MT SAN JACINTO COLLEGE
SAN JACINTO CAMPUS
1499 N STATE ST
SAN JACINTO, CA 92583

Perilous Planet Earth
Catastrophes and Catastrophism through the Ages

Perilous Planet Earth places our present concern about the threat to Earth from asteroids and comets within an historical context, looking at the evidence for past events within the geological and historical records. Professor Palmer argues that the better we understand our past, the greater the likelihood that we will be able to take appropriate action to preserve our civilisation for the future.

Two particular historical aspects are considered. Firstly, the book looks at the way in which prevailing views about modes of global change have changed dramatically over the years. The move away from support for change through relatively rare catastrophic events, toward theories of uniformity and incremental change, is charted. The author then discusses how modern theories consider both catastrophic and gradual change to be important forces in shaping the world around us. The second theme considers the way in which catastrophic events are now seen to have influenced the course of evolution in the distant past, as well as the rise and fall of civilisations in more recent times. Professor Palmer reviews a host of myths and legends that may have had their origin in actual catastrophic events, and makes a case for more research on the frequency and causes of natural catastrophes so that we are better prepared for future events.

Professional academics and students from a variety of disciplines will enjoy this book, which incorporates aspects from a wide range of subject areas, including palaeontology, geology, evolutionary biology, astronomy, social anthropology and history. Written in a clear and engaging style that avoids jargon, the book will also appeal to a wider audience of general readers.

TREVOR PALMER was awarded a Ph.D. in 1973 by the University of London, for his research into inherited medical disorders. From this emerged the two main interests of his subsequent career, enzymology and evolution, the latter stimulating a further interest in catastrophism. He moved to the Nottingham Trent University in 1974, initially as a lecturer in biochemistry, before becoming Head of the Department of Life Sciences (1987), Professor of Life Sciences (1990), Dean of the Faculty of Science (1992) and the University's Senior Dean (1998). Professor Palmer is a member of council of the Society for Interdisciplinary Studies (SIS), which provides an international platform for discussions about catastrophism and chronology. He was Chairman of the SIS from 1995 to 1998 and from 2000 to 2002. Professor Palmer has written and edited numerous other books including *Natural Catastrophes During Bronze Age Civilisations* (1998) and *Controversy – Catastrophism and Evolution: The Ongoing Debate* (1999).

Perilous Planet Earth

CATASTROPHES AND CATASTROPHISM THROUGH THE AGES

TREVOR PALMER

CAMBRIDGE
UNIVERSITY PRESS

PUBLISHED BY THE PRESS SYNDICATE OF THE UNIVERSITY OF CAMBRIDGE
The Pitt Building, Trumpington Street, Cambridge, United Kingdom

CAMBRIDGE UNIVERSITY PRESS
The Edinburgh Building, Cambridge CB2 2RU, UK
40 West 20th Street, New York, NY 10011-4211, USA
477 Williamstown Road, Port Melbourne, VIC 3207, Australia
Ruiz de Alarcón 13, 28014 Madrid, Spain
Dock House, The Waterfront, Cape Town 8001, South Africa

http: // www.cambridge.org

First published 2003

Printed in the United Kingdom at the University Press, Cambridge

Typeface Quadraat 10/14 pt   *System* LATEX 2$_\varepsilon$ [TB]

A catalogue record for this book is available from the British Library

Library of Congress Cataloguing in Publication data

Palmer, Trevor, 1944–
    Perilous planet earth : catastrophes and catastrophism through the ages / Trevor Palmer.
        p.   cm.
    Includes bibliographical references and index.
    ISBN 0 521 81928 8
    1. Asteroids – Collisions with Earth.   2. Catastrophes (Geology)   3. Evolution.   I. Title.
QE506 .P35   2003
551.3′97–dc21   2002035183

ISBN 0 521 81928 8 hardback

This book is dedicated to the members of the Society for Interdisciplinary Studies and the Cambridge Conference Network.

Contents

Contents

Acknowledgments

It would be impossible to mention all of those who, directly or indirectly, have made a contribution to this book. However, I am particularly grateful to the following (in alphabetical order), for making helpful suggestions (in response to previous writings or drafts of the present work), or for providing useful information: Mark Bailey, Alasdair Beal, Neil Bone, John Bimson, Victor Clube, Richardson (Dick) Gill, John Grattan, Joel Gunn, Anthony Hallam, Richard Huggett, Peter James, John Mccue, Carl Murray, Bill Napier, Bernard Newgrosh, Bob Porter, Brian Pyatt, David Rohl, David Salkeld, Robin Turner, Kay Warrington and the anonymous referees of Cambridge University Press, whose views were communicated to me in constructive fashion by the editor, Susan Francis. Special thanks are due to Benny Peiser, for maintaining the Cambridge Conference Network, an invaluable source of information. I acknowledge the holders of copyright for the passages quoted, full details being given in every case. I am also grateful to William Schupbach, of the Wellcome Library, London, and to Rob May and Karen Roberts, of Nottingham Trent University Library, as well as to Pauline Polkey and Lynda Byford for their general assistance. Again, for a variety of important reasons, thanks are due to my wife Jan, son James and daughter Caroline. Of course, if any mistakes have been made, they are entirely my own responsibility. Similarly, I take full responsibility for the illustrations, particularly my own drawings and photographs. Nevertheless, I am grateful to Maggie Martin for her help with the preparation of all of the illustrations. Credit for individual figures is due to the following: figures 2.1 (left), 3.1, 5.1 (right), 7.2, 8.1 and 9.1, Caroline Palmer; figure 15.3, Maggie Martin; figures 19.1, 19.2, and 19.3, NASA; figures 24.1, and 24.2, Maggie Martin; figure 28.1, Steven Robinson.

Introduction

Today, we are becoming increasingly aware of the threats to Earth from space. We are conscious of the dangers of bursts of lethal cosmic radiation arising, for example, from a nearby supernova explosion, or from some other cause. We know of asteroids and comets in Earth-crossing orbits, some large enough to cause global devastation in the event of a collision, and we can see huge craters on the surface of the Earth that have resulted from past impacts. Regardless of hazards from above, we are also at risk from below: there have been several occasions during the history of the Earth when outpourings of lava on a continental scale, or the explosion of a supervolcano, must have caused worldwide havoc.

We have evidence of at least five major mass extinction episodes, when many species became extinct. It is still uncertain whether these occurred over millions of years or over a much shorter time-scale, but they nevertheless took place. Perhaps the best-known mass extinction event was that at the end of the Cretaceous Period, when the dinosaurs were amongst the victims. Relatively large amounts of iridium, which could have come only from space or the Earth's core, have been found at the Cretaceous–Tertiary boundary at sites throughout the world. This has stimulated arguments about the possible involvement of a catastrophic event, either an extraterrestrial impact or extensive volcanic activity, in these extinctions. It is even possible that both might have been involved. There is also increasing evidence that natural catastrophes, albeit on a smaller scale, have influenced the rise and fall of civilisations in more recent times. Indeed, as we shall be discussing in detail, many people are now convinced that the courses of both biological evolution and human history have been shaped by catastrophic events.

And yet, only twenty years ago, such ideas would have seemed unthinkable. As a matter of firm principle, it was believed that, ever since life had become established on Earth, all geological and environmental changes of any significance had taken place gradually. Thus, the fields of evolution and history could only be considered within that context.

How did this gradualistic paradigm come to be so firmly established, and what brought about a re-assessment of the situation? There is a fascinating story to tell, and the aim of the first part of this book is to tell it. As we shall see, it is a saga involving honour and deceit, perception and blindness, pragmatism and dogma, careful strategies and accidental happenings. Let us start at the beginning, with the earliest written and oral traditions.

Part I

Catastrophism: the story of its decline and fall . . . and resurrection

Section A

From prehistory to 1899: catastrophism dominates for centuries, but then gives way to gradualism

1   Mythology, religion and catastrophism

Ancient beliefs

In the ancient world, deities were generally believed to intervene in human history, often in a very major way. So, for example, according to the book of *Genesis*, in the Jewish and Christian traditions, God created the world. Then, six days later, after spending the intervening time filling it with fish, birds and land animals, he breathed life into Adam and Eve, the first man and woman. Just nine generations later, corruption had become so widespread that God brought about the Flood, when 'the waters prevailed upon the earth an hundred and fifty days', and 'all the high hills, that were under the whole heaven, were covered'. However, Noah, who was an exception to the general rule of wickedness, had been given a warning about the coming deluge. That enabled his family to build a large boat, the Ark, on which to sail on the waters. In this way, they survived the Flood, supposedly the only humans to do so.[1]

These events were all believed to have taken place within the past six thousand years. Using genealogies and information about time intervals taken from the Bible, James Ussher (1581–1656), the protestant archbishop of Armagh, and an authority on Semitic languages, argued in his book *Annales veteris testamenti* ('*Annals of the Old Testament*'), published in 1650, that the Earth must have been created in 4004 B.C. For that he has become a figure of ridicule, but in fact he was only following a long tradition, using a well-established methodology. The Jewish calendar introduced by the Palestinian patriarch, Hillel II, in 359 A.D., and still widely used today, starts when the world supposedly began, at a date equivalent to 3761 B.C. Similarly, the Venerable Bede estimated in the eighth century of the Christian era that the Creation took place in 3952 B.C. Returning to the seventeenth century, the great polymath, Sir Isaac Newton, who included the chronology of the ancient world amongst his many areas of interest, agreed with the conclusions of Ussher. Indeed, a 4004 B.C. Creation became generally accepted in Britain and, for the next two centuries, dates from Ussher's chronology were often inserted in the margins of the Bible. One such date was that for the Flood, which was believed to have taken place in 2349 B.C.[2]

However, the legend of Noah does not stand in isolation. Indeed, it is just one of several hundred flood myths from around the world, many of which similarly involve a man and a woman escaping by boat. Amongst these is the one told in the Babylonian epic of Gilgamesh, where the hero, Uta-Napishtim, was warned by Ea, God of the Waters, about the coming deluge. Others include a Greek myth, where the survivors were Deucalion and his wife, Pyrrha, and one from the Aztecs of Mexico, where the equivalent figures were Coxcoxtli and Xochiquetzal.[3]

As well as legends of a catastrophic flood, there are other widespread myths where the Earth suffered near destruction by fire. According to the Aztecs, the present age (or 'sun') had been preceded by four others, each of which ended in catastrophic fashion. One of the transitions between world-ages involved (needless to say) a deluge, whereas another was brought about by fiery rain falling from the sky. Conflagrations were also a feature of the Greek tales of the battles fought by the Olympian gods against the Titans, the Giants, and the monstrous Typhoeus (or Typhon), when thunderbolts and molten rocks were hurled around as weapons. The fact that Zeus clashed with Typhoeus near Mount Vesuvius and finally trapped him under Mount Etna has suggested to some that these stories were inspired by a series of volcanic eruptions, involving an outburst of molten lava and ash from within the Earth, a process known alternatively as 'vulcanism' or 'volcanism'. These various terms were derived from the name of Vulcan, the Roman God of Fire, who was associated in legend with Vulcano, one of the Aeolian Islands off the northern coast of Sicily, between Vesuvius and Etna, and part of the same volcanically active region. Vulcano itself experienced major eruptions in 424 and 360 B.C. (and more recently in 1786 and 1888 A.D.).[4]

In Norse mythology, Odin and his fellow gods of Valhalla fought against the monstrous wolf Fenrir and the poisonous serpent Jormangard at the time of *Ragnarok* (or *Götterdamerung*), when the world-order changed, accompanied by earthquakes, tidal waves, and episodes of fire and frost. Other legends where conflicts between heroic gods and evil monsters led to environmental convulsions on a massive scale include the battles involving Marduk and Tiamat in Babylonian mythology, Feridun and Zohak in stories from Persia, and Huitzilopochtli and Coyolxauhqui in Aztec tradition. Even where there were no such clashes between supernatural rivals, the human race could sometimes be threatened with mass destruction, as in the Greek myth where Phaeton tried to drive the Sun-chariot of his father, Helios, but lost control and came too close to the Earth. People were in danger of being burned alive until Zeus cast one of his thunderbolts, diverting the chariot and causing Phaeton to fall to his death.[5]

According to the philosopher Plato (*c*. 429–347 B.C.), writing in the *Timaeus*, his distant ancestor Solon had been told by an Egyptian priest at Sais, in the Nile delta, that the Phaeton myth owed its origin to one of a series of cosmic disturbances which produced periodic catastrophes on Earth. The priest claimed, 'That story, as it is told, is in the style of a legend, but the truth of it lies in the shifting of the heavenly bodies which move around the Earth, and a destruction of many things on the Earth by fierce fire, which recurs after long intervals'.[6]

Naturally, we are under no compulsion to accept this as a statement of fact. According to the *Timaeus* and another of Plato's works, the unfinished *Critias*, a separate and more detailed story told to Solon on the same occasion concerned the destruction of the island civilisation of Atlantis by a catastrophic flood, but this is generally regarded with considerable scepticism. Even if the two books were intended as strictly historical accounts, which is by no means certain, Plato might have been misinformed. By his own admission, Plato's source, Critias, was an old man of failing memory, who had learned the entire story at the age of ten from his ninety-year-old grandfather, whose father had been told it by Solon. Critias did, however, claim that he had some of Solon's original writings in his possession.[7] But even if the transmission of the story had been accurate, it might not have been so firmly based on knowledge as the Egyptian priest supposed.

How the ancient myths and legends came into being is far from clear, and the stimulus may have been quite different from one to the next. It is likely that some myths are dramas based on the replacement of one cult by another in a particular region, whereas others could be stories associated with rituals whose purpose was to induce fertility, the succession of seasons, or a hoped-for life after death.[8] It is also possible that some legends may, to a greater or lesser extent, have been inspired by actual happenings. It will probably never be known whether any of the specific characters mentioned in Homer's *Iliad* and *Odyssey* ever lived, or whether the events took place as described, but archaeologists such as Heinrich Schliemann, Wilhelm Dörpfeld, Sir Arthur Evans and Carl Blegan found abundant traces of pre-classical civilisations at sites located from details in these stories.[9] As to the flood and fire myths, they seem to indicate at the very least that ancient societies believed in the possibility of cataclysmic events, regardless of whether they themselves had actually experienced any.

In most ancient traditions, catastrophes were associated with divine displeasure. In the book of *Genesis*, as we have seen, God caused Noah's Flood because of the increasingly wicked behaviour of humankind. Shortly afterwards, the twin cities of the Dead Sea plain were destroyed for similar reasons. As related in *Genesis*, 'the Lord rained upon Sodom and upon Gomorrah brimstone and fire . . . out of Heaven', because not even ten righteous people could be found within them.[10] Prophecies of further punishment for evil abound. According to *Psalm 11*, 'Upon the wicked he shall rain snares, fire and brimstone, and an horrible tempest';[11] whilst *Malachi* warns, 'For behold, the day cometh, that shall burn as an oven; and all the proud, yea, and all that do wickedly, shall be stubble; and the day that cometh shall burn them up, saith the Lord'.[12]

In an Egyptian myth, the sun-god, Ra, began to lose the respect of humankind as he grew older, so he loosed his 'eye' upon the Earth, causing great slaughter.[13] Similarly,

Figure 1.1   Left: Plato, based on a bust in the Fitzwilliam Museum, Cambridge, a first century A.D. Roman copy of a bronze by Silanion, probably made during Plato's lifetime. Right: Aristotle, based on a bust in the Kunsthistorisches Museum, Vienna, a Roman copy of a Greek original dating from the fourth century B.C.

in Greek mythology, Zeus often indicated his displeasure by casting thunderbolts, as in the story of Phaeton, whilst Poseidon was inclined to cause floods or storms when annoyed. So, for example, when the Trojan king Laomedon broke a promise to him, Poseidon flooded the nearby coastal plain and, for good measure, sent a sea-monster to terrorise the people.[14]

Such floods had undoubtedly occurred. By the time of Aristotle (384–322 B.C.), a pupil of Plato (figure 1.1), the evidence of marine fossils in outcrops of rock made it clear that at least part of what was now land had once been covered by sea. In his *Meteorologica*, Aristotle wrote that there were periodic transpositions of land and sea, but generally those occurred too slowly and over too long a time interval for anyone to notice them happening. Nevertheless, on rare occasions a great winter could occur, bringing protracted heavy rainfall and causing devastating floods, such as that of Deucalion.[15] Similarly, there could be a very lengthy arid period, a great summer. The Greek word for the great winter flood, *kataklysmos*, is the origin of the modern word cataclysm.

Aristotle had views very different from those of his teacher Plato who, as we have seen, accepted that sudden and violent events could take place in the heavens, with serious consequences for the people below. According to Aristotle, writing in his *De Caelo*, the stars and planets occupied a series of concentric spheres and, unlike the corrupt

Earth, which was located at the centre, the heavens were perfect and unchanging.[16] Comets and shooting stars, as he explained in another book, the *Meteorologica*, were purely terrestrial phenomena, caused by changes of temperature, coupled with friction between the Earth's atmosphere and the innermost sphere.[17]

Many of Aristotle's ideas, including his concept of a stable Universe, with the heavens segregated from the Earth, were taken up by the Christian philosopher Thomas Aquinas in the thirteenth century, and remained influential amongst scholars for many centuries afterwards.[18] At the same time, the Church maintained a belief that the world would come to an end in catastrophic fashion. As prophesied in the *New Testament*, 'the day of the Lord will come as a thief in the night; in which the heavens shall pass away with a great noise, and the elements shall melt with fervent heat, the earth also and the works that are therein shall be burned up'.[19] There was not seen to be any contradiction here: cosmic catastrophes could be brought about by the intervention of God, but not by any natural process.

The appearance of a comet in the sky was generally viewed with alarm, as it was thought to signal some coming disaster.[20] For example, the Venerable Bede, in his *Ecclesiastical History of the English People*, wrote, 'In the year of the incarnation of Our Lord 729, two comets appeared about the Sun, to the great terror of the beholders'.[21] The very word 'disaster' was derived from the Latin words '*dis*' and '*astrum*', together meaning 'evil star'. The disaster could be to the population as a whole, or to an important individual. So, in William Shakespeare's play *Julius Caesar*, written in 1599, Caesar's wife, Calpurnia, is concerned by unusual features in the sky, and warns her husband, on the night before his assassination, 'When beggars die there are no comets seen; The heavens themselves blaze forth the death of princes'.[22]

Cosmogonists: blending belief and observation

In the seventeenth and eighteenth centuries, various theories of the formation and development of the Earth were put forward by the so-called cosmogonists. Their interest was the origins of stars and planets, whereas the main concern of a different group, the cosmologists, was the nature of the Universe as it actually existed at the time. Posterity has consistently admired cosmologists such as Galileo Galilei (1564–1642) and Sir Isaac Newton (1642–1727), who followed the example of Nicolaus Copernicus (1473–1543) in rejecting Aristotle's geocentric Universe in favour of the modern system in which the Earth and planets orbit the Sun, according to natural laws. In contrast, for reasons we shall come to later, twentieth century geologists generally believed that the cosmogonists who lived around the same time as Galileo and Newton had been extremely poor scientists.

The cosmogonists were catastrophists, i.e. they believed that 'the history of the Earth has to be explained by events radically different from anything going on at the present day', which is the definition of catastrophism given in the *Cambridge Encyclopedia*.

Other reference books define catastrophism as the theory that the Earth's geological features have been fashioned by 'sudden, short-lived, worldwide events' (the *McGraw-Hill Dictionary of Scientific and Technical Terms*), by 'sudden, violent and unusual events' (the *Oxford English Reference Dictionary*) or by 'infrequent violent events' (the *Chambers Dictionary of Science and Technology*).[23]

Whichever of these precise definitions is used, it is clear that geological catastrophism, in itself, is a perfectly rational notion, regardless of whether or not it is correct. However, it has generally been supposed that a characteristic feature of seventeenth, eighteenth and nineteenth century catastrophism was an association with supernatural forces, particularly as an explanation of the replacement of one set of fossil forms by another during the course of the Earth's history. So, for example, in 1982, the University of Guelph science historian Michael Ruse wrote of the catastrophists in his book, *Darwinism Defended*: 'They argued flatly that new species of organism, including God's final creation, man, were produced miraculously by God. God wants no nonsense about unbroken laws coming between them and his handiwork. He intervenes personally.'[24] Similarly, a few years later, the Oxford University zoologist Richard Dawkins claimed in his book *The Blind Watchmaker* 'Catastrophism was an eighteenth – and nineteenth – century attempt to reconcile some form of creationism with the uncomfortable facts of the fossil record'.[25] Again, the Johns Hopkins University palaeontologist Steven Stanley wrote in *Earth and Life Through Time*, published in 1986, that, up until the early nineteenth century, many natural scientists were catastrophists who believed that 'floods caused by supernatural forces formed most of the rocks visible at the Earth's surface'.[26]

As we shall see, such statements present a false picture of the catastrophists of the seventeenth to nineteenth centuries. Whilst it is true that they were unable to separate science from religion, the same was true of all their contemporaries. There seems no justification for making critical judgements on, say, cosmogonists, by the strict application of twenty-first century standards, whilst ignoring some of the strange views of cosmologists. We have to consider both groups in the context of the times in which they were living, including the fact that they were sometimes forced to adopt orthodox views (as Galileo was compelled by Pope Urban VIII to recant his belief that the Earth moved around the Sun), or risk sharing the fate of the philosopher Giordano Bruno who was burned to death as a heretic in 1600.[27] If cosmogonists and cosmologists are looked at together, it can be seen that they had much in common, operating within the complex intellectual climate of their day.[28] Even Newton, who is justly given great credit for formulating the mathematical laws of gravity, could never accept that gravitational forces were purely materialistic phenomena. Rather, he considered them to be an expression of God's will.

That comes over clearly in an exchange of letters between Newton and Richard Bentley, Chaplain to the Bishop of Worcester, following a series of sermons on the 'evidences for Christianity' preached by Bentley at St Martins-in-the-Fields, London, in 1682. At the end of that year, Bentley wrote to Newton, 'It is inconceivable that inanimate brute matter should (without a divine impression) operate upon and affect other matter without mutual contact'. Newton replied that he agreed wholeheartedly, echoing Bentley's words and adding, 'That gravity should be innate, inherent, and essential to matter, so that one body may act upon another, at a distance through a vacuum, without the mediation of anything else, by and through which their action and force may be conveyed from one to another, is to me so great an absurdity, that I believe no man who has in philosophical matters a competent faculty of thinking, can ever fall into it'.[29]

It should not be forgotten that, for many centuries, the Church had maintained and controlled scholarship in Christian countries, everything having to be considered within a spiritual context.[30] Without question, the most reliable source of information was believed to be the Bible, which testified that the Earth was only a few thousand years old, and that there had been a single major cataclysm, the Flood in the time of Noah. Anyone who thought differently was in a difficult position, because to have said so would have incurred the wrath of the Church, and the risk of being condemned as a heretic. That continued to apply even after the Reformation in the early sixteenth century. The leading Protestants, including Martin Luther and John Calvin, emphasised the role of *The Bible* in determining 'true' doctrine. Similarly, for the Catholic Church, the Council of Trent, which met at Trento, in northern Italy, between 1545 and 1563, confirmed the belief that the Latin Vulgate Bible represented the authentic word of God.[31]

At that time, it was still generally accepted that all marine fossils found inland had been carried there by Noah's Flood, although Leonardo da Vinci (1452–1519), Girolamo Fracastoro (1483 1553) and a few others had argued that this was impossible, because the thickness of beds containing such fossils was incompatible with the short time-scale. Instead, the land must have risen in places, changing the shoreline in significant fashion, just as Aristotle had suggested.[32]

In 1669, the Danish naturalist Niels Stenson, better known as Nicolaus Steno (1638–1686), produced a theory to explain the landscape of Tuscany in which the Flood played a prominent but far from unique role. Other features included the elevation of land in some locations because of precipitation of sediments from the waters, and lowering of the land elsewhere as a consequence of the collapse of caverns under the ground. Steno appreciated that not all rocks had been formed at the same time and so, since rock formation most likely involved the deposition of sediment on existing rock, each stratum in a formation must be younger than the one below. Some rocks would have been formed from sediments precipitated from the waters of Noah's Flood, whereas earlier ones must have been part of the land which emerged from the original worldwide ocean, as described in the *Genesis* account.[33]

Later, following a similar methodology, several English cosmogonists put forward models that tried to reconcile observations of the natural world with the teachings of the Church. Thomas Burnet (1635–1715), an Anglican clergyman who became chaplain to King William III, proposed a system which had some features in common with that of Steno. However, instead of relying on rain and subsidence to cause the Flood, Burnet suggested that wide cracks appeared in the Earth's surface, allowing water to be forced upwards from underground stores. He believed this sudden release of water might have caused the Earth's axis to tilt.

Considered against the attitudes of his time, Burnet was undoubtedly a rationalist. He wrote to Newton early in 1681 to explain why he could not accept a literal interpretation of *Genesis*, which maintained that the Earth had been created in just six days. His major work was the four-volume book, *The Sacred Theory of the Earth*, published during the 1680s. As was inevitable in the seventeenth century, he started with the assumption that the Biblical record was essentially true, even if not accurate in every detail, and then sought natural explanations for the events described. Newton, as we have seen, thought that gravitational forces must be supernatural in origin, yet operated in an unchanging fashion according to physical laws. Similarly, Burnet, having little time for those who invoked miraculous interventions as an explanation for observed phenomena, believed that God had set natural laws in motion at the time of Creation. He pointed out, 'We think him a better artist that makes a clock that strikes regularly at every hour from the springs and wheels which he puts in the work, than he that hath so made his clock that he must put his finger to it every hour to make it strike'.[34]

Burnet was not prepared to accept that the waters causing the Flood had been created miraculously by God. However, they must have come from somewhere, so the interior of the Earth seemed the most likely possibility. That suggestion was strongly opposed by those who viewed with displeasure any departure from the traditional interpretation. On the other hand, whatever the cause of Noah's Flood, Robert Hooke (1635–1703), a founder-member of the Royal Society of London, was unconvinced that it could have lasted long enough to account for all the world's fossil-bearing strata. Instead, the evidence seemed to suggest that earthquakes had occasionally caused significant rearrangements of land and sea.[35]

Nevertheless, cosmogonists continued to propose systems in which Noah's Flood played a major role. In the model of John Woodward (1665–1728), a geologist based at Gresham College, London, who was a pioneer of fieldwork and a Fellow of the Royal Society, the Flood arose much as Burnet had suggested. Then, in some way, it dissipated all the rocks previously in existence. Afterwards, materials carried by or released from the waters of the Flood sedimented according to their specific gravities to form horizontal strata. These were later dislocated by depressions and elevations of unspecified origin to form the patterns which were so apparent to observers.[36]

William Whiston (1666–1753), who succeeded Isaac Newton in the chair of mathematics at Cambridge University, was another who thought that some of the waters of the Flood might have been released from the interior of the Earth, but he considered that the major proportion had fallen as rain derived from the vapours in the tail of a passing comet. He also believed that the passage of the comet brought about the diurnal rotation of the Earth and changed its orbit from a perfect circle to an ellipse. These ideas were presented in his book, *A New Theory of the Earth, from its Original, to the Consummation of all Things*, which was published in 1696.[37] Long before this, in 1577, detailed observations by the Danish astronomer Tycho de Brahe had demonstrated that, contrary to the teachings of Aristotle, comets travelled in regions far beyond the Earth's atmosphere. Whiston was also aware, from observations of the great comet of 1680 made by John Flamsteed, the Astronomer Royal, that comets moved between the region of the Sun and the outer Solar System in highly elongated orbits. Furthermore, Edmond Halley had calculated that the comet of 1682 (which subsequently took his name) had the same orbit as those of 1531 and 1607, and predicted, correctly as it turned out, that it would return in 1758. He was not so close to the mark when he deduced (quite incorrectly) that the 1680 comet was the same as the great comet of 1106, thus having a periodicity of 575 years, but this led Whiston to believe that it could have made an earlier visitation in 2342 B.C., around the time the Flood was thought to have occurred.[38]

Whiston was eventually dismissed from his post for expressing various views which gave rise to theological concern. Amongst these was his belief that global catastrophes, past and future, might be caused by natural phenomena. Halley had similar thoughts about this subject, but showed more caution about letting them become generally known. In 1694, in a lecture entitled 'Some considerations about the Universal Deluge', he had suggested to the Royal Society that the story of Noah's flood might be an account of a cometary impact, the projectile splashing into the Caspian Sea, with devastating consequences for the surrounding lands. Precisely what happened after the lecture is uncertain, but Halley went back to the Royal Society a few days later to tell the members that he had been mistaken, and his paper was not published for another 30 years.[39]

By this time it was clearly established that comets were neither atmospheric phenomena nor signs of divine displeasure. Instead, as conclusively demonstrated in Newton's *Principia Mathematica* of 1687, they were objects obeying the same laws of motion as all cosmic bodies. However, even here, in his greatest scientific work, Newton emphasised his belief that these laws had been established by a benign God. Hence, comets obeying them were far more likely to have beneficial effects than to bring disaster to the planet. Near the end of Book 3 of the *Principia*, he suggested that they formed part of some divine plan to maintain life on Earth by replenishing the planet's stores of water during a close passage, writing that 'comets seem to be required, so that from the condensation of their exhalations and vapors, there can be a continual supply and renewal of whatever liquid is consumed by vegetation and putrefaction and converted into dry earth'.[40]

Eighteenth century cosmogony and Neptunism

On the continent of Europe, Newton's laws of planetary motion were slow to become accepted, largely because of the influence of the theories of the French mathematician and philosopher René Descartes (1596–1650). In 1644, in his *Principia Philosophiae*, Descartes argued that space was full of matter which, in the beginning, had been stirred into movement by God, as part of a carefully formulated plan, and then left alone for the system to develop in purely mechanical fashion. Vortices were produced, in which the Sun and planets were able to form by condensation, the planets being whirled around the Sun by the continuing action of the vortices.[41] In Germany, another mathematician and philosopher, Baron Gottfried von Leibniz (1646–1716), accepted Descartes' view that the newly condensed Earth would have been very hot, and hence in a fluid-like state. He developed the notion that, as it cooled, a crust formed which later cracked on occasions to release flood water from within the Earth, each time depositing a layer of sediment.[42]

Around the middle of the eighteenth century, the French naturalist Georges-Louis Leclerc, Comte de Buffon (1708–1788), suggested that the 'days' of creation in *Genesis* were not meant to be taken literally. It made more sense, he thought, to regard them as periods of unspecified but great length. Buffon calculated that if, as he personally supposed, the Earth had been formed by a collision between the Sun and a comet, it could have cooled down sufficiently within 35,000 years to allow condensation of atmospheric water vapour to form a universal ocean. Further cooling over many tens of thousands of years caused cavities to appear in the Earth's surface, through which sea water drained until it reached its present level. As volcanoes began to erupt, the continents appeared and valleys were gouged out by ocean currents. More cooling took place and there was gradual erosion of the mountains, until the Earth assumed its present form.[43]

Buffon's fellow-countryman Benoît de Maillet (1656–1738) believed that erosion of the earliest mountains by the action of the ocean over a time-scale of millions of years was an important factor in producing sediment from which new mountains could be made. These ideas were expressed in a book entitled *Telliamed* (a reversal of the author's name), published in 1748 but written over thirty years earlier. In an attempt to avoid offending the Church, *Telliamed* was presented as a work of fiction, de Maillet's speculations about the history of the world being put into the mouth of an Indian philosopher. As well as having heretical views about geology, de Maillet also believed that every species alive today had originated in the primaeval ocean as a result of the natural germination of seeds which pervaded the Universe, each developing separately into modern forms as conditions changed.[44]

Theories that a universal ocean once contained in solution all the material that later formed the Earth's crust were generally labelled as 'Neptunist' (after Neptune, the Roman God of the sea). The most influential advocate of Neptunist views was

Abraham Gottlob Werner (1749–1817), a German mineralogist and geologist associated with the Freiburg Mining Academy, who developed the ideas of his fellow countrymen Johann Gottlob Lehman and George Christian Füchsel. In Werner's theory, precipitation of dissolved material took place over long periods of time, first forming primitive rocks such as granite, and then, as erosion of these began to contribute to the process, deposits such as limestones and slates. Later, when mechanical deposition became more significant than chemical precipitation, came the laying down of chalk and other fossil-rich rocks, together with basalt. Whilst all this was going on there were occasional episodes when powerful currents, associated with sudden drops in water-level, cut deep channels through the sediments. Localised uplift of rock also occurred, generally linked to volcanic activity.[45]

Werner's ideas were most certainly not determined by religion. He believed that the Earth was far older than a few thousand years, and refused to speculate, because of a lack of scientific evidence on which to do so, about where the universal ocean might have come from, and where the water went when the sea-levels fell. However, the British supporters of Neptunism, such as Robert Jameson in Scotland and Richard Kirwan in Ireland, tried to find, and then emphasise, links with the scriptures. Similarly, Jean André de Luc, who was Swiss by birth, but spent most of his working life in England, presented a Neptunist theory which was explicitly in line with the *Genesis* account. De Luc claimed that there had been a universal flood in relatively recent times, and six periods of deposition, which corresponded to the six days of creation.[46]

He was not alone. At around the same time, the second half of the eighteenth century, others in Britain such as Alexander Catcott, Patrick Cockburn, John Whitehurst and John Williams were still maintaining the old traditions, producing cosmogonies which attempted to be consistent with accumulating field evidence, whilst retaining a place for Noah's Flood. Often, these were variants of earlier models. So, for example, Whitehurst agreed with Woodward that materials had precipitated from the waters of the Flood, settling according to their densities, but differed from him by thinking that this process had been uneven, because of the gravitation effects of the Moon. However, by the end of the century it was becoming increasingly clear that, even if the Flood had occurred, it could only have been one of many factors responsible for the formation of features at the Earth's surface.[47]

2 Hutton: fact and fiction about the origins of modern gradualism

The legend: Hutton as the founder of modern geology

Let us now look at the career of James Hutton (1726–1797), who has been widely regarded as a pioneer of modern scientific practices, like Newton, who died in the year after Hutton was born (figure 2.1), and Charles Lyell, who was born in the year Hutton died. Both Hutton and Lyell were geologists who believed that major changes at the Earth's surface took place over very long periods of time, i.e. they were gradualists. Until quite recently, it was generally accepted that their gradualistic theories were derived from firm scientific evidence, unlike the fanciful speculations of the catastrophists.[1]

The establishment of that view owed much to *The Founders of Geology* by Sir Archibald Geikie, Director-General of the British Geological Survey, which was first published in 1897. In the book, Geikie condemned the 'monstrous doctrines' of the cosmogonists, and wrote of speculation running 'completely riot' in their theories. He continued, 'It was a long time before men came to understand that any true theory of the earth must rest upon evidence furnished by the globe itself, and that no such theory could properly be formed until a large body of evidence had been gathered together.'[2] Hutton, in the opinion of Geikie, was the first geologist to realise what was required, so he 'went far afield in search of facts, and to test his interpretation of them'.[3] On the basis of this extensive fieldwork, carried out over a thirty-year period, he was then able to develop his theories. Geikie wrote, 'In the whole of Hutton's doctrine, he rigorously guarded himself against the admission of any principle which could not be founded on observation. He made no assumptions. Every step in his deductions was based upon actual fact, and the facts were so arranged as to yield naturally and inevitably the conclusion which he drew from them.'[4]

That was the story which, throughout most if not all of the next century, was taught to geology students as the unquestionable origin of their subject as a scientific discipline. So, for example, Stokes and Judson's *Introduction to Geology*, published in 1968, maintained, 'Many concepts of present-day geology stem directly from observations of the Scotsman James Hutton', whilst the 1971 edition of Leet and Judson's textbook, *Physical Geology*, stated, 'Modern geology was born in 1785 when James Hutton . . . formulated the principle that the same physical processes that are operating in the present also operated in the past', and the 1973 CRM book, *Geology Today*, claimed, 'The first to break formally with religion-shrouded tradition was James Hutton'.[5]

Figure 2.1   Left: James Hutton, based on a 1787 etching by John Kay. Right: Sir Isaac Newton in old age, based on a portrait by John Van der Bank.

Similarly, in 1982, the British palaeontologist Beverley Halstead wrote in *Hunting the Past*, 'The first detailed attempt to understand the meaning of rocks was made by the Scotsman James Hutton . . . after thirty years of observation and study of the rocks in Scotland and elsewhere'.[6]

Hutton was a gentleman farmer and manufacturing chemist, who had medical qualifications from Edinburgh, Leiden and the Sorbonne, but never made use of them. He eventually became a member of a dazzling Edinburgh intellectual circle, which included David Hume, Adam Smith and James Watt amongst its members.[7] Hume, a philosopher and historian, was author of the *Treatise of Human Nature*; Smith, the first major political economist, remains famous for his *Wealth of Nations*; and Watt played a significant role in the development of the steam engine.[8] Hutton was determined not to be outdone. His ambition was to produce a mechanistical model for the Earth, much as Newton had done for the Solar System, and he presented his theory before the Royal Society of Edinburgh in 1785, giving a full account in the *Transactions of the Royal Society of Edinburgh* in 1788, after previously providing a summary.[9] The culmination of his work came with the *Theory of the Earth with Proofs and Illustrations*, published in two volumes in 1795.[10]

To contrast Hutton's views with the Neptunist ones of Werner, which were very popular at the time, they were termed 'Plutonist', since they implied that there was a source of heat deep within the Earth (Pluto being the Roman god of the Underworld). The main advance was in recognising that rocks such as basalt and granite were igneous, not sedimentary, in origin.[11]

Volcanoes were, of course, known to erupt on occasions, but that did not necessarily mean that vulcanism was of any particular geological importance, or had made any significant contribution to the formation of the Earth's rocks. The Neptunists believed that the lava thrown out by volcanoes consisted of rock which had become molten when underground beds of coal caught fire. A somewhat similar view was taken by Peter Simon Pallas (1741–1811), a German geologist who worked in Russia under the patronage of Tsarina Catherine II ('Catherine the Great'). Pallas carried out extensive surveys of the Ural and Altai mountain ranges, observing major disruption of the strata, which he attributed to the effects of volcanic eruptions, probably caused by the ignition of deposits of iron pyrites mixed with the decaying remnants of marine organisms. In France, 'Vulcanists' such as Nicholas Desmarest (1725–1815) came to the conclusion that basalt was an igneous rock produced by volcanic activity, but they could never accept that the same applied to granite. Like Pallas and the Neptunists, Desmarest believed that volcanic activity had only begun to occur relatively recently, and played no part in the formation of granite, which was considered to be a component of the Earth's original crust.[12]

The Neptunists argued that crystalline structures found in basalt and granite proved that these rocks could not be igneous in origin. However, Hutton's friend the Scottish geologist James Hall (1761–1832) was able to demonstrate in laboratory experiments that such structures could be formed if molten rock was cooled very slowly. He also advanced the Plutonist cause by showing that rocks of the basalt and granite types had much in common, and could be partially interconverted.[13]

Another of Hutton's major insights was in recognising that uplift could be more than a localised phenomenon, instead providing a general mechanism which could act in the opposite direction to erosion and decay. From this he developed a model of a self-restoring Earth, which operated on a three-stage cycle. Firstly, rain and river-water wore away rock, releasing particles which eventually found their way into the sea. Secondly, this debris from the rivers, together with material resulting from the action of tides on coastal regions, accumulated in the ocean basins, one layer on top of another. As suggested by the experiments of another of Hutton's friends from the Edinburgh circle, the chemist Joseph Black (1728–1799), each layer would eventually compact to form a stratum of rock, because of the heat and pressure induced by overlying material. This process continued over a considerable period of time until the accumulated weight was sufficient to generate enough heat to cause the rock in the lower strata to melt. In the third stage, the expanding molten rock was forced upwards, until eventually it solidified again to form new continents, displacing the oceans into low-lying regions of the former ones. However, since everything was now much as it had been previously, the end of one cycle must have been the start of another. Thus, the system could maintain itself indefinitely.[14] In the 1788 paper, Hutton summarised his views on the time-scale with the words: 'Time, which measures every thing in our idea, and is often deficient to our schemes, is to nature endless and as nothing',[15] and he concluded, 'If the succession of worlds is established in the system of nature, it is in vain to look for anything higher in the origin of the earth. The result, therefore, of our present enquiry is, that we find no vestige of a beginning, – no prospect of an end'.[16]

The reality: Hutton, the eighteenth century cosmogonist

Without question, Hutton's three-stage model was a major achievement, identifying the different processes involved in rock formation, over very long periods of time. Furthermore, as the geology textbooks claimed, he certainly used actualistic methods in his work. In other words, he studied present-day processes as a means of interpreting past events. (Actualism as an explicit concept was developed in France early in the nineteenth century, the name coming from its reliance on '*causes actuelles*', i.e. causes now in operation.) In the *Theory of the Earth*, Hutton made it clear that, in attempts to account for the history of the Earth, he wanted 'no powers to be employed which are not natural to the globe, no actions to be admitted of except those of which we know the principle, and no extraordinary events to be alleged in order to explain a common appearance'.[17] However, he was not the first to take this approach, such methods having previously been employed by Werner, Pallas and Desmarest, and even, in sporadic fashion, by cosmogonists such as Buffon.[18] Moreover, when Hutton developed his ideas about the geology of the Earth having a long time-scale, he undoubtedly coupled actualism with an a priori assumption that natural processes always take place at the same rate. In the 1788 paper, he wrote:

> In examining things present, we have data from which to reason with regard to what has been; and, from what has actually been, we have data for concluding with regard to that which is to happen hereafter. Therefore, upon the supposition that the operations of nature are equable and steady, we find, in natural appearances, means for concluding a certain portion of time to have necessarily elapsed, in the production of those events of which we see the effects.[19]

Despite the later legend of the empirical scientist, Hutton was not seen by his contemporaries as being very different from the other speculative cosmogonists of the period.[20] For example, Desmarest dismissed the notion of a cyclical Earth on the grounds that it was insufficiently based on evidence.[21] Certainly, the assumptions of seventeenth and eighteenth century philosophy can be seen clearly in Hutton's 1788 paper, when he wrote that the Earth had been 'adapted to the purpose of being a habitable world'[22] and, again, when he claimed that the 'purpose' of volcanoes was not 'to overwhelm devoted cities with destruction' but to 'prevent the unnecessary elevation of land, and fatal effects of earthquakes.'[23]

It is now established beyond question that Hutton's formulation of the model of a self-renewing Earth was based not only on the desire to restrict himself to the use of observable causes, but also on his views on the relationship between God and the Earth. He objected strongly to Buffon's theory that the planets had been formed as a result of a chance encounter between the Sun and a comet. Hutton did not doubt that the world had been carefully designed by a wise and benevolent Creator, but he was not a Christian of the traditional kind, a 'theist' who believed that God actively intervened to change the course of history. Instead, Hutton was a 'deist' who considered, rather like Isaac Newton and Thomas Burnet (see Chapter 1), that such intervention was unnecessary, because God had created a Universe which operated like clockwork. Deism, developed from the 'mechanical philosophy' of René Descartes, was supported in the eighteenth century by a number of leading intellectuals, particularly in France (e.g. by Voltaire) and Scotland. Hence, it is clear that, from his religious views, Hutton was pre-disposed towards a system which God could allow to operate without interference.[24]

The legend claimed that Hutton produced his theory of a self-renewing world as a result of logical deductions following careful field observations of granite and of geological unconformities, i.e. clear breaks between episodes of rock formation. The reality, as shown by Gordon Davies, a geomorphologist and science historian then working in Dublin, in his 1969 book, *Earth in Decay*, was that when Hutton presented a full account of his theory to the Royal Society of Edinburgh in 1785, he had seen no unconformities and had observed granite at just one location, in northeastern Scotland, which provided little information. It was not until later in 1785 that he saw a granite–limestone interface in an outcrop at Glen Tilt, in the Grampian Mountains, which appeared to have been formed by molten granite forcing its way up into cracks in the overlying limestone and greywacke. The first unconformity he saw was in 1787 at Loch Ranza, on the Isle of Arran, and this was almost immediately followed by another in the Tweed basin, in southern Scotland. The latter in particular provided clear evidence that sedimentary rocks had been uplifted, tilted into a vertical orientation and subjected to some erosion before being overlaid by further sedimentary strata. However, the most significant unconformity, at Siccar Point near St. Abbs Head, was located by Hutton in 1788. He took James Hall and John Playfair, Professor of Mathematics at Edinburgh University and an amateur geologist, to see it by boat, and they were profoundly impressed. However, although these observations provided strong support for his ideas, it is clear that they came *after*, not *before*, the formulation of the self-renewing Earth theory.[25]

Hutton had no personal experience of the great mountain chains of continental Europe, his knowledge of them being based mainly on a detailed investigation of the Alps carried out by Horace-Bénédict de Saussure (1740–1799), a Swiss physicist and geologist, who had Neptunist views.[26] James Hall, who travelled to Switzerland to observe the Alps at first-hand, could not reconcile what he saw with Hutton's gradualism. Guided by his religious beliefs, Hutton had refused to contemplate the possibility of catastrophes occurring, even during the periods of extensive uplift of land. In the final chapter of the *Theory of the Earth* he wrote, 'In whatever manner, therefore, we are to employ the great agents, fire and water, for producing those things which appear, it ought to be in such a way as is consistent with the propagation of plants and life of animals upon the surface of the earth. Chaos and confusion are not to be introduced into the order of nature'.[27] However, to Hall, the violent twisting of the strata did not seem consistent with gradual uplift, and the abundance of huge erratic boulders (ones foreign to the region in which they were found) could not be explained by the action of rain or rivers. Although Hall remained committed to his friend's model of a self-renewing Earth, he concluded that there must have been episodes of paroxysmal uplift, during which tidal waves swept over the land, disrupting glaciers and their underlying rocks, the detached ice carrying boulders over large distances.[28]

Hutton undoubtedly believed that the Earth functioned as a machine whose purpose was to sustain life, and his aim was to produce a theory consistent with that belief, not one derived from scientific observations.[29] He stated that unequivocally in his 1788 paper:

> This is the view in which we are now to examine the globe; to see if there be, in the constitution of this world, a reproductive operation by which a ruined constitution may be again repaired, and a duration or stability thus procured to the machine, considered as a world sustaining plants and animals. If no such reproductive power, or reforming operation, after due enquiry, is to be found in the constitution of this world, we should have reason to conclude, that the system of this earth has either been intentionally made imperfect, or has not been the work of infinite power and wisdom.[30]

Nevertheless, the myth persists that Hutton's theories were formulated strictly on the basis of scientific evidence. Simon Lamb and David Sington, in their 1998 book, Earth Story, wrote of Hutton and the Siccar Point unconformity, 'In 1788, he made an observation which changed for ever the way geologists thought about the Earth, opening up undreamt-of vistas of time stretching into the remote past'.[31] Yet his theory had been presented in a complete state to the Royal Society of Edinburgh three years previously.

When Hutton came to write his 1795 book, he now had this scientific evidence available to him, so made use of it. Nevertheless, even then, he was interested solely in the support it gave to the theory he had formulated by philosophical reasoning, not in any further information it could give about the history of the Earth. He did not appear to realise that the particular features of each stratum could indicate the conditions under which the rock had been deposited, and might also be used as a basis for comparing strata in different locations.[32] The eminent University of Birmingham geologist Anthony Hallam wrote in the 1989 edition of his Great Geological Controversies that he could not agree with those who believed 'that Hutton was the founder of modern geology, because of the ahistorical nature of his work, and geology is nothing if not an historical science.'[33]

One reason why later generations of geologists held such misconceptions about Hutton's work is that he was known to them not so much through his own writings, which were difficult to read, but through John Playfair's Illustrations of the Huttonian

Theory of the Earth, published in 1802, five years after Hutton's death.[34] So, for example, Charles Coulston Gillespie wrote approvingly, in his 1959 book, *Genesis and Geology*, 'Despite his impatience with unfounded speculation, Playfair's discussion of progressive scientific method did not minimize the importance of framing theories. He insisted, however, on limiting theories to the interpretation of observable phenomena.'[35] The implication, unfounded as we now know, was that Hutton did the same.

Although Playfair gave a reasonably accurate account of Hutton's theory of a self-renewing Earth, he played down those aspects of his work which would have seemed too firmly rooted in an earlier age, such as Hutton's assumption that everything had a purpose, and his failure to appreciate that rocks could be a source of historical information. At the same time as promoting the self-renewing Earth model, Playfair attacked the alternative theory of a directional process, involving a continually cooling planet. He suggested it was reminiscent of 'the wild fictions of Scandinavian mythology', and dismissed it (revealingly) as a 'dismal and unphilosophic vision'.[36] Thus began the process of creating a false dichotomy between gradualism and catastrophism, in which the scientific basis of the former was exaggerated and that of the latter denied.

3 Cuvier and Lamarck: choosing between extinction and evolution

Revolutionary geology and anatomy

Across the English Channel, in France, the dominant figure at the time Hutton was developing his theories was the Comte de Buffon. Right up to his death in 1788, the year in which Hutton's Edinburgh paper appeared in print in its complete form, Buffon maintained considerable power and influence in Paris through his post as 'keeper' of the Jardin du Roi (the present-day Jardin des Plantes). However, important changes were taking place. A new generation of naturalists was emerging, and these sought a fresh approach to science. One of the chief critics of Buffon's style was a man baptised Jean-Léopold-Nicholas-Frédéric Cuvier (1769–1832), but known as Georges, after a dead brother.[1]

Whilst most pre-nineteenth century cosmogonists, including Buffon, used rational methods and acknowledged the importance of evidence, as we saw in Chapter 1, their arguments were, without question, often speculative and influenced by prevailing assumptions. Cuvier, in contrast, guided by the belief of the English philosopher John Locke (developed by Étienne Bennôt, Abbé de Condillac) that the chief cause of human error is a compulsion to extend inquiries beyond sensible limits, was determined to keep his arguments free from unjustified speculation. He carried out much impressive work, for example in putting the classification of plants and animals onto a more scientific basis, developing the ideas of the Swedish naturalist Carl von Linné (1707–1778), better known as Carolus Linnaeus. When Buffon died, Cuvier was openly delighted, proclaiming, 'The naturalists have finally lost their leader. This time, the Comte de Buffon is dead and buried'.[2]

An era was undoubtedly coming to an end, not only in academic circles but also more generally. The Bastille prison was stormed in 1789, only a year after Buffon's death, signalling the start of the French Revolution. Nevertheless, scientists tried to carry on with their work, despite disruptions such as the execution of the chemist Antoine-Laurent Lavoisier (for reasons which had nothing to do with chemistry). The Jardin was reformed in 1793, and the Muséum d'Histoire Naturelle created in the process. Twelve chairs were made available, including two in zoology. One, for the study of worms, insects and micro-organisms, was given to Jean-Baptiste-Pierre-Antoine de Monet, Chevalier de Lamarck (1744–1829), whose career had originally been advanced by Buffon. Lamarck, although a minor aristocrat, remained free in Paris throughout the

Figure 3.1   Left: Georges Cuvier, based on a lithograph after a painting by Nicolas Jacques, included in the 1826 edition of *Discours sur les révolutions de la surface du globe*. Right: Jean-Baptiste de Lamarck, based on an 1801 portrait by C.Thévenin.

Revolution. The second zoological chair, for the study of the remaining animals, went to the young Etienne Geoffroy Saint-Hilaire (1772–1844), who promptly arranged for Cuvier to come to Paris and join the Muséum. He taught comparative anatomy in place of the ageing Antoine-Louis-François Mertrud, eventually succeeding to Mertrud's chair in 1802.[3]

In 1795, two years after the reform of the Jardin, the Académie des Sciences was dissolved and re-created as the Institut National, with Cuvier and Lamarck (figure 3.1) being amongst the first members elected to the physical, natural and mathematical sciences division. That provided Cuvier with a major opportunity to develop his career. He quickly impressed his colleagues by his ability, with the result that he was elected secretary in 1800, and then made permanent secretary three years later. After that, he controlled elections to the Institut, as well as preparing the funeral orations for members who had died. By such means, Cuvier was able to promote his own beliefs about the natural sciences, and to discredit those holding different views, often attacking a caricature of an opponent's argument, rather than the real thing.[4] Partly as a result of this, but mostly because of the positive virtues of his own investigations and theories, he soon gained a high reputation as a scientist.[5]

Cuvier was a committed catastrophist (although that particular term had not yet been coined). Not surprisingly, therefore, in line with the legend that catastrophists were driven by religious dogma, he has generally been seen as resorting to supernatural explanations for events in the history of the Earth. So, for example, the science journalist Roger Lewin wrote in his 1993 book, *Complexity*, 'At the beginning of the nineteenth century, the great French geologist and naturalist Baron Georges Cuvier proposed what came to be known as the Catastrophe theory, or Catastrophism. According to the theory, the abrupt faunal changes geologists saw in rock strata were the result of periodic devastations that wiped out all or most extant species, each successive period being repopulated with new kinds of animals and plants, by God's hand'.[6] Similarly, when discussing Cuvier in 1999, another science journalist, Austen Atkinson, claimed, 'Mindful of and committed to the Bible as the source of knowledge regarding the origin of the Earth, he believed in the conclusion of Bishop Ussher . . . that the Earth was created by God in 4004 BC'.[7] Again, although with less explicit references to supernatural intervention, Richard Fortey, a palaeontologist from the Natural History Museum, London, wrote in his 1997 book, *Life – An Unauthorised Biography*, of 'the series of successive "creations" described by the great anatomist Baron Cuvier to account for the sequence of fossil mammal faunas he was studying'.[8] What was the truth? Firstly, let us consider geological issues.

In 1808, Cuvier presented to the Institut the preliminary results of a detailed investigation of the geology of the Paris basin, carried out over many years in collaboration with the mining engineer and mineralogist Alexandre Brongniart.[9] A more detailed report followed in 1812, in Cuvier's book, *Recherches sur les ossemens fossiles*.[10] In the introduction, published separately as *Discours sur les révolutions de la surface du globe*,[11] Cuvier expressed his view that there had been several sudden advances and retreats of the sea. Alternating layers of saltwater and freshwater deposits rested on a thick bed of chalk, whilst overlying the stratified rocks in valley bottoms was a layer of loose material, which he termed 'detrital silt'. The transitions were associated with major catastrophes (which he called *révolutions*) for, on each occasion, almost all the animals and plants then living were destroyed. In the aftermath, new types emerged, according to the evidence of the fossils found in the rocks. The scale was such that the processes involved must have affected an area much more widespread than just the Paris basin. As an indication of the speed of action of the most recent of the *révolutions*, if not for the others, Cuvier drew attention to the discovery of unputrefied carcasses of large extinct mammals such as mammoths, in frozen regions to the north, reports of which had reached Paris in 1807.[12]

In contrast, Cuvier's older colleague Lamarck was a gradualist through and through, accepting the methodological principle that science could deal only with processes which were still taking place.[13] On this basis, he presented in his *Hydrogéologie* of 1802 a theory of the geological history of the Earth, in which the surface features were being constantly re-shaped by mechanisms involving the movement of water.[14] As in Hutton's theory, rain eroded the continents, with rivers carrying the debris to the sea. The rivers could also cut through plateaux, forming mountains and valleys. Because of the direction of the Earth's rotation, river-borne debris tended to be carried to the western coasts of continents, whereas the eastern coasts were the ones likely to be eroded.[15] As Lamarck explained: 'in this imperceptibly slow process, the sea is constantly breaking up, destroying, and invading the continental coasts it encounters on its path; meanwhile, on the opposite coast, the sea is constantly falling, withdrawing from the land it has raised, and forming new continents behind itself, which one day it will return to destroy'.[16]

Thus, to Lamarck, the process was a gradual one, taking place over long periods of time. As in Hutton's model, there was no way of knowing how or when it had started up, and no indications that it would come to an end in the foreseeable future. Lamarck maintained a belief in the uniformity of operation of the laws of nature, and the constancy of action of water on the continents. Although he accepted that upheavals of land could be caused by earthquakes or volcanic eruptions, he did not believe that such phenomena could make a significant contribution to large-scale transformations of the Earth's surface. To him, they always occurred on a strictly localised basis, and were due to an extremely rare combination of circumstances.[17]

As might be supposed, Cuvier took a very different view, but his reasoning was no less rational. He found it impossible to believe that the colossal disruption of rock strata observed in Alpine regions (close to his birthplace of Montbéliard) could have resulted from the everyday processes of weathering and sedimentation, whatever the time-scale, even with occasional help from volcanic eruptions.[18] Indeed, no forces known at the time could account for these features. Geologists now consider that the Alps were formed when Africa collided with Eurasia, as a result of continental drift, a concept only introduced in the twentieth century.[19]

Without doubt, Cuvier was a gifted scientist. As well as his geological work, his major contributions to the field of comparative anatomy cannot be brought into question. Cuvier showed that the anatomical characteristics of each major group of animals are unique and, according to his principle of the 'correlation of parts', the anatomical structure of every organ and bone is functionally related to all other parts of the body. On this basis, the overall characteristics of an animal, and possibly its species, could be identified from the shape of a single bone, which was very useful in the interpretation of incomplete fossils.[20] The novelist Honoré de Balzac, a contemporary, claimed, 'Is Cuvier not the greatest poet of our century? Our immortal naturalist has reconstructed worlds from blanched bones. He picks up a piece of gypsum and says to us "See!" Suddenly stone turns into animals, the dead come to life, and another world unrolls before our eyes'.[21]

Conflicts over evolution

Cuvier's deductions about previous ages were to play an important part in an impassioned debate about whether species could become extinct, and whether they could change into other species by evolution (or, as it was known at the time, transmutation). The Christian Church had very strong views on these issues. After all, *Genesis* made it clear that all types of living creature were created separately and, by implication, had remained exactly as they were since Creation. Again, to the fundamentalist believer in the literal truth of the Bible, extinction and evolution were both excluded by the words of *Ecclesiastes* 3:14, which stated, 'I know that, whatsoever God doeth, it shall be forever: Nothing can be put to it, nor any thing taken from it'. What is more, there was a long-standing belief in a 'great chain of being' (the *Scala Naturae*), linking humankind to the simplest organism. This ruled out the possibility of extinctions because they would create gaps and break the chain whereas, according to the widely accepted concept of *plenitude*, an idea which had been strongly supported by Leibniz, the chain must be continuous.[22] On the other hand, it appeared from the fossil record that extinctions of species or evolution, or both, had taken place, because animals which were completely unknown at the present day had clearly been alive in former times. Cuvier concluded there had been extinctions but no evolution, whereas Lamarck believed that there had been evolution but no extinctions.[23]

The first faltering steps towards a belief that life on Earth could undergo significant modification had been taken in the previous century, for example in the speculations of Benoît de Maillet, which we have already noted (see Chapter 1). In contrast, the core principle of the orthodox belief was that, although individuals were born, matured and died, populations remained relatively unchanged, making up the *Scala Naturae*.[24]

There were, however, two different ways of looking at the great chain of being. Following the tradition of Aristotle, the chain could be imagined as being made up of a series of quite discrete species. Christian philosophers, though, tended to concentrate on individuals rather than species and pointed out that, because of the great range of forms in existence, it would be virtually impossible to identify a point in the chain where one link ended and the next began. Thus, the founder of Methodism, John Wesley (1703–1791), writing in his *Survey of the Wisdom of God in Creation*, claimed, 'The whole progress of nature is so gradual, that the entire chasm from plant to man is filled up with divers kinds of creatures, rising one above another by so gentle an ascent that the transitions from one species to another are almost insensible'.[25]

Indeed, to the French philosopher Jean-Baptiste Robinet, writing in the 1760s, species could not exist, because there must be individuals corresponding to every possible point along the sequence, making each link infinitesimally small. Hence, the apparent identification of separate species was an illusion.[26]

To naturalists, however, it was clear that individuals tended to belong to one or another of the distinct groups known as species. In 1764, the Swiss naturalist Charles Bonnet compiled a list of species which he believed formed a linear sequence from the lowest form of life to the highest, humankind. Later, in his *Palingénésie philosophique* of 1769, Bonnet suggested that the sequence was not static, but one which showed progressive change with time. In his view, the lower forms of life were created first, and higher forms developed from these according to God's pre-ordained plan. Extinctions caused by geological catastrophes were also part of the plan, clearing space for new species.[27]

Before then, John Ray (1627–1705), an English clergyman and naturalist, had suggested in 1691 in his book, *Wisdom of God in the Creation*, that God had originally created distinct species, but variations subsequently arose in individual members as a result of changing conditions. Ray introduced the idea of a species being a group of animals or plants capable of breeding with each other, a concept still in general use today, and he began attempts at classification.[28]

This was taken much further in the middle of the eighteenth century by Linnaeus, who grouped animals and plants into species, on the basis of structural characteristics, and then further grouped species into genera. Thus each species was given two names, the first being the name of the genus (the group of similar species) and the second an adjective defining the species itself. The same classification system is used today, as in the well-known binomial for our own species, *Homo sapiens* (and has been extended, so that genera are now further grouped into families, families into orders, orders into classes, classes into phyla and phyla into kingdoms).[29]

Cuvier and Geoffroy Saint-Hilaire improved Linnaeus's original approach by formulating the principle of the 'subordination of characters'. This maintained that features essential to an organism's existence (e.g. characteristics of the nervous system in animals) should be given greater weighting in classification than more superficial features. Cuvier also broke with the concept of the chain of being by grouping animals, on the basis of patterns of body organisation, into four 'types': vertebrates; molluscs; articulates (insects, worms, etc.); and radiates (e.g. starfish). These types, with some modifications, have formed the basis of the phyla of the present day.[30]

In contrast, Buffon, an opponent of Linnaeus, had argued that many of the groups classified as separate species were really variant forms resulting from migration into areas with different environmental conditions. If they were brought back together, he was sure they would be able to interbreed, casting doubt on the whole concept of species.[31]

Lamarck, although a protégé of Buffon, was prepared to accept the existence of species, but he retained the notion of the possibility of limited change. Eventually, when he was over fifty years old, he went further and suggested that one species could actually be transformed into another as environmental conditions changed. In some ways, Lamarck's ideas can be seen as an extension of those of his mentor (and of Bonnet), allowing for progress along a chain of being. Like Buffon, Lamarck also believed that lower organisms could arise by spontaneous generation from inanimate matter, without God having to be involved in the process.[32]

Lamarck's theory, outlined in 1801 in his *Système des animaux sans vertèbres*, and developed in a variety of ways in subsequent books, had two main principles. Firstly, Lamarck believed that some internal mechanism forced evolution onwards, generally causing an increase in the complexity and size of individuals in a species over successive generations. This inherent drive has been given various names, but is generally called orthogenesis, or sometimes autogenesis. The supposed trend towards larger individuals is often called 'Cope's Law', after the American naturalist, Edward Drinker Cope, even though he lived a generation after Lamarck. The second principle in Lamarck's theory was that an organ or other feature developed by use, as a consequence of some 'need', would be passed on in its improved form to a descendant, whilst a feature which became atrophied through disuse would eventually be lost altogether. In other words, inheritance of acquired characteristics would take place. Lamarck believed that, rather than becoming extinct, a species would generally evolve into something else, a supposition which inevitably led him into conflict with Cuvier. The latter was adamant that no evolution of species could have occurred, challenging Lamarck and, later, Geoffrey Saint-Hilaire, to produce evidence of fossil forms representing transitional stages between the different species. (Note in passing that Geoffroy became receptive to the possibility of evolution as he became increasingly aware of the similarity of features of different species within the same animal 'type'. However, in contrast to Lamarck, he thought that the significant developments in response to environmental change were more likely to take place in embryos rather than adult animals.)[33]

No examples of transitional forms could be given but, in his *Philosophie zoologique* of 1809, Lamarck emphasised the general plausibility of his ideas, arguing that the observed differences between fossil species and ones still living could be completely explained by a gradual transformation of life forms as conditions slowly changed, making it unnecessary to think in terms of extinctions. Lamarck did not deny the possibility that some life forms might die out completely rather than survive through change, but he nevertheless believed strongly that such extinctions, if they ever occurred, would be just rare and unimportant exceptions to what normally happened.[34]

Cuvier and the causes of catastrophes

Regardless of arguments about evolution, Cuvier's knowledge of comparative anatomy helped to cast light on the relationship between fossils and living species, establishing beyond question that some fossilised animal bones were from species quite distinct from those still living. Another eminent French geologist, Louis-Constant Prévost (1787–1856), agreed with Cuvier that the more ancient a stratum, the greater were the differences between the fossils it contained and the species alive at the present time.[35]

Much evidence had come from the Paris basin, where the alternating layers of saltwater and freshwater deposits showed that, at intervals in the Earth's history, the area had been submerged beneath the seas, whereas at other times it had been dry land. Prévost formed the opinion that the movements of the water, backwards and forwards, had occurred in gradual fashion. In contrast, Cuvier, as we have seen, concluded that the transitions must have been rapid, forming part of more extensive catastrophic events, which caused the extinction of many species. He wrote, in his *Discours*, 'Living organisms without number have been the victims of these catastrophes. Some were destroyed by deluges, others were left dry when the seabed was suddenly raised; their races are even finished forever, and all they leave in the world is some debris that is hardly recognizable to the naturalist'.[36]

But how had each area been re-populated? It would be easy to assume, given the legend of unscientific catastrophism, that Cuvier must have been a believer in Creation by divine intervention (see, for example, the quotation from Lewin's *Complexity*, given near the beginning of this chapter). Cuvier was indeed a committed Christian, a member of the small French Protestant community. However, he never suggested the involvement of a supernatural mechanism in any of his published works, nor did he try to establish any linkage between fossil evidence and the teachings of the Church. On the contrary, he pointed out that the *révolutions*, however destructive they may have been, might not have been entirely global in scale. Hence, animals could have continued living in unaffected parts of the world, and later migrated into the depopulated areas where the catastrophes had taken place. There was therefore no necessity to have to speculate about the possibility of new acts of creation.[37] Thus, in the *Discours* he suggested that, after the most recent of the catastrophic events, 'the small number of individuals spared by it have spread out and reproduced on the land newly laid dry'.[38] The question of how these had come into existence in the first place was never addressed. Steven Stanley claimed, in *Extinction*, that Cuvier favoured the idea that they 'were species of the original biblical creation',[39] but that was nothing more than supposition.

The science historian, Martin Rudwick, wrote in his 1997 book, *Georges Cuvier, Fossil Bones, and Geological Catastrophes*, 'It is of course possible that Cuvier privately believed in some kind of supernatural causation for new species or for the origin of reptiles and mammals, but there is no historical evidence for it'.[40] Similarly, in *The New Catastrophism*, published shortly after the author's death, in 1993, Derek Ager, a geologist from University College, Swansea, called Cuvier a 'much misunderstood man' and referred to the 'false linking of Cuvier with the biblical fundamentalists'.[41] The notion that Cuvier derived his ideas from the scriptures largely arose in the English-speaking world from a translation of the *Discours* commissioned by Robert Jameson. This included a preface and extensive footnotes written by Jameson, in which he expressed his own views, but gave the impression they were those of Cuvier.[42] The latter undoubtedly used the term 'creations' in his writings, but that was common amongst the naturalists of the period, and did not necessarily imply an origin by divine intervention.[43] As already stated, Cuvier avoided speculation as a matter of principle, and attempted to limit himself to conclusions which could be justified by scientific evidence. Similarly, he was not prepared to speculate on the cause of the *révolutions*, even though he was aware of some possibilities.

In 1742, Pierre-Louis-Moreau de Maupertuis, one of the earliest French supporters of Newton's laws of planetary motion, wrote of the devastating effects which would result from the collision with a comet.[44] He claimed, 'Comets have occasionally struck the Earth, causing extinction by altering the atmosphere and oceans'.[45] Half a century later, in 1796, Pierre-Simon, Marquis de Laplace, expressed similar views in his

Exposition du système du monde, the same book in which he proposed the condensation hypothesis for the origin of the Solar System which, with some modifications, is still accepted today (see Chapter 19). Laplace, who had observed the close approach of Comet Lexell in 1770, suggested that an impact or even a glancing encounter with a comet might change the axis of rotation of the Earth and cause cataclysmic floods.[46] He wrote, 'The seas would abandon their ancient positions, precipitating themselves towards a new equator; a great portion of the human race and the animals would be drowned in the universal deluge, or destroyed by the violent shock imparted to the terrestrial globe; entire species would be annihilated; all the monuments of human history destroyed'.[47]

However, the possible dangers from comets were not taken very seriously, except by a few eccentrics. One of these was the writer on gastronomy Jean-Anthelme Brillat-Savarin who, in the middle of his 1825 book, *Physiologie de goût*, suddenly turned to the subject of cometary catastrophes, writing, 'Let us suppose . . . that some comet passes close enough to the Sun to be charged with superabundant heat, and comes close enough to Earth to cause for the space of six months a temperature of 168 degrees . . . At the end of that fatal season, all living things, both animal and vegetable, will have perished; every sound will have died away; the earth will revolve in silence, until new circumstances develop new germs'.[48] However, the mainstream view was much more sceptical about the dangers from comets. So, for example, the French scientist François Arago argued, early in the nineteenth century, that the chances of a collision with the Earth, for any particular comet, were no more than 1 in 281 million.[49]

Cuvier must have been very familiar with the ideas of Laplace, because the latter was one of the most prominent members of the Institut National in its early days, being elected Vice-President in 1795 and President in the following year. Indeed, *Recherches sur les ossemens fossiles* was dedicated to him.[50] However, despite that, Cuvier remained silent about whether cometary impacts, divine wrath or, indeed, anything else, was the likely cause of his *révolutions*. He said, with characteristic caution, during the first public session of the Muséum d'Histoire Naturelle, in 1796:

> But what then was this primitive earth where all the beings differed from those that have succeeded them? What nature was this that was not subject to man's dominion? And what révolution was capable of destroying it to the point of leaving as trace of it only some half-decomposed bones? But it is not our duty to engage ourselves in the vast field that these questions present. Let hardier philosophers undertake it. Modest anatomy, limited to detailed examination, to the meticulous comparison of objects submitted to its eyes and to its scalpel, will content itself with the honour of having opened this new route to the genius who will dare to travel along it.[51]

In 1829, three years before Cuvier's death, the mining engineer Léonce Élie de Beaumont (1798–1874) proposed a mechanism for the *révolutions* which was simply an extension of Buffon's ideas, and had no requirement for any factors, natural or supernatural, from beyond the Earth. He argued that even if the Earth was cooling slowly and gradually as Buffon had proposed, and that the reduction in volume led to mountain building, then this latter process was likely to occur in an episodic and catastrophic fashion, with upheavals of submerged land creating gigantic and destructive waves. This model had occurred to Élie de Beaumont whilst he carrying out extensive fieldwork, in partnership with Ours-Pierre Dufrénoy, for a detailed geological map of France. His observations indicated that different mountain ranges had been elevated at different times, which suggested there had been a series of convulsions of the Earth's crust.[52] Cuvier, however, remained true to his principles and never expressed an opinion on possible causal mechanisms, believing he had insufficient evidence to do so.

Politics, religion and science

According to the long-enduring myth, the nineteenth century catastrophists were all given to rash speculation, whilst the gradualists of the same period interpreted the evidence in much more sober fashion. However, if we compare the gradualist, Lamarck, with the catastrophist, Cuvier, we come to a very different conclusion. Whereas Cuvier was reluctant to put forward speculative theories, Lamarck was more than willing to do so, about a variety of topics. As we have already seen, he advanced a model for the geological history of the Earth, and also a mechanism for the transformation of species. Given the changing intellectual climate, and Cuvier's political skills, the outcome, at least in the short-term, was predictable: Cuvier was able to portray himself to his colleagues as a dedicated scientist, concerned solely with topics in which he was an acknowledged expert, whereas Lamarck was seen to be putting forward general theories in the outdated fashion of the speculative cosmogonists, and confronting the specialists in more disciplines than was credible.[53] As might be expected in the circumstances, Cuvier's reputation continued to grow, and in 1831 he was made a Baron, a rare honour for a French Protestant.[54] In contrast, Lamarck faded from the scene. When he died, he was buried in a pauper's grave, and for many years his theories were known mainly through a distorted funeral oration written by his rival, which claimed that Lamarck had written:

> It is the desire and the attempt to swim that produces membranes in the feet of aquatic birds; wading in the water, and at the same time the desire to avoid getting wet, has lengthened the legs of such as frequent the sides of rivers, and it is the desire of flying that has converted the arms of all birds into wings, and their hairs and scales into feathers.[55]

Lamarck's actual words, in the preface to his *Système des animaux sans vertèbres* of 1801, were somewhat different, as is clear even from an English translation:

> The bird attracted to water, because there it can find the prey necessary for its existence, spreads the toes of its feet when it wants to hit the water and move on

its surface. The skin connecting these digits at their base eventually becomes accustomed to stretching. Thus, over time, the large membranes that link together the toes of ducks, geese, etc., have been formed as we see them today . . . In the same way, one may perceive that the shore-dwelling bird, which dislikes swimming but must go up to the water to find its prey, will be constantly obliged to wade in the mud. Wishing to keep its body out of the liquid, it will acquire the habit of stretching and elongating its legs. As a result of the succession of generations and generations of birds which continue to live in this manner, individuals eventually find themselves elevated as if on stilts, on long bare legs . . . [56]

It cannot be denied that Lamarck wrote of an organism 'wishing' to develop a certain characteristic, which enabled Cuvier to make fun of his rival and suggest that he believed the transformation could be brought about simply by desire. However, it is clear from the context that Larmarck was just using a figure of speech to convey that the development took place through use, because of actions resulting from the organism's needs, without implying any involvement of thought in the process.[57]

In the manuscript of *Recherches sur les ossemens fossiles* is an even more distorted account of the same passage from *Système des animaux sans vertèbres*. Although removed before publication, it clearly demonstrates the author's satirical method. Here, Cuvier suggested it was Lamarck's belief that:

> ducks by dint of diving became pikes; pikes by dint of happening upon dry land changed into ducks; hens searching for their food at the water's edge, and striving not to get their thighs wet, succeeded so well in stretching their legs that they became herons or storks. Thus took form by degrees those hundred thousand diverse races, the classification of which so cruelly embarrasses the unfortunate race that habit has changed into naturalists.[58]

It was not in Cuvier's nature to miss a chance to promote his cherished theories. In consequence, in a lecture course given at the Athenée de Paris in 1805 for politicians and other prominent members of society, Cuvier did something he always avoided in his published writings, and suggested scriptural as well as scientific reasons for preferring his own ideas to those of Lamarck. According to a report produced by the Italian count Marzari Pencati Cuvier divided the history of the Earth into six main eras, paralleling the six days of Creation described in *Genesis*, with living creatures appearing only in the fifth era. This may have occurred relatively recently, with the present continents perhaps being less than ten thousand years old. Whatever the reason, *Genesis* was consistent with the geological record, for fossil fish were found in older strata than fossil mammals, and human fossils had been located in only the most recent layers. Pencati suggested that Cuvier was 'on the lookout for a cardinal's hat', which would have been a strange ambition for one of his denomination. Indeed, there is every reason to think that Cuvier had different motives. He knew that his influential audience would be receptive to arguments based on religion, so, seeing yet another opportunity to advance his scientific career, he hinted (if no more) at scriptural support for his theories. However, whilst such tactics might have been of benefit to him in the short-term, they made it easy for geologists of later generations to believe that his ideas arose from religious dogma, and so were of dubious value.[59]

In fact, as already noted, Cuvier was more than prepared to accept scientific evidence which was incompatible with Christian tradition. In particular, he believed that there had been several major catastrophes, not just one, and that many animal species had become extinct during the course of the Earth's history. Similarly, he concluded from the geological record that a very long time had elapsed before animal life first appeared on Earth, not just the 'four days' of *Genesis*. In the reports he wrote with Brongniart about the geology of the Paris basin, he made no mention whatsoever of the supposition that the Earth may be very young, refusing to speculate on the time-scale of the series of catastrophic events. Even on the occasion Pencati had referred to, when he claimed that Cuvier 'doesn't even assign them ten thousand years', the celebrated anatomist was clearly referring to the time since the present continents surfaced above the waves, not indicating the age of the Earth.[60] Similarly, when Cuvier wrote in the *Discours* that it 'cannot be dated much further back than five or six thousand years ago', he was talking in unambiguous terms about the most recent *révolution*.[61] Indeed, in a memoir of 1804, describing fossil bones of a small quadruped, found near Paris, Cuvier demonstrated beyond any doubt that he accepted the concept of a long time-scale, writing, 'They are the imprint of an animal of which we find no other trace, of an animal that, buried perhaps for thousands of centuries, reappears for the first time under the eyes of naturalists'.[62]

For all that, Cuvier remained adamantly opposed to the concept of evolution. It has generally been thought that this was because of his adherence to Christian dogma but, since his views were in significant conflict with the Bible in other respects, that cannot have been the reason. Cuvier's main opposition to the possibility of evolution arose from his knowledge of comparative anatomy, which showed that each organism possessed a high degree of structural organisation. His principle of 'correlation of parts' led to another principle, the 'prerequisites of existence', which suggested that the structural and functional relationships of all the organs in an individual were closely linked to the requirements for survival. Lamarck believed that, if these requirements changed, as they would if the environment changed, processes would be set in motion that would transform an organ, or even create a new one. Cuvier did not doubt that something along these lines could happen, but it had to be maintained within strict limits. This was because a major change in one organ would most likely disrupt the crucial balance between the various functions in the animal as a whole, making it unable to survive. Because of this important consideration, the successful evolution of one species into another, as opposed to variation within the same species, seemed to be outside the bounds of possibility.[63]

Cuvier may also have been worried about whether an acceptance of evolution would have an adverse effect on his primary interests, comparative anatomy, classification and the extinction of species. The science of comparative anatomy, for example, undoubtedly relies on the existence of a system of types with stable characteristics. Apparently Cuvier once mentioned to Prévost that he believed in evolution, but preferred not to say so in public because it would destroy all the work that had been done in classifying species.[64] He may also have thought that it would undermine his arguments that major extinctions had occurred at the times of the great geological *révolutions*.[65] The irony is that, whatever Cuvier's motives in saying what he did to Prévost, his failure to appear receptive to the possibility of biological evolution was used by later generations to discredit both him and his arguments for extinctions.

4   Natural theology and Noah's Flood: the high-water mark of catastrophism

God in nineteenth century Britain

In the previous chapter, we saw how the Frenchman Georges Cuvier made a determined effort (most of the time) to keep his science and his religion separate. Such an attitude would have been unusual at the other side of the English Channel, because of the continuing influence of a philosophy known as natural theology. This was a synthesis of science and religion, which maintained that the main purpose of studying nature was to find, in its harmony and purposiveness, evidence for the existence of a Creator. Natural theology was popular during the Renaissance, and still dominated the writings of later scientists, such as Leibniz and Linnaeus, but it faded out in continental Europe during the eighteenth century. However, in Britain, where it had become prominent in the late seventeenth century through, for example, the sermons of Richard Bentley (see Chapter 1) and John Ray's *Wisdom of God in the Creation* (see Chapter 3), natural theology continued to be influential. It helped shape the attitudes of Hutton (which we noted in Chapter 2), and remained an important consideration well into the nineteenth century.[1]

The year 1802 saw the appearance of *Natural Theology*, an influential book by William Paley, Archdeacon of Carlisle, who was also an amateur naturalist. Paley's main theme, developing one introduced by Ray, was that the design of each animal, whose component parts fitted together to form a successful, functioning organism, pointed to the existence of a Creator, just as the design of a watch was evidence of the existence of a watchmaker.[2] Several decades later, between 1833 and 1836, came the *Bridgewater Treatises*, eight works commissioned by the Earl of Bridgewater in his will, as an act of repentance for the sins of his life. They were intended to demonstrate the 'Power, Wisdom and Goodness of God as manifested in the Creation'.[3]

A significant number of prominent British scientists of the early nineteenth century were clergymen, for this was still a requirement for obtaining a senior post at either Oxford or Cambridge Universities.[4] Of the clergymen–scientists at Cambridge, John Stevens Henslow (1796–1861) and William Whewell (1794–1866) were successive occupants of the chair of mineralogy, the former subsequently becoming Regius Professor of Botany and the latter Professor of Moral Philosophy. Similarly,

Adam Sedwick (1785–1873) held the title of Woodwardian Professor of Geology at Cambridge whilst, at Oxford, William Buckland (1784–1856) was Reader in the same subject, after five years as Reader in Mineralogy.

When Charles Darwin, acclaimed by posterity as one of the greatest of all scientists, was an undergraduate student at Cambridge, seemingly destined to become a clergyman himself, he attended the lectures of both Henslow and Sedgwick, and they did much to encourage his growing interest in biology and geology. Darwin became known as 'the man who walks with Henslow', during the course of which they discussed innumerable scientific topics. Henslow also played a major role in getting Darwin a place as gentleman naturalist on H.M.S. *Beagle*'s voyage of exploration to South America and beyond, an event which was to have enormous consequences for the history of science (see Chapters 7 and 8).[5] Sedgwick was acknowledged as the leading British geologist of his day, and was President of the Geological Society of London from 1829 to 1831.[6] His colleague Whewell was a Fellow of the Royal Society, President of the Geological Society in 1837, President of the British Association four years later and eventually Vice-Chancellor of the University of Cambridge. He helped name several geological epochs, and wrote the most popular *Bridgewater Treatise*, which covered a wide range of the physical sciences. His recommended approach was that a scientist (a term which was coined by Whewell himself) should use both knowledge and intuition to formulate a hypothesis, and then test and refine it by further experiment and observation.[7] The final member of this quartet, Buckland, was a Fellow of the Royal Society, President of the British Association in 1832 and author of the *Bridgewater Treatise* on geology, before becoming the Dean of Westminster.[8]

As will be clear from the above, Henslow, Sedgwick, Whewell and Buckland were all major figures in the British scientific establishment in the first half of the nineteenth century. As clergymen, they would never have doubted the existence of an omnipotent God but, nevertheless, they took their responsibilities as scientists very seriously. Without question, they were sincere in their search for truth, and were prepared to change their opinions in the light of new evidence, provided this was strong enough. Such open-mindedness brought them into conflict with William Cockburn, John Keble and others within the Church who had what we should now call fundamentalist views, believing the testimony of the Bible in unquestioning fashion.[9] The geologists, Buckland and Sedgwick (figure 4.1), had a particularly difficult time, because of the nature of their subject area. They were keen to operate as independent scientists, yet, as a consequence of their background, they began with an expectation that fieldwork would rapidly confirm the essential features of the *Genesis* account.[10] That guided their initial interpretations of what they observed but, eventually, they came to a realisation that the evidence pointed to a somewhat different conclusion.

Figure 4.1   Left: William Buckland, based on an 1832 portrait by Samuel Howell. Right: Adam Sedgwick, based on an 1850 lithograph by T.H. Maguire.

Buckland, Sedgwick and the Flood

Early in his career, Buckland interpreted a widespread layer of loam and gravel, corresponding to Cuvier's 'detrital silt', lying on top of stratified formations, as the product of the universal deluge in the time of Noah. He was disturbed by the fact that there were no obvious traces of humans from before the Flood, as ought to have been the case from the *Genesis* account, and the immense depths of deposits beneath the loam and gravel layer suggested that the Earth must be very old, with Creation taking far longer than the six days mentioned in the Bible. Nevertheless, the evidence for the deluge itself seemed very clear. Fossils had been found in mud deposits in caves in Kirkdale, Yorkshire, and throughout Europe, and these must surely have been the remains of animals trapped by the rising flood water. In *Vindiciae geologica* ('Geology Vindicated'), his inaugural lecture as Reader in Geology at Oxford in 1819, Buckland argued, 'The grand fact of a universal deluge at no very remote period is proved on grounds so decisive and incontrovertible, that had we never heard of such an event from scripture or any other authority, geology of itself must have called in the assistance of some such catastrophe'.[11]

The case was presented in detail in his *Reliquiae diluvianae* (*Relics of the Flood*), published in 1823. As Charles Coulston Gillespie wrote during the 1950s, in *Genesis and Geology*, 'Diluvial geology . . . succeeded in winning, for a time, nearly unanimous assent both to its assumptions and conclusions, though these were practically the same thing. Buckland came out right where he went in, with the ark, but he brought with him a lot of new information picked up in the course of his circular process'. What, then, about the causes of the Flood? According to the University of Manchester Earth-scientist Richard Huggett, writing in 1997, 'Buckland and his supporters . . . all believed in God's special intervention in the regular course of Nature through geological catastrophes and the sudden rise of species'.[12]

However, *Reliquiae diluvianae* avoided any mention of miraculous events and cosmogony-style speculation, concentrating instead on evidence which appeared to demonstrate that a single, major flood had taken place. Buckland suggested that the deluge might have been associated with a change in climate, but he never even hinted that this might have been the result of an intervention by God.[13] On the contrary, he wrote, 'What this cause was, whether a change in the inclination of the earth's axis, or the near approach of a comet, or any other cause or combination of causes purely astronomical, is a discussion of which is foreign to the object of the present memoir'.[14]

Sedgwick agreed with Buckland's views on the interpretation of the evidence,[15] writing in 1825, 'The sacred record tells us – that a few thousand years ago . . . the earth's surface was submerged by the waters of a general deluge; and the investigations of geology tend to prove that that the accumulations of alluvial matter have not been going on many thousand years, and that they were preceded by a great catastrophe which has left traces of its operation in the diluvial detritus which is spread out over all the strata of the world'.[16]

However, it soon became apparent that the loam and gravel layer was restricted to northern latitudes, so it could not be taken as evidence of a universal flood. Also, as investigations continued, it became clear that the fossils in the various caves did not all come from the same period.[17] Despite the testimony of *Genesis*, Buckland announced in his *Bridgewater Treatise* of 1836 that he no longer believed in a single, universal flood. Sedgwick had done the same several years earlier, during his farewell presidential address to the Geological Society in 1831, when he acknowledged that he and his colleagues had been led astray by their expectation of finding evidence of Noah's Flood, and continued, 'There is, I think, one great negative conclusion now incontestably established – that the vast masses of diluvial gravel, scattered almost over the surface of the earth, do not belong to one violent and transitory period'.[18]

Nevertheless, although Buckland and Sedgwick came to reject the idea of a single Flood, they retained their catastrophist views, continuing to find the evidence strongly suggestive of the involvement of cataclysmic forces. It was just that they now thought these had acted on more than one occasion, just as Cuvier had supposed. Another prominent British geologist, William Conybeare (1787–1857), also concluded that there had been several catastrophic floods. He came to that view after carrying out extensive fieldwork, particularly in the Thames valley, presenting his findings to the Geological Society in London in 1829. Buckland, Sedgwick and Conybeare, like all the geologists of the period, paid a great deal of attention to the large erratic boulders found throughout much of Britain, Europe and North America, as well as to the unstratified loam and gravel deposits scattered over these same regions. In an attempt to explain the origin of these features, supposedly linked to catastrophic floods, William Whewell and others developed theories of tidal waves from the 'cooling Earth' scenario of Léonce Élie de Beaumont.[19]

As we saw in Chapter 3, Élie de Beaumont believed that, as the Earth cooled over the course of time, natural shrinkage gave rise to episodic large-scale disruptions of the crust. During each of these intermittent upheavals, mountain building took place because of 'wrinkling' of the Earth's crust, volcanoes erupted in many areas, some former continental regions were flooded by sea water, and many species became extinct.

These ideas were well-received when they reached Britain, shortly after their publication in *Annales des sciences naturelles* in 1829, and an English translation appeared in the *Philosophical Magazine and Annals* in 1831. It seemed that an explanation had been found for the series of cataclysms which the British geologists now believed to have occurred.[20] In his final presidential address to the Geological Society, Sedgwick enthused, 'I am using no terms of exaggeration when I say, that in reading the admirable researches of M. de Beaumont, I appeared to myself, page after page, to be acquiring a new geological sense, and a new faculty of induction; and I cannot express my feelings of regret, that during my recent visit to the Eastern Alps I did not possess this grand key to the mysteries of nature'.[21]

Nevertheless, only a few years later, the concept of recurrent catastrophic floods no longer found general favour. This was because Charles Lyell (1797–1875) established the belief that the only significant processes bringing about changes to the Earth's surface were the everyday ones familiar to his contemporaries, acting over very long periods of time. Lyell was the son of a wealthy Scottish landowner, and he was originally intended for the legal profession. However, he attended Buckland's lectures when he was a student at Oxford, which turned his interests towards geology. He subsequently operated largely as a self-funded geologist, supported by his writings, but nevertheless he was Professor of Geology at King's College, London, from 1831 to 1833, and was knighted for his services to the subject in 1848.[22]

At first, Lyell was prepared to find a geological role for floods, even very major ones, provided there was no suggestion that they had arisen in catastrophic fashion. In his marine erosion theory, a sub-section of his more general world-view, Lyell argued that many features of the landscape originated when seas were covering land. Migrating icebergs could have scraped against the rocks below, creating valleys both small and large, and, as the ice melted, they could have deposited boulders and gravel on whatever lay underneath. Other features could have arisen when the seas advanced or receded, which, of course, in his opinion, took place very gradually. However, despite the general success of Lyell's views (see Chapters 5 and 6), his marine erosion theory was soon abandoned, when the British geologist Joseph Beete Jukes (1811–1869) argued convincingly for the 'fluvialist' view that the action of rivers over many thousands of years was sufficient to explain the characteristics of even the deepest valleys.[23]

Furthermore, largely because of the work of Louis Agassiz (1807–1873), a Swiss naturalist who eventually moved to the United States and became Professor of Zoology at Harvard University, it quickly became accepted that the erratic boulders and drift deposits had been carried by glaciers during an 'Ice Age', not by tidal waves or floating icebergs.[24] Benjamin Silliman, of Yale University, and other American geologists, continued to maintain a belief in the reality of Noah's Flood well into the 1840s, by which time Buckland, previously the most enthusiastic British advocate of the involvement of catastrophic floods in the history of the Earth, had become a convert to the glacial theory.[25] The tide of geological opinion was turning, in a profound fashion.

5 Catastrophism, uniformitarianism and idealist philosophy

Interpreting the record of the rocks

Having focused narrowly on floods in the previous chapter, let us now take a broader look at the views of the British catastrophists and gradualists of the period, to see what they had in common, and to identify the nature of their disagreements.

During the first half of the nineteenth century, there was great interest in trying to deduce the history of past ages from the record left in the rocks. In Britain, it has been common to think that this arose largely from the work of William Smith (1769–1839), popularly known as 'Strata' Smith, an industrial surveyor by profession. However, without question, many British geologists learned their techniques from the continental tradition, as exemplified by the maps of Élie de Beaumont and others. Later British geologists were reluctant to admit these overseas influences.[1] Smith's contribution was, nevertheless, a significant one. In 1815, he produced a *Map of the Strata of England and Wales*, each stratum being characterised by its unique fossils. On the basis of the fossil evidence viewed in comparison to living organisms, it was clear that lower strata were generally older than higher ones, as Nicolaus Steno had suggested 150 years previously.[2]

The pioneering studies of Smith, Cuvier, Brongniart, Élie de Beaumont and Omalius d'Halloy stimulated a great deal of further activity, catastrophists and gradualists alike making significant contributions. Cuvier classified rocks in the Paris basin, in increasing order of age, as Recent (later termed Quaternary), Tertiary and Secondary formations, following a system derived from Werner and, more particularly, from the Italian geologist Giovanni Arduino (1714–1795).[3] Rocks older than Secondary formations were discovered elsewhere, and investigated in detail. So, for example, Adam Sedgwick and Sir Roderick Impey Murchison, a former soldier, studied and named the Silurian and Cambrian systems in Wales.[4]

In 1822, William Conybeare, together with John Phillips (1800–1874), nephew of William Smith, wrote a classic book, *Outlines of the Geology of England and Wales*. Showing how knowledge had advanced since the time of Hutton, the authors used fossils as markers to compare strata in different locations. In 1841, Phillips, then working for the British Geological Survey and later to succeed Buckland as Reader in Geology at Oxford University, eventually becoming Professor, classified fossil-bearing rocks into those of the Cenozoic Era (the age of new animal life), the Mesozoic Era

Figure 5.1   Left: William Whewell, based on an 1866 engraving by 'S.T.'.
Right: Charles Lyell, based on an 1849 lithograph by T.H. Maguire.

(the age of middle animal life) and the Palaeozoic Era (the age of ancient animal life).
These three Eras, which together came to be regarded as making up the Phanerozoic
Eon (the age of visible life) were further divided into Periods. Thus, the Cenozoic Era,
the one closest to the present day, consisted of the Quaternary and Tertiary Periods.
Before then, the Mesozoic Era (comprising the Secondary formations) was made up of
the Cretaceous, Jurassic and Triassic Periods, whilst the Palaeozoic Era (comprising
fossil-bearing rocks older than Secondary formations) included the Permian, Silurian
and Cambrian Periods.[5] Charles Lyell (figure 5.1), with assistance from Gérard Paul
Deshayes, divided up the Tertiary Period into Epochs on the basis of the percentage of
living species found in the rocks, the names of the Epochs (such as Pliocene, Miocene
and Eocene) being formulated with help from William Whewell.[6] These classifications,
all introduced before 1850, form the basis of the system which is still in general use at
the present time (figure 5.2).

As a result of the work of these various geologists, a consensus emerged about
the outlines of the Earth's history. However, there were major disputes about some
of the details, particularly concerning the transitions between geological periods.
Sedgwick, Whewell, Conybeare, Murchison and others argued that the new information supported a catastrophist view, at least at these points, for there were sharp
breaks between some geological formations, and corresponding discontinuities between fossil populations, which seemed to indicate sudden changes in conditions.
Charles Lyell, in contrast, being a convinced gradualist, believed that all geological
processes had a long time-scale, and tried to find alternative explanations for the
apparent breaks.

To subsequent generations, raised on the legend of a dichotomy between
'scientific' gradualism and 'dogmatic' catastrophism, only the work of Lyell was worthy

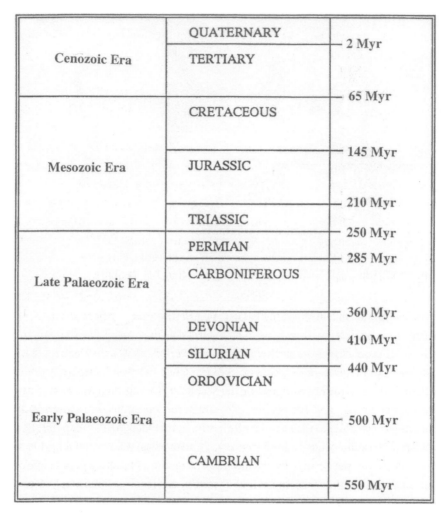

	QUATERNARY	2 Myr
	TERTIARY	
Cenozoic Era		
		65 Myr
	CRETACEOUS	
Mesozoic Era		145 Myr
	JURASSIC	
		210 Myr
	TRIASSIC	
		250 Myr
	PERMIAN	285 Myr
Late Palaeozoic Era	CARBONIFEROUS	
		360 Myr
	DEVONIAN	
		410 Myr
	SILURIAN	
		440 Myr
	ORDOVICIAN	
Early Palaeozoic Era		500 Myr
	CAMBRIAN	
		550 Myr

Figure 5.2   The geological periods, with approximate dates, as millions of years (Myr) before the present, obtained by radiometric methods. The Tertiary Period is sub-divided into Epochs, namely the Palaeocene (65–55 Myr ago), Eocene (55–35), Oligocene (35–25), Miocene (25–6) and Pliocene (6–2 Myr ago). The main part of the Quaternary Period is termed the Pleistocene Epoch, at the end of which (around 11,500 years ago) came the transition to the Holocene.

of note. So, Arthur Strahler, in the 1976 textbook *Principles of Earth Science*, wrote, 'Thanks to Lyell's efforts, stratigraphy became established on a sound basis'.[7] The contributions of the catastrophists were ignored, which was manifestly unfair. Similarly, in 1986, in *Geological Science*, Andrew McLeish wrote how Lyell 'unified the work of Hutton, Werner, Hall, Cuvier, Smith and Brongniart'.[8] Yet, Lyell, like Hutton, was far from being a purely dispassionate investigator. That does not diminish his achievements, for he undoubtedly made major contributions to geology. Were it not for the legend, it would hardly seem necessary to point out that, like any human being, he had strengths and weaknesses. However, whatever his limitations as a strictly objective scientist, Lyell was unrivalled as a persuasive advocate of a particular philosophy. The catastrophists were heading for trouble.

Prominent amongst the catastrophists was William Whewell, of Cambridge University (see figure 5.1). The British science historian Robert Young, writing in 1985 in his book, *Darwin's Metaphor*, claimed that 'Whewell was the most articulate and sophisticated interpreter of the providential view which was associated with the belief that God intervened in the course of nature with catastrophic alterations of its geology and its complement of species'.[9] That was consistent with the legend. Nevertheless, it was highly misleading, certainly as far as geology was concerned, to suggest that Whewell believed in wholesale and direct interventions by the Creator. Young should have known better. Featured on the frontispiece of the first edition of Charles Darwin's *On the Origin of Species*, a major source of material for Young's own book, was a quotation from Whewell's *Bridgewater Treatise*, written in the mid-1830s, which stated, 'But with regard to the material world, we can at least go as far as this – we can perceive that events are brought about not by insulated interpositions of Divine power, exerted in each particular case, but by the establishment of general laws'.[10]

Indeed, as we noted in the previous chapter, he and many other major British catastrophists had, several years earlier, adopted Élie de Beaumont's model in which the shrinking of a naturally cooling Earth caused intermittent episodes of turmoil and destruction, of decreasing intensity. At first, it was widely thought that catastrophism had been given what it had hitherto lacked, a mechanism to provide a convincing, scientific answer to many geological problems. However, that situation was short-lived, as gradualism quickly replaced catastrophism as the ruling paradigm.

The *Principles of Geology*

The advance of gradualism in the English-speaking world was accomplished largely through the influence of Lyell's *Principles of Geology*, a long argument for the scientific importance of uniformity, which was published in three volumes between 1830 and 1833. Taking the traditional view of the history of the period, Charles Coulston Gillespie wrote in *Genesis and Geology* in 1959, 'If Buckland feared that without catastrophism there was no God, Lyell was as fundamentally apprehensive lest, without uniformity, there be no science'.[11] However, the situation was nothing like as straightforward as that, either for Buckland (who, as we have seen, moved from a belief in a single Universal Deluge to multiple floods and then to Ice Ages, as more evidence came to light) or for Lyell (who confused the establishment of his personal beliefs with the objective pursuit of science). Indeed, in the *Principles of Geology*, the author made good use of his legal training, promoting his views with a deviousness worthy of Cuvier.

Lyell's most successful device for establishing gradualism as the ruling paradigm, as pointed out from the 1960s onwards by science historians such as

Reijer Hooykaas and Martin Rudwick, followed by Roy Porter, Stephen Jay Gould and Anthony Hallam, was to use the term 'uniformity' in a way which conflated two quite distinct meanings. The first of these was a methodological principle which appealed to most scientists, whilst the second was a substantive world-view which was more controversial.[12] However, by this stratagem, Lyell managed to persuade the scientific community that the second meaning followed automatically from the first. The methodological principle, the 'Principle of Uniformity', was alluded to in the sub-title of the first edition of the *Principles of Geology*, an 'attempt to explain the former changes in the Earth's surface, by reference to causes now in operation'. It had two components, a 'uniformity of law' and a 'uniformity of process', maintaining that the same laws and processes operated throughout time. The Principle of Uniformity was essentially a formal statement of the actualistic method, employed in the previous century (albeit without use of the term) by Pallas, Desmarest, Saussure and Hutton, and more recently by Lamarck, Cuvier, Prévost, Playfair and others. Substantive uniformitarianism, on the other hand, was a belief in the 'uniformity of rate'. It maintained that, since all the changes taking place at the Earth's surface at that time appeared to be gradual ones, then all past changes must also have been gradual. As we shall see in the next section, William Whewell pointed out that the substantive aspect of uniformitarianism was no more than an assumption because, at infrequent intervals, terrestrial processes might operate in a more violent fashion than they do at the present time. Nevertheless, once a link had been forged (however inappropriately) between the two meanings, such arguments fell on deaf ears.[13]

In playing his part in the establishment of the legend, Sir Archibald Geikie wrote of Lyell, in *Founders of Geology*, 'With unwearied industry he marshalled in admirable order all the observations that he could collect in support of the doctrine that the present is the key to the past. With inimitable lucidity he traced the operation of existing causes, and held them up as the measure of those which have acted in bygone time'.[14]

However, as well as advancing his case by such admirable methods, Lyell also attempted to discredit his opponents in ways that were far less worthy. In the first volume of the *Principles of Geology*, Lyell attacked the catastrophist cosmogonists of earlier times for their supposed reliance on speculation and mysterious forces, contrasting their broad philosophical approach with his own more narrowly focused, scientific study of geology, and claiming that 'geology differs as widely from cosmogony, as speculations concerning the creation of man differ from history'.[15] He suggested that Burnet's writings were reminiscent of *Paradise Lost*, the epic by John Milton, and commented, 'Even Milton had scarcely ventured in his poem to indulge his imagination so freely in painting scenes of the Creation and Deluge, Paradise and Chaos, as this writer, who set forth pretensions to profound philosophy'.[16] Lyell had no greater regard for Whiston and his ideas about cometary-induced catastrophes, implying that his mathematical arguments were little more than devices 'to throw an air of plausibility over the most improbable parts of his theory'.[17] Whiston, who, like Newton, had been

Lucasian Professor of Mathematics at Cambridge University, was finally dismissed by Lyell in the following words, 'Like all who introduced purely hypothetical causes to account for natural phenomena, he retarded the progress of truth, diverting men from the investigation of the laws of sub-lunary nature, and inducing them to waste time in speculations on the power of comets to drag the waters of the ocean over the land'.[18]

In Volume 3 he denounced the catastrophists of his own day in similar fashion, writing, 'We hear of sudden and violent revolutions of the globe, of the instantaneous elevation of mountain chains, of paroxysms of volcanic energy... We are also told of general catastrophes and a succession of deluges, of the alternation of periods of repose and disorder, of the refrigeration of the globe, of the sudden annihilation of whole races of animals and plants, and other hypotheses, in which we see the ancient spirit of speculation revived, and a desire manifested to cut, rather than patiently to untie, the Gordian knot'.[19] Lyell then went on to claim that the arguments given by Élie de Beaumont for his cooling Earth model were 'mysterious in the extreme, and not founded upon any induction from facts'.[20]

His references to the contemporary British catastrophists were more circum-spect than those to their predecessors, or to their colleagues overseas, but he was, nevertheless, more than ready to accuse them of poor judgement in accepting spec-ulative ideas without properly assessing them against the available evidence. Thus, he commented, 'Professor Sedgwick has declared his adhesion to the opinions of de Beaumont; but we are not aware that he had maturely considered them in all their bearings'.[21]

There is a considerable irony here, with Lyell insisting that his ideas, unlike those of the catastrophists, were determined by the evidence at hand. On the important issue of the transition between geological periods, it was undoubtedly the catastrophists who were the more willing to take evidence at face value. They interpreted the sharp breaks between formations as an indication that sudden changes had taken place at the Earth's surface at those times. Lyell, however, cleverly reversed the situation by accusing the catastrophists of being unable or unwilling to exercise proper critical judgement. Had they been less blinkered in their approach, they would have been able to understand that, where there was no evidence for a gradual transition, this was simply because the evidence had gone missing, as a result of erosion and other effects over countless millennia. As should have been evident to everyone, the geological record gave a far from complete picture of what had really happened.[22]

Roy Porter, a historian of science from Cambridge University, pointed out in 1976 how, in the *Principles of Geology*, 'caricature heroes and villains loom out of the empty backdrop of history'. He went on to comment that Lyell was 'not interested in the subtle discrimination of their thoughts or in the fine-textured relationship of men to their times'.[23] Indeed, building on the foundation laid by John Playfair (see Chapter 2), Lyell's denigration of the approach of nineteenth century catastrophists was counter-balanced by a glorification of the work of James Hutton. He depicted

Hutton as the ideal scientist, producing theories solely on the basis of observations made during apparently endless periods of laborious fieldwork:

> His application was unwearied, and he made frequent tours through different parts of England and Scotland, acquiring considerable skill as a mineralogist, and constantly arriving at grand and comprehensive views in geology. He communicated the results of his observations unreservedly, and with the fearless spirit of one who was conscious that love of truth was the sole stimulus of all his exertions.[24]

In this way, Lyell tried to create a heroic counterpart to the 'villainous' catastrophists. Even so, he was critical of some aspects of Hutton's work, particularly his failure to use the rocks as a primary source of historical information, and to maintain a true gradualistic position. For example, Lyell disagreed with Hutton's view that huge land-masses could arise in a single, albeit non-catastrophic, process by uplift of molten rock, as a result of heat generated by the weight of accumulated sediments. He accepted that some rocks, such as granite, had been formed by the solidification of molten material. Uplift of the Earth's surface could also occur, but this must always be limited in both intensity and scale.[25] Drawing attention to observations that an earthquake in Chile had raised the coast by a height of about one metre over a considerable distance, Lyell argued that a series of such shocks at the rate of one or two per century could produce a mountain chain within a few thousand years.[26]

However, whatever the shortcomings of the *Principles of Geology* as an objective piece of writing, it must be stressed that the scientific credentials of its author were genuine. Although he had no formal scientific qualifications, Lyell put a great deal of effort into his field studies, learning much, both about actualist methods and gradualist interpretations, from Louis-Constant Prévost and the British geologist and politician George Poulett Scrope (1797–1876).[27] As part of a detailed study of the volcanic rocks of Europe, Lyell visited Mount Etna in Sicily, where he found evidence of innumerable lava flows, the older ones not appearing to be any more extensive than those which origi-nated in more recent times. Purely on the basis of the evidence, therefore, he was able to conclude that the process of mountain-building, at least in this particular location, had involved a series of relatively small events, taking place over a very long time-scale.[28]

The uniformitarianism–catastrophism debates

Without question, and despite the personal prejudices of the various partic-ipants, the debates between British geologists in the 1830s were about the relative merits of two scientific models, one of them catastrophist and the other gradualist. Indeed, Henry de la Beche (1796–1855), shortly before becoming the first Director of the Geological Survey, wrote, in 1831, 'The difference in the two theories is in reality not very great; the question being merely one of intensity of forces, so that probably, by uniting the two, we should approximate nearer the truth'.[29] Contrary to what might be supposed from the legend, even catastrophists had positive things to say about the *Principles of Geology*. This was the case, for example, in a copy of the book annotated by Louis Agassiz (see Chapter 4), a catastrophist who had been a pupil of Cuvier. As might be expected, there are a few passages criticising Lyell's substantive uniformitarianism, i.e. his insistence on gradualism, but Agassiz then went on to comment that the book as a whole was 'the most important work' of geology known to him. Similarly, Alcide d'Orbigny, another former student of Cuvier, who, in making major contributions to stratigraphy, saw evidence of even more major catastrophes in Earth history than his mentor, complimented Lyell for his methodological principle, but pointed out that it was already well-established in France.[30] The British catastrophists of the time were also operating in line with the principle, according to William Conybeare, in a series of papers which appeared in *Philosophical Magazine and Annals of Philosophy* during 1830 and 1831, and Adam Sedgwick, in his final presidential address to the Geological Society in 1831.[31]

William Whewell attacked Lyell's substantive world-view, but not his methodological principle, in a discussion of the second volume of *Principles of Geology* which appeared in the *Quarterly Review* in 1832. In the process, Whewell fell into Lyell's trap and used the terms 'uniformitarians' and 'catastrophists' to label the two sides participating in the debate. Indeed, he actually coined these terms, writing:

> Have the changes which lead us from one geological state to another been, on a long average, uniform in their intensity, or have they consisted of epochs of paroxysmal and catastrophic action, interposed between periods of comparative tranquillity? These two opinions will probably for some time divide the geological world into two sects, which may perhaps be designated as the *Uniformitarians* and the *Catastrophists*. The latter has undoubtedly been of late the prevalent doctrine, and we conceive that Mr. Lyell will find it a harder task than he appears to contemplate to overturn this established belief. Indeed, we think it ought to be so.[32]

Unfortunately for Whewell, he had just made Lyell's task much easier. The use of the single word, 'uniformitarianism', to describe the anti-catastrophist doctrine made it almost inevitable that people would link Lyell's belief in the 'uniformity of nature' (as he described his world-view) with the Principle of Uniformity.[33]

Whewell later expanded his own views in his book *History of the Inductive Sciences, from the Earliest to the Present Time*, which was published in 1837. In this, using purely rational arguments, he attacked Lyell's belief that the Earth's surface had been shaped solely by forces of a type and intensity witnessed in recent times. On what basis, Whewell asked, could we exclude the possibility that immense, violent events had occurred in the past? He concluded his argument in these words:

> In order to enable ourselves to represent geological causes as operating with uniform energy through all time, we must measure our time by long cycles, in which repose and violence alternate; how long must we extend this cycle of

change, the repetition of which we express by the word *uniformity*? And why must we suppose that all our experience, geological as well as historical, includes more than *one* such cycle? Why must we insist upon it, that man has been long enough an observer to obtain the *average* of forces which are changing through immeasurable time.[34]

Whewell accepted that it would be invalid for him to make an arbitrary assumption that catastrophes had taken place, but it would be equally wrong for Lyell to exclude the possibility of catastrophes, for 'the degree of uniformity and continuity with which terromotive forces have acted, must be collected, not from any gratuitous hypothesis, but from the facts of the case'. Similarly, when Lyell saw merit in rejecting any difference between the intensity of forces acting on the Earth's surface at the present time and in the past, he was just as wrong as someone taking the opposite view. As Whewell pointed out, 'The effects must themselves teach us the nature and intensity of the causes which have operated; and we are in danger of error, if we seek for slow and shun violent agencies further than the facts naturally direct us, no less than if we were parsimonious of time and prodigal of violence'.[35]

As well as the confusion between substantive and methodological uniformitarianism, there was another major reason why the terms 'catastrophism' and 'uniformitarianism' do not fully convey the views of the two camps. In addition to believing that great catastrophes had taken place, the catastrophists also held a *directional* view of history. As they saw the situation, from the perspective of the cooling-Earth model, geological forces decreased in intensity as time passed and climates became milder. Although this was a natural process, they regarded it as part of a divine plan, enabling more complex life-forms to be brought into existence when conditions became suitable for them. Lyell, on the other hand, wished to establish not only a strict gradualistic view, a 'uniformity of rate', but also a 'uniformity of state', a scenario of *non-progression*. He believed in a *steady-state* Earth, whose features could fluctuate to some degree, but which would never change in an irreversible fashion. He thought that, as environments gradually changed, different forms of life came into being, which were ideally suited to the conditions of the time. As conditions reverted to previously existing ones, so old forms of life would come back into existence.[36] He wrote: 'Then might those genera of animals return, of which the memorials are preserved in the ancient rocks of our continents. The huge iguanodon might reappear in the woods, and the ichthyosaur in the sea, while the pterodactyle might flit again through umbrageous groves of tree-ferns'.[37]

Apart from such fluctuations, the Earth, considered as a whole, could not have changed very much in appearance or behaviour throughout its long history. In Volume 1 of the *Principles of Geology*, Lyell summarised his views about the uniformity of forces acting on the Earth as follows:

> There can be no doubt, that periods of disturbance and repose have followed each
> other in succession in every region of the globe, but it may be equally true, that the
> energy of the subterranean movements have been always uniform as regards the
> *whole earth.* The force of earthquakes may for a cycle of years have been invariably
> confined, as it is now, to large but determinate spaces, and may then have
> gradually shifted its position, so that another region, which had for ages been at
> rest, became in its turn the grand theatre of action.[38]

Lyell had no wish for anything to be allowed to threaten his vision. He went on to
argue, rather strangely (particularly in view of the legend of the prime importance of
evidence), that even if it could be shown that the whole Earth had, at some time in the
past, been subjected to forces of different intensity from those of the present day, his
theory would not be affected, claiming:

> But should we ever establish by unequivocal proofs, that certain agents have, at
> particular periods of past time, been more potent instruments of change over the
> entire surface of the earth than they now are, it will be more consistent with
> philosophical caution to presume, that after an interval of quiescence they will
> renew their pristine vigour, than to regard them as worn out.[39]

Without question, natural theology helped to formulate Lyell's ideas, just as much as
those of his opponents, even though they had very different views of God. Most of
the British catastrophists, including Whewell, belonged to the Anglican Church, and
believed in a divinity who took an active, personal interest in Earthly events. In contrast,
Lyell's sympathies were with the Unitarian Church, and its deist-like view of a God who
was disinclined to intervene in what was happening on Earth. As we have seen, Lyell envisaged species appearing and disappearing as environmental conditions fluctuated,
whereas the catastrophists believed that new species were created as the Earth underwent progressive development.[40] However, by what mechanism was it supposed that
the new species came into being? In the *History of the Inductive Sciences,* Whewell argued:

> We conceive it undeniable (and Mr Lyell would probably agree with us) that we see
> in the transition from an earth peopled by one set of animals, to the same earth
> swarming with entirely new forms of organic life, a distinct manifestation of
> creative power, transcending the known laws of nature: and, it appears to us, that
> geology has thus lighted a new lamp along the path of natural theology.[41]

To anyone retaining an unquestioning belief in the legend of the dichotomy between
uniformitarianism and catastrophism, the idea that Lyell might have agreed with
Whewell on this issue must seem absurd. In 1993, in *Complexity,* Roger Lewin wrote
of the catastrophists' belief in the creation of new species, and then hammered home
the point that 'Lyell rejected so nonscientific a hypothesis'. That was precisely what

the legend said about the situation. Nevertheless, as Whewell knew full well, Lyell's beliefs about the succession of biological species through time were not very different from those of the catastrophists. All were committed creationists and, as such, they were completely opposed to the possibility of evolution. Above all, none of them was prepared to contemplate any evolutionary origin for humankind. Lyell may have been a rational scientist (as were his opponents) when thinking about geological matters, but not when addressing the question of the origin of new species.[42]

By the 1830s, it was becoming clear that Cuvier's suggestion about extinct species being replaced in a region by ones moving in from elsewhere could not always be true. In some instances, species appeared which had not existed, anywhere in the world, before that time. There was no avoiding the crucial question: had the new species arisen by creation or by evolution? Without doubt, not many people, whether in Britain or elsewhere, were prepared to believe that biological evolution could occur. Because of the fossil evidence now available, it was no longer possible for scientists to accept a single act of Creation, as in the *Genesis* account, so they generally believed that separate creation events must have taken place at intervals throughout Earth history. Precisely what had happened was never discussed. The only certainty was that the process could not have occurred by the natural transformation of previously existing forms of life. Instead, it must have involved the 'First Cause', i.e. God, either directly or through the operation of intermediate (or secondary) causes. Because of this, there was no point in speculating further, or trying to carry out a scientific investigation to determine the details. Lyell's views, as well as those of the British catastrophists, fitted into this general picture. However, whereas the catastrophists were happy to envisage the direct involvement of the First Cause, Lyell, in keeping with his philosophy of life, preferred to think of intermediate causes being at work. Also, to fit in with his gradualistic stance, he envisaged new species arising on an individual basis, as conditions became appropriate for them, rather than as part of a mass turnover at the beginning of a new geological period, as believed by the catastrophists.[42]

Lyell discussed the evolutionary ideas of Lamarck in the second volume of the *Principles of Geology*, attempting to discredit them by utilising Cuvier's arguments about the limitations of variation (see Chapter 3). Thus, Lyell maintained:

> first, that the organization of individuals is capable of being modified to a limited extent by the force of external causes; secondly, that these modifications are, to a certain extent, transmissible to their offspring; thirdly, that there are fixed limits beyond which the descendants from common parents can never deviate from a certain type; fourthly, that each species springs from one original stock, and can never be permanently confounded, by intermixing with the progeny of any other stock; fifthly, that each species shall endure for a considerable period of time.[43]

In line with the traditions of natural theology, he continued, 'We must suppose, that when the Author of Nature creates an animal or plant, all the possible circumstances in which its descendants are destined to live are foreseen, and that an organization is conferred upon it which will enable the species to perpetuate itself and survive under all the varying circumstances to which it must be inevitably be exposed'.[44]

Several chapters later, when discussing the replacement of one species by another, Lyell again referred unequivocally to creation, writing of his assumption 'that single stocks only of each animal and plant are originally created, and that individuals of new species do not suddenly start up in many different places at once'.[45] Nevertheless, in Lyell's view, God did not intervene directly to create new species. He had, instead, set up 'laws of creation', operating through intermediate causes, by which new forms of life appeared at appropriate times, without further intervention being necessary.

Idealist philosophy

Similar ideas to those of Lyell, whereby the creation of new species did not require the intervention of a deity, yet was subject to some form of divine control, dominated 'idealist' philosophy in early nineteenth century Germany. Idealist philosophy attracted romantics who deplored the tendency to try to explain the world in mechanistic terms, which had become increasingly common since the Newtonian revolution. It emphasised the difference between appearance and reality, and maintained that the only access to reality was through the human mind.

Unlike Lyell, the idealists, taking their initial ideas from Immanuel Kant (1724–1804) and Johann Gottfried von Herder (1744–1803), believed the history of the Earth to have been directional, and generally thought that the developments were teleological, occurring in line with the intentions of God. The form of idealism known as *Naturphilosophie*, which was associated particularly with the name of Lorenz Oken (1779–1851), maintained that everything that had happened in the world was designed to lead to the eventual emergence of humankind. There may have been some creations which were not a direct part of this process, but the idealists did not regard these as being of any importance. In 1821, Johann Friedrich Meckel (1781–1833) wrote about a supposed 'parallelism' between the development of the human embryo and the main sequence of species leading to humankind. Of course, the appearance of new species owed nothing to evolution, being created instead as part of a plan devised long ago by God.[46]

The naturalist Louis Agassiz was one of many who came under the influence of Oken. Agassiz first made his reputation as a scientist with a study of fossil fish, and later went on to establish that many of the geological features thought to be evidence of a great deluge had in fact been produced by an Ice Age, as we have noted in the previous chapter. Yet, right up to his death in America in 1873, Agassiz maintained some of the beliefs of the idealist philosophers.[47] In 1842, whilst still at Neuchâtel, Switzerland, he wrote, 'The history of the Earth proclaims its Creator. It tells us that the object and term of creation is man. He is announced in nature from the first appearance of organized beings; and each important modification in the whole series of these beings is a step towards the definitive term of the development of organic life'.[48]

Agassiz thought that life on Earth had become extinct on many occasions, to be replaced by more advanced forms. He believed in progression, whilst denying any that there was any continuity in the process. To him, and other idealists, it simply demonstrated the working out of the divine plan, with a species becoming extinct, to be replaced by one with more advanced characteristics.[49]

However, this was only a single conceptual step away from the view that the process might be continuous, involving the pre-ordained transmutation of one form into the next. Robert Chambers, a member of the Scottish publishing family, of *Chambers's Encyclopaedia* fame, took this very step in his anonymous *Vestiges of the Natural History of Creation*, but, unfortunately, his grasp of some aspects of the subject matter was poor. As a result, his arguments were easy to discredit by the many who were opposed to any form of continuous development, regardless of whether it was steered by God or by nature.[50]

In 1828, Karl Ernst von Baer (1792–1876) started an alternative form of the idealist philosophy movement, when he challenged the concept of parallelism, claiming that Oken, Meckel and others had overstated the similarity between the early stages of the human embryo and the adult structures of lower forms of life. He argued that all vertebrate embryos are much the same in their early stages, but they subsequently differentiate and specialise in ways unique to the species. One of von Baer's followers was the British anatomist Richard Owen (1804–1892) who, after holding professorships at the Royal College of Surgeons and the Royal Institution, eventually became the first Director of the Natural History Museum in South Kensington, London. In 1848, Owen proposed the concept of *homologies*, showing that all adult vertebrates had essentially the same body plan. This could be developed in different ways, depending on the life-style of the particular animal. It is probable that, like Chambers, Owen came to believe in a form of evolution, controlled by God through the agency of intermediate or secondary causes. This belief is suggested in his *On the Nature of Limbs*, which was published in 1849. Such a scheme, proceeding in a pre-determined way, in step-wise fashion, would result in new species arising at appropriate times by transmutation. For example, in the case of animals, females of an existing species might give birth to the founding members of the succeeding species. However, Owen always remained vague about his precise views on the subject of evolution. This continued to be an unmentionable topic for anyone anxious to maintain a good reputation.[51]

6 Lyell triumphant: gradualism dominates geology

Arguments against Earth-centred and cometary catastrophes

In geology, by the middle of the nineteenth century, the balance was changing inexorably in favour of the uniformitarian case. Furthermore, that owed more to observation and to the limited outlook of the catastrophists of the time, than to any further deviousness on the part of Lyell. Accumulating field evidence offered little support for the belief that periodic large-scale convulsions of the Earth's surface could have caused massive extinctions of species. It emerged that many disruptions of the strata were extremely localised, for considerable areas of Central Europe and North America had remained flat and undisturbed throughout the period in which the Alps were formed. Also, it was not generally possible to explain uplift of mountain ranges in terms of single events, and no evidence could be found to suggest that volcanic activity had, as a rule, been greater in former times than in more recent ones. For these various reasons, the cooling Earth model at the centre of mid nineteenth century catastrophism fell increasingly out of favour.[1]

That left no plausible alternative to the growing uniformitarian consensus. There was no reason to think that, on very rare occasions, volcanic eruptions could take place on a much larger scale than those normally experienced, and very few people took the prospect of cosmic catastrophes at all seriously. One who had done so, earlier in the century, was David Milne. In his 'Essay on Comets', which won an astronomy prize at Edinburgh University in 1828, Milne acknowledged that the heads of comets were not the massive bodies supposed by some who had previously written about cometary catastrophes. He continued, 'It is very true . . . that the masses of comets are usually small; and for this reason we might be disposed to imagine that the result of a collision would be trivial. But if a comet, moving with the prodigious velocity which it acquires near its perihelion, should chance to strike a planet, as for instance the Earth, then coming in an opposite direction, the consequences would be truly disastrous . . . The waters of the ocean, now attracted by the close approach and next driven from their ancient beds by the contact of the comet, would sweep over the face of the globe . . . Whole species of plants and animals, existing in different quarters of the Earth, would, by this cataclysm, be at once overwhelmed and annihilated'.[2]

Concerning possible attempts to avert such a disaster, the poet Lord Byron suggested to Thomas Medwin in 1822, 'Who knows whether, when a comet shall approach this globe to destroy it, as it often has been and will be destroyed, men will not tear rocks from their foundations by means of steam, and hurl mountains, as the giants are said to have done, against the flaming mass?– and then we shall have traditions of Titans again, and of wars with heaven'.[3]

Around this time, several writers followed the example of Plato, after a gap of more than two millennia, and suggested that the Phaeton myth owed its origins to an extraterrestrial event. One of these was Johann Wolfgang von Goethe, writing in 1821.[4] Like Byron, Goethe was best known as a poet (although originally destined for the legal profession) but, influenced by Herder (see Chapter 5), he devoted much of his time to a study of science.[5]

In 1823, the German linguist Johann Gottlieb Radlof argued that a large planet orbiting between Jupiter and Mars might have been smashed into pieces in historical times by a cometary impact. Four objects (Ceres, Pallas, Juno and Vesta), classified as 'asteroids', had recently been discovered in this region of the Solar System, and these could be remnants of the former planet. Radlof suggested that another piece of the shattered planet had encountered Mars and been thrown into an almost circular orbit around the Sun, as the planet we know as Venus, whilst a further fragment collided with the Earth, giving rise to myths describing battles in the sky between various gods.[6]

Similarly, Ignatius Donnelly, an American lawyer and politician, who served for eight years in the House of Representatives, wrote in his 1883 book, *Ragnarok: The Age of Fire and Gravel*, of a supposed close encounter between the Earth and a comet, traces being left behind in the form of the gravel deposits spread throughout much of the northern hemisphere. (As we noted in Chapter 1, Ragnarok was the event in Norse mythology characterised by celestial battles, fiery catastrophes and terrible, prolonged winters, during which many of the old gods died.[7]) Moreover, according to Donnelly, that encounter (which he also associated with the Phaeton myth) was only one in a series of cometary catastrophes evident from the geological record, and it was not even the most recent or the most devastating of them. It seemed to him that 'many times have comets smitten the earth, covering it with *débris*, or causing its rocks to boil, and its waters to ascend into the heavens'.[8]

However, all of this was generally regarded as fantasy. The existence of comets was, of course, well known, but most scientists maintained they presented little threat to the Earth. In 1879, the American astronomer Simon Newcomb (1835–1909) wrote in his *Popular Astronomy* about the devastation which might theoretically be caused in the event of a collision between the Earth and the nucleus of a comet, but continued, 'Happily, the chances of such a calamity are so minute that they need not cause the slightest uneasiness.' Although it had been largely forgotten by this time, philosophical and religious traditions had played a significant role in shaping this belief. So, for example, Thomas Dick, a populariser of science, wrote about the dangers of cometary impacts in his book *The Sidereal Heavens*, published in 1840, but added that one would never take place without the 'sovereign permission' of God. Because no such event was listed amongst the prophecies given in the Bible, it could only happen after all the ones which were mentioned had been accomplished, that is, when every Jew had become a Christian, all wars had ceased, and the kingdom of the Messiah extended over the entire Earth. Thus, readers could be re-assured that there would be no cometary impacts for a very considerable period of time.[9]

Even Donnelly, despite writing of the horrors of cometary catastrophes in *Ragnarok: The Age of Fire and Gravel*, concluded the book by giving it a moral dimension. He wrote, 'So far as we can judge, after every cataclysm the world has risen to higher levels of creative development . . . Nothing died but that which stood in the pathway of man's development . . . Surely, then, we can afford to leave God's planets in God's hands'. Nevertheless, he continued, 'There may be even now [a comet] coming from beyond Arcturus, or Aldebran, or Coma Berenices, with glowing countenance and horrid hair, and millions of tons of *débris*, to overwhelm . . . all the ant-like devices of man in one common ruin'. However, if humankind became truly righteous, then 'from such a world God will fend off the comets with his great right arm, and the angels will exult over it in heaven'.[10]

Other writers, such as the clergyman Thomas Milner, in his 1860 book, *Gallery of Nature*, supposed that comets were so insubstantial that even if one collided with the Earth, there would be no harmful effects.[11] The notion that comets were innocuous objects was consistent with what happened after Comet Biela began to disintegrate in 1846. The breakdown products gave rise to an impressive display of meteors (shooting stars), known as the Andromedid shower, in 1872 and on subsequent occasions, but that was all.[12]

Meteorites and changes of tilt

By the middle of the nineteenth century, it was generally agreed that meteorites (rocks of unknown origin found at the surface of the Earth) had come from space, but there had been much resistance to that notion. When a group including Antoine-Laurent Lavoisier, acting on behalf of the Académie des Sciences, investigated a 1768 incident in which it was claimed that a large stone had fallen from the sky near Lucé, they concluded that the object was simply an ordinary piece of Earth-rock which had been struck by lightning. All the witnesses to its arrival from above must either have been mistaken or were lying, because the event could not possibly have happened as described.[13] Much the same happened again in 1790, when a shower of stones was reported to have fallen in Barbotan, in the Gascony region of southwestern France. In the *Journal des Sciences Utiles*, Pierre Berthelon ridiculed those who claimed to have seen the stones fall.[14]

In contrast, a report written in 1794 by the German physicist Ernst Chladni (1756–1827) concluded that several large masses of iron had probably come to the

Earth from cosmic space. These included the 680 kg Krasnojarsk iron, discovered in Siberia in 1749, and investigated 20 years later by Peter Simon Pallas (see Chapter 2), as well as the even larger Tucuman iron, found in South America in 1783. However, Chladni's conclusions were dismissed out of hand by most scientists of the time, one critic writing, 'He contradicts the entire order of things and does not consider what evil he is causing for the moral world'.[15]

Almost immediately afterwards, there was a shower of stones near Siena, Italy, whilst a 25 kg stone fell from the sky close to Wold Cottage, near Scarborough, in Yorkshire, just missing a man who was working there. These falls of stones could not be disputed, but Edward King, writing about them in 1796 in the *Gentlemen's Magazine*, suggested that they were of volcanic origin. The object which almost hit the Yorkshire labourer had possibly been thrown from Mount Hekla, in Iceland, whereas the Siena stones had probably been ejected by Vesuvius, which was located less than 400 km to the southeast, and had erupted a few hours previously. However, when another meteorite shower struck Benares, India, in 1798, material from this and the other recent falls were investigated by Edward Howard, who reported to the Royal Society in 1802 that the mineral content in each case, particularly the high nickel concentration, suggested an extraterrestrial origin.[16]

A report on the Benares shower by Johann Friedrich Blumenbach convinced Goethe, who had previously rejected the authenticity of the 140 kg stone which many claimed to have seen fall at Ensisheim, in Alsace, in 1492, that meteorites were of extraterrestrial origin. That led to his suggestion, mentioned earlier, that a similar event had been the stimulus for the Phaeton myth.[17]

Many were still sceptical, particularly in France, but that changed with the arrival of several thousand stones out of a clear sky near L'Aigle, Normandy, in 1803. The Académie investigators, who included Jean-Baptiste Biot (1774–1862), a distinguished physicist, as well as Pierre-Simon Laplace, had little option but to conclude that they had come from space, possibly being ejected from volcanoes on the Moon.[18] Nevertheless, when Thomas Jefferson, President of the U.S.A., who included palaeontology amongst his many interests, was told in 1807 that Benjamin Silliman, from Yale, together with a colleague, was claiming that meteorites had recently struck the Earth at Weston, Connecticut, he is said to have replied, 'It is easier to believe that two Yankee professors would lie, than that stones should fall from the sky'.[19]

In any case, such events were on too small a scale to do more than cause a little damage and localised panic. A fall of stones on a frozen lake at Hessle, Sweden, in 1869 did not even manage to break the ice. One of the Siena meteorites hit a child's hat, but caused no harm to the child who had been wearing it. There was no shortage of incidents. Buildings were reported to have been damaged by falling stones, many of them glowing with heat, in China in 1661, 1670, 1845 and 1850, in France in 1759, 1761, 1835 and 1858, and in England in 1801 and 1803, the last-mentioned event involving the 'White Bull' tavern in East Norton. Similar happenings also occurred in Eastern Europe in 1803, 1847 and 1868, at Waseda, Japan, in 1823 and at Macao, Brazil, in 1836. However, in not one of these incidents was anyone seriously hurt.[20]

Some oxen apparently died during the Macao event, and a young horse was killed by a meteorite at New Concord, Ohio, in 1860 but, despite various claims, there was no confirmed case of any human dying in similar fashion. According to some reports, an Italian monk may have died in an incident at Cremona in 1511 and another at Milan in 1650, whilst two Swedish sailors were supposedly killed at sea by a meteorite in 1674. Several people were rumoured to have died when a large stone fell in a Chinese market in 1639, and a farmer may have been killed in Gascony during the 1790 meteorite shower. However, firm evidence of human fatalities was lacking in every case.[21]

Hence there seemed to be nothing, either on the ground or in the skies, to challenge the uniformitarian consensus. Catastrophist theories were still occasionally put forward, but these were generally of a speculative nature, so were not taken seriously. For example, Alfred Wilks Drayson, who had been a major-general in the army, presented in his 1873 book, *On the Cause, Date and Duration of the Last Glacial Epoch of Geology*, the hypothesis that the tilt of the Earth's axis of rotation varied much more than was supposed, changing from its present value of 23°, relative to the perpendicular to the orbital plane, to one of around 35° and back over a 30,000 year cycle. In the context of uniformitarian geology, that was a very short time-scale for a large effect. At times of maximum tilt, winters would be much colder than at present and summers much hotter, particularly at intermediate latitudes, and the transition from winter to summer could cause catastrophic floods, with the rapid melting of ice. However, in the absence of any convincing evidence, this hypothesis was almost completely ignored, as the uniformitarian consensus became ever more dominant.[22]

This consensus nevertheless incorporated elements of catastrophism as well as of gradualism, and the steady-state aspect of Lyell's views was conveniently forgotten. Ice Ages, despite the association with Agassiz, were admitted on the supposition that environmental changes associated with them, although major and widespread, occurred in a gradual fashion. The advance and retreat of the glaciers was thought, from ideas suggested by the Scottish 'independent thinker' James Croll (1821–1890), who worked as a museum caretaker in Glasgow before joining the staff of the Geological Survey, to be related to a slow fluctuation in the eccentricity of the Earth's orbit, acting in concert with the precession of the equinoxes, which resulted from a slight wobble in the Earth's rotation.[23] Paroxysmal events such as earthquakes and volcanic eruptions were known to occur, so they could hardly be ignored. However, they were thought to be too localised in their effects to be a direct cause of species extinction. Hence, neither the existence of such events, nor anything else, was seen to challenge the general belief that the terms 'uniformitarianism' and 'gradualism' overlapped considerably in their meaning.[24]

7 Darwin and evolution

Influences on Darwin

As we discussed in Chapter 3, the concept of a static *Scala Naturae*, or great chain of being, was modified during the eighteenth and early nineteenth centuries to allow for movement along it, as environmental conditions gradually changed. These developments were eventually synthesised into a theory of evolution by Jean-Baptiste de Lamarck. However, Lamarck's arguments for gradual evolutionary change were challenged by his influential contemporary Georges Cuvier, whose catastrophist ideas, allied to his knowledge of stratigraphy and comparative anatomy, led him to reject both gradualism and evolution.

In 1831, two years after Lamarck's death, Charles Darwin (1809–1882) left England for South America, as gentleman-naturalist on board H.M.S. *Beagle*. Although his main influence up to this point had been the botanist John Henslow (see Chapter 4), a man of traditional views, he had been introduced to speculations about evolution by the writings of his grandfather Erasmus Darwin (1731–1802), a medical practitioner with a great appetite for food and, even more so, for knowledge.[1] Erasmus Darwin was a member of the Lunar Society of Birmingham, a dining club formed by a small group of intellectual giants, including the chemist Joseph Priestley and the engineers Matthew Boulton and James Watt (who introduced his fellow Scot James Hutton to the other members).[2]

In his *Zoonomia*, which first appeared in 1794, the elder Darwin wondered, 'Would it be too bold to imagine, that in the great length of time since the earth began to exist, perhaps millions of ages before the commencement of the history of mankind . . . that all warm-blooded animals have arisen from one living filament'. In the same book, he wrote, in a way reminiscent of Lamarck, 'Some birds have acquired harder beaks to crack nuts, as the parrot. Others have acquired beaks adapted to break the harder seeds, as sparrows. Others for the softer seeds of flowers, or the buds of trees, as the finches . . . All which seem to have been gradually produced during many generations by the perpetual endeavour of the creatures to supply the want of food'.[3]

Again, in *The Temple of Nature*, published after his death in 1803 he wrote, this time in verse:

Organic life beneath the shoreless waves
Was born and nursed in Ocean's pearly caves;

First forms minute, unseen by spheric glass,
Move on the mud, or pierce the watery mass;
These, as successive generations bloom,
New powers acquire, and larger limbs assume;
Whence countless groups of vegetation spring,
And breathing realms of fin, and feet, and wing.[4]

The evolutionary views of Erasmus Darwin, after being ignored to start with, were savagely attacked towards the end of his life, and did not stimulate any positive interest amongst his British contemporaries.[5] He died before his grandson, Charles, was born, but the younger Darwin would have been aware of his theories through his books, and family tradition. Charles Darwin was also influenced by the zoologist Robert Grant (1793–1874) when he was a reluctant and ultimately unsuccessful medical student at Edinburgh University. Grant was sympathetic towards evolution, having a great admiration for both Lamarck and Geoffroy Saint-Hilaire.[6] After failing his medical course, Darwin went to Cambridge University to study theology, again without any great enthusiasm, although this time going on to complete his degree. In Cambridge he learned a great deal about geology from Sedgwick (whom he found much more stimulating than Jameson in Edinburgh), and about biology, as already stated, through his friendship with Henslow.[7]

The evolutionary theory which Charles Darwin eventually produced differed from the Lamarckian view in maintaining that inheritable variations arose in a population by factors which were not necessarily linked to environmental change. Then, those which gave individuals an advantage in terms of survival or mating over other members of the population would be, as a consequence of natural selection, the most likely to be passed on to subsequent generations.[8] However, as the Harvard University evolutionary biologist Ernst Mayr pointed out in 1982 in his book *The Growth of Biological Thought* Darwin in fact produced a cluster of five interacting theories. These were that biological evolution had occurred; that every group of organisms had evolved from a common ancestor; that new species arose in populations, so a number could evolve separately from each existing species; that the process was driven by the action of natural selection upon variant forms; and (of particular importance to our story) that it was a gradual one.[9]

As the *Beagle* crossed the Atlantic and sailed down the South American coast towards Tierra del Fuego, Darwin read with enthusiasm the first two volumes of Lyell's *Principles of Geology*. He was particularly stimulated by the second of these, picked up in Montevideo, the central sections of which described how the survival of existing groups of plants and animals could be threatened by environmental changes, or the introduction of new species into the regions they occupied.[10] This was argued from an anti-evolutionary point of view, accompanied by attacks on Lamarck, as we have seen. However, the visit of the *Beagle* to the Galápagos Islands, off the west coast of South America, in 1835, proved a key event in turning Darwin's thoughts towards the possibility that evolution might have occurred. When the birds he collected in the Galápagos were brought to Britain and classified (mainly by the ornithologist, John Gould), they showed that species living on the various islands were slightly different from each other, and from those on the mainland. This caused Darwin to think it possible that species of birds arriving from the mainland to each of the separate islands had then remained effectively in geographical isolation and had evolved in slightly different ways. The alternative, that God had created slightly different finches and mockingbirds for each of the individual small islands in the group, seemed less plausible. Similar considerations also applied to other animals, such as tortoises, which showed small differences of detail from island to island.[11]

The fact that the differences between the species were very slight was suggestive of gradualism, but there were also other reasons why Darwin became a gradualist. Even before that time, during the *Beagle* voyage, the arguments of Lyell's *Principles of Geology* had been influential, as we have noted, and that was not all. In 1832, as he sailed past St. Jago, in the Cape Verde Islands, Darwin was struck by an almost continuous band of compressed shells in the rocks above his head, the regularity of which suggested a gradual elevation of the land rather than convulsive uplift. Years later, as he developed his theory of evolution, he envisaged it as something which took place in populations rather than individual organisms, which also seemed more suited to gradual rather than sudden change.[12]

Darwin may have hit upon the idea of natural selection as a mechanism for evolution through observing artificial selection being carried out by breeders of 'fancy' pigeons (an activity in which he was later to become personally involved). However, the full potential of natural selection only became apparent to Darwin in 1838, when he read the sixth edition of the *Essay on the Principle of Populations*, written by the clergyman Thomas Malthus. The principal argument in the *Essay* was that populations tended to increase as a geometrical progression, whereas, even by bringing more land under cultivation, food supply could only increase arithmetically. Thus, Malthus warned that human populations might easily expand to the point where there was a shortage of food, which would lead to desperate struggles for whatever resources were available.[13]

Patrick Matthew and natural selection

In fact, Darwin was not the first to advocate evolution by natural selection, as he himself acknowledged.[14] Even his grandfather Erasmus Darwin (figure 7.1), had hinted at such a process. In his *Zoonomia*, he wrote about the competition between male animals for females and continued, 'The final cause of this contest amongst the males seems to be that, the strongest and most active animal should propagate the species, which should thence become improved'. Erasmus Darwin also showed himself to be aware of the pressures of overpopulation, perhaps as a result of reading the first edition of Malthus's classic book in 1798.[15] However, he never explicitly incorporated these ideas into a coherent theory of evolution.

Figure 7.1   Left: Erasmus Darwin, based on an 1801 portrait by James Rawlinson, now in the Erasmus Darwin Building, Nottingham Trent University. Right: Patrick Matthew, based on an early photograph.

The naturalist William Wells suggested the idea of natural selection in 1813, though only to explain variation within a species, not as anything to do with an evolutionary process. Again, in 1835, another naturalist, Edward Blyth, considered but rejected it as a mechanism by which new species might be formed. Blyth, a dedicated believer in the fixity of species, was happy to conclude that it could do no more than maintain the perfection of the type.[16] However, the Scottish botanist Patrick Matthew (1790–1874) outlined all the important aspects of the theory which would later become known as Darwinism, in an appendix to his book *Naval Timber and Arboriculture*, published in 1831, when the *Beagle* was just leaving England on its voyage of discovery. Towards the middle of the 6500-word appendix, which was entitled *Accommodation of Organized Life to Circumstance, by Diverging Ramifications*, Matthew (see figure 7.1) pointed out that successive geological formations generally contained quite different fossils.[17] In his view, there could be only two possible explanations:

> We are therefore led to admit, either of a repeated miraculous creation; or of a power of change, under a change of circumstances . . . The derangements and changes in organized existence, induced by a change of circumstances from the interference of man, affording us proof of the plastic quality of superior life, and the likelihood that circumstances have been very different in the different epochs, though steady in each, tend strongly to heighten the probability of the latter theory.[18]

From that, it is clear that Matthew preferred evolution to special creation as an explanation for the progression of biological forms. Such a preference was hardly surprising, given what we know about Matthew, for he held radical views on a variety of subjects.[19]

Unlike Darwin, he undoubtedly saw evolution taking place within the context of a series of large-scale catastrophes, for he wrote, 'The destructive liquid currents, before which the hardest mountains have been swept and comminuted into gravel, sand and mud, which intervened between and divided these epochs, probably extending over the whole surface of the globe, and destroying [nearly] all living things, must have reduced existence so much, that an unoccupied field would be formed for new, diverging ramifications of life'.[20] Matthew went on to explain the likely mechanism of evolution in a way which anticipated Darwin, many years later. Even though he did not use the term 'natural selection', it is clear from the following passage that he envisaged such a process acting on variants within a population:

> The self-regulating adaptive disposition of organized life may, in part, be traced to the extreme fecundity of Nature, who, as before stated, has, in all the varieties of her offspring, a prolific power much beyond (in many cases a thousandfold) what is necessary to fill up the vacancies caused by senile decay. As the field of existence is limited and preoccupied, it is only the hardier, better suited to circumstance individuals, who are able to struggle forward to maturity, these inhabiting only the situations to which they have superior adaptation and greater power of occupancy than any other kind; the weaker, less circumstance-suited, being prematurely destroyed.[21]

Matthew concluded, again with strange pre-echoes of Darwin, that, given the 'unremitting operation of this law acting in concert with the tendency which the progeny have to take the more particular qualities of the parents', a change in circumstance would result in 'the breed gradually acquiring the very best possible adaptation' to the new conditions.[22]

It has generally been supposed by later generations that Matthew must have been an obscure figure to his contemporaries, but he was far from that. Like Darwin, he was a product of a wealthy family, who failed to complete a medical degree at Edinburgh University (in Matthew's case, he left to run his mother's estate at Gourdiehill, Perthshire, when his father died). He travelled widely in Europe, and took an active interest in international politics, as well as in trying to improve conditions for the working class in Britain.[23] *Naval Timber and Arboriculture*, the first of several books written by Matthew, was widely reviewed at the time of its publication, and the contents of the appendix were specifically mentioned in two of the reviews.[24] However, Darwin was out of the country until 1836 so it is likely, as he subsequently maintained, that he remained unaware of Matthew's ideas until they were pointed out to him in 1860.[25]

The development of Darwin's theory

When Darwin arrived home after the *Beagle* voyage, it was thought likely he would seek a profession within the Church, but he never did. Indeed, devastated by the death of his ten-year-old daughter in 1851, he eventually became an agnostic. Darwin first took a house in London, where he married in 1839 and was Secretary to the Geological Society from 1838 to 1844. However, city life was not to his liking, so, in 1842, he and his wife moved to Downe, in Kent. Thereafter, Darwin lived in the style of a country gentleman, on income from the investment of money he had inherited. He was frequently ill, suffering from headaches and dizziness, accompanied by skin and digestive problems. Possibly he could have contracted a tropical disease during his time in South America but, more likely, as he himself suspected, the illness was inherent in his family, the symptoms being exacerbated by stress. Despite these problems, he worked very hard on his scientific studies, whenever he could, and gained a high reputation, initially as a geologist, as a result of his writings.[26]

The development of Darwin's theory of evolution is a complex subject, and needs to be considered within the circumstances of the time. Science, religion and politics were all relevant factors in the catastrophist/gradualist and creationist/evolutionist conflicts. During the time when Darwin was a student at Edinburgh and Cambridge, the Tories, the party of the aristocracy and the Anglican church (but not all of its clergymen) had been in power. Under their firm control, the relatively relaxed intellectual atmosphere of the closing decades of the previous century, when Erasmus Darwin's speculations about evolution could be tolerated (at least for a time, even if never supported), had long since disappeared. The French Revolution had, of course, shown what could happen if there was a collapse in the social order, and the British electorate (which was far from representative of the total population) reacted by giving support to conservative, orthodox values. Inevitably, that led in some quarters to increasing demands for change, especially when another revolution, bloodless on this occasion, took place in France in 1830. At the time of the *Beagle*'s departure, the political situation in Britain was highly unstable.[27]

When Darwin returned to England, freedom of thought was still not encouraged, but some aspects of life had altered. Power had shifted to the Whigs, whose values in general were not vastly different from those of the Tories, but whose membership included people with a wider range of backgrounds and beliefs. Of particular concern to the Tories were the activities of the Dissenters, who sought to remove the privileges associated with membership of the Anglican Church. However, Whigs and Tories alike were alarmed by the riots which followed the introduction of the New Poor Law in 1834, a measure inspired by the ideas of Malthus, which forced the poor to compete successfully for limited resources, or be forced into workhouses. This was the context in which Darwin began his work.[28]

Darwin soon made it clear that his geological observations on the *Beagle* voyage supported the gradualistic ideas expressed in the *Principles of Geology*. Because of this and other common interests and attitudes (both having relatively liberal views), he and Lyell became friendly. Another of Lyell's friends was Charles Babbage, inventor of a calculating machine, a forerunner of the modern computer. Babbage was also author of the unofficial *Ninth Bridgewater Treatise*, in which he argued that an apparent miracle could be brought about through the operation of a 'higher' law of nature. In the context of the succession of species, he therefore portrayed God as a programmer who devised laws by which new species were created at intervals throughout history, as required, without any further intervention being required. By arguments such as this, Lyell and his circle sought to present themselves as objective scientists, unlike their catastrophist opponents, who retained a belief in the possibility of divine intervention. During 1837, Darwin was a regular visitor to the homes of Lyell and Babbage, where he heard this claim repeated again and again.[29]

Although Darwin was now thoroughly assimilated into the Lyell camp, all was not well. His thoughts were already turning towards evolution, which, had Lyell known, would have appalled him just as much as it would have appalled the catastrophists.[30] Despite Lyell's eagerness to condemn interventionism, and his claim to believe in the operation of natural laws, he remained a creationist at heart. The 'higher laws' he spoke about in relation to the creation of species were very different from Newton's laws. As we have seen, Newton may have believed that the forces of gravity were maintained by metaphysical powers, but his law about the behaviour of bodies under gravitational attraction operated continuously and was therefore capable of being verified by carrying out a scientific experiment. In contrast, the 'laws of creation' operated only intermittently and, for that reason, they could never be tested. As the science historian Peter Bowler wrote in *Evolution – The History of an Idea*, 'Lyell had decided that species could not be modified by natural causes and in addition was convinced that adaptation of every species to its environment reflects the Creator's wisdom and benevolence. In modern terms, it sounds as though he was trying to have his cake and eat it: the creation of new species was not a miracle, yet it occurred in distinct acts embodying design'.[31]

Lyell may have rejected the involvement of the First Cause, and made a virtue of his belief in the operation of 'intermediate causes'. However, as Ernst Mayr pointed out in the notes to *The Growth of Biological Thought*, when Lyell was asked to explain exactly what he meant by them, he always included in his answer some reference to the arrival of a new species as being 'pre-ordained', 'pre-determined' or 'appointed'. Mayr concluded, 'Lyell's beliefs, his own disclaimers notwithstanding, amounted to rank special creationism'.[32]

So, Darwin said nothing, but agonised in private over deductions he had made from the species collected in the Galápagos islands. To him, an evolutionary process seemed to be indicated, but Gould, who had classified the birds, and Owen, who had helped with the classification of the other specimens, clearly did not think so. Moreover, a good natural theologian would have had no difficulty in accepting the small variations as evidence of the care which God had taken to ensure that every species was ideally suited to its environment.[33]

An even greater problem for Darwin was that he could find no convincing reason for excluding the origin of humankind from the overall process of evolution.

That meant that humans must have the same common ancestor as everything else. Darwin knew that, if he tried to say so, he would receive hostile opposition from everyone, including those he regarded as his friends. Within his social class, it was generally appreciated that the moral accountability of humankind to the Creator was an important component of the glue which held society together, with social inequalities being accepted as a deliberate part of God's plan.[34] This belief is clearly expressed in a verse from the 1848 hymn *All things bright and beautiful*, which is now usually omitted. It states, 'The rich man in his castle, the poor man at his gate: God made them high and lowly, and ordered their estate'.[35] If such an understanding was shattered, then the consequence might well be a social revolution.

For these reasons, Darwin continued to work alone on the detail of his theory, keeping detailed notes, but not showing them to anyone. Eventually, in 1844, he brought his various thoughts on evolution together in a two-hundred page essay.[36] However, since he had no wish to be involved in major controversy, publication was out of the question. That point was underlined by the hostile reception given in the same year to the *Vestiges of the Natural History of Creation*, which was referred to in Chapter 5. *Vestiges*, influenced by idealist philosophy and also by the *Ninth Bridgewater Treatise* of Babbage, attempted to construct an argument in favour of the possibility of divinely directed transmutation, which was viewed as evolution, and so a matter for condemnation. It seemed clear to Darwin that he would provoke an even greater outcry if he proposed a theory of evolution with a natural mechanism. Hence, he revealed the contents of his essay in the first instance only to the botanist Joseph Hooker (1817–1911), later to become Director of Kew Gardens, who had much in common with Darwin, both having sailed around the world to study nature in different regions. Even so, Darwin said that it was 'like confessing a murder'. In fact, at the time, Hooker was far from convinced by Darwin's arguments, but he encouraged his friend to develop his ideas further and discuss them with him, without fear of disclosure.[37]

Another possible reason why Darwin was reluctant to publish at this time could have been that, regardless of other considerations, he believed that his thoughts on evolution would never be taken seriously until he could establish a similar reputation as a biologist to the one he already had as a geologist. He was devoting a great deal of time to this, carrying out a detailed study of barnacles. The results were published in a two-volume book in 1854, to great acclaim. These barnacle studies revealed to Darwin the full range of individual variability within a single species so, as well as helping to build his reputation, they also contributed to the development of his evolutionary ideas.[38]

The *Origin of Species*

Darwin's barnacle book made a great impression on the biologist Thomas Henry Huxley (1825–1895), bringing him into contact with the author. Huxley (figure 7.2) was yet another former ship's naturalist, having sailed the South Seas for four years

Figure 7.2   Left: Charles Darwin in 1854. Right: Thomas Huxley. Both drawings are based on early photographs.

as assistant surgeon on H.M.S.*Rattlesnake*, following medical studies at Charing Cross Hospital, London. During his travels he improved on Cuvier's system of classification, and built a scientific reputation by sending detailed observations back to England for publication. He was elected to the Royal Society in 1851, and became its President from 1881 to 1885.[39]

Both Huxley and Lyell eventually came to know of Darwin's evolutionary beliefs. Huxley, who was looking to create a new type of science, free from the constraints of the past, took them in his stride. Lyell, in contrast, was shaken by the revelation but, nevertheless, he remained friendly and supportive towards Darwin. His own views, however, remained unchanged, largely because of the implications of evolution for humankind. He believed that humans were unique in having an immortal soul but, in any case, for practical reasons, our special status in creation had to be maintained, or civilisation could fall apart.[40] Thus, Lyell, in his 1851 Presidential Address to the Geological Society, drew attention to the lack of evidence in the fossil record for a 'gradual advance towards a more perfect organization . . . resembling that of man.' He went on to stress that 'the superiority of man, as compared to the irrational mammalia, is one of kind rather than of degree, consisting in a rational and moral nature . . . and not in the perfection of his physical organization'. As to how new species came into being, Lyell was not willing to speculate, apart from pointing to the involvement of God, saying, 'Whether such commencements be brought about by the direct intervention of the First Cause, or by some unknown Second Cause or Law appointed by the Author of Nature, is a point upon which I will not venture to offer a conjecture'.[41]

Lyell could not have disagreed more strongly with Darwin's views on evolution. Nevertheless, he urged him to publish them, to establish ownership. However, Darwin continued to delay for, throughout the 1850s, he could see publication causing more controversy than he was prepared to face. In 1857, the appearance of the first gorilla in a travelling menagerie encouraged social revolutionaries to try to de-stabilise society by proclaiming our monkey origins. This particularly inflamed Owen, who, like Lyell, was concerned about the social consequences of humankind losing its special status. Although, by this time (as we saw in Chapter 5), he was prepared to accept some kind of guided evolution, this could never have included the formation of *Homo sapiens* from an ape.[42]

The situation was suddenly brought to a crisis point when Alfred Russel Wallace, a naturalist living in the East Indies, sought Darwin's help in finding a publisher for a paper he had written. As was immediately apparent, Wallace, like Darwin, had come to believe in evolution by natural selection after reading what Malthus had written about populations.[43] Darwin's initial reaction was apparently to let Wallace take all the glory, and all the abuse. However, Lyell and Hooker (who by this time had tested Darwin's theories against his observations in various parts of the world, and had become convinced of their general validity) persuaded him to allow his 1844 essay, together with a summary of his ideas on natural selection which had been enclosed in a confidential letter written in 1857 to the American naturalist Asa Gray, to be read alongside Wallace's paper at a meeting of the Linnean Society in July 1858. His best-known book, *On the Origin of Species*, was published in the following year.[44]

In fact, as Darwin stated in the introduction, he regarded the *Origin* as an abstract of a larger book, *Natural Selection*, which he had been working on for several years. Since he still intended to complete and publish *Natural Selection* at a later stage, he gave few references to his sources. *Natural Selection* never reached publication during his lifetime, and much of the material was used in other books.[45] That left the *Origin* as a somewhat detached work, leading to much controversy. However, the copious notes which Darwin left behind have enabled scholars to prepare an edition of *Natural Selection* (which was published in 1975) and also to chart the development of his theory of evolution. Although most of the essential features were present by 1838, it is clear that significant modifications were made throughout the next twenty years.[46]

Inevitably, Darwin was a product of his environment. Early in his career, he was influenced by the writings of William Paley, and his belief, as one of his evidences of God (see Chapter 5), that all species were perfectly adapted to their habitat. The replacement of one species by another in the fossil record indicated that a climatic change had taken place, after which God, directly or indirectly, removed the old species and replaced it by one which was ideally suited to the new environment. The change from one species to another must have been saltational (i.e. it involved a sudden jump). For this reason, although Darwin was committed to geological gradualism, his earliest evolutionary theories included saltational changes. Even after he came upon natural selection as a mechanism for evolution in 1838, removing the need to involve God in the process, Darwin's belief in perfection of adaptation prevented him from discarding saltational evolution. Thus, in the 1844 essay, Darwin argued that organisms would generally be ideally suited to their habitat, so would not change in form unless there was some alteration in environmental conditions. For this reason evolution must be an intermittent, rather than a continuous, process.[47]

However, during the 1850s, Darwin rejected the concept of perfection of adaptation and, with it, saltational evolution, moving to a complete acceptance of gradualism. There were several reasons for this. One was the continuing influence of Lyell. Another was that Darwin's barnacle work had demonstrated the possibility of significant variation within a species, which was inconsistent with perfection of adaptation. It also led to the notion that a variant form with advantageous characteristics, however slight, would slowly but continuously make up an ever increasing proportion of the population, as a result of natural selection. Evolution as he saw it was simply an extension of this process, so it must be a gradual one.[48] Possibly not all Darwin's reasons for adopting gradualism were based in science, for he made much use of a popular dictum of the natural theologians, attributed to Linnaeus, which stated, 'Natura non facit saltum' ('Nature does not make jumps'). Furthermore, Darwin's nervous disposition, coupled with the fear of revolutions prevalent throughout his social class, may have made him receptive to a biological system where change could only occur in an imperceptible fashion.[49]

Without question, Darwin could not have been led to a belief in evolutionary gradualism by the fossil record. This, as we have already seen, tells a very different story, if read in straightforward fashion. For this reason, Darwin was forced to stress the incomplete nature of the fossil evidence when defending evolutionary gradualism. In the final chapter of the *Origin*, after acknowledging several problems facing his theory, he wrote, 'I can answer these questions and grave objections only on the supposition that the geological record is far more imperfect than most geologists believe . . . That the geological record is imperfect all will admit; but that is imperfect to the degree which I require, few will be inclined to admit.'[50]

Nevertheless, he made a determined effort to justify the strongly gradualistic nature of his conclusions, going on to claim that: 'If we look to long enough intervals of time, geology plainly declares that all species have changed; and they have changed in the manner which my theory requires, for they have changed slowly and in a graduated manner. We clearly see this in the fossil remains from consecutive formations invariably being much more closely related to each other, than are the fossils from formations distant from each other in time'.[51]

Darwin's argument was unsatisfactory, because these observations were just as compatible with an evolutionary mechanism proceeding via a series of discrete steps as with one involving imperceptible transitions. Moreover, even then, there was some positive evidence against gradualism. For example, as a result of Cuvier's work, it was known that three genera of elephants, one including the African elephant, one the Indian elephant and one the mammoth, had appeared more or less simultaneously from the ancestral genus *Primelephas*, and the two surviving species had changed very little from the time of their first appearance (now known to be around 4 million years ago) to the present day.[52] On the eve of publication of the *Origin*, Thomas Huxley had written to Darwin, to express his concern about the issue, warning, 'You have loaded yourself with an unnecessary difficulty in adopting *Natura non facit saltum* so unreservedly'.[53] Nevertheless, Huxley continued to offer full support for Darwin's theory as a whole and, in order not to provide ammunition for the opponents of evolution, he made no public criticism of the gradualistic model his friend had chosen to adopt.[54]

8 After the *Origin*: the triumph of evolutionary gradualism

The initial reaction

The *Origin* made an immediate impact, provoking reactions both for and against the author's arguments. The scientific clergymen showed less hostility that Darwin had expected but, not surprisingly, they could not agree with his views.[1] Henslow, Darwin's former mentor, came to stay with him early in 1860, and they discussed their differences amicably enough, without coming any closer together.[2] As Henslow explained in a letter to his brother-in-law, the Rev. Leonard Jenyns, he considered the book to be 'a marvellous assemblage of facts and observations ... but it pushes *hypothesis* (for it is not real *theory*) too far'.[3] Whewell also remained reasonably friendly, writing to tell Darwin, 'I cannot, yet at least, become a convert. But there is so much of thought and of fact in what you have written that it is not to be contradicted without careful selection of the ground and manner of the dissent.' Surprised by this mild reaction, Darwin exclaimed to his friends, '[He] is not horrified with us'. Afterwards, Whewell concentrated mainly on arguing that, even if biological evolution had taken place, it must have been a process initiated and controlled by God. He pointed out that Darwin had not provided any actual proof in his book of a species originating by natural selection. Furthermore, there was 'an inexplicable gap at the beginning of his series" which 'leaves room for, and requires, a supernatural origin'.[4]

The more emotional and straightforward Sedgwick expressed his views to Darwin 'in a spirit of brotherly love', exclaiming, 'If I did not think you a good-tempered and truth-loving man, I should not tell you that ... I have read your book with more pain than pleasure'. Whilst he admitted there were parts of the book which he 'admired greatly', other parts were 'utterly false and grievously mischievous'. As befitting someone who was steeped in natural theology, he went on to accuse Darwin 'of trying to sever the link between material nature and its moral meaning'.[5]

In America, Agassiz, still adhering to the principles of idealist philosophy, was more openly critical of Darwin's theory than the Anglican catastrophists, writing that it was 'a scientific mistake, untrue in its facts, unscientific in its methods, and mischievous in its tendency'.[6]

Some scientists, including Owen and Gray, were prepared to accept many of Darwin's arguments, but they still clung to the belief that there must be some form of divine control over selection. The difference was fundamental: Darwin was committed to an evolutionary mechanism operating completely through natural processes, whilst the others could not accept any mechanism which excluded the involvement of God. A model of evolution whose course was determined by variants arising through chance and surviving or dying as a result of natural selection, termed the 'law of higgledy-piggledy' by the astronomer Sir John Herschel, was of no interest to those looking for a mechanism through which there could be an unfolding of a divine plan. In contrast, Darwin could not see how the evidence of cruelty in nature was compatible with the providential view.[7] In a letter to Gray, in 1860, he wrote, 'I cannot persuade myself that a beneficent and omnipotent God would have designedly created the Ichneumonidae with the express intention of their feeding within the living bodies of Caterpillars, or that a cat should play with mice'.[8]

Whilst most of those committed to natural theology and idealist philosophy remained antagonistic to evolution, particularly as a natural process, they were generally in the twilight of their careers and, indeed, of their lives. Darwin's evolutionary views became increasingly popular, particularly amongst the younger generation of scientists. However, even so, almost no-one was convinced by his arguments about natural selection. In the annual report of the Linnean Society for the period which included the presentations of Darwin and Wallace, the President, Thomas Bell, wrote that the year which had just passed had not been marked 'by any of those striking discoveries which at once revolutionize, so to speak, the department of science on which they bear'.[9] Even Huxley was not so much concerned with establishing a particular mechanism for evolution, as in destroying natural theology.[10] In his review of the *Origin* in *The Times* in December 1859, he emphasised the distinction between Darwin's approach and the ideals of natural theology, writing, 'Mr Darwin abhors mere speculation as nature abhors a vacuum. He is as greedy of cases and precedents as any constitutional lawyer, and all the principles he lays down are capable of being brought to the test of observation and experiment'.[11]

Prominent amongst those who were determined to resist any attempt to separate science from religion was Samuel Wilberforce, Bishop of Oxford (1805–1873). In an attempt to turn the tables on his opponents, Wilberforce invoked Lyell's uniformitarianism to attack natural selection in a review of the *Origin* in the *Quarterly Review*, writing, 'But why should Nature, so uniform and persistent in all her operations, tend in this instance to change? Why should she become a selector of varieties?'[12] For Wilberforce, like many others inside and outside the Church, the particular problem was Darwin's inclusion of humankind in the evolutionary process.[13] To minimise controversy, Darwin had avoided any explicit mention of this in the *Origin*. However, near the end of the book, when speculating on the outcomes of further research, he wrote, 'Light will be thrown on the origin of man and his history'.[14] That was sufficient to give the game away: Darwin believed that humans had evolved by natural processes from apes. On hearing this from her husband, the wife of the Bishop of Worcester is said to have cried out, 'Descended from apes! My dear, let us hope it is not so; but if it is, that it does not become generally known'.[15] The fear that such knowledge would destabilise society was still strong.

Since even Wallace found it impossible to accept that the process of natural selection alone could have given rise to the human brain, the opposition of a reactionary bishop such as Wilberforce to the idea of apes evolving into humans can easily be imagined.[16] In his review of the *Origin*, he wrote that 'such a notion is absolutely incompatible not only with single expressions in the word of God on that subject of natural science with which it is not immediately concerned, but, which in our judgement is of far more importance, with the whole representation of that moral and spiritual condition of man which is its proper subject matter.'[17]

Such antagonism probably cost Darwin a knighthood. A proposal to award him one was being considered at the time of publication of the *Origin*, but this was turned down by Queen Victoria on the advice of her ecclesiastical advisors, one of whom was Wilberforce.[18]

At the British Association for the Advancement of Science meeting held in Oxford, in 1860, Wilberforce took part in a famous debate against Darwin's supporters, including Huxley and Hooker, the former becoming known as 'Darwin's bulldog' because of his robust support for evolution. Both sides claimed victory, but the verdict of posterity has generally been in favour of the Darwinians. Wilberforce had ended his speech by asking Huxley whether it was through his grandfather or his grandmother that he claimed his descent from a monkey. Huxley murmured, 'The Lord hath delivered him into mine hands', before replying that if he was offered the choice of having an ape for a grandfather, or a man who misused his talents by introducing ridicule into what was meant to be a serious scientific discussion, then without hesitation he would state his 'preference for the ape'. This caused uproar, during which one person (Lady Brewster) fainted. Then, after the chairman (Henslow) had restored order, Hooker made another telling contribution on behalf of the Darwinian camp, pointing out that Wilberforce had shown a poor grasp of his material (even questioning whether the bishop had read the *Origin*), whereas he himself, after initially being opposed to evolution, had studied the subject for many years and had eventually been persuaded, on the basis of the evidence, that Darwin was on the right track.[19]

In the late nineteenth century, as at other times, there were religious fundamentalists such as Wilberforce whose minds were not in any way receptive to scientific arguments. However, it would be misleading to regard the debates over evolution, any more than those concerning geology, as being ones between objective scientists and religious bigots. Indeed, in Britain, following a long-established tradition, science and religion were thought for much of the nineteenth century to be inseparable, and were blended together in a variety of ways by different individuals. Darwin adopted much the same public stance as many of his scientific contemporaries and predecessors, back to the times of the cosmogonists and cosmologists, suggesting that God had devised the system, set it in motion, and was now letting it operate without interference. Near the end of the *Origin*, he wrote, 'Authors of the highest eminence seem to be satisfied with the view that each species has been independently created. To my mind it accords better with what we know of the laws impressed on matter by the Creator that the production and extinction of the past and present inhabitants of the world should have been due to secondary causes, like those determining the birth and death of the individual'. However, what set Darwin completely apart from what had gone before was his willingness to discard the notion that the system was teleological, operating according to some pre-determined plan towards a definite goal.[20]

Owen, one of those who clung to teleological ideas, had briefed Wilberforce for the Oxford debate, even though he was much more open-minded than the fundamentalist bishop. Huxley, in turn, did not accept all aspects of Darwin's theory. He was never a committed gradualist, and was far from convinced that natural selection provided an adequate evolutionary mechanism. To him and to others, including the Irish physicist John Tyndall, the detail was of secondary importance, the crucial point being that Darwinism provided a lever for separating science from religion.[21]

Darwin and Lyell

What were Lyell's thoughts at this time? Without doubt, he had proved a good friend of Darwin and given him much support, but he remained distant from some of the key components of Darwin's evolutionary theory. Lyell never accepted natural selection as a possible mechanism for evolution, being convinced that God must have retained control of the process. Above all, he always denied that humankind could be regarded as an animal which had evolved from other animals.[22]

In 1863, four years after the appearance of the *Origin*, Lyell's *The Geological Evidences of the Antiquity of Man* was published.[23] It detailed many reasons for thinking that humankind had been on Earth for much longer than the 6,000 years suggested by biblical chronology, and it also included a short section outlining arguments in favour of evolution. However, Lyell never suggested in the book that he accepted these arguments, or believed they had any relevance to the origin of humankind.[24] Darwin was not impressed, telling Hooker, 'I must say how much disappointed I am that he has not spoken out on species, still less on man. And the best of the joke is that he thinks he has acted with the courage of a martyr of old'.[25] Without doubt, Lyell still maintained a belief that God operated above the 'normal' laws of nature. Thus, events of an occasional nature, which had major implications, could be seen as evidence of the unfolding of a divine plan. He made his views on this subject clear in the *Antiquity of Man*, writing:

> In our attempts to account for the origin of species, we find ourselves still sooner brought face to face with the working of a law of development of so high an order as to stand nearly in the same relation as the Deity himself to man's finite

understanding, a law capable of adding new and powerful causes, such as the moral and intellectual faculties of the human race, to a system of nature which had gone on for millions of years without the intervention of any analogous cause. If we confound 'Variation' or 'Natural Selection' with such creational laws, we deify secondary causes or immeasurably exaggerate their influence.[26]

However, in the tenth edition of his *Principles of Geology*, which was published in 1867, Lyell, then aged seventy, tried to be more openly supportive of Darwinism than he had been in the past. For the first time, he committed himself to an unequivocal statement of his belief that biological evolution had taken place, and that the fossil record showed indications of progression. Unfortunately, although Darwin was pleased by this, he was profoundly disappointed by Lyell's discussions about the origin of humankind, which he found to be 'too orthodox, except for the beneficed clergy'.[27] To make matters worse, Lyell told Darwin in 1869 that he had recently written to Wallace to congratulate him on his views on human evolution, adding, 'I rather hail Wallace's suggestion that there may be a Supreme Will and Power which may not abdicate its function of interference, but may guide the forces and laws of Nature.'[28] The two former friends had little further contact with each other. When Lyell died in 1875. Darwin was asked to be a pallbearer at the funeral in Westminster Abbey, but made excuses on the grounds of ill health. Seven years later, in 1882, Darwin himself was to be buried in Westminster Abbey.[29]

Darwin's evolutionary theory has often been seen as a natural extension of Lyell's uniformitarianism. As Peter Bowler wrote, in the 1989 edition of *Evolution – The History of an Idea*, 'It has been fashionable to trace the ancestry of Darwinism back to Charles Lyell's uniformitarian geology, in which all changes occurred gradually through the action of observable causes. Darwin was seen simply to have applied this element to the organic as well as the physical world'.[30] Moreover, that 'fashion' continued to be in vogue. So, science journalist Roger Lewin wrote in 1993 in *Complexity*, 'In his major work, *Principles of Geology*, Lyell said that abrupt transitions in the geological record would one day be shown to be erroneous, when transitional strata were discovered. Gradualism superseded catastrophism. Darwin's and Lyell's worldviews were therefore perfectly complementary'.[31]

However, for reasons we have already discussed, that is far too simplistic a picture. Certainly, Darwin took the 'principle of uniformity' from Lyell, and also followed him in becoming a convinced gradualist. On the other hand, evolution had more in common with the catastrophist's ideas of progression than with the steady-state world-view of Lyell. Without question, Darwin's evolutionary theory owed something to both the uniformitarian and the catastrophist traditions.[32] Thomas Huxley pointed this out, and at the same time expressed his concerns about giving undue emphasis to gradualism, in his 1869 Presidential Address to the Geological Society, saying,

> To my mind there appears to be no sort of necessary theoretical antagonism
> between Catastrophism and Uniformitarianism. On the contrary, it is very

conceivable that catastrophes may be part and parcel of uniformity . . . Still less is there any necessary antagonism between either of these doctrines and that of Evolution, which embraces all that is sound in both Catastrophism and Uniformitarianism, while it rejects the arbitrary assumptions of the one and the, as arbitrary, limitations of the other.[33]

As Ernst Mayr wrote, in *The Growth of Biological Thought*, 'No matter how great Darwin's intellectual indebtedness to Lyell was, uniformitarianism . . . was actually more of an hindrance to the development of his evolutionism than a help. Gradualism, naturalism, and actualism were the prevailing concepts from Buffon to Kant and Lamarck.'[34]

Darwin and gradualism

Darwin's commitment to gradualism was made very clear in the *Origin* and, furthermore, his interpretation of gradualism could hardly have been more extreme. Thus, in the fourth chapter of the first edition, he wrote, 'Natural Selection can act only by the preservation and accumulation of infinitesimally small inherited modifications, each profitable to the preserved being . . . That natural selection will always act with extreme slowness I freely admit . . . Nothing can be effected unless favourable variations occur, and variation itself is apparently always a very slow process.'[35]

Two chapters later, he re-emphasised these points, after asking a rhetorical question: 'Why should not Nature have taken a leap from structure to structure? On the theory of natural selection, we can clearly understand why she should not; for natural selection can act only by taking advantage of slight successive variations; she can never take a leap, but must advance by the shortest and slowest steps'.[36]

In Darwin's view, the living world was always in a state of severe overpopulation. The consequence of this was a system under considerable strain, characterised by intense competition between individuals for survival. Darwin described the situation in forthright terms in the third chapter of the *Origin*:

> In looking at Nature, it is most necessary . . . never to forget that every single organic being around us may be said to be striving to the utmost to increase in numbers; that each lives by a struggle at some period of its life; that heavy destruction inevitably falls either on the young or old, during each generation or at recurrent intervals. Lighten any check, mitigate the destruction ever so little, and the number of the species will almost instantaneously increase to any amount. The Face of Nature may be compared to a yielding surface, with ten thousand sharp wedges packed close together and driven inwards by incessant blows, sometimes one wedge being struck, and then another with greater force.[37]

In such a dynamic system, individuals with advantageous characteristics not only survived, but tended to form an increasing proportion of the population, eventually constituting a new species. Several new species could arise from a single ancestral species, each consolidating a different advantageous variation, so a highly branched pattern of evolution would be expected to emerge. Species which were less successful than rival species would eventually become extinct, just as individuals who were unsuccessful in the struggle for life would die without leaving offspring. So, Darwin wrote of evolution proceeding 'slowly and progressively . . . like the branching of a great tree from a single stem', whilst those species which became extinct were simply compared to branches that had 'decayed and dropped off'.[38] As he explained in the fourth chapter of the book, 'I think it inevitably follows that as new species in the course of time are formed through natural selection, others will become rarer and rarer, and finally extinct. The forms which stand in closest competition with those undergoing modification and improvement, will naturally suffer most'.[39]

In Victorian England, it was understandable that such an idea, with progress or failure being determined through competition, should have struck a chord which was attractive to many. Indeed, the philosopher and man of letters Herbert Spencer argued for an even more explicitly progressive view of evolution. The phrase, 'survival of the fittest', which was coined by Spencer, was incorporated by Darwin into later editions of the *Origin*.[40] In the first edition, Darwin discussed evolutionary progress in the tenth chapter, writing, 'The inhabitants of each successive period in the world's history have beaten their predecessors in the race for life and are, in so far, higher in the scale of nature; and this may account for that vague yet ill-defined sentiment, felt by many palaeontologists, that organisation on the whole has progressed'.[41]

However, this idea was not invented by Darwin. As we have seen, progression, albeit arising by a non-evolutionary mechanism, was an integral feature of the beliefs of the catastrophists (in England and also in France, where it featured strongly in the writings of Alcide d'Orbigny) and it also formed an important part of the Lamarckian model of evolution. To Darwin, of course, evolutionary progression did not occur in a direct fashion. Instead, it emerged as an overall trend from complex, branched patterns. These various themes of natural selection, gradualism, extinction and progress were brought together in the final chapter of the *Origin*, when Darwin wrote:

> As natural selection acts solely by accumulating slight, successive favourable variations, it can produce no great or sudden modification; it can act only by very short and slow steps. Hence the canon of '*Natura non facit saltum*', which every fresh addition to our knowledge tends to make more strictly correct, is on this theory simply intelligible . . . The extinction of species and of whole groups of species, which has played so conspicuous a part in the history of the organic world, almost inevitably follows on the principle of natural selection; for old forms will be supplanted by new and improved forms . . . Thus, from the war of nature, from famine and death, the most exalted object which we are capable of conceiving, namely, the production of the higher animals, directly follows.[42]

Darwin and human evolution

What were Darwin's precise views on the 'production of the higher animals' and, in particular, humankind? At the time when he was working on the *Origin*, some evidence of a prehistoric, stone-age culture was being uncovered at sites in England and continental Europe. The best known of these was at Crô-Magnon, in western France, discovered in 1868. However, the only skulls found near the stone tools seemed identical to those of modern humans, members of the species *Homo sapiens*.[43]

Nevertheless, more isolated fossil evidence of a different human type, the Neanderthals, was starting to emerge. These had skulls with slightly more ape-like features, such as a sloping forehead, no chin, and ridges over the eyebrows. The first Neanderthal skull was unearthed at Gibraltar in 1848, but little attention was paid to it until a similar skull was found in 1856, together with other parts of a thick-boned skeleton, buried in a cave in the Neander Valley, near Düsseldorf, Germany. This location gave the type its name, 'valley' being 'thal' in the German of the time, now modified to 'tal'. A report of the find by the German anatomist Hermann Schaffhausen appeared in 1858, the year before the publication of the *Origin*, but another two years passed before there was an English translation. In 1864, the Irish anatomist William King classified the finds as *Homo neanderthalensis*. That, however, was controversial, for not everyone accepted that the Neanderthal man was a member of an extinct species, separate from *Homo sapiens*. Some, including the distinguished German anthropologist Rudolf Virchow, believed it to be the remains of a modern human with skeletal deformities.[44]

As we have seen, Darwin avoided specific mention of human evolution in the *Origin*, but Thomas Huxley, needless to say, was not afraid to force the issue. In 1863, in his *Evidence as to Man's Place in Nature*, he argued that apes and humans must have a close evolutionary relationship, because of their many anatomical similarities. Huxley appreciated that the Neanderthal skull had features which made it distinct from those of modern humans but, as it contained a modern-sized brain, it could not be regarded as belonging to a species intermediate between apes and humans. In his view, fossils of such a species remained to be discovered.[45]

Another who was quick to speculate about human evolution was the German zoologist Ernst Haeckel (1834–1919), who coupled Darwinism with some of the theories of the idealist philosophers, envisaging evolution as an orthogenetic process, showing a progression towards more advanced forms. Unlike his idealist predecessors, however, he viewed this in entirely naturalistic terms.[46]

Haeckel, in his *Generelle Morphologie der Organismen* of 1866, and in later books, gave versions of a 'tree of life' which had labels depicting a variety of organisms, some on the main trunk and some on the side branches. *Menschen* ('Man') was located at the uppermost point of the tree, with the present-day apes (gorillas, chimpanzees, orang-utans and gibbons) on branches to one side or the other. Haeckel argued that investigators would have to look back to the Tertiary Period, beyond the time of the

Neanderthals, who lived during the Pleistocene, the first epoch of the Quaternary Period, for the common ancestor of modern apes and humans. He assigned this postulated 'missing-link' to the genus *Pithecanthropus* ('Ape-like Man').[47]

Darwin's silence on the subject of human evolution came to an end in 1871, with the publication of his book *The Descent of Man*.[48] Fossil evidence was still very scarce so, inevitably, the arguments were largely theoretical, following on from his previous writings. Not surprisingly, Darwin believed that natural selection must have played a very important role. He also thought that sexual selection, the desire to find mates with characteristics which seemed attractive, but which might not give any survival advantage against the environment, could have been a significant factor. Furthermore, he was prepared to give serious attention to the possible involvement in the evolutionary process of the acquisition and elimination of characteristics through use and disuse.[49] However, whatever the specific details, the process of human evolution must have been a natural one. He wrote:

> We have now seen that man is variable in body and mind; and that the variations are induced, either directly or indirectly, by the same general causes, and obey the same general laws, as with the lower animals . . . The early progenitors of man must also have tended, like all other animals, to have increased beyond their means of subsistence; they must therefore occasionally have been exposed to a struggle for existence, and consequently to the rigid law of natural selection. Beneficial variations of all kinds will thus, either occasionally or habitually, have been preserved, and injurious ones eliminated.[50]

Despite the impression given by some of his critics, Darwin was not arguing that *Homo sapiens* had evolved from any of the species of ape alive at the present time. Instead, like Haeckel, he believed that modern apes and humans must have had a common ancestor. As to the point of divergence, he had little doubt that the acquisition of bipedalism, i.e. the ability to walk on two legs, was an important development in the line leading to humankind. Explaining this, Darwin wrote:

> As soon as some ancient member in the great series of Primates came, owing to a change in its manner of procuring subsistence, or to a change in the conditions of its native country, to live somewhat less on trees and more on the ground, its manner of progression would have been modified: and in this case it would have to become either more strictly quadrupedal or bipedal . . . Man alone has become a biped; and we can, I think, partly see how he has come to assume his erect attitude, which forms one of the most conspicuous differences between him and his nearest allies. Man could not have attained his present dominant position in the world without the use of his hands which are so admirably adapted to act in obedience to his will.[51]

However, the acquisition of bipedalism could not have occurred in isolation. The gradual development of the ability to walk upright must have had associated effects, changing the shape of the pelvis, the curvature of the spine and the position of the head. After the hands started to become free, and undergo modifications to allow more effective use of tools and weapons, there would have been less need for large canine teeth and strong jaws. Hence, these ape-like features would have slowly disappeared. As these changes would inevitably have resulted in a gradual loss of physical strength and mobility, so increased intelligence and better social organisation must have been developed at the same time to enable the bipeds to survive. As Darwin put it, the transitional forms 'might have existed, or even flourished, if they had advanced in intellect, whilst gradually losing their brute-like powers''. Therefore, as the physical sign of this advance in intellect, the brain must have increased in size, as the ability to walk upright was being acquired.[52]

Without question, given the information available at the time, Darwin presented a plausible scenario in *The Descent of Man*. This, as we have seen, envisaged modern humans evolving from ancestral apes through a gradual process which involved the co-ordinated development of bipedalism, the use of tools, social organisation and an enlarged brain. That last aspect, brain size, came to be seen as a particularly important marker of progress along the evolutionary path to humankind.

Darwinism and uniformitarianism

Both with regards to human evolution and also more generally, it still remained unclear as to what gave rise to the variations on which natural selection could act. In the first edition of the *Origin*, Darwin thought that use-and-disuse might play a part, but was sceptical about the possible involvement of environmental factors such as climate, although not dismissing them altogether. In later editions, as in *The Descent of Man*, Lamarckian factors such as these became more central to Darwin's ideas, after the intervention of Fleeming Jenkin, an engineer by profession. Jenkin had argued that if, as was generally supposed, inheritance consisted of an even blending of characteristics from each parent, then any new characteristic, however advantageous, would quickly be diluted to insignificance as a result of breeding with members of the vast majority of the population who lacked it.[53] Was there any way in which a novel characteristic could be preserved in a lineage without constant reinforcement through a Lamarckian mechanism?

As it happened, Johann Mendel (1822–1884), a Moravian monk who adopted the Christian name of Gregor when he joined the Augustinian order, had already suggested an answer to that question. Seven years of studying the characteristics of peas had convinced him that inheritance involved the transmission of specific factors, which later became known as genes. Mendel (figure 8.1) concluded that each of the traits he investigated was determined by two factors of inheritance (one from each parent) and, if these were different, the expression of one could be dominant over the other, so preventing blending. His findings were published in detail in the *Proceedings of the Brünn* (now Brno) *Society for the Study of Natural Science* in 1866, and copies of his

Figure 8.1   Gregor Mendel, based on a portrait in the Mendelianum, Brno, painted after Mendel's death.

paper sent to 40 eminent botanists and to more than one hundred scientific institutions, including the Royal Society and the Linnean Society in London. However, his work was ignored until the beginning of the twentieth century.[54]

Despite uncertainties concerning the mechanism, Darwinism, with its emphasis on gradualism, continued to attract support as the best scientific explanation for the development of life on Earth, just as Lyell's uniformitarianism was widely accepted as the *only* legitimate interpretation of the sequence of rocks. Without question, as the nineteenth century drew to a close, catastrophism seemed to have vanished without trace, and gradualism was established as the ruling paradigm, in both geology and evolution.[55]

Catastrophism, of course, could accommodate a short time-scale for the history of the Earth, even if nineteenth century catastrophists did not generally advocate a starting date of anything like as recent as 4004 B.C., as often (but erroneously) supposed. Georges Cuvier, for example, wrote in 1804 (as we noted in Chapter 3) about a fossil he had investigated which had been 'buried perhaps for thousands of centuries'. Similarly, William Whewell, in his *History of the Inductive Sciences* of 1837, referred (as we saw in Chapter 5) to 'forces which are changing through immeasurable time'. However, although catastrophism could afford some flexibility over the time-scale, gradualism had no alternative but to require a very long one. So Darwin, in the ninth chapter of the *Origin*, estimated that erosion of the Weald in Kent must have been going on for something in the region of 300 Myr (million years), if present-day features were to be explained by forces currently acting.[56]

From considerations of erosion and sedimentation rates, and estimates of the thickness of the Earth's crust, John Phillips calculated in 1860 that the crust must have been formed over a period of something like 96 Myr. Nineteen years later, in the light of additional information, this value was adjusted to 600 Myr by the engineer and amateur geologist T. Mellard Reade. Although a time-scale such as this would not exclude the possibility of major catastrophes having occurred, it was *essential* for the gradualists. Hence, there was some concern in 1883 when the eminent physicist Lord Kelvin, formerly William Thompson (1824–1907), calculated, on the basis of cooling processes, that the Earth was not likely to be more than 400 Myr old. He went on to reduce this figure to 50 Myr in 1876 and to 24 Myr in 1897. However, it then became apparent that the omission of a factor that Kelvin knew nothing about, heat from radioactive decay, invalidated his calculations. When Ernest Rutherford (1871–1937) came to the Royal Institution in London in 1904 to make the announcement, Kelvin was in the audience, and Rutherford was not looking forward to telling the elderly scientist of his mistake. However, when the moment came, he had the inspiration to suggest that, when Kelvin had said that his conclusions were unchallengeable provided no new source of heat was discovered, he was, in fact, predicting the discovery of radioactivity. As Rutherford later wrote, 'Behold, the old boy beamed upon me'.[57]

Studies of the degree of breakdown of unstable radioactive isotopes in rocks, coupled with calculations based on the uniformitarian assumption that radioactive decay occurred at a steady rate, gave a time-scale for the Earth which was at least as long as the estimates of the geologists. It became clear, from reviews by the British physicist/geologist Arthur Holmes, in 1913, and the American geologist Joseph Barrell, in 1917, that the age of the Earth was to be reckoned in hundreds, if not thousands, of millions of years. Eventually it became generally accepted that the Earth is about 4,600 Myr old, and that organisms have been living on the planet for more than 3,000 Myr. This had implications for the interpretation of transitions in the geological or fossil records, for an apparently abrupt change could have taken place over what, from the point of view of living organisms, was a very long period of time, i.e. approaching a million years.[58]

Although precise dates had still to be worked out, the general truth of that was clear from the very early years of the twentieth century. Hence, even disregarding the undoubted imperfections of the fossil record, it seemed to most people that there was no longer any reason to even contemplate the possibility that an apparently abrupt transition might have been caused by a genuinely sudden catastrophe. Uniformitarianism could explain everything, so catastrophism, having been thoroughly discredited, could now be thrown onto the rubbish-tip of history, together with other false doctrines, and remain there forever.

Section B
From 1900 to 1979: gradualism reigns supreme

9   Neo-Darwinism: the Modern Synthesis

The development of genetics

To most biologists, the twentieth century, from its very beginning, was the century of genetics. In 1900, Gregor Johann Mendel's work on the genetics of peas, ignored by his nineteenth century contemporaries, was re-discovered by three botanists, Hugo de Vries of Holland, Carl Correns of Germany and Erich Tschermak of Austria, operating independently of each other. The contribution of de Vries (1848–1935) was particularly significant in establishing that biological characteristics were inherited through the transmission of discrete factors (which were named 'genes' in 1909). However, it quickly became apparent that Mendelian inheritance could not, in itself, explain evolutionary change. Godfrey Harold Hardy, of Cambridge University, and Wilhelm Weinberg, of Stuttgart, both pointed out that any particular genetic trait would be expressed in a population in exactly the same proportions from one generation to the next, unless equilibrium was disturbed by natural selection or some other factor, a conclusion subsequently known as the Hardy–Weinberg principle.[1]

In a series of significant experiments carried out at Columbia University, New York, from 1910 onwards, Thomas Hunt Morgan (1866–1945) and Hermann Joseph Müller (1890–1967) demonstrated that variations in some of the characteristics of fruit flies (Drosophila) could be associated with structural changes ('mutations') in their genes. At first it seemed that these new developments in genetics posed a challenge for Darwinism, for new mutations might be able to disrupt the Hardy–Weinberg equilibrium and give rise to evolutionary jumps. This was indicated, for example, by the sudden appearance of distinctive new traits in the evening primrose, Oenothera lamarckiana, which had been noted by de Vries during a twenty-year period of observation.[2]

On the other hand, it soon became clear that most mutations had only slight consequences, suggesting a possible mechanism for the production of Darwin's small variations. Evidence for this became particularly noticeable when the focus of attention changed from experiments with pure stocks carried out in laboratories or gardens, towards investigation of the natural world. However, although Darwinism was seen to fit well within the bounds of possibility, it did not necessarily follow that it was of any great significance in evolutionary terms. Indeed, for some years after the realisation that most genetic changes were small ones, it was still generally believed that evolutionary progress owed more to the appearance of new mutants than to natural selection.[3]

There were also a number of people who maintained a belief in the Lamarckian theory of the inheritance of acquired characteristics, despite the fact that no evidence of it was emerging from the genetics studies. Some found Lamarckism attractive because they believed it to differ from Darwinism in being something other than a soul-less, mechanical process. However, to imagine the involvement of thought or desire in Lamarckian evolution was to misunderstand Lamarck's ideas, as we discussed in Chapter 3. Another factor which owed little to biology was the observation that a form of Lamarckian inheritance operated at the social level. Every day it was possible to observe acquired knowledge and acquired possessions being passed on from generation to generation, and some were tempted to think that the same thing must happen in physical evolution.

A more science-based reason for supporting the involvement of the theory of inheritance of acquired characteristics in evolutionary processes was that the development of certain inherited features, such as the long neck of the giraffe, seemed to be explained in a more straightforward fashion by Lamarckism than by natural selection. It was difficult to see how the fact that a particular giraffe had a neck which was a millimetre longer than other giraffes would give it any survival advantage over the rest, and for this process to lead to significant changes over the course of time, whereas the stretching of a giraffe's neck by striving to reach otherwise inaccessible food had a satisfying if superficial logic to it.[4]

In contrast, Darwinian logic could sometimes appear rather strained, as with the so-called Baldwin Effect, put forward to explain features such as the inherited calluses on the rump of the ostrich, on which the bird sits. According to the Baldwin Effect (named after James Mark Baldwin, who proposed the general idea in 1896, one of several to have done so at around the same time), a genetic change arising by accident could turn out to provide evolutionary advantages, fortuitously replacing or complementing non-inheritable acquired characteristics such as calluses formed by friction on an ostrich's rump. Perhaps not surprisingly, some found this proposal difficult to accept, as it seemed to depend on unlikely coincidences.[5] On the other hand, there was no clear-cut experimental evidence to prove the involvement of Lamarckism in evolution, or even to eliminate the possibility of alternative Darwinian explanations in any particular instance. Furthermore, not only did geneticists fail to find anything which could offer positive support to Lamarckism, but it began to appear, in the light of growing knowledge about genes, that such a mechanism might even be physically impossible.

Back in 1885, the German zoologist August Weismann (1834–1914) had suggested that the germ cells of an animal (those cells whose genetic content is passed on to the next generation) were completely separated from the somatic cells (those of the rest of the body). If he was correct, environmentally induced changes in somatic cells could not be transmitted to germ cells, so they could never be passed on to subsequent generations.[6] In an attempt to put his theory to the test, Weismann cut off the tails of twenty-two successive generations of mice, without this resulting in any mouse being born without a tail. He also drew attention to the fact that the circumcision of boys over more than a hundred successive generations had failed to produce any who lacked foreskin. Of course, neither of these 'experiments' really tested the 'use–disuse' principle of Lamarck, in the sense that no feature was lost in natural fashion as a result of disuse, so they could not be considered conclusive. Nevertheless, no-one was able to disprove the existence of the 'Weismann barrier' between germ cells and somatic cells, which remained a major problem for Lamarckism. An even greater problem became apparent when it was confirmed that there were indeed distinct differences between somatic cells and germ cells.[7]

Although the amphibian breeding studies of the Austrian biologist Paul Kammerer appeared to offer evidence of the inheritance of acquired characteristics, his work fell into disrepute when it was revealed that some of the evidence had been enhanced artificially. Kammerer subsequently committed suicide in 1926. However, that was not a straightforward admission of guilt, for it seems unlikely that he was involved personally in the fraudulent activities, and he was under great stress at the time for a variety of reasons. He had lost almost everything during the First World War, including breeding populations which had taken many years to accumulate, and his mistress was refusing to go with him to Russia, where he had been offered a new post.[8]

However, regardless of the issues of fraud and guilt, Kammerer's results were capable of being explained in a Darwinian way. His most famous, or notorious, study involved the midwife toad, *Alytes obstetricans*. This normally breeds on land but, when bred in water, it was found that male toads developed seemingly inheritable horny patches on the hands to aid copulation. Although these patches had been injected with ink to make them more conspicuous, they could have been genuine. They are found in some other species of toads, so could have been present in ancestral midwife toads. The success rate in breeding midwife toads in water was extremely low, so it was perfectly plausible to argue in a Darwinian fashion that those variants in the population which retained this atavistic feature to the greatest degree would be the most successful in surviving and producing offspring under the conditions of the experiment. Even so, Kammerer's results could not be confirmed when the same experiments were carried out by others.[9]

After the Kammerer scandal, Lamarckism was further discredited when the Soviet biologist Trofim Lysenko, supported by Stalin, attempted to use its principles in association with those of Marxism to produce increased yields of crops. In fact, the results were disastrous, which had serious consequences for the Soviet economy.[10]

Nevertheless, even if Lamarckism was making no progress, it was still generally believed that the force driving evolution forward was not natural selection but mutational pressure, the occurrence of new mutations. Thus, contrary to what Darwin had supposed, it seemed that evolution might be essentially a saltational process. Amongst the main champions of this point of view were de Vries and Morgan, together with the British geneticist William Bateson (1861–1926) and the man who coined the term, 'gene', the Danish botanist/geneticist Wilhelm Johannsen (1857–1927).[11]

However, there was a move back towards Darwinism in the 1920s, when several geneticists, including the Britons Ronald (R.A.) Fisher and John (J.B.S.) Haldane, the Russian Sergei Chetverikov and the American Sewall Wright, began to link genetics to the study of variation in populations.[12] Fisher's *The Genetical Theory of Natural Selection*, published in 1930, explained the new thinking.[13] As Reginald (R.C.) Punnett, a former student of Bateson, wrote in an unsympathetic review of the book for the journal *Nature*, Fisher saw the evolutionary process as 'a very gradual, almost impalpable one in spite of the discontinuous basis upon which it works'. This was because the overall trends of evolution were envisaged as being determined by changes taking place by natural selection in populations, not by mutations in individuals.[14]

The Genetical Theory of Natural Selection did not prove particularly influential at the time. In 1932, in *The Scientific Basis of Evolution*, Morgan re-iterated his long-held views that natural selection was only a minor factor in evolution, being far less important than mutational pressure.[15] Even amongst the population geneticists, it was not unanimously believed that all significant evolutionary change was adaptive, i.e. determined by the fact that variant forms which were better adapted to their environment than others would, through the operation of natural selection, be the ones most likely to survive, breed and pass on their genes (and hence characteristics) to future generations.[16] Sewall Wright (1899–1988) argued in 1931 that effective breeding populations were often small enough to allow significant changes to take place by 'genetic drift', i.e. by random fixation of particular gene combinations, whereas, in larger populations, these same changes might be eliminated by natural selection. A year later, he developed the concept of an 'adaptive landscape', with peaks representing those sub-populations within a species that had particular favourable combinations of genes. According to Wright, genetic drift might push a sub-group off an adaptive peak, which would lead to extinction if it could not return, which would generally be the case. However, in rare circumstances, natural selection might pull it rapidly to another peak. As each sub-group generally remained in contact with others of the same species, a favourable new characteristic which had emerged in this way in a small cluster of individuals, by a mixture of genetic drift and natural selection, should be able to spread through the entire population.[17]

Despite the arguments of Wright, Fisher continued to be convinced of the supreme importance of natural selection. In his view, the number of individuals making up a species would generally be high enough to negate any possible effects of random genetic drift. Thus, in 1936, he claimed, 'Evolution is progressive adaptation, and consists of nothing else'.[18]

Figure 9.1   Two of the founding fathers of the Modern Synthesis. Left: Theodosius Dobzhansky. Right: Ernst Mayr.

The birth of the Modern Synthesis

Over the next fifteen years, a variety of geneticists and other biologists went on developing these various ideas, combining Darwinism and population genetics to produce the 'Modern Synthesis' of evolution. This was also generally referred to as neo-Darwinism, although that particular term had an earlier origin, being coined by George (G.J.) Romanes in 1896 to describe a form of Darwinism which excluded any possible involvement of the inheritance of acquired charateristics.[19] The first major step in the establishment of the Modern Synthesis came in 1937, with the publication of *Genetics and the Origin of Species*, by the Russian-American geneticist, Theodosius Dobzhansky (1900–1975).[20] Taking his lead from Fisher, but including the results of more recent studies, particularly those of Wright, Dobzhansky (figure 9.1) argued that everything that was known about population genetics showed it to be entirely consistent with Darwinism. He acknowledged that he could 'partly follow' Wright's lead towards a scheme involving both genetic drift and natural selection, because 'species frequently differ in characteristics to which it is very hard to ascribe any adaptive value'. Thus, he accepted that genetic drift in a small sub-population could throw up a new variant for natural selection to act upon in the main population. Nevertheless, regardless of whether genetic drift was involved, Dobzhansky maintained that the overall process must be a gradual one.[21]

Wright's idea of an adaptive landscape was modified when it was incorporated into the Modern Synthesis, each peak coming to be regarded as the representation of a separate species, characterised by a favourable combination of genes and also a harmonious interaction with the environment. When environmental conditions were stable, natural selection would tend to eliminate those individuals who were not to be found near the top of a peak, maintaining species as stable entities. Hence, for evolutionary change to take place without the involvement of genetic drift, there would first have to be an environmental change, moving the positions of the peaks in the adaptive landscape. Evolutionary success would then depend on the ability of a population to adjust to the environmental change which, it was assumed, would occur gradually, over a long period of time.[22] It is interesting to note that Darwin, whilst rejecting the need for any major change, accepted that a gradual drift in environmental conditions could produce new habitats for evolving species to occupy. He explained his thinking in the fourth chapter of the *Origin*, writing:

> Not that, as I believe, any extreme amount of variability is necessary; as man can certainly produce great results by adding up in any given direction mere individual differences, so could Nature, but far more easily, from having incomparably longer time at her disposal. Nor do I believe that any great physical change, as of climate, or any unusual degree of isolation to check immigration, is actually necessary to produce new and unoccupied places for natural selection to fill up by modifying and improving some of the varying inhabitants.[23]

Thus, via Wright, these ideas also became part of Dobzhansky's thinking. Dobzhansky's Darwinian view of evolution, with gradual changes taking place in populations by a process involving natural selection, perhaps assisted by a slow drift in environmental conditions, was reinforced in 1942 by the British biologist Julian Huxley (1887–1975), grandson of Darwin's 'bulldog', Thomas, in his *Evolution – The Modern Synthesis*, a title which gave the developing theory a name. Huxley stressed, in the same way as Dobzhansky (and Darwin), that 'most evolutionary change will be gradual, to be detected by a progressive shifting of a mean value from generation to generation'.[24] In the same year, the Harvard University zoologist Ernst Mayr (see figure 9.1), who was born in Germany in 1904, showed, in his *Systematics and the Origin of Species*, that current knowledge about animal diversity, geographical variation and speciation fitted into the picture described by Dobzhansky.[25] Further support was soon to come from others, including the German zoologist Bernhard Rensch (1900–1990), and the American plant geneticist George Ledyard Stebbins (1906–2000).[26]

No proper definition of the Modern Synthesis was ever issued. Indeed, the leading figures disagreed about some matters of fine detail, and their precise views changed over the course of time. Nevertheless, in general terms, the main principles of the Modern Synthesis have always been clearly understood. The notion that mutational pressure could be the driving force of evolution was firmly rejected, that argument being strengthened when it emerged that the appearance of new types of *Oenothera lamarckiana* observed by de Vries resulted from the inheritance of different combinations of existing genetic factors, in a complex system, rather than the production of new genes. Similarly, there was no place in the Modern Synthesis (unlike the original form of Darwinism) for the Lamarckian features of orthogenesis or the inheritance of characteristics acquired by an organism in response to its environment. According to the Modern Synthesis, evolutionary change was essentially adaptive, genetic variants with advantageous characteristics spreading through a population by natural selection. Furthermore, in line with Darwin's own views, evolutionary processes were seen as gradual ones, giving rise to more advanced forms in a non-linear fashion.[27]

There was no clear understanding of what was meant by evolutionary progress. Various suggestions were put forward, including the notion that progress involved a change towards more complex organisms, or ones which were less dependent upon the environment.[28] However, there was a general appreciation of what had happened. So, Mayr wrote, in *The Growth of Biological Thought*, that, whatever the details, the 'series of morphological and physiological innovations that have occurred in the course of evolution can hardly be described as anything but progress'. He went on to insist, of course, that this was not teleological, involving change towards a pre-determined outcome, but the result of 'random variation and natural selection'.[29]

Use of words such as 'random' and 'chance' in connection with Darwinism has led to some misunderstandings. Darwin himself wrote, in the fifth chapter of the *Origin*, 'I have hitherto sometimes spoken as if the variations . . . had been due to chance. This, of course, is a wholly incorrect expression, but it serves to acknowledge plainly our ignorance of the cause of each particular variation'.[30] By the time of the Modern Synthesis, even though the precise relationship between the genotype (the genetic make-up) and the phenotype (the observable characteristics) of an organism remained uncertain, a great deal more was known about the nature of genes than had been the case in Darwin's day. Nevertheless, the confusions persisted with some, mainly from outside the ranks of biologists, interpreting 'random' and 'chance' in too literal a fashion, implying that everything was equally possible.[31] In fact, it was generally accepted by geneticists that, from a particular starting point, there were limitations to the number and type of variants which could arise by mutation, and some mutations would be more likely to occur than others. However, within those constraints, it was certainly believed by adherents to the Modern Synthesis that mutations arose by chance, in the sense that they were not in any way directed by external agencies, such as environmental pressure (as in Lamarckism) or the will of God.[32]

At the heart of the Modern Synthesis was the principle that no two members of a population, other than identical twins, have exactly the same genotype. Hence, some individuals have a built-in advantage over the others in their capacity for survival and reproduction, and these will be the ones most likely to pass their genes on to the next generation. However, that process is far from straightforward, its complexity ensuring the continual production of a wide range of variant forms.

Except in simple single-celled organisms such as bacteria, two types of cell are present, as already discussed. Somatic cells, the typical cells of the body, generally contain two equivalent sets of chromosomes (i.e. discrete assemblies of genes), one inherited from the mother and one from the father. Germ cells, which participate in sexual reproduction, are formed from somatic cells by meiosis, a process of chromosome separation. This results in each germ cell having only one set of chromosomes, produced by random mechanisms. Crossovers may occur between pairs of chromosomes during meiosis, so that each chromosome in a germ cell is made up of a chance combination of sections from corresponding maternal and paternal chromosomes. When sexual reproduction takes place, one of these germ cells fuses with a complementary germ cell from another individual, producing once again a somatic cell with two equivalent sets of chromosomes. This new cell will then start to multiply, giving rise to an individual with millions of cells of the same genetic make-up as the one produced by sexual reproduction. In all probability, the genes of both parents would have included some which had previously been modified by mutation, as well as the many that were normal for the species. Hence, it can be seen that sexual reproduction, involving the processes of meiosis and chromosome recombination, will give rise to an individual with a unique combination of normal and mutated genes, to face the next round of the ongoing process of natural selection within the population.[33]

In 1944, Oswald Avery demonstrated that genes are made of DNA (deoxyribonucleic acid)[34] and, nine years later, James Watson and Francis Crick elucidated its double-helix structure.[35] Once this was established, it immediately became clear that genetic information is stored as the sequence of nucleotide bases making up the DNA, and expressed by the synthesis of proteins whose component amino acid sequences are determined by the DNA structure. A complete DNA molecule consists of many discrete sections (the genes), each of which can be transcribed into an RNA (ribonucleic acid) molecule of related structure, and each RNA in turn can be translated into a protein molecule by use of the 'genetic code'. The sequence of bases in the DNA (and hence RNA) is in fact sub-divided into units of three, each 'triplet' specifying a particular amino acid in the protein being synthesised.[36] In 1958, Crick formulated the so-called central dogma of molecular biology, which states that DNA can replicate itself or be used to synthesise RNA and protein, the flow of information being from nucleic acid to protein and never in the reverse direction.[37]

The central dogma was seen to be entirely consistent with Darwinism, for changes in the genotype would result in changes in the phenotype, which might or might not confer an advantage in evolutionary terms. Accumulating information about molecular genetics never challenged the essentials of the Modern Synthesis, whereas these developments revealed no mechanism which could support Lamarckism.[38] Although it emerged that the RNA of some viruses could specify DNA synthesis, by reverse transcription, it was far from clear whether this had any relevance to evolution.[39]

Furthermore, the inheritance of acquired characteristics required changes in the genotype to result from changes in the phenotype, so reverse translation as well as reverse transcription would be needed for Lamarckian evolution to take place, and this was never observed. In contrast. a neo-Darwinian mechanism of evolution required a mutation in DNA to result in the synthesis of a protein of a different structure, thus possibly affecting the survival chances of the individual, and evidence for this was accumulating all the time.[40]

10 Phyletic gradualism

What exactly is a species?

Amidst all the certainties, one topic that continued to be a problem for the Modern Synthesis was speciation. In spite of calling his major work *On the Origin of Species*, Darwin had great doubts whether the concept implied in the title actually had any real meaning.[1] From his broad knowledge of biology, and what he had learned through his active involvement in pigeon breeding during the five years before the publication of the *Origin*, Darwin did not doubt that different 'varieties' could exist within what was generally regarded as a single species. Nevertheless, he argued that, if looked at from an evolutionary perspective, 'a well-marked variety may be justly called an incipient species'. To support this claim, he wrote, 'Certainly, no clear line of demarcation has as yet been drawn between species and sub-species . . . or, again, between sub-species and well-marked varieties, or between lesser varieties and individual difference. These differences blend into each other in an insensible series; and a series impresses the mind with the idea of an actual passage'.[2] So, to Darwin, the difference between species and varieties was not a fundamental one, merely a matter of degree.

Without question, Darwin's concept of a species was somewhat vague whereas, partly because of Cuvier's anatomical studies, the nineteenth century catastrophists, as well as Lyell, believed that a species was something which could be clearly defined, and this belief helped to determine their anti-evolutionary views.[3] Four years after the *Beagle* returned to England, William Whewell wrote, in his *Philosophy of the Inductive Sciences*, 'Species have a real existence in nature, and a transition from one to another does not exist'.[4] In these circumstances, it must have seemed self-evident to Darwin that he could not advance a theory of biological evolution without denying that species were distinct entities. In the *Origin*, he drew a hypothetical family tree to demonstrate that branching leads to increased diversity, pointing out how 'the line of succession is broken at regular intervals by small numbered letters marking the successive forms which have become sufficiently distinct to be regarded as varieties. But these breaks are imaginary, and might have been inserted anywhere, after intervals long enough to have allowed the accumulation of a considerable amount of divergent variation'. Then, extending the time-scale, the diagram illustrated the steps by which the small differences distinguishing varieties were 'increased into the larger differences distinguishing species'.[5]

Darwin argued on many occasions in the *Origin* that evolution could only proceed by slight, successive steps giving rise to numberless intermediate varieties, and this would be clear to us were it not for the fact that the fossil record was far from complete. After giving reasons why the formation and preservation of fossils might be unrepresentative of the history of life, Darwin went on to write:

> All these causes taken conjointly, must have tended to make the geological record extremely imperfect, and will to a large extent explain why we do not find interminable varieties, connecting together all the extinct and existing forms of life by the finest graduated steps. He who rejects these views on the nature of the geological record, will rightly reject my whole theory.[6]

In Darwin's time, as now, it was understood that fossilisation occurred as an exception rather than a rule, requiring certain uncommon circumstances to bring it about. Shortly after death, an animal had to be buried in sediment or volcanic ash to protect its body from being broken down by oxygen or scavengers. Then, if remaining undisturbed, bones and teeth could absorb minerals from their surroundings and be converted into stone over a period of thousands of years. Because of this, it was obvious that the fossil record could not show a continuous and completely representative picture of the history of life on Earth.[7]

Even so, the species problem remained. With the coming of the Modern Synthesis, the pre-Darwinian concept of a biological species being a reproductively isolated group of individuals was accepted from the very beginning. Thus, in 1937, in the seminal *Genetics and the Origin of Species*, Theodosius Dobzhansky stated that new species arose when 'the once actually or potentially interbreeding array of forms becomes segregated in two or more separate arrays, which are physiologically incapable of interbreeding'.[8] Consistent with that, Ernst Mayr, in *Systematics and the Origin of Species*, published in 1942, defined species as 'groups of actually or potentially interbreeding natural populations, which are reproductively isolated from other such groups'.[9] At any given time, therefore, the only issue was whether similar but not identical groups of individuals in different locations were variant forms of a single species or members of different species. From an evolutionary point of view, however, the adherents to the Modern Synthesis, like Darwin, doubted whether species really existed as distinct entities over periods measured in tens of thousands of years or more.

Despite the incorporation of the 'biological species concept', it was thought that species were subject to constant, gradual change.[10] Dobzhansky's *Genetics and the Origin of Species* stated, 'Species is a stage in a process, not a static unit'.[11] Similarly, Hermann Müller, whose work on mutations helped to lay the foundations for the Modern Synthesis, wrote in 1940 that 'speciation represents no absolute stage in evolution, but is gradually arrived at, and intergrades imperceptibly into racial differentiation beneath it and generic differentiation above'.[12] Any fossil species which seemed to be a distinct entity over a period of geological time could only appear so because of the imperfections of the record. So, for example, Mayr wrote in *Systematics and the Origin of Species*:

> The species of each period are the descendants of the species of the previous period and the ancestors of those of the next period. The change is slight and gradual and should, at least theoretically, not permit delimitation of definite species. In practice, the fossil record is fragmentary, and the gaps in our knowledge make convenient gaps between the 'species'.[13]

That view continued to hold sway. Thus, the British zoologist John Maynard Smith claimed in the 1975 edition of his book *The Theory of Evolution* that 'any attempt to group all living things, past and present, into sharply defined groups, between which no intermediates exist, is foredoomed to failure'.[14]

Without question, then, the founders and the supporters of the Modern Synthesis all envisaged evolution as a process of continuous, gradual modification.

Discord between palaeontologists and geneticists

Regardless of what ought to have been the case, if the Modern Synthesis was correct, it was an undeniable fact that the fossil record did not provide positive support for the idea of gradual change. The population geneticists and others who studied living organisms brushed this aside with references to the imperfections of the record, but the palaeontologists, who spent their working lives studying fossils, took a very different view, and queried the gradualist assumptions. For much of the first half of the twentieth century, palaeontologists generally believed that the driving force of evolution was either mutational pressure or orthogenesis. Very few thought that natural selection had much to do with the process, or that the work of the population geneticists, which provided the main theoretical basis for gradual change, was of any relevance.[15]

During the 1940s, the divergence of the evolutionary disciplines of genetics and palaeontology became a cause for concern, and attempts were made to reverse the process. Following a suggestion made at a meeting of the Geological Society of America, the Society for the Study of Evolution was formed in 1946, and its journal, *Evolution*, had the explicit aim of encouraging interdisciplinary communication. This brought palaeontology into the Modern Synthesis, but it did not initially bring the views of the geneticists and the palaeontologists any closer together. Steven Stanley referred to the situation in his book *The New Evolutionary Timetable* as 'a shotgun wedding'.[16]

Nevertheless, at least one palaeontologist, George Gaylord Simpson (1902–1984), tried to be positive. In 1944, his *Tempo and Mode in Evolution* was published.[17] This book attempted to build bridges between palaeontology and genetics by giving examples of instances where the fossil evidence of evolutionary trends was consistent with the ideas generated by those investigating living populations. One such example was the evolution of the horse, which showed the branched pattern typical of a Darwinian mechanism, rather than the linear development expected from orthogenesis. However, Simpson went further than just looking for support for Darwinism, for he also used the fossil record as a primary source of evidence. Thus, he sought an evolutionary explanation for its discontinuities, rather than simply dismissing them as artifacts. Because of this, there was less emphasis on gradualism and the control of adaptation by natural selection than in the works of the geneticists.[18]

Simpson went along with the conventional view that most evolutionary change was gradual, but he nevertheless developed the belief that accelerated evolution could sometimes take place within existing phyletic lines (i.e. in linear fashion, without branching). This he termed *quantum evolution*, defining it as 'the relatively rapid shift of a biotic population in disequilibrium to an equilibrium distinctly unlike an ancestral condition'. Simpson was undoubtedly influenced by Sewall Wright's suggestion that a rapid evolutionary change could result from genetic drift in a small sub-population, followed by its spread through the population as a whole by the action of natural selection (see Chapter 9). Given the localised nature and short time-scale of the key events in such a process, it was hardly to be expected that traces would be found in the fossil record.[18]

The concept of quantum evolution was severely criticised in reviews of *Tempo and Mode in Evolution*, but the book as a whole was well-received, and Simpson became a respected figure in the Modern Synthesis movement. Despite that, he and other palaeontologists had little influence on the population geneticists, whose views continued to dominate. Indeed, attitudes hardened even more towards a strict insistence on evolution being characterised by gradual transformations involving adaptation driven by natural selection, and nothing else. Thus, the 1951 edition of Dobzhansky's *Genetics and the Origin of Species* was markedly narrower in outlook than the original 1937 version. Similarly, in the revised edition of Simpson's *Tempo and Mode in Evolution*, published in 1953 with the new title *The Major Features of Evolution*, there was more emphasis on gradualism than previously, with the inclusion of extra sections on genetics, and a three-fold reduction in the space devoted to quantum evolution.[19]

Apart from the erratic nature of fossilisation, which we mentioned earlier, interpretation of the fossil record of extinct animals could be very difficult, for there was no way of knowing whether a slight difference in characteristics between fossils was evidence of variation within a species or of evolutionary change. Also, the precise sequence of events within a period of a few hundred thousand years was unlikely to be apparent, for this is a very short period of time, in geological terms. Further uncertainties could arise because of erosion or possible disturbances of the strata, and the likely necessity of having to try to determine trends using fossils from different sites, which introduced problems of comparative dating and possible regional variation. For these various reasons, when the population geneticists, zoologists and botanists were creating the Modern Synthesis, with its emphasis on gradualism, the absence of any direct support from the fossil record could be ignored without any undue concern, whilst attention focused instead on studies involving living organisms. Because of the obvious limitations of time, these could give little more than hints about how new species appeared under natural conditions, but nevertheless these hints seemed to support a gradualistic view.

Microevolution and macroevolution

One of the best-known examples of an investigation that appeared to demonstrate natural selection in action involved the peppered moth, *Biston betularia*. Two forms were recognised, which were members of the same species: one form was speckled, black and grey, whereas the other was all black, or melanic. This melanic form was first observed in 1846, by which time the speckled form was already well known. Later, during the 1950s, it was shown by Bernard (H.B.D.) Kettlewell (1907–1979) that the speckled form predominated in rural areas, because there the melanic form was more conspicuous to predators, standing out against relatively light backgrounds. Conversely, in industrial areas, the melanic form was the predominant one, because the speckled form could be seen more easily against a dark background, and so was a more likely prey. As industrial pollution advanced and receded, the proportions of the two forms of *Biston betularia* in local populations were observed to vary in the way expected.[20]

Of all the possible changes to the phenotype of the black-and-grey moth which could be imagined as occurring by a chance mutation, the modification to an all-black form seemed one of the most likely. Furthermore, it was easy to imagine that such a small change could be brought about by a 'point mutation', i.e. one involving a change in a single base in the DNA. Industrial pollution would have helped the survival and proliferation of the melanic form, but it could not, according to the Modern Synthesis, have *caused* the mutation which brought about the change in colour. That must have been an accidental event.[21]

Evolutionary biologists fully accepted the principle that *microevolution*, i.e. variation within a species, such as that observed with *Biston betularia*, occurred mainly by the processes of natural selection acting on a population which included variants introduced by the chance occurrence of point mutations. However, there was no such consensus about *macroevolution*, which deals with the origin and extinction of species.

To the followers of the Modern Synthesis, the gradual accumulation of small mutations with advantageous phenotypic effects, however slight, within a population would, under the continuous pressures of natural selection, eventually result in the formation of a new species. This even-paced development of an existing species into a completely new one over a period of a hundred thousand years or more was termed phyletic gradualism.[22]

According to this view, two or more separated populations of the same species would probably evolve slightly differently, because of different selection pressures in the different localities. After a period of time, therefore, they might no longer be able to interbreed, if they were brought back together. If this was the case, they would be regarded as having formed different species, with the lineage having branched. Evidence of such a situation was provided by what were termed 'ring-species', where imperceptible transitions could be observed around the perimeter of a large island such as Australia, in hills enclosing a central valley, or around the globe at a fixed latitude. So, for example, when travelling from the United Kingdom towards the east, it could be seen that the British lesser black-backed gull (*Larus fuscus*) was replaced in the first instance by the Siberian lesser black-backed gull; then, in turn, by Heuglin's gull, Birula's gull, the Vega gull and the American herring gull; and finally, as the circuit of the Earth was completed, by the British herring gull (*Larus argentatus*), which was quite different in both appearance and behaviour from *Larus fuscus*, with which it co-existed.[23] Nevertheless, both were still members of the same genus, so it was far from certain that this type of gradual process could lead to the origin of something fundamentally different.

Although it was generally thought that birds had evolved from reptiles (possibly early dinosaurs), the question was whether this could this have happened by a mechanism involving phyletic gradualism. Some fossils of *Archaeopteryx*, a creature from the Late Jurassic having some characteristics of birds and some of reptiles, had been known for some time, but it was still far from clear whether *Archaeopteryx* lay on the direct line of descent from reptiles to birds.[24] Even if it did, there was no evidence of a continuum of forms linking it to reptiles in one direction and birds in the other. Some argued that genetic macromutations, i.e. mutations having a large effect, could be responsible for such transitions.[25] Thus, in 1940, in *The Material Basis of Evolution*, the geneticist Richard Goldschmidt (1878–1958), a refugee from Germany who was working in America, suggested that a macromutation could produce a 'hopeful monster', thereby giving rise to the first member of a new species.[26] For this, Goldschmidt was subjected to much abuse. It seemed to most people that such a monster would have to remain hopeful: how could it be expected to find a mate and have fertile offspring, let alone become the founder of a species?[27] Nevertheless, a similar idea was put forward by the German palaeontologist Otto Schindewolf (1896–1972), in 1950. Schindewolf suggested that a new family or order of animals might arise in this way with, say, the first bird hatching out of a reptile egg.[28]

To most people, however, processes involving phyletic gradualism appeared much more likely, despite the lack of supporting fossil evidence. From the earliest days of the Modern Synthesis, its followers believed phyletic gradualism to be far more significant than speciation. Julian Huxley set the tone, in 1942, when he wrote in *Evolution – The Modern Synthesis*, 'Species-formation constitutes one aspect of evolution; but a large fraction of it is in a sense an accident, a biological luxury, without bearing upon the major and continuing trends of the evolutionary process'.[29] Later, Dobzhansky, in his book *Mankind Evolving*, explained how the steady build-up of favourable small mutations in a population as a result of natural selection must eventually give rise to new species, writing that 'species arise gradually by the accumulation of gene differences, ultimately by summation of many mutational steps which may have taken place in different countries and at different times'.[30]

The Modern Synthesis went from strength to strength, and the neo-Darwinian views of its founders became firmly established. In 1958, one hundred years after the first publications on evolution by Darwin and Wallace, Gavin de Beer wrote, for an exhibition at the Natural History Museum in London, 'With the same confidence as it accepts Copernicus's demonstration of the movement of the earth round the sun and Newton's formulation of the laws of this movement, science can now celebrate the centenary of the first general principle to be discovered applicable to the entire realm of living beings'.[31] By this time, palaeontologists had generally assimilated the beliefs of the population geneticists and, like many belated converts to a cause, they came to be numbered amongst its staunchest advocates. So, by the end of the 1950s, phyletic gradualism ruled almost unchallenged. In line with this, de Beer, in his guide to the centenary exhibition, explained that 'species are not necessarily the end-products of finite changes, but stages in a continual process of potentially infinite change in which they may be at any level'.[32]

Reductionism and holism

The supporters of the Modern Synthesis became extremely reductionist in approach, paying little attention to the biology of whole organisms. Darwin himself had spent much of his time studying whole organisms, but he was quite prepared to believe that significant evolutionary progress could be achieved by the piecemeal accumulation of advantageous characteristics.[33] Although he appreciated the difficulties in trying to explain the evolutionary origin of complex organs such as the eye, admitting to Asa Gray in 1860, 'The eye to this day gives me a cold shudder',[34] Darwin had swept his doubts aside in the sixth chapter of the *Origin*, writing, '[If] the eye does vary ever so slightly, and the variations be inherited, which is certainly the case; and if any variation or modification in the organ be ever useful to an animal under changing conditions of life, then the difficulty of believing that a perfect and complex eye could be formed by natural selection, though insuperable to our imagination, can hardly be considered real'.[35] Similarly, adherents to the Modern Synthesis concentrated on the spread by natural selection of advantageous individual features arising from genetic mutations, without worrying about the complexities of overall body plans.

There was, however, an alternative, more holistic, approach to organisms and to evolution. The nineteenth century 'rational morphologists', such as Geoffroy Saint-Hilaire and Georges Cuvier, saw similarities in the structure of different organisms as evidence that some as yet unknown organising principles played an important role in determining overall body plans.[36] Following in this tradition, the British biologist D'Arcy Thompson (1860–1948), in his 1917 book, *On Growth and Form*, argued that the change in body plan from one species to another might be governed by mathematical laws.[37]

Also, it was clear that an evolutionary development, such as the extending of the giraffe's neck, involved not only the lengthening of the bones, but also corresponding changes to the circulatory, nervous and other systems, in order to maintain a viable organism. It was considerations such has these which had caused Cuvier, with his great knowledge of anatomy, to consider it unlikely that evolution from one species to another could ever take place (see Chapter 3). If it nevertheless did, then those who were unconvinced of the sufficiency of the reductionist approach pointed out that no-one had yet satisfactorily addressed the issues of how it was possible for body plans to change in significant fashion and still retain viability. Since the precise links between the characteristics of the whole organism and those of the genetic material were still far from clear, it was possible that genes might not be the only factor in determining overall structure, function, organisation and development.[38]

Some even maintained the vitalist tradition of Aristotle, Leibniz and the idealist philosophers, arguing that life could never be fully explained in terms of physical substances and processes, no matter how complex these might be. In the early 1900s, Hans Driesch proposed a vitalist theory in which some aspects of living organisms could be explained mechanistically, but a factor called 'entelechy' (a term taken from Aristotle) co-ordinated the activity of the physical processes.[39]

During the 1920s, several biologists, especially Hans Spemann, Alexander Gurwitsch and Paul Weiss, argued that morphological development was directed by a complex but unspecified mechanism, operating through what eventually became known as 'morphogenetic fields'. Cells in the same field would all develop in the same way, whereas ones in a different field would develop differently. This notion was as vague as that of entelechy but, because of the replacement of the vitalist 'life force' by a mechanism dependent on physical forces and processes, it was more scientifically acceptable.[40] For the next three decades, the morphogenetic field concept was developed by the British embryologist Conrad (C.H.) Waddington (1905–1975). Waddington also introduced the idea of 'chreodes' or discrete developmental pathways, to describe how cells with an identical genetic component could develop in the embryo into different specific end-products, e.g. a liver or a kidney, but not into anything between. As an aid to visualising what he thought was happening, he created 'epigenetic landscapes', using the analogy of a ball rolling down a hillside into one of a series of interconnected valleys. If given a push (analogous to being subjected to environmental stress), a ball could be forced up a ridge between valleys, from which it could either roll back into its original valley, or have enough momentum to be propelled over into the next one. In the early 1970s, René Thom described this mathematically as part of his 'catastrophe theory', which we shall discuss in Chapter 15.[41]

Early in the twentieth century, the American geneticist, Thomas Hunt Morgan (see Chapter 9), suggested that worms and other animals which could regenerate their bodies did this by setting up chemical gradients, controlling by such means the growth and specialisation of individual cells. Later, in the 1950s, the British mathematician Alan Turing proposed that gradients of such chemicals, which he termed 'morphogens', might play an important role in the development of embryos. Later still it was thought that morphogens, such as retinoic acid, might act by stimulating receptor molecules in the membranes of cells, causing them to send signals to the genome. Depending on which particular genes were activated in this way, different cells could develop in different ways from an identical starting point, giving rise to a complex organism made up of specialised components.[42]

Back in the nineteenth century, as we have noted (see Chapter 3), Geoffroy Saint-Hilaire suggested that evolutionary change might be brought about by an environmental stress acting during the embryonic stage of development. More recently, examples of environmental factors with the ability to affect epigenetic development (i.e. the development of an embryo from undifferentiated cells) have been discovered, e.g. the drug thalidomide, which prevents the 'normal' growth of limbs. In between, some environmentally induced effects had been demonstrated in the fruit fly, *Drosophila*, by Waddington.[43]

In his 1957 book, *The Strategy of the Genes*,[44] Waddington presented holistic ideas that never became part of the Modern Synthesis, stressing three particular themes. These were the amplification of gene expression through epigenetic mechanisms, the capacity of embryos to self-regulate, and the creation of more advantageous pathways of development by means of genetic assimilation. The last-mentioned term referred to something similar, but not identical, to the Baldwin Effect, which, as we saw in Chapter 9, relied on the appearance of a favourable new mutation. In contrast, genetic assimilation was concerned with the interplay between environmental factors and existing genetic variants. It described the situation where a particular variant might switch from its normal pathway of embryo development to one with more advantageous consequences, as a result of the action of a chemical or some other external factor. As a consequence, this variant would suddenly start to spread through the entire population.[45]

Perhaps because of the involvement of environmental factors, even though in a way that was clearly non-Lamarckian, genetic assimilation was generally viewed with suspicion by strict adherents to the Modern Synthesis. In *Adaptation and Natural Selection*, written in 1966, George Williams, an ecologist from the State University of New York at Stony Brook, brushed aside the concept, questioning whether it had any importance as an explanation of adaptive evolution.[46] His own book, as he made clear from the start, was 'based on the assumption that the laws of physical science plus natural selection can furnish a complete explanation for any biological phenomenon'. He hoped that it would 'help to purge biology of . . . unnecessary distractions that impede the progress of evolutionary theory', one of these clearly, in his mind, being genetic assimilation. Because of the adequacy of what could be explained by the simplest form of natural selection, that of selection in a population between variant forms of the same gene, Williams argued that adaptation 'should be attributed to no higher a level of organization than is demanded by the evidence'.[47]

Another influential defence of the reductionist approach came from the French molecular biologist and Nobel laureate Jacques Monod. In his 1971 book, *Chance and Necessity*, Monod used his profound knowledge of molecular genetics to argue against the possibility of teleological mechanisms, maintaining that the structure of macro-molecules, and therefore, by extension, evolution, was nothing more than 'random chance caught on the wing'.[48]

Throughout that period, phyletic gradualism continued to be accepted in widespread fashion. So, in 1970, in *Populations, Species and Evolution*, Ernst Mayr explained that those evolutionary biologists who, like himself, were supporters of the Modern Synthesis, believed that 'all evolution is due to the accumulation of small genetic changes, guided by natural selection', and that macroevolution is simply 'an extrapolation and magnification of the events that take place within populations and species'.[49] Similarly, in the textbook *Evolution*, which boasted Dobzhansky as the first author, and was published (after Dobzhansky's death) in 1977, we read:

> The process of natural selection, acting upon the sources of genetic variability that reside in the gene pools of species, is clearly adequate to produce, preserve, and accumulate the sorts of changes that lead from one species to another. There is a voluminous body of theory and evidence to explain the origin of species through microevolution.[50]

In the 1978 edition of another multi-author textbook, *Life on Earth*, it was claimed that, as a result of work over the preceding decades, biologists had been able to observe the 'beginnings of evolutionary change' in a variety of animals and plants, and thus confirm some of the assumptions made by the founders of the Modern Synthesis. The question was then asked as to whether a continuation of processes like that observed in the peppered moth, *Biston betularia*, could eventually give rise to a major innovation, such as the production of birds from reptiles. In other words, could a series of simple microevolutionary shifts lead to macroevolution? The authors of *Life on Earth* suggested that the answer was probably 'Yes', and they continued:

> One must keep in mind the enormous difference in time scale between the observed cases of microevolution and macroevolution. Under natural conditions the nearly complete substitution of the melanic gene of the peppered moth took 50 years. Evolution of the magnitude of the origin of birds usually, perhaps invariably, takes many millions of years. As paleontologists explore the fossil record with increasing care, transitions are being documented between increasing

numbers of species, genera, and higher taxonomic groups. The reading from these fossil archives suggests that macroevolution is indeed gradual, paced at a rate that leads to the conclusion that it is based upon hundreds or thousands of allele substitutions no different in kind from the ones examined in our case histories.[51]

Without doubt, the belief that phyletic gradualism was the predominant feature of evolution was still widespread during the 1970s.

11   Gradualist perceptions of human evolution

Fossil finds in Asia and Europe

In 1871, as we saw earlier (in Chapter 8), Charles Darwin proposed a scheme for human evolution in which several important characteristics were developed in gradual and related fashion. At about the same time, Ernst Haeckel suggested that humans and modern apes had evolved from an unknown common ancestor, to which he assigned the name, *Pithecanthropus*. This motivated a Dutch anatomist, Eugène Dubois (1858–1940), to resign in 1887 from his lectureship at the University of Amsterdam, to search for traces of the 'missing link' in the East Indies, where modern apes still lived in the wild. Four years later, on the island of Java, Dubois found a fossilised tooth and skullcap, near to a thigh-bone which showed clear evidence of an upright stance. Examination of the skullcap showed that it had contained a brain much smaller than a Neanderthal's, but larger than that of any species of modern ape. After some hesitation, Dubois classified the specimen as *Pithecanthropus erectus*. The fossil subsequently became famous as 'Java Man' but, at the time, there was a marked reluctance to acknowledge it as a direct ancestor of modern humans. This was because it was thought that a larger brain should have developed by the stage of human evolution in which upright walking was a clear feature.[1]

During the time these arguments about *Pithecanthropus* were taking place, more Neanderthal fossils were being discovered in Europe. Later, in 1908, a jawbone found in a sand quarry at Mauer, near Heidelberg, appeared more apelike than those of the Neanderthals, so it was assigned to a new species, *Homo heidelbergensis*. However, although the Mauer skull may have been a little older than the Neanderthal fossils, all were from the Pleistocene Epoch, the first of the Quaternary Period.[2]

Aleš Hrdlička (1869–1943), an American palaeoanthropologist (investigator of fossil humans), of Czech and German origin, was for many years a strong advocate of the view that Neanderthals were the ancestors of modern humans. Instead of regarding them as a distinct species, he referred to 'the Neanderthal phase of man', emphasising that the overall process was one of continuous, gradual development.[3]

In contrast, in 1912, the French palaeoanthropologist Marcellin Boule (1861–1942) concluded from his studies of specimens found in his own country that the Neanderthals formed an evolutionary dead-end, and did not give rise to humankind.[4] Later, after some initial doubts, he went along with many others, particularly in

England, in believing that a more likely human ancestor was a species represented by fragments of skull bones found in Pleistocene deposits at Piltdown, Sussex. These had been uncovered over a number of years, between 1908 and 1915, by Charles Dawson and the Jesuit priest Pierre Teilhard de Chardin, both of whom were amateur palaeontologists, and by Arthur Smith Woodward, Keeper of Geology at the Natural History Museum, London. In contrast to the jawbone, which resembled an ape's, the reconstructed cranium was similar in both size and shape to one belonging to a modern human. An enlarged brain was exactly what many had anticipated finding in a creature developing human characteristics. However, as further discoveries of fossils were made around the world, particularly in Africa, it became increasingly difficult to fit the Piltdown specimen into the picture of human evolution which the new finds were revealing. It was not altogether surprising, therefore, when, in 1953, Joseph Weiner and Wilfred Le Gros Clark of Oxford University, together with Kenneth Oakley from the Natural History Museum, showed the Piltdown Man skull to be a fake. Its component parts were of a jaw from an orang-utan, some ape teeth which had been modified with a file, and pieces of a braincase from a thick-boned member of our own species. The technique of fluorine dating, based on the rate of uptake of fluorine into bones from surrounding deposits, showed the jaw to be from the present day, and the cranium only slightly older. Both had been stained with iron and chromium salts to make them look like genuine fossils.[5]

Meanwhile, teeth found in 1927 at Zhoukoudien, China, were investigated by Davidson Black, a Canadian who was then Professor of Anatomy at the Union Medical College in Peking (now Beijing). Black eventually assigned them to a new species, *Sinanthropus pekinensis* ('Chinese Man of Peking'), generally known as 'Peking Man'. A cranium very similar to that of Java Man was later found at the same site, followed by more fossil bones and teeth, all appearing to belong to close relatives of Peking Man.[6] On the island of Java itself, more specimens were found by the German–Danish palaeoanthropologist Ralph von Koenigswald (1902–1982) and others. In 1939, von Koenigswald and Franz Weidenreich (1873–1948), Black's successor in Peking, worked together to compare the various specimens from Java and China. They concluded that all were very likely to be of the same species, the differences being due to racial variations, except perhaps in one case, the so-called 'Solo Man', or *Homo soloensis*, a specimen with features that were slightly more modern than the rest, found close to the Solo River in Java. All the others were eventually assigned to *Pithecanthropus erectus*, later to be re-named *Homo erectus* as its close evolutionary relationship to present-day humans became clear.[7]

These findings led Weidenreich to develop a 'multi-regional hypothesis' of human evolution, which maintained that evolutionary pressures led to similar development taking place independently in different areas of the world. According to this scheme, the people represented by Java Man developed, via Solo Man, into Australian aborigines, whereas the line of Peking Man eventually gave rise to modern Mongoloid races. Modern Africans could have originated via 'Rhodesian Man', a skeleton found in 1921 at Broken Hill (now Kabwe), which was Neanderthal-like, but had lighter bones. Modern Europeans, in contrast, had probably developed from true Neanderthals, which may have arisen in the first place from the Heidelberg people.[8]

Even if the multi-regional hypothesis was correct, these sequences covered just the final stages of human evolution. Dubois may have been justified in regarding *Pithecanthropus* (later *Homo*) *erectus* as a 'missing link' between humans and ancestral apes but, nevertheless, its characteristics were too close to modern humans for it to have occurred near the beginning of the process. Fossils dating from before the start of the Pleistocene Epoch would be required to show how the earliest human ancestors differed from their ape cousins.

African hominids

Apart from the Piltdown forgery, the first claims of another 'missing link' came in 1925, when the Australian anatomist Raymond Dart (1893–1988), then at Witwatersrand University, South Africa, carried out studies on the fossil skull of a child discovered in a lime quarry at Taung, northwest of Johannesburg. The cranium was far smaller than that of a modern infant, but the forehead was high and rounded, without eyebrow ridges, indicating some human-like characteristics. Also, the canine teeth were small, whilst the jaw was short and delicate. Furthermore, from the way that the spinal cord linked to the brain, it seemed that the Taung child must have walked upright. The species was classified by Dart as *Australopithecus africanus* ('Southern Ape of Africa'). At the time, the evidence from a single, immature specimen was insufficient to enable Dart to convince many people that *Australopithecus africanus* was a possible human ancestor. However, a similar fossil, but of an adult, was found in 1936 at Sterkfontein, near Pretoria, by the Scottish-born anthropologist Robert Broom, then working for the Transvaal Museum. These fossils appeared to date from the second half of the Pliocene Epoch, which preceded the Pleistocene.[9]

In 1938, Broom investigated fragments of another fossil skull at the nearby site of Kromdraai. There were many points of similarity, but this skull was bigger and much heavier in appearance than the gracile ones from Taung and Sterkfontein. In particular, the jaw was more powerful, and the teeth were larger. Broom gave the Kromdraai fossil the name *Paranthropus robustus* ('Robust Near-Man').[10]

In the same year, William Gregory and Milo Hellman of the American Museum of Natural History visited South Africa to make a detailed study of the Taung, Sterkfontein and Kromdraai fossils. Their conclusion was that the South African specimens represented genuine intermediates between extinct apes of the Miocene (the epoch before the Pliocene) and modern humans, and they placed them all within the same sub-family, the Australopithecinae. Furthermore, they classified these australopithecines as hominids, i.e. members of the same taxonomic family as modern humans, the Hominidae. Prevailing opinion was changing, and this became obvious to all when two British anatomists, Wilfred Le Gros Clark and Sir Arthur Keith, who had previously been hostile to the notion that australopithecines were ancestral to humans, both announced that they now thought differently.[11] In the case of Le Gros Clark, this change of mind came in 1947 after he had opportunity to investigate the fossils at first hand.[12] Keith, once an advocate of the evolutionary importance of Piltdown Man, explained his conversion in his book *A New Theory of Human Evolution*, which was published in 1948. However, he took pains to make clear that he did not regard australopithecines as actually having achieved human status. He specified a brain size of 750 cubic centimetres as a 'cerebral Rubicon' for the ape–human lineage, with small-brained apes remaining on one side, whilst humans were those who had been able to cross to the other and develop larger brains. Thus, whereas *Pithecanthropus* and the Neanderthals qualified as humans by virtue of brain size, all australopithecines remained on the wrong side of the divide.[13]

In 1959, freelance palaeoanthropologists Louis Leakey (1903–1972) and his wife, Mary (1913–1996), discovered the first australopithecine fossil from outside of South Africa in the Olduvai Gorge, Tanzania. This specimen was named as *Zinjanthropus boisei* ('Boise's East African Ape'), after the Leakeys' financial backer, Charles Boise. However, Broom's assistant, John Robinson, quickly pointed out that the East African fossil was not sufficiently distinct from the robust australopithecines of South Africa to justify being regarded as a distinct genus. On the other hand, Robinson objected to the increasing tendency to group the robust and gracile australopithecines together in a single genus, *Australopithecus*.[14]

Meanwhile, from fossils of common animals found in the same stratum, Louis Leakey deduced that *Zinjanthropus* lived in the Early Pleistocene. It was still far from clear what that meant in terms of years, but then a new radiometric technique, potassium–argon dating, became available. This was applied to rocks close to where *Zinjanthropus* had been found, and indicated an age of 1.75 Myr.[15]

The potassium–argon method was used again during the 1970s, when fossils of bipedal hominids with even smaller brains than the gracile australopithecines were discovered in the Afar Triangle of Ethiopia and at Laetoli, Tanzania. Donald Johanson, of the Cleveland Museum of Natural History, together with Tim White, who had previously worked with the Leakeys, named the species *Australopithecus afarensis*. Results of the potassium–argon dating studies, confirmed later by other techniques, indicated that *Australopithecus afarensis* had existed around 4 Myr ago, with the gracile *Australopithecus africanus* emerging about a million years later.[16]

Preconceptions of gradual, progressive development

When leading figures of the Modern Synthesis, including Theodosius Dobzhansky and Ernst Mayr, turned their attention towards human evolution in the 1940s and 1950s, they came to the conclusion that humankind was probably a single, evolving population. Accordingly, only one hominid species ought to be identifiable from the fossil record at any particular period of history.[17] The 'single-species hypothesis' continued to be strongly supported, its main advocate being Loring Brace, accompanied from the late 1960s by Milford Wolpoff. These two palaeoanthropologists, both from the University of Michigan, saw hominid evolution as a gradual, linear progression from australopithecines to modern humans. During this process, exactly as envisaged by Darwin in *The Descent of Man*, brain size and social organisation slowly increased, whilst teeth and jaws became progressively weaker. It was only because of the imperfections of the fossil record that distinct hominid species appeared to have lived at different times.[18] The suggestion that Neanderthals might have died out by conflict or competition with modern humans, rather than simply being an ancestral form, was dismissed by Brace as being 'hominid catastrophism', and hence unscientific.[19] In the 1977 book, *Human Evolution*, Brace and his co-author, Ashley Montagu, summarised their views in clear fashion, writing: 'we recognise four stages or phases in human evolution: Australopithecine, Pithecanthropine, Neanderthal, and Modern. In reality, these merely represent arbitrary stages in what in fact is a continuum'.[20]

Wilfred Le Gros Clark, writing in 1970 in *History of the Primates*, claimed that, although the fossil record of human evolution was still fragmentary, there was some clear evidence of 'gradual and progressive modifications in a definite time sequence'.[21] For the best-documented period, the last quarter of a million years, Joseph Weiner, who had helped to expose the Piltdown forgery, wrote in *Man's Natural History* in 1971 that the hominid fossil record formed a 'morphological continuum'.[22] Three years later, the gradual, continuous nature of the process was also stressed by Bernard Campbell, a palaeoanthropologist from the University of California, Los Angeles, in the second edition of his book *Human Evolution*. Campbell claimed, 'Fossil species are properly called chronospecies, which implies that they have a dimension of time as well as spacial and morphological variability. They represent a segment of an evolving lineage, a biological species extended through a long period, with the variation due to evolution added to what we would expect of any population at any given point in time'.[23]

Much the same picture was presented to the general public by Clark Howell, a palaeoanthropologist from the University of Chicago, in *Early Man*, a 1970 offering from the Time-Life library. In this book, Howell wrote of the 'stages of man's long march from ape-like ancestors to *sapiens*', accompanied by an illustration which depicted the march through time in literal fashion. *Homo sapiens*, upright and alert, led the procession, with every succeeding figure stooping slightly more than the one in front and looking increasingly ape-like (figure 11.1). Finally, at the end, came figures representing ancestral apes of the Miocene Epoch.[24]

By this time, a more comprehensive understanding of the hominid fossil record was emerging, at least in general terms. Although the precise details of human evolution were still far from certain, the preconception which had made many people receptive to the Piltdown hoax, that the development of a large brain must have been

Figure 11.1   Sketch based on a section of the portrayal of 'man's long march from ape-like ancestors to *sapiens*', included in the 1970 Time-Life book, *Early Man*, and depicting (from left to right) *Dryopithecus*, *Oreopithecus*, *Ramapithecus*, gracile *Australopithecus* and robust *Australopithecus* (or *Paranthropus*). The original illustration contains 15 figures, starting with 25 Myr old *Pliopithecus* and ending with modern *Homo sapiens*.

an early and essential feature, was now completely discredited. Without any doubt, ape-like gracile australopithecines, who had small brains but walked erect and had human-like jaws, lived in South Africa during the Pliocene Epoch.[25] Yet, to a significant extent, Darwin's assumptions of a link between bipedalism, increasing brain size, social development and use of tools, with consequent changes to jaws and teeth, continued to guide the interpretation of new fossil finds. The only difference from before was that culture, rather than an enlarged brain, was now regarded as the most significant human characteristic.

The Olduvai bed in which the skull of *Zinjanthropus* (later *Australopithecus*) *boisei* had been found also contained crude stone tools, but Louis Leakey refused to link these to an ape-like robust australopithecine. Similarly, he could not accept that either the robust or the more gracile australopithecines could be ancestral to humans. Indeed, he had searched since the 1930s for a species of *Homo* contemporary with the australopithecines and had made claims, later withdrawn, that he had found what he was looking for at two Kenyan sites, in the form of some skulls from Kanjera and a mandible from Kanam.[26] So, in 1960, when fragments of a jaw and cranium of a gracile hominid (known as 'Jonny's Child') were found by his son, Jonathan, at Olduvai in the same bed as the tools, Leakey quickly came to believe that this fossil represented the people who had made and used them, and who could therefore be the missing species of *Homo* he was searching for. He also identified a roughly circular pile of stones as a windbreak constructed by these early humans.[27]

After the cranium had been assembled from the various fossil fragments, it could be seen that the brain capacity was around 680 cubic centimetres. Although this was below Keith's 'cerebral Rubicon', it was higher than the brain capacities of a number of gracile australopithecines from South Africa which had been investigated by the anatomist Phillip Tobias, the successor to Raymond Dart at the University of Witwatersrand. Furthermore, the Olduvai individual may not have been an adult. The bed representing the next period of time at Olduvai then yielded up two fossil skulls (known as 'Cindy' and 'George') which, although incomplete, had somewhat similar characteristics.[28] Leakey and Tobias, together with John Napier (1917–1987), of the University of London, concluded that the three fossils were of the same species, and in 1964 named this *Homo habilis* ('Handy Man'), because of the association with the tools.[29] However, others considered that the fossil from the earlier bed fell within the likely range of variations of the gracile australopithecines, whilst the later fossils, if not australopithecines, may have been representatives of *Homo erectus*.[30]

In 1972, a group led by Leakey's younger son, Richard, uncovered hundreds of fragments of a hominid skull at Koobi Fora, Kenya. This was carefully reconstructed by Richard Leakey's wife, Maeve (also an anthropologist), together with Alan Walker (a former postgraduate student of John Napier, then at the University of Nairobi), revealing *habilis*-like features and a braincase between 750 and 800 cubic centimetres in volume, crossing Keith's Rubicon. Although initial estimates gave it an age of around 2.5 Myr, it was eventually acknowledged, after much argument (the 'KBS tuff controversy'), that this had to be reduced by over half a million years, making the Koobi Fora skull about the same age as the supposed *Homo habilis* from Olduvai, and hence a likely member of the same species.[31] After this, the classification of *Homo habilis* became generally accepted. However, nagging doubts remained, and not even Richard Leakey was entirely convinced that the Koobi Fora fossil belonged to the same species as the Olduvai hominids.[32] The only certain conclusion was that the original naming of *Homo habilis* had owed more to pre-conceived assumptions about the nature of human evolution than to information derived from the fossils.[33] Moreover, at around the same time, there was another significant instance of speculation being driven by gradualistic dogma and, in consequence, leaving the evidence far behind. This concerned fossils from an earlier epoch, the Miocene, investigated by the palaeoanthrapologist Elwyn Simons.

Soon after moving to Yale University from Oxford in 1960, Simons became intrigued by two fragments from the jaw of a Miocene ape, which had been collected a quarter of a century earlier by Edward Lewis in the Siwalik Hills of northern India, and placed in the Peabody Museum of Natural History. Lewis had assigned the specimens to a new genus, *Ramapithecus*, claiming that the features of the jaw were less typically ape-like than those in the well-known Miocene ape *Dryopithecus*. This made *Ramapithecus* a likely ancestor of humankind. After investigating the specimens, Simons agreed with these conclusions.

British-born David Pilbeam then joined Simons at Yale, initially as his postgraduate student, and together they produced a detailed scenario from the original jaw fragments and similar ones discovered more recently. The arrangement of the teeth seemed more akin to that in humans than in modern apes, making it likely that the face of *Ramapithecus* had been relatively flat and humanlike. Furthermore, the chewing teeth were larger and flatter, and the canines and incisors smaller, than those found in the jaw of a modern ape. The chewing teeth also had thick enamel coatings, as in *Homo* and the australopithecines, but not the chimpanzee or gorilla.[34] So, in the mid-1960s, Simons and Pilbeam concluded that, by the time of *Ramapithecus*, the hominid line, leading to modern humans, had already separated from the lineage which gave rise to the apes of today.[35] Simons, in a 1967 article in *Scientific American*, provided a figure which showed the hominid line proceeding via *Ramapithecus* and the ape line via *Dryopithecus*, the split between the two taking place over 25 Myr ago, before the Oligocene–Miocene transition,[36] and Pilbeam gave essentially the same figure in his 1970 book, *The Evolution of Man*.[37] Shortly afterwards, Wilfred Le Gros Clark, in the third edition of *The Antecedents of Man*, similarly depicted the ape–hominid separation occurring sometime around the start of the Miocene Epoch.[38]

Clearly influenced by Darwin's gradualist, progressive scenario, Simons and Pilbeam argued that if significant changes to the teeth and jaws had occurred within the hominid line by the time that *Ramapithecus* came into existence, then other major changes must also have taken place. Accordingly, they imagined that *Ramapithecus* had the ability to walk erect, even though it may have done so only on occasions. Furthermore, it could use tools and had introduced some elements of social organisation. In *The Evolution of Man*, Pilbeam wrote, 'The small canines are important and imply that certain basic behavioural changes had already occurred by the time of *Ramapithecus*, and these behavioural changes may have had something to do with tool- and weapon-use'.[39]

That picture of *Ramapithecus* came to be generally accepted. Thus, Clark Howell wrote in *Early Man* in 1970 that 'one simply does not find a jaw like *Ramapithecus*'s on the body of a quadrupedal animal, any more than one would find grasping toes associated with the fossil remains of a horse'.[40] In the same year, Wilfred Le Gros Clark, discussing *Ramapithecus* in his *History of the Primates*, commented that 'it has been argued with good reason that it should be included in the family *Hominidae*'.[41] Similarly, Herbert Wendt, in *From Ape to Adam*, published in 1972, referred to *Ramapithecus* as 'the first demonstrably human type',[42] whilst Bernard Campbell, in the 1974 edition of *Human Evolution*, called *Ramapithecus* 'the oldest genus of the Hominidae'.[43]

Various apes lived in Africa during the Miocene, including *Kenyapithecus*, which Simons regarded as being identical to *Ramapithecus*. However, almost no African fossils had been found for the period covering the Miocene–Pliocene transition, between 8 and 4 Myr ago, perhaps simply because of the scarcity of strata of that age in accessible locations.[44] Whatever the reason, it was particularly unfortunate for an understanding of hominid evolution because by the end of the fossil gap, in the Middle Pliocene, the East African apes known from the Miocene had disappeared to be replaced by the bipedal australopithecines. However, the absence of evidence did not curtail speculation about human origins. One popular theory was put before the general public in the 1960s by the writer Robert Ardrey, based on the theories of Raymond Dart. In line with Darwin's ideas about the related development of a variety of human characteristics, Ardrey and Dart argued that the relative success of hominids in a supposed time of hardship was due to their abilities as 'mighty hunters'. This was based on assemblages of animal bones being found in association with those of australopithecines at various sites in South Africa, the animals having apparently been butchered by the hominids.[45]

So, well into the 1970s, Darwin's general picture of human evolution as a progressive, gradual process, still remained intact. Fossil finds had simply served to add detail to his imaginings.[46]

12 Heretical catastrophists

Crustal displacements and projectiles from space

Possible catastrophist scenarios, often speculative in nature, continued to be put forward, without any of them receiving much attention from mainstream academics.[1] Some considered the possibility that, on occasions, the geographical poles might move rapidly to new positions, causing major climate changes and cataclysmic floods. So, for example, Hugh Auchincloss Brown (1879–1975), an engineer who graduated from Columbia University, New York, proposed in a private publication of 1948 that the spinning Earth could suddenly tumble sideways, in catastrophic fashion, the instability being a consequence of the weight of ice at the poles. Nineteen years later, these ideas received a wider circulation in Brown's book *Cataclysms of the Earth*.[2]

Earlier in the century, it had been suggested that the poles might change their positions (in more gradual fashion) because of slippage of the Earth's crust over the interior, caused by a slowing of rotation through tidal friction. However, geologists such as Harold Jeffreys, of Cambridge University, argued that tidal friction could not possibly produce effects of this magnitude. Nevertheless, in 1952, an American electrical engineer, Karl Pauly, presented evidence in *The Scientific Monthly* which seemed to suggest that crustal displacements had taken place.[3] The concept was then developed by Charles Hapgood (1904–1982), a science historian at Keene State College, University of New Hampshire, who, working in collaboration with the engineer, James Hunter Campbell, suggested that the slippage of the crust was triggered in catastrophic fashion by the weight of polar ice. The causal mechanism was the same as in the theory of Hugh Auchincloss Brown, but Hapgood considered Brown's overall model to be less plausible than his own, since it involved displacement of the entire Earth, not just its outer skin. Hapgood presented his ideas in 1958 in *Earth's Shifting Crust*, and twelve years later in *The Path of the Pole*.[4]

In 1953, Hapgood sent a preliminary draft of *Earth's Shifting Crust* to the great physicist Albert Einstein, who replied, 'I find your arguments very impressive and have the impression that your hypothesis is correct. One can hardly doubt that significant shifts of the crust of the earth have taken place repeatedly, and within a short time. The empirical material you have compiled would hardly permit another explanation'.[5] Einstein went on to write a sympathetic foreword to the American edition of the book, expressing doubts only about whether displacements could be brought about by the weight of the polar ice.[6] The author subsequently accepted this point (which was also made by the Harvard University geologist Kirtley Mather in the foreword to the British edition) and, when revising his theory in *The Path of the Pole*, suggested that 'the forces responsible for shifts of the crust lie at some depth within the Earth rather than at its surface'. However, as to the details of how the process might take place, he acknowledged that 'at the present time there is no satisfactory explanation'.[7] Hence, despite the support from Einstein, Hapgood's ideas received little consideration in mainstream scientific circles.

Scenarios based on extraterrestrial impacts fared no better. That was, perhaps, hardly surprising, given that the starting point in many cases was the assumption that some myths and legends were based on catastrophes of cosmic origin, or that the story told by Plato of the destruction of the ancient island civilisation of Atlantis was essentially a factual account. One of those who interpreted a myth as an actual event was Franz Xavier Kugler (1862–1929), a Jesuit priest who was an acknowledged authority on Babylonian and Hebrew astronomy, as well as mythology and chronology. In 1927, Kugler wrote a short book entitled (when translated from the original German) *The Sybilline Battle of the Stars and Phaeton seen as Natural History*.[8] Making use of ancient sources, Kugler argued that Phaeton had been a very bright celestial object, which made its entrance onto the scene several hundred years before the founding of Rome. Then, not long afterwards, it fell to Earth as a shower of large meteorites, causing catastrophic fires and floods in Africa and elsewhere.[9]

Another believer in cosmic catastrophes was the British journalist Comyns Beaumont, who presented an original view of comets in 1932 in his book *The Mysterious Comet*, suggesting they were inhabited planets displaced from their natural orbits. In Beaumont's opinion, cometary heads could disintegrate to form meteors. However, their usual fate was to crash into the Sun, giving rise to sunspots, although sometimes the Earth got in the way, with catastrophic consequences for humankind. Beaumont believed that the interior of the Earth was cold and inert, so all earthquakes and volcanic eruptions had to be the result of impacts by comets. In this scenario, episodes of mountain building could not have been caused by the continued cooling of the Earth's crust, as generally assumed by the British catastrophists of the previous century, so the origin of mountains must have had an extraterrestrial dimension. Following the example of Ignatius Donnelly (see Chapter 6), Beaumont saw evidence of a cometary impact in the loam and gravel deposits widely spread throughout northern regions, and he associated the same impact-event with the Phaeton myth, the floods of Noah and Deucalion, and the destruction of Atlantis. The sixth century Spanish historian Paulus Orosius had placed the Deucalion Flood 810 years before the founding of Rome, so, on that basis, Beaumont suggested a date of around 1560 B.C. for the time when the comet (or comets) concerned had struck.[10]

Figure 12.1   Left: Immanuel Velikovsky. Right: Claude Schaeffer.

Velikovsky and planetary catastrophism

In 1950, Immanuel Velikovsky (1895–1979), a Russian-born psycho-analyst living in the U.S.A., launched what was to become a comprehensive assault on the bastions of gradualism and orthodoxy, putting forward in his book *Worlds in Collision* a highly controversial catastrophist scenario which had some features in common with those of Kugler and Beaumont, and of the nineteenth century writers, Radlof and Donnelly. Velikovsky (figure 12.1) derived his theory of 'planetary catastrophism' not from scientific evidence but on interpretations of ancient myths and records from various parts of the world. On the basis of stories such as the Greek myth in which Pallas Athena emerged from the head of Zeus, he argued that the most recent of a series of global catastrophes of cosmic origin began when Venus was ejected from the core of Jupiter as a body with a substantial comet-like tail, and passed very close to the Earth. As our planet moved into the tail of Venus, poisonous red dust and molten rocks showered down on the Earth's surface, and inflammable hydrocarbons were introduced into the atmosphere. Immense electrical discharges took place between the bodies, and gravitational forces gave rise to massive earthquakes and tidal waves, as well as causing the Earth to turn over completely, giving a North–South reversal. There was also an East–West reversal for, from this time onwards, the Sun was observed to rise in the East, whereas previously it had apparently risen in the West. On the basis of biblical reckoning, Velikovsky dated this close encounter at around 1450 B.C., believing it to have caused the plagues of Egypt, which enabled Moses to lead the Israelites out of the country, as described in the book of *Exodus*. Venus and the Earth then moved apart but, 52 years later, they came together again, causing similar catastrophes to the previous occasion, but no further geographical reversals.[11]

According to Velikovsky, Venus then moved away from the Earth once more and came into contact with Mars, losing its tail in the process. As a result of this encounter, Venus acquired its present orbit, keeping it well clear of other planets. Mars, however, was propelled into an orbit which brought it very close to the Earth on several occasions between 747 B.C. and 687 B.C., afflicting further large-scale catastrophes on our planet. The final encounter between Mars and the Earth resulted in both planets taking up their present orbits, thus restoring a measure of stability to the Solar System. In this scenario, therefore, the orbits of the inner planets as we know them today were established only a few thousand years ago, not many millions of years in the past, as generally supposed. Velikovsky argued that this was possible because, contrary to orthodox opinion, interactions between cosmic bodies involved not only gravitational forces, but also electrostatic ones.

Before the publication of *Worlds in Collision*, Albert Einstein, who from 1921 to 1924 had been co-editor with Velikovsky of the *Scripta Universitatis atque Bibliotechae Hierosolymitarum*, from which the Hebrew University of Jerusalem was to grow, wrote to Velikovsky in much the same way as he did to Hapgood:

> There is much of interest in the book which proves that in fact catastrophes have taken place which must be attributed to extraterrestrial causes. However, it is evident to every sensible physicist that these catastrophes can have nothing to do with the planet Venus and also that the direction of the inclination of the terrestrial axis towards the ecliptic could not have undergone a considerable change without the total destruction of the entire Earth's crust.[12]

That view undoubtedly prevailed. Many scientists were infuriated by the ideas expressed in *Worlds in Collision*, believing them to be unworthy of serious consideration. However, Velikovsky's attacks on orthodoxy were only just beginning. In 1952, in *Ages in Chaos*, he enraged another group of academics by using references to catastrophes in the records of different civilisations as markers to construct a revised chronology for the ancient world. The stimulus for this was the Papyrus Ipuwer, now in Leiden Museum, Holland, which told of terrible suffering amongst the Egyptians: 'The river is blood . . . walls are consumed by fire . . . The land is without light'. Velikovsky became convinced that this referred to the plagues of Egypt described in the Old Testament, even though the Papyrus Ipuwer could not have been written later than the end of the Middle Kingdom, generally regarded as coming several centuries before the time of the Exodus. Once he had started to question accepted chronologies, he quickly came to the conclusion that the Dark Ages of Greece and other countries, thought to have lasted several hundred years between 1200 B.C. and 800 B.C., were just figments of scholars' imaginations.[13]

Revolutionary though this may have seemed, Velikovsky was in fact taking much the same stance as the British archaeologist Cecil Torr, over half a century earlier, in defence of what was then the traditional view. Torr, in his *Memphis and Mycenae* of 1896, maintained that the start of the Archaic period in Greece in the eighth century B.C. followed immediately after the Mycenaean age. In doing this, he was opposing the arguments then being put forward by William Flinders Petrie and others, that the Mycenaean period must have ended in the twelfth century B.C., much earlier than previously supposed. Petrie's arguments were based on finds of Greek pottery in Egypt, which enabled them to be linked to Egyptian chronology. As his views became firmly established, a Dark Age was inserted to fill the gap created by moving the Mycenaean period into an earlier time-slot.[14] Velikovsky was now trying to reverse that process.

Three years after the publication of *Ages in Chaos* came the next book, *Earth in Upheaval*, in which Velikovsky presented what he saw as strong geological evidence for global catastrophism. Again, though, his arguments failed to stimulate any serious debate amongst professional scientists, being either ignored or rejected as being self-evidently flawed.[15]

However, perhaps because of an attempt by some American academics to suppress Velikovsky's writings, they stimulated considerable interest amongst the general public in the subject of global catastrophes affecting the Earth, as well as possible chronological revisions.[16] That applied particularly to young people. In an address given in 1953 to the Graduate College Forum of Princeton University, and included as a supplement to *Earth in Upheaval*, Velikovsky repeatedly urged the students to 'dare' to formulate their own independent views. He assured them that they could be greater than their teachers, if they had the ability to see things in a different light. Not surprisingly, many were receptive to the suggestion that they should not accept orthodox opinions without challenge.[17]

Eventually, a number of writers, including scientists and other scholars with University posts, made a serious effort to assess Velikovsky's work. Some participated in debates on the subject organised by the Student Academic Freedom Forum, based in Portland, Oregon. The Forum published a journal, *Pensée*, between the years 1972 and 1975.[18]

In Britain, the Glasgow University archaeologist Euan MacKie wrote in *New Scientist* in 1973 that, regardless of whether Velikovsky's scenario seemed plausible, he had formulated hypotheses which should be tested in the normal way.[19] In the same year, he suggested in *Pensée* that radiocarbon dating might provide the evidence for a test of Velikovsky's theories of global catastrophes and chronological revisions.[20] MacKie went on to become a founding member of the Society for Interdisciplinary Studies (SIS), an organisation designed to provide a forum for open-minded discussions of all aspects of catastrophism and chronology. From 1975, debates took place regularly at SIS meetings, and in the pages of the Society's journal, the *SIS Review*, later re-named the *Chronology and Catastrophism Review*. To avoid possible misunderstandings, it was made clear right from the start that the Society had been formed to *examine* the ideas of Velikovsky and other catastrophists, not to promote any particular point of view.[21] Nevertheless, although some reputable scholars took part in these activities, the academic community in general continued to shun catastrophism, in whatever form it was presented.

Controversies over possible Bronze Age catastrophes

During the 1970s, the gradualist paradigm was supremely dominant, as it had been throughout the previous hundred years, and any attempts to suggest catastrophist mechanisms for events in geology or evolution were viewed with great suspicion and generally ignored. Exactly the same applied to catastrophist explanations for events in ancient history, particularly ones in the Middle East. Rightly or wrongly, such arguments were generally seen as moves to provide support for a literal interpretation of the Bible.[22]

When the British archaeologist Sir Leonard Woolley excavated the ancient Sumerian city of Ur, in what is now southern Iraq, between 1928 and 1934, he found a three-metre-thick layer of alluvial silt on top of the levels of the Ubaid Period (dated to around 4000 B.C.) and beneath the first traces of the succeeding Uruk Period. To some, including Woolley himself, this seemed like evidence for the flood of Noah. However, no other sites were found to show similar alluvial deposits during the Ubaid Period. On the other hand, at the nearby city of Shuruppak (the modern Fara), there was evidence of a flood during the Early Dynastic Period, around 2750 B.C., and an alluvial deposit dating from around the same time was found at another Sumerian site, the city of Kish. Although it became generally accepted that some localised event in the region, caused by the Tigris and/or Euphrates bursting their banks, might have been the stimulus for both the Uta-Napishtim and the Noah stories, no serious investigation took place as to whether there had been a more widespread flood in Sumer during the Early Dynastic Period, as this would have smacked of unfashionable 'Biblical Archaeology'.[23]

Similarly, there was a hostile reaction in 1980 when Walter Rast and Thomas Schaub began to suggest that five Early Bronze Age sites they had been investigating near the southeastern corner of the Dead Sea since 1975 might be the five 'cities of the plain', Sodom, Gomorrah, Admah, Zeboiim and Zoar, mentioned in *Genesis* 14:2 (see Chapter 1). One of these sites, at Bab ed-Dra, had first been discovered by a team led by the American archaeologist William Albright, in 1924. However, the excavations of Rast and Schaub revealed that it was far more extensive than had been realised, supporting a large population, before being overwhelmed by fire towards the end of the Early Bronze Age, as was the nearby settlement of Numeira, 10 km to the south. After these catastrophes, whatever their cause (and the assumption was that natural processes were involved, not the wrath of God), both sites were abandoned. However, the suggestion that there might have been some truth in the *Genesis* account of the destruction of Sodom and Gomorrah was too much for some scholars to contemplate.[24]

Even where there was no obvious biblical agenda, there was still widespread suspicion of catastrophist ideas. So, silence greeted the publication in 1948 of

Stratigraphie comparée et chronologie de l'Asie occidentale,[25] by Claude Schaeffer (see figure 12.1) who argued that Bronze Age civilisations had, on more than one occasion, been overwhelmed by large-scale natural catastrophes. That was of course contrary to orthodox opinion, which maintained that civilisations could come to an end as a result of gradually changing circumstances, or because of conquest by invading armies, or a combination of the two. It was a matter of firm dogma that natural catastrophes could never be involved, except on a very localised scale (e.g. an earthquake making a particular city susceptible to capture by an enemy). Hence, very little notice was taken of Schaeffer's book, despite the eminence of the author, who at various times occupied chairs at the École de Louvre and the Collège de France. Schaeffer's main professional achievement was the excavation of an ancient artificial mound (or 'tell') at Ras Shamra on the Mediterranean coast of Syria, which he was able to identify as Ugarit, an important city throughout the period from the Neolithic (the New Stone Age) to the end of the Bronze Age.[26]

On the basis of findings here and at other sites throughout the Middle East, Schaeffer (1898–1982) claimed that there had been at least five occasions in the Bronze Age when catastrophic destructions occurred in widespread fashion, often with evidence of earthquakes and/or fire. Two of these were in the Early Bronze Age, the first around 2300 B.C., co-incident with the end of the Old Kingdom in Egypt, involving sites in Syria (including Byblos and Ugarit), Palestine (particularly Beth Shan) and Anatolia (e.g. Alaça Hüyük and Troy). The second Early Bronze Age catastrophe occurred perhaps 200 years later, affecting many of the same locations as the first, together with others such as Jericho in Palestine and Tell Brak in Mesopotamia. About 500 years after that, the end of the Middle Bronze Age was marked by destructions at many sites, including Ugarit in Syria, Jericho, Hazor and Lachish in Palestine, Alaça Hüyük and Boghazköy in Anatolia and Tepe Gawra in Mesopotamia. This was also the time the Hyksos invaded Egypt. Schaeffer further claimed that there were two episodes of widespread catastrophic destruction in the Late Bronze Age. The first of these occurred around 1365 B.C., the time of the Amarna Period in Egypt, and affected locations in Syria (including Ugarit), Palestine (e.g. Beth Shan, Megiddo and Lachish), Anatolia (Boghasköy, Tarsos and Troy) and Mesopotamia (Chagar Bazar and Tell Brak). The other was dated around 1200 B.C., bringing to an end some Bronze Age cultures, with destructions at most of the same sites in Syria, Palestine and Anatolia as in the previous wave.[27]

Schaeffer was convinced that these catastrophic destructions were the result of natural events, rather than human activity. However, he was undecided as to the precise causes, although undoubtedly favouring the involvement of earthquakes. He did not consider the possibility of the involvement of extraterrestrial factors, a point picked up by the Belgian mathematician, engineer and amateur geologist René Gallant (1906–1985), in his 1964 book, *Bombarded Earth*.[28] Gallant argued that the seismic activity and climate changes which, according to the evidence provided by Schaeffer, occurred at the times of the destructions, were both likely to have resulted from large meteoritic impacts. *Bombarded Earth*, however, received even less attention than Schaeffer's major work had done.[29]

With the establishment of a forum to discuss such ideas, Euan MacKie followed up his earlier suggestion of using radiocarbon dating to test for possible correlations between catastrophic events in different locations by carrying out a survey of published data. His tentative conclusion, published in SIS *Review* in 1979, was that the end of the Old Kingdom in Egypt, which Schaeffer had included as part of the first wave of Early Bronze Age catastrophes in the Middle East, could also have been contemporaneous with the end of the Chalcolithic (a copper-using culture) in the western Mediterranean regions, the fall of the Harappan civilisation in the Indus Valley, northwestern India, and the end of the Neolithic in northwestern Europe.[30]

Velikovsky's ideas considered

In 1978, the SIS, in collaboration with the Extra-Mural Department of Glasgow University, organised a conference, entitled *Ages in Chaos?*, to discuss the chronology issues raised by Velikovsky. The consensus which emerged at this Glasgow conference was that there were indeed problems with the conventional chronologies but, equally, there were major difficulties with Velikovsky's proposed revisions.[31] Afterwards, several historians with SIS associations, including Gunnar Heinsohn (writing in *Ghost Empires of the Past*), Peter James (in *Centuries of Darkness*), David Rohl (in *A Test of Time* and *Legend*) and Emmet Sweeney (in *The Genesis of Israel and Egypt* and *The Pyramid Age*), went on to propose revised chronologies different from those of Velikovsky. However, they were also different from each other.[32]

At a meeting of the SIS in London in 1982, the British astronomer Victor Clube, then at the Royal Observatory, Edinburgh, and later at Oxford University, presented an address entitled, 'Cometary catastrophes and the ideas of Immanuel Velikovsky'. In his talk, Clube acknowledged that Velikovsky may have been on the right track in thinking that some myths were inspired by cosmic catastrophes but, if so, these were far more likely to have involved comets than wandering planets. In his interpretation of ancient writings, Velikovsky may have been misled by reports of close encounters of the Earth with Comet Encke in the fifteenth century B.C. (the time of the supposed Venus event) and with Comet Halley (rather than Mars) in the seventh. Clube concluded, 'So, in looking again at Velikovsky, we can dispense with his quaint astronomical "theory" but we must pay serious attention to all his "facts"'. Similarly, at another SIS meeting in London in 1984, René Gallant (then aged 78) explained why major catastrophes in the ancient world had probably been caused by large extraterrestrial impacts, rather than by close encounters with another planet.[33]

Whilst some considered Velikovsky's scenario from the perspective of events on Earth, others examined his ideas about a Solar System where electromagnetic and electrostatic forces are of great importance, and where planets have changed their orbits in significant fashion during the time humankind has been living on the Earth. Observations from the Earth failed to detect any electrical charges on other members of the Solar System but, according to Ralph Juergens, an American civil engineer, this did not disprove Velikovsky's theory. As Juergens pointed out in *Pensée*, it is a known fact that interplanetary space is a plasma, not a pure vacuum. In other words, ionised gas is present. Hence, each planet would be surrounded by a space-charge sheath, which would insulate the planet's electrical charge. Only on a close approach, when space-charge sheaths would inevitably be disrupted, would electrical interactions between planets take place. Hence, if Venus had brushed past the Earth, as Velikovsky believed, there might have been electrical discharges between the planets. At other, more normal, times, planets would interact solely by gravitational attraction, exactly as observed.[34]

However, calculations made on the basis of gravitational forces indicated that Venus must have been in its present orbit for a very long time. In articles in the *Chronology and Catastrophism Review* and elsewhere, the British engineer Eric Crew confirmed by computer simulations that it was extremely unlikely that Venus could have followed the suggested path from Jupiter and reached its present position solely as a result of gravitational forces, no matter what encounters with other planets it had on its way. However, the possibilities were increased if unscreened electrostatic forces played a part. Assuming that the Sun has a stable positive charge, and that the body ejected from Jupiter might also have had a positive charge, but one which declined with time, an orbit around the Sun close to the present position of Venus could be attained.[35] Yet, even so, such an orbit would almost certainly be more elliptical than the almost circular one which Venus currently possesses. Crew concluded that, if an object had indeed been ejected from Jupiter and travelled through the Solar System as Velikovsky supposed, it must have been a much smaller body than Venus, and its only possible link with that planet was that it eventually collided with it.[36]

As to catastrophes in earlier times, Velikovsky suggested, in an unpublished book, and then in the journal, *Kronos*, in 1979, that because myths often refer to a Golden Age associated with the figure known in Roman mythology as Saturn, the Earth might originally have been a satellite of the planet bearing that name, its subsequent escape from Saturn's influence being linked with the event we call Noah's flood.[37] This speculation was developed, in the pages of the American journal *Aeon* by Dwardu Cardona, David Talbott and Ev Cochrane. Led by myths and ancient inscriptions, they concluded that, in ancient times, Saturn was a very prominent object, occupying a fixed position in the northern sky, with Venus and Mars lying between the Earth and Saturn. The largest planet, Jupiter, could not be seen from the Earth in this period, because it was obscured by Saturn. The reason for this 'polar configuration' was that the five planets Jupiter, Saturn, Venus, Mars and the Earth orbited the Sun as a single linear unit, which rotated about a point close to Saturn, before its break-up at the end of the Golden Age.[38] However, regardless of the difficulties of reconciling such a hypothesis with the laws of physics, it was never explained how humans could have remained alive on Earth throughout cosmic turmoil of this magnitude.

Not surprisingly, considerations of Velikovsky's ideas largely took place away from the mainstream of academic debate. Even disregarding the Saturnian extension, the essential features of the Velikovsky scenario were firmly rejected by professional scientists, including ones who were convinced that the Earth had indeed suffered catastrophes in recent times. This was despite the fact that Velikovsky made certain predictions which turned out to be correct. He suggested, for example, that, contrary to what was then generally supposed, the surface of Venus would be found to be very hot, and that Jupiter would be shown to emit radio waves.[39] Because of these correct predictions, Albert Einstein believed that Velikovsky's ideas should not be dismissed out of hand.[40] Most scientists, however, took a very different view. For example, the Cornell University astronomer Carl Sagan wrote in his 1979 book, *Broca's Brain*, 'To the best of my knowledge, in *Worlds in Collision* there is not a single correct astronomical prediction made with sufficient precision for it to be more than a vague lucky guess – and there are . . . a host of demonstrably false claims'. Moreover, whilst Sagan deplored the 'disgraceful' attempts to prevent the publication of *Worlds in Collision*, he nevertheless considered Velikovsky's approach to his work to be 'shoddy, ignorant and doctrinaire'.[41]

Similarly, Velikovsky's arguments for a revised chronology were rejected by professional Egyptologists. For example, at the 1978 *Ages in Chaos?* conference, Michael Jones, of the University of Glasgow, discussed the wealth of information obtained from discoveries at the Deir el-Medineh site, near Luxor, and commented, 'The manipulation of dynasties undertaken by Dr Velikovsky overlooks the continuity of archaeological and linguistic development which finds comfortable accommodation within the framework of the existing chronology'. He added that Velikovsky's proposals 'display blatantly' a variety of shortcomings.[42]

Likewise, professional geologists were unimpressed by the arguments for global catastrophes which Velikovsky presented in *Earth in Upheaval*. That could have been because he was seen as an outsider, lacking any appropriate qualifications or experience but, as in the areas of physics and ancient history, there were undoubtedly other reasons. For example, he showed a lack of discrimination in his readiness to interpret finds, however localised, as proof of worldwide catastrophes; indeed, he even saw evidence of such catastrophes in facial expressions and body contortions of individual fossils. Then there was the fact that *Earth in Upheaval* gave the impression of being out of date even at the time of its publication, most of the cited references being from before 1940. Furthermore, it was Velikovsky's misfortune that he pointedly rejected continental drift as a possible explanation for any of the geological features considered, just as a revolution was about to take place which brought this hitherto minority view into general favour, as we shall discuss in Chapter 14. In his 1978 book, *Ever Since Darwin*, the Harvard University palaeontologist and science historian Stephen

Jay Gould expressed considerable sympathy for Velikovsky because of the way he had been treated by some professional academics, but pointed out that 'a man does not attain the status of Galileo merely because he is persecuted; he must also be right'. Although Gould acknowledged that 'many fundamental beliefs of modern science came as heretical speculations advanced by nonprofessionals', he continued, 'Unfortunately, I don't think that Velikovsky will be among the victors in this hardest of all games to win'.[43]

In 1975, Velikovsky's ideas were strongly opposed at a meeting of the American Association for the Advancement of Science.[44] Without question, none of his specific proposals, whether concerning chronology, or planetary encounters and their effects, had much of an influence on the mainstream of academic thought. Indeed, it could be argued that they served to reinforce existing beliefs within establishment circles about the supposedly unscientific nature of catastrophism. On the other hand, his writings, and attempts to suppress them, stimulated much critical examination of orthodox views, and creative searches for alternatives. Above all, they made many people start to think seriously, for the first time, about the possibility that catastrophic events had influenced the history of life on Earth.

13 Atlantis: rational and irrational theories of a 'lost' civilisation

Plato's island civilisation

As we noted in the previous chapter, many catastrophists have been interested in the theory that a major civilisation existed in ancient times on an island called Atlantis. According to Plato, writing in the *Timaeus*, an elderly priest in the Temple of Neith at Sais, in northern Egypt, said to the Athenian, Solon:

> It is told in our records how your City once repulsed a mighty power which, starting from a point in the Atlantic Ocean, made an unprovoked attack on the whole of Europe and Asia [Minor]. For in those days the Atlantic was navigable and, in front of the straits known by you as the pillars of Heracles, was an island [Atlantis] which was larger than Libya [North Africa] and Asia [Minor] put together, and was the way to other islands, and from these you might pass to the whole of the opposite continent which surrounded the true ocean.

The Athenian army, in driving back the Atlantean invaders, liberated all those within the pillars of Heracles who had been enslaved by them. However, as the Egyptian priest told Solon, 'at a later time, there occurred great earthquakes and floods and, in a single day and night of misfortune, the whole body of your warriors sank into the Earth, and the island of Atlantis in like manner sank beneath the waves and disappeared'. These events were said to have taken place 9,000 years before that time, i.e. around 11,500 years ago.[1]

Despite Plato's claim that all of this was true, his pupil Aristotle regarded it as fiction, as did succeeding generations of Greeks, including the great geographer and historian, Strabo.[2] Over a thousand years later, around the beginning of the seventeenth century, Sir Francis Bacon wrote, in *The New Atlantis*, of an island in the South Seas inhabited by Atlanteans, who had moved there from their original home in the Americas, after its destruction by floods. This was probably just a device to allow Bacon to explore ideas about an ideal society. Nevertheless, in a map produced by Athanasius Kircher in 1665, Atlantis was shown as a large island in the centre of the Atlantic Ocean.[3]

In the late nineteenth century, Ignatius Donnelly argued for the reality of Plato's Atlantis, in his books *Atlantis – The Antediluvian World* (1882) and *Ragnarok: The Age of Fire and Gravel* (1883) (figure 13.1).[4] Donnelly placed Atlantis on the

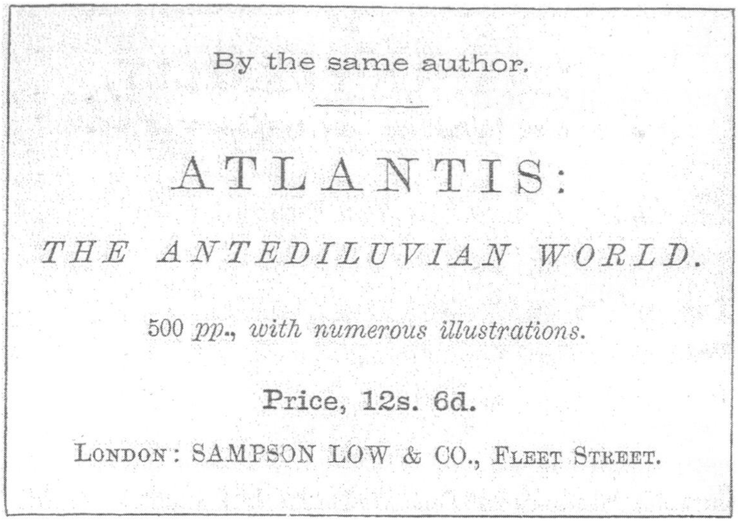

> By the same author.
>
> ---
>
> # ATLANTIS:
>
> ## *THE ANTEDILUVIAN WORLD.*
>
> *500 pp., with numerous illustrations.*
>
> Price, 12s. 6d.
>
> LONDON: SAMPSON LOW & CO., FLEET STREET.

Figure 13.1   Advertisement for Ignatius Donnelly's 1882 book, *Atlantis – The Antedeluvian World*, included in the same author's *Ragnarok – The Age of Fire and Gravel*, published in the following year. Note that the price of the book was more than would have been earned in a week by some in full-time employment.

mid-Atlantic ridge, with the present-day island of the Azores being the mountain peaks of the sunken continent. In his view, the cometary catastrophe which gave rise to the widespread gravel deposits (see Chapter 6) occurred about 30,000 years ago, driving the Atlanteans from North America to their new island home. Then, at the time described by Plato, a further catastrophe, associated with the event we know as the flood of Noah (and also that of Deucalion), caused this to sink beneath the waves of the Atlantic Ocean. Apart from myths and legends, his arguments were based on geological evidence, and supposed linguistic and cultural similarities (e.g. pyramid building) on both sides of the Atlantic, which suggested a common origin on the island of Atlantis. The French physician and explorer Augustus Le Plongeon (1826–1908), a contemporary of Donnelly's, also saw similarities between the cultures of the Old and New World, claiming that Maya hieroglyphics had much in common with Egyptian ones. He followed his compatriot the Abbé Charles-Etienne Brasseur (1814–1874) in supposedly identifying a reference in Maya writings to Mu, a land destroyed by earthquakes and floods, which might have been Atlantis. Le Plongeon and Brasseur's belief that, like the languages of Europe, the Maya language was based on an alphabet was eventually shown to be incorrect but, even at the time, their claims of decipherment failed to convince many other scholars.[5]

Atlantis became an obsession with some individuals and groups far away from the mainstream of scholarly activity. In consequence, theories became increasingly far-fetched. Various occultists believed in not one but two sunken continents, each of which had been the homes of advanced civilisations. These were Atlantis and Lemuria, the latter being located in the East, either in the Indian Ocean (where its existence had been proposed by naturalists before there was any knowledge of continental drift to explain the distribution of lemurs and closely related lorises in Africa, Madagascar and India) or in the Pacific. One of the most influential of the occultists was Helena Petrovna Blavatsky (1831–1891), who was born in Russia, but lived at different times in various parts of the world. Blavatsksy's 1882 book, *Die Geheimlehre (Secret Doctrines)*, supposedly based on an Atlantean record, described cities with statues of precious stones, and a terrible flood which drowned the sinners but not the good people.[6] Moving forward into the twentieth century, Karl Georg Zchaetzsch claimed, in his 1922 book, *Atlantis – die Urheimat der Arier (Atlantis – the Original Home of the Aryans)*, that, following the destruction of Atlantis by a comet, the only survivors were Wotan, his daughter and his pregnant sister. This was argued mainly on the basis of Norse legends, and 'racial memory'.[7]

In 1913, Hans Hörbiger, an Austrian engineer and mystic, claimed that there had been several previous advanced civilisations on Earth, each of which had been destroyed by a cosmic catastrophe. The last such cataclysm, when Atlantis was destroyed, occurred when the Earth captured its present Moon and lost its previous one, both of which were made of ice.[8] Hörbiger's ideas, which suggested a series of cleansing catastrophes and the periodic re-establishment of civilisation by an Aryan 'master-race', gave rise to a series of novels (in German) by Edmund Kiss, and appealed to the prejudices of Adolf Hitler and other Nazis, who accepted them as being essentially correct. However, even some who were persecuted by the Nazis claimed mystical knowledge of Atlantis. One of these was Rudolf Steiner, who maintained that Hitler was influenced by occultists who practised 'black magic', thus releasing evil into the world, whereas he himself tried to use 'white magic' for the benefit of all humankind. Steiner, in his *Unserer Atlantischen Vorfahren (Our Atlantean Ancestors)* of 1928, claimed on the basis of his intuition that the Atlantis civilisation had existed 80,000 years ago, and that the Atlanteans used 'organic energy' to make their vehicles float above the ground.[9]

However, dubious works about Atlantis were not confined to continental Europe. So, William Scott-Elliott, an English merchant banker, who was a disciple of Helena Blavatsky, claimed to have had occult revelations which enabled him to publish maps of the world, showing the locations of Atlantis and Lemuria, in 1896. He described the Lemurians as being over twelve feet tall, with an extra eye for seeing backwards, who were originally egg-laying hermaphrodites.[10] Again, in 1936, Hans (H.S.) Bellamy, an Anglo-Austrian writer, supported Hörbiger's theories about the catastrophic destruction of Atlantis, in his *Moons, Myths and Man*.[11] During the same decade, James Churchward claimed in a series of books that, when he had been with the British Army in India, a Hindu priest taught him how to decipher some stone tablets, which he proceeded to do, aided by intuition. The tablets told of two advanced civilisations in the remote past, one of which was Atlantis whilst the other, located in the Pacific Ocean, was called Mu (the name incorrectly deciphered from Maya writings by Brasseur and Le Plongeon).[12] Wilder still, the American psychic Edgar Cayce claimed that Atlantis had suffered a catastrophe before that described by Plato, when the inhabitants used crystals to harness the rays of the Sun, and blew themselves up. The survivors then developed technologies such as televisions and anti-gravity machines, and set up bases in Mexico, Egypt and northern Spain, before their island home sank beneath the Atlantic Ocean. In 1933, Cayce prophesied that part of an Atlantean temple would soon be discovered in the sea near Bimini, in the Bahamas and, in 1940, suggested that portions of Atlantis would 'rise again' in 1968 or 1969.[13]

More rationally, in 1924, in The Problem of Atlantis, and two years later, in History of Atlantis,[14] the Scottish anthropologist and journalist Lewis Spence argued that the Atlanteans were the Crô-Magnon people, who migrated into western Europe towards the end of the Pleistocene Ice Age, bringing with them a distinctive Stone Age culture, typified by cave paintings at Lascaux and elsewhere. At about this time, the original landmass of Atlantis split into two, as a result of geological disturbances, the northern part being the island described by Plato, which eventually sank beneath the waves. The southern part, which Spence called Antilla, survived this catastrophe, but eventually fragmented into the islands we now know as the West Indies. It was from Antilla that the civilisations of Central and South America developed.[15]

No physical evidence of the lost civilisation had come to light. There was excitement when an article appeared in the New York American in 1912, which referred to items found at Troy, bearing the inscription, 'From King Cronos of Atlantis'. The article was written by someone who claimed to be Paul Schliemann, grandson of Heinrich Schliemann, the excavator of Troy, but, alas, it turned out to have been a hoax: Heinrich Schliemann did not have a grandson named Paul, and no evidence of the Atlantean artifacts was ever produced.[16] Nevertheless, the subject of Atlantis had become very popular, with new ideas continuing to be put forward. Innumerable possible locations were suggested for Atlantis, including South America, Mexico, Britain, Greenland and Spain, the various authors taking considerable liberties with Plato's account.[17] Without question, most of this served to reinforce the view in orthodox scholarly circles that 'catastrophism' was more-or-less synonymous with 'unscientific nonsense'.

There were, nevertheless, some Atlantologists who tried to reconcile the precise details of Plato's story with theories based on purely scientific evidence. One of these was the Austrian engineer Otto Muck (1892–1956), who expressed his ideas in Atlantis, die Welt von der Sinflut (Atlantis, the World of the Sin-Flood).[18] An edited version appeared in English in 1978 (long after Muck's death) as The Secret of Atlantis. In this book, Muck agreed with Donnelly that the Azores were the remnants of Atlantis, which sunk after an extraterrestrial impact.[19] The Polish Atlantologist L. Zajdler also attributed the destruction of the island civilisation to an impact, in his book Atlantyda, which was published in 1972.[20]

In Muck's scenario, a 10-km-diameter asteroid entered the Earth's atmosphere at the end of the Pleistocene, and began to break up before striking the planet. Some of the smaller debris produced the Carolina Bays, a cluster of water-filled depressions found between Maryland and northern Florida, whilst the two largest fragments landed in the Atlantic near Puerto Rico, where there are still two large depressions in the ocean floor. According to Muck, the main impactors both punctured the Earth's crust, causing vulcanism along the line of the mid-Atlantic ridge on a scale far greater than the huge Krakatoa explosion of 1883. A tidal wave would have struck Atlantis, whilst poisonous gases enveloped the island, and hot pumice rained down from above. So much light pumice would have landed in the Atlantic Ocean, floating on the surface for perhaps thousands of years afterwards, that it would have appeared like a shoal of mud, as described by Plato in the *Timaeus*.[21]

Smaller particles thrown up by the volcanic activity would have spread around the Earth, eventually to fall as the Loess belt, stretching from northern Europe to China. Similarly, torrential and sustained rainfall resulting from the immense clouds produced by the clash of fire and water in the Atlantic gave rise to widespread Flood myths, whilst the asteroid was the origin of the Phaeton story. Furthermore, the impacts caused a crustal slippage and a 2° shift of the pole of rotation, bringing much colder temperatures to Siberia. The mammoths and other large animals living there were first asphyxiated by the gases, then buried by mud when rain and ash fell from the sky in torrents, and finally frozen solid as temperatures plummeted, to be found intact thousands of years later, as noted by Cuvier. Atlantis itself sank, over a 24-hour period, because the magma on which it rested was released and hurled into the air as a result of the asteroid impact. Eventually, only the mountain peaks, now known as the islands of the Azores, showed above the surface of the ocean. A consequence of this was that the gulf stream, previously deflected to the south by Atlantis, could now carry on and bring warmer climates to the lands bordering the northeastern Atlantic. Finally, Muck suggested that the Maya of Central America counted their dates from the day the asteroid struck. In the calendar generally used in the western world at the present time, that would be equivalent to 5 June, 8498 B.C.[22]

Unfortunately, the phenomenon of sea-floor spreading, which involves the mid-Atlantic ridge, was largely unknown at the time Muck was writing, so his geological arguments (like those of Donnelly before him) were based on false assumptions.[23] This was soon to become apparent (see Chapter 14), destroying many of the key features of Muck's case.

Regardless of that, a major difficulty for all those who maintained that Plato was presenting an accurate account of a civilisation around 11,500 years ago was that he described it as making extensive use of metal weapons, for which there was no archaeological evidence whatsoever. Of course, Atlantis itself might have sunk beneath the waves, but traces should have remained of its influence around the Mediterranean and elsewhere. However, nowhere at all were there any indications of a metal-using culture until thousands of years after the supposed destruction of Atlantis. Muck acknowledged this, following Spence in identifying the Altanteans as the people who brought the Stone Age Crô-Magnon culture into Europe, rather than being the users of metal described by Plato.[24]

The Russian chemist Nicolai Zhirov also located the lost continent in the middle of the Atlantic Ocean, near the Azores, in his 1964 book, *Atlantis* (which appeared in an English translation in 1970).[25] Zhirov accepted that the Atlanteans were unable to produce bronze and other alloys by smelting, but believed that they made use of meteoritic iron, as well as metals and alloys such as tin, copper and brass, which could be obtained directly from ores.[26]

Some, including Adolf Schulten, writing in *Pettermanns Geographische Mitteilungen* in 1927, saw a link with Atlantis in the silver-rich state of Tartessos, situated in southwestern Spain around the Guadalquivir river.[27] Tartessos was known to the Greeks from about 650 B.C., and to the Phoenicians from possibly even earlier, for it seems that they built their base of Cadiz adjacent to Tartessos territory. However, there was little evidence of a Tartessos civilisation as such, no city having been located.[28] As to the earlier history of the Iberian peninsula, there was carbon-14 and thermo-luminescence evidence that the first megalithic culture of Europe arose there some 6,500 years ago. This was initially based purely on stone tools although, at a later stage, copper was used, and then bronze. It was apparent that farming developed in Iberia about 8,000 years ago, but this was still some time short of the date of Plato's Atlantis.[29]

Atlantis and Thera

Many writers concluded that, if there really had been an Atlantis, it was much smaller and more recent than Plato stated. Indeed, apart from the supposed date, everything that Plato wrote about Athens in the Atlantis account fitted better into a Bronze Age rather than a Stone Age setting.[30] Velikovsky wrote in *Worlds in Collision* that there must be one nought too many in the age of Atlantis given by Plato, which would place its destruction within the period of his postulated (but widely rejected) Venus-induced catastrophes, around 1500 B.C.[31] Comyns Beaumont, as we saw in Chapter 12, came to similar conclusions. Although he challenged Plato's date, Beaumont still located the lost continent within the Atlantic Ocean.[32] In his 1946 book, *Riddle of Prehistoric Britain*, he argued that the legendary sunken land of Lyonesse, to the southwest of England, was part of Atlantis.[33] However, others looked elsewhere. In particular, some linked the story of the destruction of Atlantis to the Bronze Age eruption of Thera, in the Aegean Sea. Much of the volcano disappeared during the eruption, leaving behind the group of islands now known as Santorini, the largest of which is still called Thera.[34]

The first to make the connection between Thera and Atlantis seems to have been Louis Figuier, in his 1872 book, *La terre et les mers* (*The Earth and the Seas*). Evidence of a pre-Greek Bronze Age settlement had been found on Santorini during the nineteenth century, during the quarrying of ash for use in the construction of the Suez Canal, and traces of a similar culture were later found at Knossos, on the much larger island of Crete. When the latter site was excavated by Sir Arthur Evans, from 1900 onwards, it became clear that it must have been the heart of a major Bronze Age civilisation, which was given the name 'Minoan'. As details emerged, K.T. Frost, of the Queen's University, Belfast, in an article in *The Times* of 19 February 1909, entitled 'The Lost Continent', and the Egyptologist, James Baikie, in his 1910 book, *The Sea Kings of Crete*, both pointed out some parallels with the details of Atlantis given by Plato in the *Critias*. For example, bull cults were a feature of the religion of Minoan Crete just as, according to Plato, they were of Atlantis. Again, from Crete, as from Atlantis, it was possible to pass, via a series of smaller islands, to the mainland opposite. By this time, knowledge of the Minoan finds under the volcanic ash of Santorini had faded from the public consciousness, so neither Frost (who was killed in action in the First World War) nor Baikie suggested a link between the fall of the Minoan civilisation of Crete and the eruption of Thera.[35]

That situation changed, however, when the Greek archaeologist Spyridon Marinatos found volcanic pumice and beach sand amongst Minoan ruins on Crete.[36] In 1939, in the journal *Antiquity*, he proposed that the Minoan civilisation of Crete had been brought to an end by the eruption of Thera and its accompanying tidal wave, thus giving rise to the legend of Atlantis.[37] Marinatos went on to excavate a Minoan settlement near Akrotiri, on Santorini, which turned out to be far more extensive than anyone had expected (figure 13.2).[38] In consequence, the seismologist Angelos Galanopoulos argued that Thera, not Crete, was the location of the capital city and religious centre of Atlantis.[39] Galanopoulos, in collaboration with the archaeology journalist Edward Bacon, presented his case in the 1969 book, *Atlantis – The Truth behind the Legend*.[40] In the same year, two further books were published which supported the association between the destruction of Atlantis and the eruption of Thera. One of these was *The End of Atlantis*, by John (J.V.) Luce, Reader in Classics at Trinity College Dublin; the other was *Voyage to Atlantis*, by James Mavor, an engineer from the Woods Hole Oceanographic Institution, who collaborated with Galanopoulos in carrying out an underwater geological survey around Santorini, and with Marinatos during the early stages of the Akrotiri excavations.[41]

Many were unconvinced. Despite some points of similarity, both Crete and Thera were too small, as well as being located in the wrong place, to fit Plato's account of Atlantis, and the Minoan civilisation was far too recent. In an attempt to overcome that final problem, Galanopoulos took the same line as Velikovsky, arguing that Plato had made a mistake in claiming that the destruction of Atlantis took place 9,000 years before the time of Solon, for the figure should really have been 900 years. However, it was pointed out that such an error could not have arisen easily, because neither the Greek nor the Egyptian numerical systems involved zeros and, in each case, the symbols for 'hundreds' and 'thousands' were very different.[42]

Figure 13.2   'West House, Triangular Square'. This building, where several high-quality frescoes (now in the National Archaeological Museum, Athens) were found, was part of the Minoan town near Akrotiri, Santorini, destroyed by the eruption of Thera in the Bronze Age.

Nevertheless, it was becoming clear that natural catastrophes had undoubtedly struck the eastern Mediterranean in relatively recent times. Thera had experienced one major eruption before 54,000 B.C. and another around 18,300 B.C. Then, in the Bronze Age, during the second millennium B.C. came the one we have been discussing, comparable in scale with the Krakatoa eruption of 1883. A tidal wave struck Crete and tephra fell across the Mediterranean all the way to the Nile delta. Regardless of the issue of the origin of the Atlantis story, the Minoan civilisation must have suffered a severe battering.[43]

14 Evolutionary mass extinctions and neocatastrophism

Were mass extinctions real events?

Returning to the subject of the rate and nature of evolutionary change, the particular problem for palaeontologists continued to be the transitions between geological periods. Some boundaries, if not others, seemed to indicate times when a mass extinction of species had occurred on a worldwide basis. This was particularly true of the boundaries between the Palaeozoic, Mesozoic and Cenozoic Eras, which was hardly surprising, because John Phillips (see Chapter 5) had defined the main geological eras on the basis of the marked discontinuities between them.[1] It subsequently emerged that around half of all families of marine organisms, including most of the ammonoids, brachiopods and crinoids, died out at the end of the Permian, the final period of the Palaeozoic Era, together with many families of land animals. That was the largest of the mass extinction episodes, but some others were only slightly smaller in scale. At the end of the Cretaceous, the final period of the Mesozoic Era, something like a quarter of existing families became extinct, including all the dinosaurs, who had dominated the land for the previous hundred million years. Amongst the other victims at the end of the Mesozoic were the flying reptiles (pterosaurs), the large sea-reptiles (ichthyosaurs, plesiosaurs and mosasaurs), the last of the ammonoids and many bivalve molluscs, fish, gastropods and plankton. Furthermore, whilst the proportions of families which became extinct at the Palaeozoic–Mesozoic and Mesozoic–Cenozoic boundaries were high, extinctions at the level of species and genera on each occasion were even greater.[2]

However, there was a great deal of argument about the time-scale of these extinctions. Some regarded mass extinction episodes as genuine crises of short duration, involving environmental catastrophes. In contrast, others accepted the arguments of Darwin and Lyell and made allowances for likely artifacts causing gaps in the record, or argued that the geological time-scale was such that a 'sudden' transition meant one that had taken up to a million years to accomplish.[3]

The first group, who included Harry Marshall, in 1928, and Edwin Hennig, four years later, proposed that the mass extinctions at the end of the Cretaceous Period and at other times were the result of catastrophic events such as an excess or deficiency of cosmic radiation. So, for example, Marshall, of the University of Virginia, argued in the *American Naturalist* that a reduction in the amount of ultra-violet radiation reaching the surface of the Earth from the Sun should be considered as a cause of animal extinctions at times when plants appeared to be relatively unaffected, such as the Cretaceous–Tertiary and Triassic–Jurassic boundaries. He thought that the disappearance of animal species 'ought to occur rapidly – within a few generations – if the species is vitally dependent upon ultra-violet radiation'. More generally, Marshall also suggested that an ultra-violet deficiency might be 'a cause of those extinctions [including the Permian–Triassic ones] which were widely spread over the face of the earth, which occurred at times of cool or cold climates, and under conditions favoring selective reduction of the shorter rays from the sun'.[4] Other scientists of the period, however, thought that mass extinctions had occurred solely as a consequence of far less dramatic environmental changes, over a much longer period of time.[5]

Without question, the latter view was the one that generally prevailed amongst evolutionary biologists. Julian Huxley's major work, *Evolution – The Modern Synthesis*, which appeared in three editions between 1942 and 1974, contained no explicit reference to mass extinctions at the end of the Permian and Cretaceous Periods (and only referred indirectly to events at these times in two or three very brief passages), making it clear by default that Huxley considered mass extinctions to be of little importance in the overall scheme of evolution. That was also the situation in books on the history of evolutionary theory from the 1980s, such as *The Growth of Biological Thought*, by Ernst Mayr, which devoted only a single paragraph in almost one thousand pages to the topic of mass extinctions, and commented that, although catastrophist scenarios might seem appealing at first sight, they raised 'numerous unanswered questions'. Peter Bowler, in *Evolution – The History of an Idea*, similarly covered the subject in just one paragraph.[6] Furthermore, this came in a section headed, 'Exotic Alternatives', just before a brief discussion of the theories of Erich von Däniken, who had argued in *Chariots of the Gods* and other books that ancient civilisations had been created by aliens from space.[7] Even at this time, it was still far from certain whether mass extinctions were artifacts of the fossil record or real events.

Neocatastrophism: supernova explosions and impacts of cosmic bodies

In 1954, Otto Schindewolf, who had already provoked controversy by suggesting an important evolutionary role for macromutations (see Chapter 10), proposed that the major mass extinction event at the end of the Permian Period had been caused by waves of cosmic radiation resulting from nearby supernova explosions. Schindewolf had previously supposed that species, like individuals, have life-spans that are limited by intrinsic rather than external factors. However, eventually he became convinced that mass extinctions could not be explained on the basis of the coincidence of many species just happening to reach the end of their natural life-spans at a particular time. It was far more plausible to think that there must have been a common, external, cause.

Schindewolf's views could not be dismissed as those of an uninformed amateur. He was for many years Professor of Palaeontology at Tübingen University, Germany, and he had carried out detailed studies of the Permian–Triassic boundary in the Salt Range of Pakistan and elsewhere. His departure from the path of orthodoxy was not made lightly. When he put forward his supernova hypothesis, suggesting that waves of radiation were the cause not only of the mass extinction, but also of macromutations which could produce successors to the extinct species, he admitted that it was a 'desperate move', because he could not explain the discontinuities in the fossil record in any other way.[8]

In 1963, Schindewolf discussed the idea again in a paper entitled 'Neokatastrophismus?', but this had little impact, having to wait until 1977 before it appeared in an English translation.[9] Nevertheless, its title subsequently led to Schindewolf being strongly associated with the term 'neocatastrophism'. However, he had always been concerned about the use of the name (which was not his invention), thinking that it might suggest a link with the catastrophists of earlier times and their supposedly unscientific attitudes. As we saw in Section A of this book, these catastrophists were not, as a general rule, influenced by their religious beliefs to any greater extent than their more highly regarded contemporaries. However, whatever misconceptions there might have been about the past, no-one could claim that Schindewolf was driven by religious dogmatism, and that was also true, in general, of other twentieth century neocatastrophists.

The supernova hypothesis was accepted as a serious possibility by a few others, including H. Liniger of Basel, and the Canadians Dale Russell and Wallace Tucker. Liniger wrote (in German) in the journal, *Leben und Umwelt*, in 1961, 'The "sudden" geological extinction of the dinosaurs is a proven fact; in contrast to the abundance of fossils in the lower layers of the rocks, these are entirely missing in the next highest layer. It seems highly unlikely that local influences could at the same time lead to the extinction of both centimetre-long gastropods and huge reptiles. The only events that could explain the worldwide dying of animals, both on land and in the oceans, have to be extraterrestrial: solar eruptions or supernova explosions, with subsequent nuclear radiation, the close passage of a comet, or similar events, that leave no signs in the geological record, but make procreation impossible for certain families of animals'.[10] Similarly, in a 1971 paper in *Nature*, Russell and Tucker argued that the biostratigraphic record provided no compelling support for considering gradual terrestrial processes as the cause of the extinctions at the end of the Cretaceous Period. Instead, it suggested that 'the extinctions were of unusual magnitude and geologically brief duration, and may have been accompanied by a thermal drop'. Russell and Tucker pointed out that 'a nearby supernova explosion might produce climatic effects so drastic as to cause the extinction of many animals, including the dinosaurs'. However, such arguments had little effect on mainstream scientific opinion.[11]

As well as neocatastrophist scenarios involving supernova explosions, ones involving extraterrestrial impacts were also put forward. For example, geochemist Allan Kelly and amateur astronomer Frank Dachille argued in 1953, in their self-published book, *Target Earth*, that encounters with asteroids or comets might have caused extinctions, thus affecting the course of evolution. They produced evidence to suggest that the huge craters visible on the surface of the Moon had not been formed by volcanic activity, as generally supposed at the time, but were in fact impact structures. That gave an indication of what must have happened to our own planet. Hence, it was likely that, on several occasions, the Earth had suffered impacts by objects large enough to have changed the tilt of the Earth's axis, triggered a crustal displacement and caused widespread floods, earthquakes and outbreaks of vulcanism.[12]

Three years later, in 1956, the American palaeontologist Max de Laubenfels presented a similar view, albeit on a less apocalyptic scale, suggesting in *The Journal of Paleontology* that the extinction of the dinosaurs at the end of the Cretaceous Period could have been impact-related. He summarised his ideas as follows: 'Attention is called to the great destruction that resulted from a meteorite impact in Siberia in 1908. A larger impact would cause more widespread destruction. Several larger impacts may have occurred in geologic times. The survivals and extinctions at the close of the Cretaceous are such as might be expected to result from intensely hot winds such as would be generated by extra large meteoritic or planetesimal impacts'.[13] However, despite the evidence of the great explosion in the Tunguska region of Siberia in 1908, which could not be explained by any terrestrial mechanism, these impact scenarios received even less attention during the 1950s and 1960s than Schindewolf's supernova hypothesis.[14]

Whatever de Laubenfels may have believed, the cause of the Tunguska event (which we shall discuss in Chapter 20) was, at the time, still a long way from being established as a blast caused by the arrival of a cosmic object. The existence of comets was, of course, well known, but they did not seem to pose a physical threat to the Earth, as far as could be ascertained from two centuries of scientific observation. Two hundred years was not a long time in relation to the age of the Earth, but the reassuring conclusions from this brief period could easily be extrapolated in view of the prevailing uniformitarian paradigm. In 1910, concerns were expressed by some people about the forthcoming passage of the Earth through the tail of Halley's comet, but no adverse effects were observed when the event took place. As to other possible causes of impacts, there had been two or three fleeting observations of what appeared to be asteroids of not-insignificant size passing by at a close distance, but there was no certain knowledge of asteroids in Earth-crossing orbits until 1952, when Apollo was discovered by the Heidelberg astronomer Karl Reinmuth.[15]

An increasing number of explosion craters and cryptoexplosion structures, which were possibly remnants of ancient craters, had been identified at the surface of the Earth, and these were taken by some to be evidence of large impacts. From 1886 onwards, a great deal of attention was paid to the 1.2-km-diameter crater at Coon Butte in Arizona (figure 14.1). However, although some small metallic meteorite fragments had been found in the vicinity, giving the first main investigator, Grove Karl Gilbert, an initial inclination towards an impact origin, he eventually concluded that the crater

Figure 14.1   Barringer Crater, Arizona, a structure 1.2 km in diameter, produced less than 50,000 years ago by the impact of an asteroid with a diameter of no more than 100 metres (drawing based on a U.S. Geological Survey photograph).

had been formed by an explosion linked to volcanic activity. The main problem for the impact theory was that there was no evidence within the crater of a buried meteorite of appropriate size. Also, the crater was roughly circular (in fact, almost square, with rounded corners), whereas it was thought that an oval one would have been created by a missile arriving at any angle other than strictly vertical. Indeed, an oval depression 17 metres in length near Haviland, Kansas, was shown in 1915 to be an impact crater because it contained a meteorite weighing 39 kg, as well as several smaller fragments. It was not realised at this time that a larger projectile would maintain a fast speed through the atmosphere, and so explode on impact, producing a circular crater of much greater diameter than that of the impacting object, and leaving few meteoritic traces.[16]

Daniel Barringer, the next major investigator of the Arizona crater, concluded that it had been caused by an impact, even though he could produce little evidence to support this belief. The structure has been named after Barringer, but it is also commonly known as 'Meteor Crater'.[17] As well as this, other similar explosion craters (e.g. Odessa, Texas; Boxhole, Central Australia; and Wolf Creek, Western Australia) were investigated, together with larger-scale cryptoexplosion structures (such as the 24-km-diameter Ries Basin in southern Germany), but the causal mechanism could not be established for certain in any of them. For example, Walter Bucher thought that several American craters had been formed by the explosive release of gases from the Earth, whereas Claude Albritton and John Boon argued that they could just as easily have resulted from impacts.[18]

Eventually, around 1960, geologists such as Edward Chao, Robert Dietz and Eugene Shoemaker finally established that explosion craters and cryptoexplosion structures generally had an impact origin. They showed that the overall geology of the sites, together with points of detail such as 'shatter cones' in the rock, fitted the impact hypothesis far better than any other explanation. Furthermore, this conclusion was in line with the outcomes of simulation experiments, and also the presence of mineral forms such as shocked quartz grains, coesite and stishovite, whose formation required very high temperatures and pressures.[19]

Paradigm shift: mass extinctions, falling sea-levels and continental drift

Initially, the demonstration of the impact origin of terrestrial craters made very little impression on palaeontologists. Nevertheless, mass extinctions were pushed into greater prominence in 1962 when Norman Newell, of the American Museum of Natural History, made them the subject of his retiring lecture as President of the Paleontological Society. Newell argued that mass extinctions were real events which marked boundaries between stratigraphic units. Although they were certainly not linked to crustal upheavals and mountain building, as proposed in the long-discredited catastrophist model of Élie de Beaumont, there was also no indication that the orthodox Darwinian process of direct competition between species could be anything more than a contributory factor.[20]

Some investigators had suggested changes in atmospheric oxygen as a possible cause of mass extinctions, but there did not seem to be any significant correlation between extinctions of animals and changes involving oxygen-producing plants. The supernova hypothesis of Schindewolf also seemed unlikely to Newell, because sea creatures, which would have received more protection from radiation than land animals, were affected to a similar, and sometimes greater, extent. There was also a marked delay in the appearance of new forms after an extinction, which hardly suggested that these had been produced by radiation-induced macromutations. So, in Newell's opinion mass extinctions were probably caused by significant environmental changes resulting from regression of the seas away from the continents, since a clear correlation had been established between some palaeontological boundaries and widespread falls in sea-level. His address was eventually published as a paper in the *Journal of Paleontology*, together with a summary, which concluded:

> The most conspicuous paleontological breaks lie at the top of the Devonian, Permian, Triassic and Cretaceous systems. These and many lesser paleontological boundaries coincide with obscure paraconformities. This relationship suggests a universal physical control, such as eustatic changes in sea level. The boundaries of many paleontological zones are recognizable on two or more continents and, for all practical purposes, are the same age throughout their extent.[21]

By the time he came to write an article for *Scientific American* in 1963, Newell had identified two further times which were particularly critical in the history of animals, making the full list the ends of the Cambrian, Ordovician, Devonian, Permian, Triassic and Cretaceous Periods (see figure 5.2).[22] Four years later, in a Geological Society of America publication, he predicted that palaeontology would eventually incorporate some aspects of both catastrophism and uniformitarianism, whilst rejecting others. He thought that each mass extinction episode might have lasted anything between a few hundred and a few million years but, whatever the time-scale, falling sea-level still appeared to be the most likely cause of the extinctions. Concerning those at the end of the Mesozoic Era, it was known by this time that over a third of what is now land was under water during the Late Cretaceous Period whereas, by the start of the Tertiary, the first period of the Cenozoic Era, the seas had retreated significantly. As a consequence, the increased size of continental land-masses would have resulted in increased seasonality (i.e. hotter summers and colder winters), whereas the space available to marine organisms, particularly those living in shallow seas, would have shrunk. As an additional factor, new land connections would have been established when sea-levels fell, bringing previously isolated groups of animals into competition with others. So, for a variety of reasons, it was not surprising that extinctions took place at the end of the Cretaceous, and at other times when circumstances were similar.[23]

All that was missing was a mechanism for the fluctuations in sea-level, and one soon appeared, from developing ideas about continental drift. In 1915, the German meteorologist and geophysicist Alfred Wegener (1880–1930) had presented evidence which supported a suggestion made in the previous century, that the continents could change their positions relative to each other. Wegener showed that the continents could be fitted together, in a rough fashion, to produce a single supercontinent which he called Pangaea. However, over the next fifty years, very few people took the hypothesis of continental drift seriously. Even though it became increasingly clear that there had been occasions in the past when life-forms spread from continent to continent, over what seemed like formidable barriers of water, scientists preferred to explain this by developing theories of temporary land bridges, rather than consider the possibility that the continents had once been much closer together. After all, no-one could think of a plausible mechanism by which continents could change their positions by skating over or ploughing through the Earth's crust, although the Durham University geologist Arthur Holmes (1890–1965) argued in 1929 that convection currents in the mantle underneath the crust could assist the process.[24]

Then, in 1962, geologist Harry Hess (1906–1969), of Princeton University, proposed that the whole of the Earth's crust might be in motion, not just the continents. However, Hess was not suggesting that it moved as a single unit, as in the crustal displacement theory discussed in Chapter 12. That, of course, would not change the configuration of the continents. In his view, new crust was formed from rising lava in the mid-ocean ridges and, accompanied by the underlying upper mantle, it then spread out in both directions. As the amount of lithosphere (crust and upper mantle) did not appear to be increasing, the formation and spread of new surface material must be accompanied by an opposing process, subduction, which took place in deep-sea trenches. There, old lithosphere descended into the interior of the Earth.[25]

A year later, Fred Vine and Drummond Matthews, of Cambridge University, suggested that as new crust solidified in the vicinity of the mid-ocean ridges, the magnetic polarity existing at that time would be imprinted upon it. Crust formed after a geomagnetic reversal would have the opposite polarity from that formed before, so, if Hess's model was correct, stripes of alternating polarity should be seen in crust spreading out from the ridges. The same idea was expressed, independently, by Lawrence Morley of the Canadian Geological Survey, but without any great enthusiasm. In the absence of supporting evidence, even the originators of the theory remained sceptical.

However, the situation continued to develop. In 1965, Tuzo Wilson, a Canadian geophysicist, introduced the concept of plate tectonics, and this was taken further by others, including Jason Morgan, of Princeton. According to the plate tectonic hypothesis, the Earth's surface consisted of an interlocking group of rigid plates, each made up of the lithosphere between a mid-ocean ridge and a corresponding subduction zone. The continents, which were areas of thick, low-density crust, could be carried along by the moving plates, but they were safe from subduction, because of their lightness.

The crucial breakthrough towards the establishment of this theory came in 1966, when Walter Pitman and other graduate students at the Lamont Geological Observatory of Columbia University showed, as part of an unrelated study, that the profile of magnetic alignments across the Pacific–Antarctic Ridge unquestionably fitted the predictions of Vine, Matthews and Morley. As a result, continental drift, together with the linked concepts of sea-floor spreading and plate tectonics, quickly became assimilated into orthodox geological theory (figure 14.2).[26]

Three decades later, the eminent American geologist and science historian William Glen argued that the rapid acceptance of continental drift, many years after the idea was first proposed, illustrated how significant changes in scientific perspective often occurred. The details, in fact, supported philosopher Thomas Kuhn's view, expressed in his 1962 book, *The Structure of Scientific Revolutions*,[27] that a paradigm shift, i.e. an overwhelming switch in support from one major hypothesis to another, could be an extremely complex affair, involving fundamental changes in cognitive values.[28]

Despite the acknowledged brilliance of the great names of science, the pursuit of scientific truth was supposed to be straightforward and objective. According to the accepted principles of the scientific method, hypotheses were formed on the basis of observation, and then experiments, capable of being repeated by others, were devised to test whether or not each hypothesis was valid. Indeed, Sir Karl Popper, in his frequently referred-to 1934 book, *The Logic of Scientific Discovery* (which appeared in an English translation in 1959),[29] argued that a hypothesis could only be regarded as a scientific one if it was capable of being tested and shown to be false, should that be the case. Even if a hypothesis was found to be consistent with the results of these initial experiments, the situation ought to be kept under constant review, as fresh information came to light. Thus, science should be in a constant state of flux.

Kuhn, however, believed that paradigms had a far more constraining influence on thought than generally realised, resulting in long periods of stasis. Fine details might be modified on a regular basis, but the larger picture remained essentially the same. Whilst there could be valid reasons for this, it seemed that another relevant factor was human nature. Scientists, after all, were not machines but human beings, with similar passions, hopes and ambitions to other people. Therefore, commitment to a

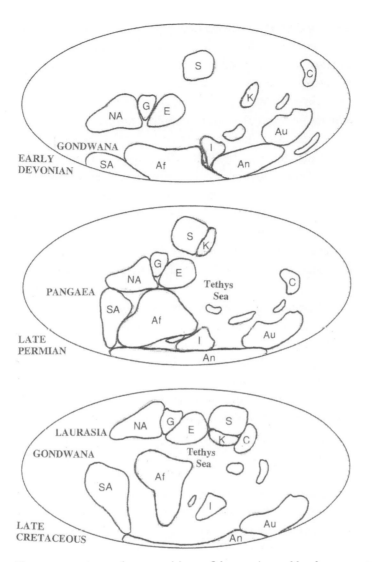

Figure 14.2   Approximate positions of the continental land-masses at various times in the past, according to the generally accepted theory of continental drift. Af = Africa, An = Antarctica, Au = Australia, C = China, E = Europe, G = Greenland, I = India, K = Kazakhstan, NA = North America, SA = South America, S = Siberia. It should be appreciated, however, that significant parts of the continental land-masses would have been covered by water.

particular point of view often went beyond an impartial assessment of the evidence. For reasons which could owe as much to prejudice as dispassionate logic, scientific paradigms tended to determine which issues could legitimately be investigated, which methods of investigation were acceptable, and which observations were most reliable or significant. Thus, anomalous facts could be disregarded, until a large number had accumulated. Then, some scientists (generally, ones of the younger generation) would start to develop an alternative paradigm which, should it provide a better explanation for the totality of facts than the existing one, would eventually replace it.[30]

However, this would not necessarily be a rapid process. As Max Planck, the celebrated German physicist, wrote in his autobiography, shortly before his death in 1947, 'A new scientific truth does not triumph by convincing its opponents and making them see the light, but rather because its opponents eventually die, and a new generation grows up that is familiar with it'.[31] That was certainly true in the case of evolution, following the Darwinian revolution. Even when scientists have undergone a change of mind, as a result of new information coming to light, they have often been reluctant to acknowledge that they have done so. After all, human beings in general rarely like to admit in public that they have made mistakes.

In the specific case of the acceptance of continental drift, a survey carried out by Glen himself in the 1970s revealed frequent inconsistencies between respondents' accounts of their intellectual postures prior to the paradigm change, and what others had told him about the stances taken by those same respondents.[32]

Of course, no paradigm change could take place without the finding of evidence to support the new theory, for, without that, not even the most brilliant ideas could ever gain acceptance. Nevertheless, this could happen in a variety of ways. As Glen pointed out, 'In the case of drift, conversionary proof came not from those like Wilson who were addressing the drift question, but instead from graduate students at Lamont Observatory who *inadvertently* found the magnetic data that served instantly to confirm seafloor-spreading theory in 1966'.[33]

In this way, by a combination of accident and perception, the concept of sea-floor spreading became established during the 1960s. However, there were reasons to think that sea-floors did not always spread at a constant rate. When tectonic activity was at its greatest, mid-ocean ridges would grow, causing sea-levels to rise, whereas when there was little activity, mid-ocean ridges would subside, leading to falls in sea-level. Therefore if, as Newell claimed, mass extinctions were associated with regressions of the sea, they must have taken place at times when there was very little sea-floor spreading. That clearly separated extinctions from episodes of mountain-building (orogeny), for these would be expected to be associated with increased tectonic activity, when sea-levels were high. At times of low sea-level, when there was minimal tectonic activity, erosion could actually have caused a lowering of mountain ranges, allowing groups of animals to move out of isolation and into competition with each other.[34] This would, of course, have been a slow-acting process, just like the others. So, it was apparent that, by the use of arguments such as these, mass extinctions could easily be fitted into a scenario which was essentially gradualistic.

Moreover, there were reasons for thinking that the biggest mass extinction episode of all, that at the end of the Permian Period, took place at a time when sea-floor spreading had slowed down or even stopped. Indeed, it coincided with a unique event, the merging of all the individual continents (which had split apart over 300 Myr previously) into the supercontinent, Pangaea (see figure 14.2). As this happened, the plates carrying the various continents must have locked together, preventing sea-floor spreading. Because of this, there would have been a reduction in sea-level around the continents, and a draining of flooded lowland regions, the so-called epicontinental seas. Furthermore, as the continents merged, water would have been eliminated from between them, leading to a considerable decrease in the overall length of coastline. At the end of the Permian, therefore, for both of these reasons, there would have been a significant reduction in the total area of shallow seas. In consequence, these would no longer have been able to support all the life they formerly contained, so extinction was inevitable for many species.[35]

However, the sea was not the only place where mass extinctions occurred during the Late Permian. Freshwater species survived with little damage, but large land animals, including the dominant group, the mammal-like reptiles called therapsids, suffered major losses. Many families of therapsids died out, including the carnivorous gorgonopsians and herbivores such as aulacephalodonts. The plate tectonic model, as we have seen, explained this on the supposition of a decreased environmental diversity in the continents, caused by the draining of the epicontinental seas and the possible erosion of mountain ranges. Whilst the total area available to land animals would have increased, there would have been fewer geographical barriers and a reduction in the diversity of habitats. Also, seasonal extremes of climate would have been magnified at the end of the Permian Period because of the formation of the supercontinent, Pangaea. As if that was not enough, this supercontinent stretched almost from pole to pole, drifting northwards, and, during the course of its movement, there were significant climatic fluctuations. During most of the Permian Period and the preceding Carboniferous, Antarctica, the most southerly of the land-masses which became incorporated into Pangaea, lay over the South Pole, as it does at the present day, facilitating the formation of a large ice-cap in that part of the world. However, the northward drift of Pangaea eventually carried Antarctica clear of the polar regions, initiating a slow thaw in the south. In contrast, the approach of the opposite tip of the supercontinent to the North Pole brought about an intense spell of glaciation in the northern hemisphere.[36]

So, purely on the basis of known tectonic movements, the Late Permian Period must have been a time of major climatic and environmental changes. These would have presented a great challenge to existing animal life.

Arguments for the involvement of asteroids or comets

Without question, plate tectonics provided a plausible explanation for 'the great dying' at the end of the Permian Period. Indeed, fossil evidence suggested that extinctions took place at different times at different locations, consistent with a scenario involving slow but significant changes of habitat and fluctuations of climate.

Figure 14.3   Left: René Gallant. Right: Sir Fred Hoyle.

However, that is not to say that tectonics was established as being sufficient to account for all the events of the Late Permian, let alone the mass extinctions at other times. Although discussions generally focused on gradualistic, Earth-centred mechanisms, there were still a few people who insisted that there was no reason to exclude the possibility of neocatastrophist explanations which involved extraterrestrial factors.

One of these neocatastrophists was the Estonian astronomer Ernst Öpik (1893–1985) who, in 1958, when he was Director of the Armagh Observatory in Northern Ireland, pointed out that a cometary impact could cause devastation on a large enough scale to give rise to mass extinctions of species, even though the geological effects might not be apparent throughout the world.[37] Another was René Gallant (see Chapter 12) who, writing in 1964 in *Bombarded Earth*, commented, 'It is not logical to reason that, because no world-wide catastrophes have taken place during the last 2,600 years (the period for which we have reasonably reliable records), they cannot have occurred earlier'. Gallant (figure 14.3) drew attention to the fact that the Barringer crater was, by this time, established as an impact crater, and (like Allan Kelly and Frank Dachille, mentioned above) pointed to much larger circular structures on both Earth and the Moon, giving reasons why these might have had a similar origin. Gallant went on to argue that discontinuities in the geological record could be genuine features, resulting from earthquakes, vulcanism, climatic changes and extinctions caused by impacts of large meteorites. He continued, 'When cosmic catastrophes occur, the whole world is in violent upheaval. Terrific tidal waves may wash away existing sedimentary layers and deposit new ones. The species which are wiped out may be massively and suddenly embedded in these new layers, thus ensuring their fossilization'.[38]

However, almost no-one paid any attention to the views of Öpik and Gallant. Interest in the possibility of impact-related catastrophes had certainly not increased in the few years since de Laubenfels had tried to argue along similar lines. No specific evidence to link mass extinctions to impacts had been produced and, rightly or wrongly, claims of major catastrophes were still associated in the minds of most scientists with biblical fundamentalism, poetic fantasy, uninformed speculation and irrational beliefs about Atlantis.

The next event of note came in 1970 when Digby McLaren, of the Geological Survey of Canada, followed the example of Norman Newell in making use of a Presidential Address to the Paleontological Society to draw attention to mass extinctions. McLaren argued that it was difficult not to come to the conclusion that 'something happened' at certain geological boundaries. One of these was between the final two stratigraphic stages of the Devonian Period, the Frasnian and the Famennian. The great Late Devonian mass extinction episode actually occurred at this time, not at the very end of the Period. Communities living in shallow seas were particularly affected, with many species of brachiopods, corals and trilobites disappearing, to be replaced by unrelated groups. Newell's favoured mechanism for mass extinctions, falling sea-level, did not seem to be the answer on this occasion, because there was evidence that seas had risen around the time of the Frasnian–Famennian transition. On the other hand, there was no reason to implicate a massive burst of radiation, as in Schindewolf's supernova hypothesis. Nevertheless, it appeared to McLaren that the phenomenon was 'too sudden to allow any but a catastrophic interpretation', so he suggested that a large meteorite may have landed in the Pacific Ocean at the close of the Frasnian, exploding with a great release of energy. As he explained, he had 'become increasingly convinced that we must look for more than everyday happenings to explain many geological features'. Moreover, he did not think that the Frasnian–Famennian extinction episode was unique in having a catastrophic origin, going on to write of his belief that 'the type of boundary afforded by a sudden extinction is one of the most valuable geological tools we possess, and that there are others in the geological column that have been caused catastrophically, although not necessarily by the same direct cause'. He concluded, 'Thus a major impact on continental areas could trigger tectonic or geomagnetic changes which could have enormous secondary effects on life'.[39]

Despite McLaren's status, his address failed to stimulate any debate about the involvement of impacts and other catastrophes in the course of evolution. Gradualism was so firmly established that there seemed no point in considering alternatives, either on this occasion, or again when another eminent scientist, Harold Urey (1893–1981), winner of a Nobel Prize in Chemistry, wrote a paper for *Nature* which suggested that cometary impacts could be responsible for the termination of geological periods. Harvey Nininger, an American meteorite expert, had made a similar suggestion (involving asteroids rather than comets) many years earlier, in 1942, but without producing any evidence to support it.[40] In contrast, Urey pointed out that the generally accepted dates for the stratigraphic boundaries over the last 36 Myr (from the Eocene to the

Pleistocene Epoch) approximated to the ages of various groups of tektites or smaller microtektites (glassy beads of fused rock, which were likely products of Earth–comet collisions). Urey also predicted that similar findings would eventually be made at the Cretaceous–Tertiary boundary, dated at approximately 65 Myr before the present.[41] However, the evidence he produced was insufficient to make anyone think what was then considered unthinkable.[42]

Again, there was no significant reaction when the eminent British astrophysicist Sir Fred Hoyle (1915–2001) (see figure 14.3), together with Chandra Wickramasinghe of University College, Cardiff, suggested in 1978 that the extinctions at the end of the Cretaceous Period had been associated with the close passage of a large comet, during which the Earth acquired a veil of dust which dimmed sunlight for perhaps 100,000 years.[43] The efforts of Hoyle and Wickramasinghe, like those of Urey, McLaren, Öpik, Gallant and de Laubenfels, made little impression against the all-powerful gradualistic paradigm.[44]

Yet, whilst attracting negligible support, it became increasingly apparent during the 1970s that neocatastrophism, introducing scenarios of events of cosmic origin, provided a genuine alternative to the gradualistic view of the Earth's history, just as the catastrophism of Cuvier and Whewell had done during the nineteenth century. MacLaren, in his 1970 paper, did not question the prevailing notion that geology had been 'liberated as a science by Hutton and Lyell', but he continued, 'Following the overthrow of catastrophism, however, there has been a natural tendency to over-compensate and to avoid catastrophic interpretations even when the evidence called for one'.[45]

Since neither catastrophism nor neocatastrophism was dependent upon any particular model, the only reason for distinguishing between them was the belief that catastrophism was derived from dogma rather than science. In fact, as we have discussed, nineteenth century catastrophism was not inspired by religious beliefs to any greater extent than the uniformitarianism of the same period and, as this became increasingly clear to historians of science, it could be seen by those who were aware of the fact that the distinction between neocatastrophism and catastrophism served no useful purpose. As a result, the two terms eventually began to be used in an interchangeable fashion in the context of twentieth century ideas.

Mass extinctions and the Modern Synthesis

Despite the arguments put forward by its supporters, catastrophism, even if called neocatastrophism, made negligible progress towards general acceptance during the 1960s and 1970s. Books written at this time by the founders of the Modern Synthesis generally included some reference to events at the end of geological eras. However, the authors avoided direct use of the expression 'mass extinctions' and stressed gradualistic aspects, whilst rejecting catastrophist theories. So, for example, in his 1970 book, *Populations, Species and Evolution*, Ernst Mayr, whilst admitting that some parts of geological periods 'have witnessed far more extinctions than others', then went on to write:

> Cosmic events, like the passing of the earth through a radioactive cloud are sometimes invoked as explanations, but the fact that the great periods of faunal turnover on land and in the seas do not coincide deprives the cosmic theories of all probability. Climatic events, orogeny, and other geophysical processes that produce fluctuations of the sea level on the continental shelves are far more likely causes.

In traditional fashion, Mayr maintained that individual species became extinct because of 'an inability of their genotype to respond to new selection pressures'. Even so, he was quite prepared to acknowledge the role played by environmental factors, for it seemed clear that each major episode of extinction coincided with a major environmental upheaval. Nevertheless, Mayr remained convinced that such upheavals must have taken place over long periods of time.[46]

Julian Huxley, in the 1974 edition of *Evolution – The Modern Synthesis*, accepted that there had been occasions when 'some groups once dominant have become wholly extinguished', and that 'the rise of the new type and the downfall of the old was accompanied and facilitated by world-wide climate change'.[47] However, competition remained a major feature throughout. Concerning the events at the termination of the Cretaceous Period (when the Mesozoic Era also came to an end), Huxley wrote, 'The worsening of the climate at the end of the Mesozoic reduced the general adaptiveness of the dinosaurs, pterosaurs, and other reptilian groups, while increasing that of the early mammals and birds'.[48]

A similar view was taken by Theodosius Dobzhansky, Francisco Ayala, Ledyard Stebbins and James Valentine, in their 1977 book, *Evolution*. They acknowledged the existence of a number of extinction episodes, without appearing to see anything special about them, in the following passage:

> There are a number of major waves of extinctions, each commonly followed closely by a wave of diversification. The Cambrian extinction wave is even accompanied by a high rate of diversification. Obviously, an excess of extinction over diversification lowers the diversity level, just as the opposite combination raises it. The great wave of extinction recorded in the late Permian is neither accompanied nor closely followed by significant recorded diversification, so that early Triassic diversity is low.[49]

Dobzhansky and his colleagues outlined the various aspects of the tectonic explanation for what happened at the end of the Permian Period, which we have already considered, and concluded that these 'well-established historical events' would have led to multiple extinctions. In their view, the immediate cause of extinction would have varied from lineage to lineage, generally involving different outcomes of competition as habitats changed.[50] Similarly, for events at the end of the Cretaceous Period, they attributed the extinction of the dinosaurs at that time to a climatic change and ensuing secondary effects.[51]

Again, in *How and Why*, published in 1980, George Gaylord Simpson gave a brief description of the 'turnovers' which occurred in the Late Devonian, Late Permian and Late Cretaceous Periods (see figure 5.2), and continued:

> It should . . . be emphasized that these episodes are of long continuation and that they are not sharply defined in time, even as to their beginnings. It has, for instance, already been stressed that the essential features of the Cretaceous–Tertiary crisis cannot really be localized just as the boundary between those periods. The episodes are part of a long and essentially continuous process.

As to the causes of the extinctions, Simpson thought it was reasonable to suppose that physical factors had been involved, but he warned, 'The conclusion that theirs was the decisive and necessary part is not supported by the biological evidence and may be gratuitous'.[52]

For yet another example of the views prevalent at the time, let us note Ledyard Stebbins' *Darwin to DNA, Molecules to Humanity*, which was published in 1982, but maintained the attitudes of the late 1970s. The closest this book came to a discussion of mass extinctions was when Stebbins acknowledged that evolution had proceded faster on some occasions than others, and continued:

> Most evolution, particularly of strikingly new adaptive types, occurs in quantum bursts that are triggered by challenges of a changing physical and biotic environment. When such a challenge occurs, the populations exposed to it respond in one of three ways. Most populations become extinct; some adjust to the new environment with minimal change in their hereditary makeup and thus persist with little evolution over millions of years; a few populations respond by evolving entirely new adaptive mechanisms.[53]

So, according to the founders of the Modern Synthesis, the same processes had operated, in the same way, throughout the course of evolution. As conditions gradually fluctuated, it was inevitable that there would have been times when the turnover of species was greater than at other times. That was all there was to mass extinctions. There had been no significant jolts to ongoing processes by external factors.

15 Punctuated equilibrium: a new evolutionary perspective

Uncertainties about evolutionary mechanisms

Even if the founders of the Modern Synthesis were wrong, and catastrophist mechanisms were indeed involved as causal agents of mass extinction episodes, there was no reason to think that 'sudden, violent and unusual events' (to quote from the *Oxford English Reference Dictionary* definition of catastrophism[1]) were common in the long intervals between them. Hence, such processes were unlikely to have exerted a continuous influence on the course of evolution. Indeed, if catastrophist mechanisms had played a part, an episodic rather than continuous involvement would be expected. The history of the Earth, like the proverbial life of a soldier, might have consisted of long periods of boredom interspersed by short episodes of terror.[2] However, the *catastrophe theory* of the French mathematician René Thom could have been of more general relevance. As envisaged by Thom, a system placed under stress would tend to resist change, provided that its components were arranged in a way which was stable, but if the stress was gradually increased, there might be a sudden adjustment to a different stable arrangement of components, without the appearance of any less-stable intermediate forms.[3] That could have been the case in evolution, with 'rapid' transitions from one species to another.

As the mechanisms of molecular genetics were elucidated during the 1950s, and shown to operate on the principle that each gene carries in its structure the information for the synthesis of a specific protein, it was seen that they were entirely consistent with Darwinism (see Chapter 9). However, further discoveries, whilst not changing that conclusion, appeared to complicate matters. An example was the identification of jumping genes (transposons) which, for unexplained reasons, could change their position on a chromosome. Also, it was discovered that eukaryotic cells contained large sequences (termed 'introns') which were never expressed, in addition to those (the 'exons') which were translated into proteins. Furthermore, after a protein had been synthesised, it could then become involved in other aspects of molecular genetics. The protein might, for example, be an enzyme which could catalyse a particular chemical reaction, or it might regulate the action of another gene.[4] As this information was assimilated, it became clear that the primary expression of each gene was simply one stage in a complex, interacting system essential for the existence and functioning of that organism. The external features of an organism were a long way removed from the expression of individual genes, for there was no obvious link between a particular part of the body (e.g. a leg or an eye) and any single gene.[5] Molecular genetics was far more complicated than had been realised and, to some extent, all that evolutionary biologists could do to account for some of its overall aspects was to speculate on the basis of fragmentary knowledge. Because of the nature of the systems under study, firm evidence was hard to acquire.

It was still uncertain whether overall evolutionary trends were determined by natural selection acting primarily at the level of species or of individuals, or even at some other level. In his 1976 book, *The Selfish Gene*, and again in *The Extended Phenotype*, published in 1982, Richard Dawkins, Reader in Zoology at Oxford University (later to be awarded a chair for the Public Understanding of Science), argued that it was appropriate to regard the gene as the actual unit of selection, with the individual serving as the gene's 'survival machine'. That would explain, for example, why animals (including humans) were often prepared to die to save the life of a younger blood relative, thus ensuring the continued survival of many genes they had in common. However, to Dawkins, focusing on the importance of the gene was only a shift of perspective, for he continued to go along with the mainstream followers of the Modern Synthesis in thinking that natural selection took place at the level of the individual. In this view, those mutated genes which conferred some advantage on an individual (such as black pigmentation in moths when trees were dark with industrial pollution) were likely to accumulate within a species over successive generations, whilst those mutations which were disadvantageous would be eliminated. Big changes would arise in the same way as small ones, but over a longer period of time, with the accumulation of a variety of small changes.[6] Provided there was a large genetic mix to start with, and a time-scale long enough for many new mutations to arise in different individuals, it was perfectly reasonable to suppose that mutations with advantageous, novel consequences could occur and spread through a population. At the time, most biologists accepted that evolution took place in precisely this way.

Nevertheless, a few believed that macroevolution was *not* simply a continuation of microevolution, but involved some extra factors. There was very little evidence from living organisms as to how macroevolution might occur and, impressive though advances in molecular genetics had been in the 1950s, the molecular basis of speciation was still a matter of supposition rather than certainty. It remained perfectly possible that the speciation process might involve chromosomal rearrangements or mutations of regulatory genes, rather than just the continued accumulation of mutations in ordinary structural genes.[7]

And even if macroevolution really was just a continuation of microevolution, the fossil record might not necessarily show evidence of even-paced progression, regardless of its supposed imperfections. As the palaeontologist Hugh Falconer suggested to Darwin, on the basis of a study of fossil evidence which he completed in 1863, and which revealed little change in characteristics over long periods of time, the important steps in evolution could take place rapidly in small populations, and so leave few traces.[8] Almost a century later, in 1942, Ernst Mayr argued that a sub-group which became completely separated from the main population would most likely have a different genetic mix, since the full range of the species' genes would not be represented in the 'founder population' of the small group. This could have major implications for the subsequent evolution of the two populations. The isolated sub-group might evolve differently from the main group, particularly if it occupied a quite distinct habitat, and could soon give rise to a new species.[9] This process was termed *allopatric speciation* (meaning 'speciation in a different place'). Alternatively, if the new species was regarded as having 'budded' from the main species as a result of peripheral isolation, it could be called *peripatric speciation*. Regardless of terminology, Mayr pointed out in the 1954 multi-author work *Evolution as a Process* that if the new species thrived and quickly expanded its territory, its appearance in the fossil record would be an abrupt one. As he put it, 'Rapidly evolving peripherally isolated populations may be the place of origin of many evolutionary novelties. Their isolation and relatively small size may explain phenomena of rapid evolution and lack of documentation in the fossil record hitherto puzzling to the paleontologist'.[10]

Mayr was quick to point out that 'rapid' in this context did *not* mean instantaneous, but a time-scale of thousands of years, which was a very short period in geological terms. Nevertheless, despite Mayr's eminence, this aspect of his thinking was largely ignored by other supporters of the Modern Synthesis, who continued to stress that all evolutionary change must take place by small increments in an even-paced fashion. Indeed, much effort was put into trying to find examples of this in the fossil record. Many believed that some gradual transitions had already been documented, particularly in the evolution of the Cretaceous sea-urchin, *Microcaster*, the Jurassic oyster, *Gryphaea*, and the horse, from *Hyracotherium* in the Eocene Epoch to the present-day *Equus*.[11] However, if a few more clear-cut examples of gradualistic evolution could be demonstrated, then that should surely be enough to bring into line the few biologists who still had doubts about the Modern Synthesis.

Eldredge and Gould

Of the attitude of palaeontologists in the 1950s and 1960s, one of their number, Niles Eldredge, of the American Museum of Natural History, New York, wrote many years later, in his 1985 book, *Time Frames*: 'we have proffered a collective tacit acceptance of the story of gradual adaptive change, a story that strengthened and became even more entrenched as the synthesis took hold. We paleontologists have said that the history of life supports that interpretation, all the while really knowing that it does not'.[12]

In that period, the many stretches of the record where nothing much seemed to happen were generally dismissed as being unimportant. However, during the 1970s, together with the Harvard University palaeontologist Stephen Jay Gould (1941–2002),

Figure 15.1   Left: Niles Eldredge. Right: Stephen Jay Gould.

Eldredge began to point out that 'stasis is data.' In 1972, Eldredge and Gould (figure 15.1), both of whom were former postgraduate students of Norman Newell, ar-gued that the fossil record had provided little evidence of phyletic gradualism, and the examples generally cited (see above) were unconvincing. Instead, taken as a whole, fossil evidence suggested that the dominating feature of evolution was not phyletic gradualism but what they termed punctuated equilibrium (figure 15.2). In their view, species tended to appear in a very short period of time (in geological terms) and then, during their subsequent history, they either did not change in any appreciable way, or showed a slight fluctuation in their characteristics. Finally, species tended to disappear rapidly, and be replaced by others, sometimes in just one location and sometimes more generally.[13]

    In the case of Eldredge, this view emerged from an investigation of Devonian trilobites (figure 15.3). In *Time Frames*, he described how he had been influenced by the observation of a sudden transition some 380 Myr ago between the trilobite species *Phacops milleri* and *Phacops rana* in the midwest of the U.S.A. These trilobites lived in a shallow epicontinental sea and were probably members of the same lineage. *Phacops milleri* (with 18 columns of lenses in its eyes) has only been found in rocks of the Hamilton Group corresponding to the first half of an 8 Myr period, and *Phacops rana* (with 17 columns of lenses) in the second half, separated by a short interval when the epicontinental sea dried up completely. To the east, where the epicontinental sea had persisted throughout the 8 Myr period, and the transition from *Phacops milleri* to *Phacops rana* occurred slightly earlier than in the midwest, Eldredge found a quarry which appeared to reveal the transition itself. During a period of 1,000–10,000 years, some trilobites had 18 columns of eyes, some 17 columns, and there were also some

Figure 15.2   The process of evolution, according to (left) phyletic gradualism, and (right) punctuated equilibrium. The passing of time is indicated by the vertical axis and morphological change by the horizontal one.

intermediate forms with varying numbers of lenses in the 18th column. Eldredge saw this as a clear example of the occurrence of allopatric speciation, as outlined by Mayr. The speciation event giving rise to *Phacops rana* had occurred, in rapid fashion, in the east, whilst *Phacops milleri* continued to prevail in the midwest, until it was destroyed by the disappearance of its habitat. So, when the shallow seas returned to cover the midwest, *Phacops rana* was able to extend its territory rapidly, in the absence of any competing species.[14]

Gould came to similar punctuational conclusions when he investigated fossils of the Bermudan Land Snail, *Poecilozonites bermudensis*, and found there had been

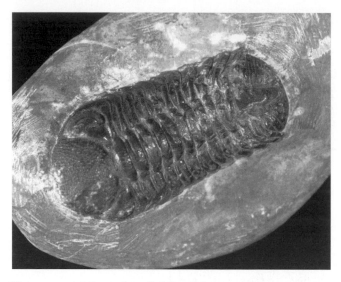

Figure 15.3 A Devonian trilobite of the genus *Phacops*, about 400 Myr old, from the Moroccan Sahara.

abrupt transitions from sub-species to sub-species during the Pleistocene Epoch. Once again, allopatric speciation seemed a likely explanation.[15]

However, Eldredge and Gould departed from the traditional Darwinian view in more ways than one. As well as seeing the transition from one species to an-other in the fossil record as being much more suggestive of punctuated equilibrium than phyletic gradualism, they also found very little evidence that forms diverged in a slow and gradual fashion. Darwin and the founders of the Modern Synthesis had maintained that the 'tree of life' gradually developed more and more branches (i.e. a greater diversity of organisms would be found as evolution progressed), yet it seemed clear to Eldredge and Gould that the 'tree' had been highly branched at a very early stage of its growth. They noted, for example, that more than twenty classes of echinoderms (sea urchins, starfish, etc.) were present during the Ordovician Period, over 500 million years ago, all but five of which disappeared during the succeeding Silurian and Devonian Periods.[16]

Yet another controversial suggestion from Eldredge and Gould was that macroevolution might be uncoupled from microevolution, with speciation being far more important than previously realised. As they saw it, new species might arise in ways not directly related to adaptive changes taking place within populations, and then have to compete for survival with other species, this process determining overall evolutionary trends. In 1980, Gould argued in an article in the journal *Paleobiology* entitled, 'Is a new and general theory of evolution emerging?', that recent develop-ments in molecular genetics made it unnecessary to think that macroevolution must be a straightforward continuation of microevolution. He suggested that a hierarchical system might exist, with the translation of nucleic acid base sequences into protein being a low-level process, and control mechanisms operating at a higher level. Certainly, new knowledge about jumping genes and insertion elements demonstrated that the genome operated in a much more complex fashion than had been suspected. It was now clear that gene sequences could be introduced, removed or transferred, and not simply translated. In any case, as Gould pointed out, gene sequences corresponding to structures of proteins generally formed less than half of the mammalian gene, and the function of the rest was uncertain. It was therefore quite possible that 'structural gene substitutions control most small-scale, adaptive variation within local populations, while disruption of regulation lies behind most key innovations in macroevolution'.[17]

As Eldredge and Gould were happy to acknowledge, very few of their ideas were totally original. In the late eighteenth century, Erasmus Darwin suggested that evolutionary progress would be made at times of environmental change, but not during periods of stability. His grandson Charles Darwin initially took a similar punctuational view, as we discussed in Chapter 7, but abandoned it before presenting his theory to the public in the *Origin of Species*. Even then, in later editions of the *Origin*, whilst continuing to stress gradualism and the imperfections of the fossil record, Darwin acknowledged the work of Falconer (mentioned above) and commented that 'the period during which each species underwent modification, though long as measured by years, was probably short in comparison with that during which it remained without undergoing any change'. More recently, as we noted in Chapter 9, Sewall Wright suggested that genetic drift, as well as natural selection, could contribute to evolution in small populations. Then, in the present chapter, we saw how, according to Ernst Mayr's 'founder principle', the small number of individuals in a group which became isolated from the rest could carry only a fraction of the total genetic variation of the species, so might evolve in a different fashion from the main population. From this, Mayr developed his theory of allopatric (or peripatric) speciation, arguing that new species could be formed relatively quickly in small, isolated groups by ordinary adaptive processes, largely independently of trends in the main populations. Later, in 1967, as a contribution to the multi-author book *Mathematical Challenges to the Neo-Darwinian Interpretation of Evolution*, Wright suggested that, just as mutations are unrelated to trends in microevolution, so speciation might be largely unrelated to trends in macroevolution.[18]

For that reason, Gould, in the 1980 *Paleobiology* article, gave the name 'Wright break' to what he saw as the discontinuity between speciation and overall evolutionary trends. Mayr had suggested in 1963, in *Animal Species and Evolution*, and again in its abridged version, *Populations, Species and Evolution*, published in 1970, that selection at the species level could be of great importance in macroevolution. However, this assumed that species had a historical as well as a geographical dimension, which most evolutionary biologists thought unlikely.[19] During the 1970s, the biologist and historian Michael Ghiselin and the philosopher David Hull, then at the Universities of Utah and Wisconsin, respectively, both argued that species were distinct entities in time as well as in space, but that view still remained a minority one.[20]

The theories of Eldredge and Gould may have been based on ideas previously put forward by respected biologists, but these proposals were not, in general, ones that had become part of mainstream thinking.[21] In any case, punctuated equilibrium clearly challenged long-established beliefs about the evolutionary process. Although it could not be said to be anti-Darwinian, for natural selection still played a major role, it presented a different view of the history of life from the gradual, continuous process envisaged by Darwin and the followers of the Modern Synthesis. For these reasons, the hostile reception given to punctuated equilibrium was predictable.

A different way of seeing

Since aspects of punctuated equilibrium had been derived from Mayr's own writings, it was hardly surprising that the eminent biologist offered some support to the theory. In *The Growth of Biological Thought* he acknowledged that his theory of allopatric speciation had been 'ignored by the paleontologists until used by Eldredge and Gould.' However, in one respect he believed that these two had views fundamentally different from his own, writing, 'They maintain that punctuated equilibria are produced by discontinuities of such size that they correspond to Goldschmidt's hopeful monsters'.[22]

Eldredge described this as 'the most seditious and irksome misconstrual of our ideas'.[23] In *Time Frames*, he explained that, in contrast to Goldschmidt (see Chapter 10), neither he nor Gould ever thought that the apparently abrupt appearance of a new species in the geological record was anything like instantaneous in real terms, writing, '5,000 to 50,000 years became the comfortable yardstick, compatible with the admittedly rather coarse level of resolution in our own paleontological data, and compatible as well with the time scales required in most theories of the speciation process'. Also, in answer to other critics, who tried to make out that there were only superficial differences between the ideas of Eldredge and Gould and those of Darwin and his followers, Eldredge pointed out that 'Darwin had effectively said that adaptive change was the origin of species. No one had ever claimed just the opposite – that most (if not all) anatomical change in evolution, adaptive though it may be, happens not throughout the bulk of a species' history, but rather at those rare events when a new reproductively isolated species buds off from the parental species'.[24] Thus, there was a fundamental difference between punctuated equilibrium and traditional Darwinism.

Eldredge's collaborator, Gould, achieved more general fame through his essays on evolutionary topics which appeared every month in the journal *Natural History* between 1974 and 2001, and were collected into a series of ten books, starting with

Ever Since Darwin (1978) and ending with *I Have Landed* (2002). In one of these essays, which was included in the 1980 book, *The Panda's Thumb*, Gould predicted that 'Goldschmidt will be largely vindicated in the world of evolutionary biology'. However, that was *not* because of any likelihood that new species could arise overnight, for Gould left his readers in no doubt that he would find 'insuperable difficulties' in advocating evolution by sudden macromutations. Instead, he argued that the supporters of the Modern Synthesis had caricatured Goldschmidt's ideas by maintaining that his theory consisted of nothing but the 'impossible' idea of sudden transitions arising from macromutations. The reality was that Goldschmidt had also made a study of embryonic development, and found that small differences in the action of 'rate genes' early in growth could result later on in major changes to the adult animal. In *The Material Basis of Evolution*, Goldschmidt had specifically invoked rate (or, as we would say, regulatory) genes as a potential maker of hopeful monsters. As Gould went on to argue, if a significant modification to the adult form arose from a small genetic change, then problems of genetic incompatibility with other members of the species would not arise, and the advantageous characteristic could spread through the population in Darwinian fashion.'[25]

Gould's 1977 book, *Ontogeny and Phylogeny*, had been, by his own admission, a 'long argument for the evolutionary importance of *heterochrony* – changes in the relative time of appearance and rate of development for characters already present in ancestors'. That in itself, he pointed out, could explain some of the relatively abrupt changes seen in the fossil record.[26]

Gould examined other aspects of Goldschmidt's writings in his 1980 *Paleobiology* article, noting that, contrary to his expectation, he had found that the heretical geneticist formulated his theories on the basis of a considerable knowledge of geographic variation. Gould continued, 'Goldschmidt concludes that geographic variation is ubiquitous, adaptive, and essential for the persistence of established species. But it is simply not the stuff of speciation; it is a different process'. Whilst Gould could not accept everything that Goldschmidt had written, he was convinced that the latter had been correct to stress the discontinuity between microevolution and speciation, and Gould termed this the 'Goldschmidt break'.[27]

By the time of this article, another palaeontologist, Steven Stanley, had, for several years, being arguing along similar lines to Eldredge and Gould, maintaining that evolution was punctuational, and denying that macroevolution was simply an extension of microevolution. Stanley had arrived at his punctuational views largely through a detailed study of the evolution of bivalve molluscs, such as clams.[28]

As he explained in 1979 in his book *Macroevolution*, and again in 1981, for a more general readership, in *The New Evolutionary Timetable*, Stanley believed that new species usually originated rapidly as a result of branching off from an evolutionary line, rather than by the slow, progressive development of previously existing species.

Like the other punctuationists, he made it clear that he intended 'rapidly' to be taken in a geological context, meaning 'in less than a few thousand years or at most a few tens of thousands of years'.[29] Whilst he had much sympathy with George Gaylord Simpson's concept of quantum evolution, he differed from Simpson in linking the rapid phases of evolution with speciation. Above all, Stanley dismissed the generally held view that overall trends result primarily from phyletic evolution, arguing instead for the involvement of a separate process, acting at a different level. In *Macroevolution*, he wrote, 'In a punctuational framework, macroevolutionary trends become analogous to microevolutionary trends: Species take the place of individuals, speciation and extinction substitute for birth and death, and speciation generates variability in the way that mutation and recombination do within a population'.[30] Furthermore, he emphasised that the role of extinction was of special importance, a view shared with Eldredge and Gould.[31]

Like them, Stanley also believed that there was no reason to think that new species would differ from their parental species in ways which would be in line with evolutionary trends. In fact, he went further than Eldredge and Gould had done and wrote explicitly of 'species selection'. A new species which happened to be well-suited to its environment would have a relatively high probability of surviving for a long time, and of giving rise to descendant species, before going extinct. Hence, species of this type would be the ones most likely to influence the overall direction of evolution. As to the biological details, Stanley maintained that a reductionist approach to living systems (studying them at the lowest possible level, as discussed in Chapter 10) would never be successful in explaining evolutionary processes, because a more complete explanation would also need to incorporate *emergent* qualities (those requiring study at more than one level).[32]

Since Stanley was a palaeontologist, the evidence for his theories came largely from the fossil record, but not exclusively so. For example, he drew attention to the fact that a salamander can be produced from its alternative and very different form, the axalotl (which is essentially an extension of the larval state into adulthood) by administration of the hormone thyroxine, whose natural synthesis could have been prevented by a mutation in a single gene. This demonstrated that small genetic changes could have major effects. Nevertheless, Stanley thought that speciation probably required several key mutations in a small population over a period of time, rather than arising from just one. For example, the process by which giant pandas arose from bears could have involved just two or three modifications to regulatory genes.[33] Stanley pointed out that there was already clear evidence for the occurrence of rapid changes in small populations because, although the tiny African lake Nabugabo had been cut off from Lake Victoria by a sand barrier only about four thousand years ago, it already contained five species of cichlid fish unique to this location. Furthermore, similar findings had been made in other African lakes.[34]

Eldredge and Gould, as we have already noted, originally pointed to the likely role of allopatric speciation in the rapid appearance of new species. Afterwards, however, both they and Stanley became more receptive to the idea that speciation could be *sympatric*, i.e. taking place without geographical isolation.[35] An example of how sympatric speciation might operate was provided by geneticist Guy Bush and his colleagues at Michigan State University, as a result of observations on the fruit fly *Rhagoletis pomonella*, and related species. Different types of fly used different fruits as hosts, adults mating around the particular trees and depositing eggs so that the larvae could subsequently feed on the fruit. In this way, it seemed that what had originally been populations of a single species became reproductively isolated from each other within a single geographical area. Eventually they formed separate species, with, for example, *Rhagoletis mendax* being associated with the blueberry, whilst *Rhagoletis pomonella* itself lived around apple trees. Another possible mechanism of sympatric speciation has been suggested in birds, where different populations within the same geographical area may start to become reproductively isolated from each other because of the use of different songs during mating.[36]

As the debate over punctuated equilibrium continued, some argued that even if the fossil record was predominantly punctuational, it still supported a gradualist scenario. This was because the 'rapid' phases of evolution were acknowledged to have a time-scale of several thousand years, instantaneous transitions not being part of punctuational models. So, it became increasingly common for evolutionary biologists to make a distinction between the terms 'rapid' and 'saltational'. For example, in the 1991 book *One Long Argument*, Mayr (by this time appreciating that Gould and Eldredge were not suggesting that new species arose overnight from a single macromutation) wrote:

> Eldredge and Gould [stated] that most major evolutionary events take place during short bouts of speciation and that successful new species, after they become widespread and populous, enter a period of stasis, lasting sometimes for many millions of years, during which they show only minimal change. Such speciational evolution, because it occurs in populations, is gradual in spite of its rapid rate and therefore is in no conflict whatsoever with the Darwinian paradigm.[37]

Regardless of that, punctuated equilibrium was clearly different from the process of stately growth illustrated by Darwin's 'tree of life' analogy. Throughout the *Origin*, Darwin insisted that evolution proceeded 'by the shortest and slowest steps', with 'numberless intermediate varieties'.[38] Equally, punctuated equilibrium had little in common with the phyletic gradualism of later times. As we have seen, the founders and followers of the Modern Synthesis consistently adopted the same ultra-gradualistic stance as Darwin, portraying evolution as a 'continual process of potentially infinite change', in which most species-formation was 'without bearing upon the major and continuing trends'.[39]

Thus, punctuated equilibrium, with its associated notions of hierarchies, was a different way of seeing evolution, compared with what had gone before. Gould used the phrase, 'a different way of seeing' in the title of a 1982 *New Scientist* article which, later in the same year, was included in the collection *Darwin up to Date*. In this article, Gould suggested that, when a fully developed theory was eventually put together from the ideas of the punctuationists, it would 'embody the essence of Darwinian argument'. Nevertheless, as he made clear, it 'would not be entirely Darwinian'.[40]

16 Human evolution: gradual or punctuational?

Questions raised by the molecular clock

As we saw in Chapter 11, it was generally thought in the 1960s and early 1970s that the Miocene ape *Ramapithecus* had characteristics which indicated that it was a direct ancestor of modern humans, and that the hominid line had already diverged from the line leading to modern apes by this time. That view owed much to the work of the Yale palaeoanthropologists Elwyn Simons and David Pilbeam. However, new discoveries began to show that there was nothing particularly unusual about *Ramapithecus*.[1] Indeed, *Sivapithecus* and *Gigantopithecus*, other Miocene apes from the same region of northern India, were found to have similar features in their jaws and teeth. Like *Ramapithecus*, both had large molars, capped with thick enamel, with *Gigantopithecus* also having relatively small canines. Furthermore, it had to be acknowledged that even the orang-utan, a modern ape from Asia, has thick linings of enamel on the molars, a feature supposedly characteristic of the hominids.[2]

As the 1970s progressed, Pilbeam accepted that *Ramapithecus* was not unique amongst Miocene apes in showing some hints of hominid characteristics, so he classified *Ramapithecus*, *Sivapithecus*, *Gigantopithecus* and the European *Oreopithecus* together as 'ramapithecids', and simply suggested that the hominid line must have emerged from within this group, rather from amongst the other Miocene apes, the 'dryopithecids'.[3]

Sherwood Washburn, an anthropologist from the University of California, Berkeley, wrote a 1978 *Scientific American* article on the subject of human evolution in which he referred to *Ramapithecus* as 'the Miocene–Pliocene ape commonly believed to be an ancestor of the hominid line'. However, he then went on to make clear that he had strong doubts both about this and also the belief that the separation of the ape and hominid lines had occurred by the time of *Ramapithecus*. He explained that, although some of his colleagues would undoubtedly disagree, the fossil and other evidence now available made it 'highly unlikely that the ape and human lines separated . . . more than 10 million years ago'.[4]

By this time, investigations in northeastern Africa had demonstrated that one particular hominid, *Australopithecus afarensis*, had walked upright around 4 Myr ago, long before the appearance of large brains or the use of stone tools. Over the whole period of those investigations, however, whilst palaeoanthropologists attempted to piece together an improved picture of human evolution in the light of these findings, they almost totally ignored the information being revealed by biochemical studies. Similarly, the debates between palaeoanthropologists as to the possible hominid status of *Ramapithecus*, and what that suggested about the date of the ape–human split, were based almost entirely on the fossil evidence, even though this was extremely insubstantial. So, why was the biochemical evidence being set aside in this way?[5]

In the early 1960s, at Berkeley, biochemists Vincent Sarich and Allan Wilson had begun to study the relationship between the genes of different living species, at the suggestion of Washburn, and it quickly became clear that their results, if they had been interpreted correctly, would force a major re-assessment of key aspects of hominid evolution. Following the example of Emile Zuckerkandl and Linus Pauling, they argued that the regular accumulation of point mutations within a population must provide a kind of 'molecular clock'. Since it would be expected that different point mutations would accumulate in different populations once they had separated, the greater the differences which were observed between the genes of any two species at the present time, the earlier should be the time of their evolutionary divergence.

For practical reasons, Sarich and Wilson had to approach this indirectly, by investigating the structure of proteins, whose amino acid composition is specified by the sequence of bases in the DNA of the genes. These studies showed that modern humans and the African apes (the gorilla, chimpanzee and pygmy chimpanzee) had genes whose base sequences were 98–99% identical. This extremely close genetic relationship led to a suggestion that the Pongidae family (the great apes) should be included within our own Hominidae family, but it was rejected. As was obvious to the eye, there were considerable differences between humans and African apes, which could not be explained purely on the basis of the minor variations in DNA structure between the different species.[6]

It was when Sarich and Wilson applied the concept of the molecular clock to their data that controversy erupted. To calibrate their method, they used the generally accepted assumption that the Old World monkeys split from the apes 30 Myr ago, as indicated by fossil evidence. On this basis, they calculated that the African apes must have split from the hominid line only 5 Myr ago, in contrast with the figure of 20 Myr or more which was widely accepted by palaeoanthropologists. The orangutan, from Asia, was found to have genes which differed slightly more in structure from those of humans, so it must have split away somewhat earlier than the African apes.[7]

However, these biochemical conclusions were widely dismissed by the palaeoanthropologists, on the grounds that they were based on a series of unproven assumptions, particularly that mutations accumulated at a steady rate, for which there was no evidence. Nevertheless, whilst they were attacking the biochemical data, the palaeoanthropologists revised their estimates of the date of the ape–hominid split to immediately before the time of *Ramapithecus*. Thus, in the late 1970s, their general view was that it occurred between 12 and 15 Myr ago.[8]

As we have seen, Washburn considered that figure to be too high and, undoubtedly, the biochemical evidence was an important factor in causing David Pilbeam to question his original interpretation of the *Ramapithecus* fossils. The debate was about far more than just dates. In 1985, in *The Emergence of Humankind*, John Pfeiffer commented that it made a great difference whether the ape–human split occurred 5 Myr ago or around 15 Myr ago, and he went on to explain:

> The long time scale suggests a gradual transition from ape to hominid, and is thus in line with the general tendency of evolution to proceed by little steps . . . The short time scale has different implications. For one thing, it raises the serious possibility that the appearance of hominids may have been a sudden phenomenon, that is, sudden in evolutionary terms – say, within half a million years or so, which is very sudden indeed considering all that has happened since.[9]

What was at stake, therefore, was the future of a key aspect of the ultra-gradualistic evolutionary paradigm.

Towards a new consensus of hominid evolution

Although, in the first instance, sequences of bases in genes had to be inferred from protein structures, it eventually became possible to determine them in direct fashion, without this leading to any significant change in the biochemical conclusions about ape evolution. Whilst there could be some dispute about relationships within the human–chimpanzee–gorilla group, these three species were seen to be more closely related to each other than to the other apes and, by applying the molecular clock, the biochemists were generally agreed that the final separation of hominids from African apes had come well within the last 10 Myr.[10]

There was still no experimental proof for the notion that the ticking of the molecular clock, the accumulation of mutations in genetic DNA, occurred in a strictly regular fashion in all organisms. Indeed, it soon became clear that the mutation rate was different in different species. Nevertheless, for comparisons in closely related species over limited periods of evolutionary time, it became accepted that the molecular clock could provide useful information. Hence, with that proviso, the conclusions which had been drawn from molecular data about the evolutionary relationships of humans and apes could be considered valid.[11]

When palaeoanthropologists began to look again at the fossil evidence in the light of the biochemical data, it became clear that they could produce nothing to demonstrate conclusively that their view of evolution, rather than that of the biochemists, was correct. Furthermore, as new fossil evidence came to light, the case for the hominid status of *Ramapithecus*, the main feature of the argument for the 'long-time-scale' model of hominid evolution, became impossible to sustain. Finds in Pakistan and in Turkey convinced many palaeoanthropologists, including Pilbeam, that there were so many similarities between *Ramapithecus* and *Sivapithecus* that all the representatives of the two genera should be grouped together within the single genus, *Sivapithecus*. Eventually, Pilbeam came to the conclusion that *Sivapithecus* and the other apes who lived in the regions linking Turkey and northern India from 12 to 7 Myr ago were distinct from Miocene apes living in Africa and Europe over the same period. One of these Asian apes, possibly *Sivapithecus* itself, was the ancestor of the orang-utan, whereas chimpanzees, gorillas and humans emerged from the Afro-European group.[12]

Inevitably there were disagreements about detail, but a consensus view eventually emerged amongst palaeoanthropologists and biochemists alike. This was very much in accord with Pilbeam's scheme, i.e. that the lines leading to the modern Asian and African apes had split over 10 Myr ago, and then the hominid line separated from that of the African apes between 5 and 10 Myr before the present day.[13]

Consistent with the implications of this shorter time-scale, Niles Eldredge and Ian Tattersall, a palaeoanthropologist who, like Eldredge, worked for the American Museum of Natural History, New York, argued in a 1975 paper and again, in more detail, in their 1982 book, *The Myths of Human Evolution*, that the hominid fossil record showed punctuational characteristics, rather than any evidence of phyletic gradualism. According to their interpretation, hominid species appeared rapidly, underwent little progressive change for a million years or more, and then disappeared as rapidly as they came.[14] A supposedly opposite view was taken by the British palaeoanthropologist John Cronin and his colleagues in a review article which appeared in *Nature* in 1981. However, despite their stated failure to find evidence of punctuated equilibrium, they accepted that the hominid fossil record showed evidence of some episodes of slow change and some of rapid change.[15] The interpretation of the 'slow' phases was inevitably subjective, by both groups, and by others debating the same issue, since the hominid fossil record was far too poor to be able to establish the presence or absence of stasis with any degree of certainty. To complicate matters further, normal variations of characteristics within a species, age differences, geographical variations and possible sexual dimorphism (differences in size between males and females, a common feature of modern apes) all needed to be taken into account.[16] Hence, no definite conclusions about stasis could be reached.

More generally, but in keeping with a punctuational view rather than the traditional one, Stephen Jay Gould argued, in *Ontogeny and Phylogeny*, that heterochrony (see Chapter 15), and in particular, neoteny, a change in development rates resulting in the continuation of juvenile characteristics into adult life, could have been an important factor in human evolution. As pointed out in 1926 by the Dutch anatomist Louis Bolk (1866–1930), adult humans have many features more like those of baby apes than adult apes, including hairlessness, a flat face, domed head and a large brain relative to body size. Indeed, the very high brain-to-body ratio of modern humans is achieved only

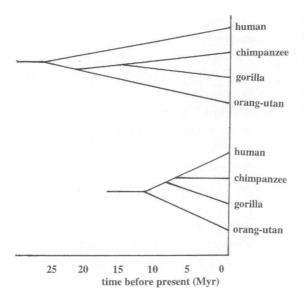

Figure 16.1 The divergence of apes and humans. Top: according to a typical scheme of the 1970s. Bottom: according to a typical scheme of the present day.

because the brain continues to grow after birth to a far greater extent than in apes, the brain size at birth being constrained by the size of the pelvic girdle.[17]

Another aspect of the traditional view which came under challenge was the notion that, since evolution was assumed to be continuous and progressive, only one hominid species must have been alive at any particular time in history (see Chapter 11). A crucial finding came in 1975 when Richard Leakey and Alan Walker discovered a *Homo erectus* skull at Koobi Fora in the same formation in which a robust australopithecine fossil had previously been uncovered. After this, even the main advocate of the single-species hypothesis, Loring Brace, retreated somewhat, acknowledging that there could be overlap between old and new forms.[18]

Regardless of these issues, and the arguments about whether there had been any periods of actual stasis, it was becoming increasingly clear that hominid species had appeared and disappeared rapidly. It was also becoming difficult to deny that these appearances and disappearances had been part of a process whose time-scale, following the ape–human split, was much shorter than previously supposed (figure 16.1). Throughout the 1960s and early 1970s, as we have seen, palaeoanthropologists generally believed that hominid evolution had involved more than 20 Myr of continuous and steady development from the last common ancestor of apes and humans through to *Homo sapiens*, via *Ramapithecus*. However, by 1980, that view could no longer be maintained with any confidence. Over a period of just a few years, the long-time-scale model for hominid evolution was abandoned.[19]

This fundamental change in perspective amongst palaeoanthropologists is illustrated by the differences between the 1974 and 1985 editions of Bernard Campbell's *Human Evolution*. In the 1974 edition, Campbell devoted several pages to the traditional view of *Ramapithecus* as an early hominid, following an ape–human separation between 15 and 25 Myr ago, and thought it unnecessary to spend more than a few lines noting the existence of an alternative theory based on biochemical evidence. In contrast, in the 1985 edition, he wrote, 'Today even the most prejudiced paleontologist, who had little time for a consideration of the biochemical evidence . . . must now agree that the common ancestor of gorilla, chimpanzee and human may not be older 6 million years'.[20] Similarly, in *Origins*, published in 1977, Richard Leakey and Roger Lewin referred to *Ramapithecus* as 'the first representative of the human family – the hominids', whereas in their later book, *Origins Reconsidered*, they presented a considerably different view, and admitted, 'We anthropologists were forced to admit that we had been wrong and that Sarich and Wilson were closer to the right track than any of us had ever imagined'.[21]

Without question, in the field of human evolution, and also more generally, some of the pillars supporting the ultra-gradualistic paradigm had started to crumble by the late 1970s. Furthermore, the issue of mass extinctions still lurked in the shadows, awaiting an opportunity to become the centre of attention, when alternative views could be debated in proper fashion. Year by year, it was becoming increasingly apparent that the paradigm which had prevailed for a century or more, in supremely dominant fashion, was no longer beyond serious challenge.

Section C
From 1980 to the present day: catastrophism strikes back

17 Evolution evolving

One hundred years after Darwin

There is nothing like an anniversary for stimulating discussion about where a subject stands. So, several books concerned with the current state of evolutionary theory were published in 1982, to commemorate the centenary of Darwin's death. In one of them, the multi-author work *Perspectives on Evolution*, John Campbell, of the University of California, Los Angeles, looked back to 1959, the centenary of the publication of *On the Origin of Species*, and commented, 'The uniformity of outlook and sense of achievement in 1959 contrasts markedly with the multiplicity of views and sense of anticipation in today's volume'.[1] That was undoubtedly true.

In 1959, the Modern Synthesis of neo-Darwinism, with its tightly focused view of evolutionary change, seemed as secure as any scientific theory could be. In aspects where the original form of Darwinism took a strict and narrow stance, the Modern Synthesis more than matched it as, for example, in maintaining that all evolutionary change was gradual. Similarly, adherents to the Modern Synthesis, just like Darwin himself, believed that all evolutionary change could be explained by a reductionist approach. There were, however, differences between the old and new forms of Darwinism. In particular, in areas where the original form was pluralistic and broad, the Modern Synthesis remained strict and narrow. So, whereas Darwin wrote, as the final words to the Introduction of the first edition of the *Origin*, 'I am convinced that natural selection has been the main but not exclusive means of modification',[2] followers of the Modern Synthesis regarded it as the *only* mechanism of any significance. As to the variations on which natural selection could act, Darwin was happy to accept that they could include characteristics acquired by an individual in direct response to environmental pressures, whereas the Modern Synthesis specifically excluded the possibility of such Lamarckian processes. By this time, Darwinism and Lamarckism were seen as mutually exclusive ways of trying to explain evolution. All these issues were discussed in detail in Chapters 9 and 10.

During the 1950s, it was generally considered that any remaining doubts about the fundamental nature of the evolutionary process had been dispelled when it became clear that the newly discovered mechanisms of molecular genetics were consistent with Darwinism but not Lamarckism. Similarly, there was a widespread belief that it had been firmly established that life itself arose by a Darwinian mechanism when Stanley Miller, working with Harold Urey at the University of Chicago in 1952, passed an electrical discharge through a simulation of the Earth's primordial atmosphere and found that some of the building blocks of proteins and nucleic acids had been synthesised during the course of this simple experiment.[3] In the early stages of the Earth's history, natural selection could have determined which of the many molecules and pre-biotic chemical systems that arose by random mechanisms and interactions over a long period of time prevailed to give rise to further systems. Thus, larger molecules would have developed from smaller molecules, and gone on to interact in ever more complex fashion, until eventually living cells, capable of replicating themselves, were produced. In this view, biological evolution followed chemical evolution, both operating according to the same principles of gradualistic Darwinism.[4]

These certainties were maintained throughout the 1960s. In the 1970s, however, as we have already noted, some of the rigid views associated with the Modern Synthesis were disputed, leading to changes in interpretation. For example, many became convinced that the formation of new species could take place rapidly, perhaps in small, isolated groups, explaining the discontinuous appearance of the fossil record. However, since these rapid bursts of speciation were thought to involve processes taking place in populations over a time-scale of thousands of years, rather than overnight changes in individuals, they were still regarded by the upholders of the Darwinian tradition as being gradual (see Chapter 15).

Whatever changes of emphasis may have occurred, it remained the firm conviction of those working within the field of evolutionary biology that natural selection, the essential characteristic of Darwinism, played an important role in evolution. Looking back to the ideas of Sewall Wright (see Chapter 9), many evolutionary biologists were also prepared to accept that genetic drift, the random increase or decrease of a particular variant in a small population, could make a contribution, if only a minor one. More controversially, the University of Cambridge geneticist Gabriel Dover, who later moved to the University of Leicester to take up a professorial position, argued in 1982 that there was a third force in evolution, which he called molecular drive. He suggested that this provided an internal mechanism, independent of natural selection, which could cause genetic changes arising from mutations to become fixed within a population. It could, for example, explain characteristics of eukaryotic genomes (i.e. the genetic material of cells of higher organisms), which contain substantial numbers of multiple-copy families of genes and non-coding sequences. Members of each family are largely identical within and between individuals of a species, but different from members of corresponding families in other species. Dover postulated that molecular drive provided a mechanism which enabled a mutation in one member of a family to be copied by all other members of the same family within an individual, regardless of whether they were on the same chromosome or different ones. Because of chromosome mixing during sexual reproduction, the mutation could then spread through a population far more quickly than would be possible by normal Mendelian inheritance.[5] However, although it remained a controversial idea, molecular drive was still regarded, like genetic drift, as something which might add an extra element to Darwinism, rather than present a fundamental challenge to the generally accepted view of evolution.

Another refinement of Darwinism was proposed in 1983 by the Japanese geneticist Motoo Kimura, in his book, *The Neutral Theory of Molecular Evolution*. Kimura's starting point, as with considerations about molecular clocks, was that point mutations accumulated in genes over the course of time. In general, it seemed to him that the rate of accumulation was too rapid to be attributable to adaptive evolution, for that would have required an impossibly high mutation rate coupled with an impossibly high birth rate, to provide sufficient variants for selection in the time available. It was therefore more plausible to think that most of the point mutations which accumulated within a species would be neutral, i.e. ones giving neither advantageous nor disadvantageous characteristics, becoming fixed by random genetic drift rather than natural selection. As evidence in favour of this, Kimura cited protein electrophoresis studies, which had shown that many variant forms of proteins could be found within a population, without there being any obvious signs that some were more advantageous than others. Although a mutation with harmful consequences would be eliminated by natural selection, the relative proportions of the various neutral mutations in the population would rise and fall by genetic drift. However, what was neutral under the original conditions might not be so if the environment changed. Hence, a particular combination of point mutations which had accidentally accumulated within a population in a neutral fashion over a long period of time might suddenly become the cause of extinction or survival in a significantly altered climate. In Kimura's view, that would explain why the fossil record was predominantly punctuational, the punctuations appearing at times of environmental change.[6]

Evolution, origins and creation science

Throughout the 1980s and 1990s, the old Darwinian principles continued to be debated in the light of new ideas, such as those suggested by Kimura, with little obvious progress being made. Much the same considerations also applied to the question of the origin of life.[7]

The arguments that life had arisen by chemical evolution continued to seem plausible when expressed in general terms, but, despite the stimulus provided by the experiments of Miller and Urey in the 1950s, further work over the ensuing years, right up to the present day, has failed to produce any convincing detailed scheme as to how living cells could have evolved from simple organic molecules. In particular, there is what has been called the 'chicken and egg problem': which came first, nucleic acid or protein? As we saw in Chapter 9, genetic information is stored as the sequence of nucleotide bases in DNA and expressed as the amino acid sequence of protein. The process by which information passes from DNA to protein, via RNA, requires catalysis by enzymes, which are proteins themselves. Enzymes are also necessary for the synthesis of DNA, whilst proteins cannot be synthesised without the presence of the DNA containing the information for their structure. A characteristic feature of living cells is replication, and in no organism can it take place without the involvement of both DNA and proteins. Many ingenious proposals have been put forward to solve the problem of the origin of this interdependence, without any of them gaining general acceptance.[8]

There have been proposals involving early genetic systems based, not on DNA, but on RNA, pyranosyl RNA (pRNA), peptide nucleic acid (PNA) or silicate units, which might have been able to replicate themselves without the presence of protein catalysts, and so provide a starting point for the evolution of the more complex process we know today. Alternatively, some have suggested that the protein–nucleic acid system developed in a co-ordinated fashion from the very beginning.[9]

Other origin-of-life arguments have focused on the location of the first living cells. Were they formed in ponds or open seas, or in the vicinity of hot sulphur springs? These have generally been assumed to be the main possibilities, but perhaps we need to look elsewhere. Did living cells arrive from space, possibly carried by comets? All that is certain is that we are still lacking a fully convincing explanation for the origin of life on Earth.[10]

Similarly for biological evolution, the Modern Synthesis has experienced much debate from the inside, and challenge from the outside, without any clear resolution being reached on the outstanding issues. One of the challenges from outside the ranks of professional biologists came from the so-called creation science movement, which maintained that what was written in *Genesis* and, indeed, the rest of the Bible was correct in every particular. The movement's use of the word 'science' in its title was highly misleading, as it made no pretence of following Karl Popper's dictum (see Chapter 14) that science operates by formulating hypotheses which can be tested and shown to be false, should they be so, in which case they would be modified or discarded. Creation science, in contrast, started with a belief founded upon religious dogma, and then selectively sought evidence to support its pre-determined conclusions, and to discredit other points of view.

One common method of approach was to search the biology literature for any statement it could use for its own purposes, usually when taken out of context. This had the unfortunate effect of inhibiting public debates about developments in evolutionary theory, because evolutionary biologists were conscious that arguments over fine detail could easily be made to seem like disagreements over fundamental principles, with the implication that 'the scientists' were in disarray. Another approach of creation science was to present a caricature of evolutionary theory, and then attack this. So, for example, it branded the essence of Darwinism as being a tautology, simply saying that those which survived were those which survived. Whilst it was true that survival by natural selection was indeed part of Darwin's theory, and of the Modern Synthesis, it was only a part. Natural selection was seen by Darwinians as operating within a specified context. It acted on variants which arose in ways not involving a direct response to environmental pressures, and it produced survivors which could pass on to future generations those factors which led to survival.[11]

Of course, it is undeniable that some scientists adopt an approach which is far from the idealised one, and not dissimilar to that used by creationists. Where that occurs, though, it is due to human weakness, not principle. Despite the action of such individuals, science continues to move forward, however imperfectly, in its search for the truth, whatever that may eventually turn out to be. Creation science, on the other hand, thinks it already knows the truth, so has nothing of significance to learn from fresh discoveries. In this philosophy, the fact that nineteenth century scientists such as Buckland and Sedgwick had no greater wish than to confirm the story told in *Genesis*, yet were increasingly unable to do so as further geological evidence came to light and changed their ideas accordingly (as discussed in Chapter 4), means absolutely nothing.

Nevertheless, Arkansas State Act 590 decreed that as much time should be devoted to teaching creation science as to teaching evolution. Although this was successfully challenged by the American Civil Liberties Union in 1981, the beliefs of creation science remained popular amongst the public at large. A 1982 Gallup poll showed that 44% of the American people accepted the *Genesis* account of creation, i.e. that God had created humans within the past 10,000 years, whereas only 9% believed that humans had evolved from other forms of life by natural processes over millions of years. Very similar figures were obtained when the poll was repeated in 1991 and again in 2001.[12] As recently as 1999, the Kansas State Board of Education removed evolution from its science curriculum (a decision reversed two years later), whilst the Kentucky Department of Education announced that evolution as such would no longer feature in state examinations.[13]

Radiometric dating methods indicate a time-scale of millions of years, not the few thousand expected from *Genesis*, but creation scientists have argued that this is meaningless, since each method is based on a series of assumptions, not all of which can be proved to be correct. However, whilst it is true that no method is free from assumptions, different dating procedures, based on different assumptions (together with the one on which all science depends, that processes operate with characteristics which are fundamental and unchanging), give very similar results when applied to samples from the same stratum.[14]

It cannot be denied, however, that problems have been encountered with some radiometric dating procedures, for example the potassium–argon one, and apparently anomalous findings have been seized upon with glee by creationists. Potassium–argon dating depends on the conversion of potassium-40 to argon-40 by radioactive decay in sealed rocks, the amount of argon formed, related to the total potassium content of the sample (almost entirely in the form of the stable isotope, potassium-39), giving a measure of its age. Naturally, if some argon was present when the process started, the date obtained would be too high. Conversely, if damage to the rock enabled argon to escape, the date would be too low. Furthermore, potassium and argon have to be analysed in different samples, providing further opportunities for error.[15] Not surprisingly, therefore, results obtained by this technique can be more variable than those generated by some others. For example, potassium–argon dating of the Late Cretaceous lava flows of the Deccan traps in India has given ages ranging from 80–30 Myr, whereas argon-40/argon-39 dating, a newer and more reliable technique (in which potassium-39 is determined after conversion by neutron irradiation to argon-39, allowing this and argon-40 to be analysed in a single crystal), produced a much narrower age range of 68–64 Myr.[16]

A second method generally acknowledged to be of uncertain accuracy, if taken in isolation, is radiocarbon dating. This procedure can be applied to biological materials from the past 70,000 years. It is based on the decay of carbon-14 to nitrogen-14 in the tissues after death, but results also depend on the concentration of carbon-14 in the atmosphere when the animal or plant was alive, which is a likely variable. However, radiocarbon dates for the most recent few thousand years have been calibrated by linkage with dendrochronology, which operates on the principle that each ring in the trunk of a tree represents the growth of a separate year. The radiocarbon date obtained for a particular ring can thus be compared to the date obtained by counting the number of annual rings from that one to the present day. Some rings are broad and others narrow, depending on conditions in the year of growth. This has enabled calibration curves to be extended back beyond the time of those trees which are still alive, using overlaps to link sequences of ring-spacings found in them to ones in wood preserved in peat bogs and river gravel[17] Calibration against dendrochronology should give reliable results for radiocarbon dating, although some people (not all of them creationists) have urged caution in placing complete trust in the outcomes, as mistakes could be made (and, indeed, have occasionally been made) when trying to link sequences in dead and living trees.[18]

Nevertheless, the various radiometric dating procedures, taken together, have undoubtedly provided a coherent framework for considering the history of the Earth. There may be problems of detail, and doubts over the accuracy of individual values (errors could be introduced, for example, by the accidental contamination of a sample with material from another stratum). However, there seems every reason to think that the time-scale now generally accepted for the course of evolution is a reasonable approximation to the true one, in marked contrast to that supported by the adherents of creation science.

Neo-Lamarckian arguments

Another external challenge to Darwinism concerned Lamarckism, the theory that characteristics acquired by an individual as a direct response to environmental pressure may be passed on to subsequent generations. This had no convincing theoretical or experimental basis, so was generally rejected by biologists as a possible mechanism of evolution. Lamarckism had some enthusiastic advocates, but these tended to be non-biologists, such as the eminent literary figure Arthur Koestler (1905–1983).[19] However, a controversial speculation was put forward during the 1980s by Rupert Sheldrake, once a Cambridge University biochemist, and then a plant physiologist at the International Crops Research Institute, Hyderabad, India, before becoming a freelance writer. Sheldrake suggested that inheritance had a second main component, in addition to genetics. Whilst he accepted that genes ensured production of the basic materials, he believed that phenomena called 'morphic fields' supplied the blueprint for the new organism. In his view, this interplay could provide a mechanism for the inheritance of acquired characteristics.

Sheldrake introduced these ideas in 1981 in *A New Science of Life*,[20] together with his hypothesis of formative causation. This maintained that the characteristics of a process were influenced by the morphic fields, which were set up the first time the process occurred and reinforced every time it was repeated. Such considerations applied just as much to processes associated with living organisms, including morphological development, as to inanimate processes like crystallisation. In this way, the past could influence the present by 'morphic resonance', for the ease with which a particular event could take place would be related to the number of times it had occurred previously, regardless of whether there was any direct link involved.

Seven years later, in *The Presence of the Past*,[21] Sheldrake attempted to place the hypothesis of formative causation in its historical and philosophical context. From the time of Descartes in the seventeenth century, all plants and animals (with the exception of humans, who had freedom of action), were generally regarded as machines which, like the components of their physical environment, operated according to the eternal law of God. As time passed, the concept of God began to intrude less and less, leaving scientists free to concentrate fully on the behaviour of systems influenced by physical forces, through the operation of unchanging laws. However, from the advent of quantum mechanics and new theories of cosmology in the twentieth century, physicists began to realise that the laws of nature might not always have been the same as they are now (although it should be realised that they were talking about changes under very exceptional conditions, such as those immediately following the supposed 'Big Bang', which gave rise to the Universe).[22]

Ever since the acceptance of the concept of evolution, biologists had always worked on the assumption that it took place by processes involving fixed and unchanging laws. That was equally true of those who took a reductionist approach and those who believed in holism, introducing additional concepts such as morphogenetic fields (see Chapter 10). Some of those who wrote about morphogenetic fields envisaged them (whatever they were) as operating under genetic control, whereas others believed that they had an existence independent of genes. However, Sheldrake was the first to suggest they could act through time. Because of this additional characteristic, he preferred to use the term 'morphic' rather than 'morphogenetic' when discussing his own theory. As Sheldrake saw it, these morphic fields acted independently of genes in both evolution and development. They were strongly influenced by the past, but they could nevertheless be modified by more recent happenings. Hence, morphic fields provided a possible mechanism for Lamarckism, if they actually existed.

Sheldrake's profound interest in religion was made very clear in *The Presence of the Past* and other books.[23] Nevertheless, he consistently maintained that his theory was scientific, because it was capable of being investigated. Sheldrake acknowledged that there was no direct evidence for the existence of morphic fields, so, because of that, everything depended upon the results of tests for formative causation. Several testing programmes were devised, some in collaboration with Sheldrake himself. These involved, for example, investigating whether it was easier to solve a crossword puzzle if thousands of other people had attempted it previously, or whether it made it easier to learn a poem in a foreign language if millions had already done so in the country of origin. Unfortunately for Sheldrake, although some early results appeared encouraging, these tests ultimately provided no reason for continuing to give serious attention to his ideas involving morphic fields.[24]

Nevertheless, the arguments continued about whether some element of Lamarckism needed to be incorporated into evolutionary theory. One particular area of interest was reverse transcription, the flow of information from RNA to DNA.

In 1964, Howard Temin of the University of Wisconsin argued that it was possible for reverse transcription to occur, subsequently providing evidence that the RNA of a virus could specify the synthesis of DNA within a host cell to produce more copies of the virus.[25] From this starting point, Ted Steele, an Australian immunologist working first at the Ontario Cancer Institute, and then at the Medical Research Council's Clinical Research Centre, in England (before eventually moving to Wollongong University, near Sydney), developed ideas which he expounded in his 1979 book, *Somatic Selection and Adaptive Evolution*, and elsewhere. In Steele's view, advantageous mutations in somatic cells could accumulate by natural selection, and these modified genes might then be transferred in the form of RNA to germ cells, to become inheritable. In association with a colleague, Reg Gorczynski, he claimed to have demonstrated that immunological tolerance developed in the somatic cells of mice could be passed on to subsequent generations. However, these results could not be confirmed by Leslie Brent and others from the Clinical Research Centre and St Mary's Medical School when they repeated the experiments under similar (although not identical) conditions.[26] Hence, the claim that immunological tolerance could be inherited was generally regarded with suspicion.

Yet even if it *had* been shown that RNA was capable of crossing the 'Weissmann barrier' between somatic cells and germ cells, that would still not have been evidence for Lamarckism and against Darwinism, as some had supposed at the time. The fact that the mutation arose in the first instance in the genome of a somatic cell rather than a germ cell was not the crucial point. Of far greater significance, there was no suggestion that it had arisen as a direct response to environmental pressure, or that reverse translation (the flow of information from protein to DNA) might have been involved. Hence, despite the complicating feature of the crossing of the Weissmann barrier by viral RNA, Steele's proposed mechanism was not Lamarckism as generally understood. Regardless of the outcomes of the immunological tolerance experiments, it remained highly unlikely that a new genotype could be synthesised in a way specified by structural changes in the phenotype.[27]

Many who considered themselves supporters of the Modern Synthesis were prepared to accept the possibility of some input from the environment, without regarding it as a move towards neo-Lamarckism, since they saw no reason to think that mutations with favourable consequences could be specifically produced in this way. Indeed, it had long been accepted that the small mutations which are a central feature of traditional neo-Darwinism may be caused, in 'random' fashion, by environmental factors such as radiation, such processes giving rise to a wide range of variant forms.

In the 1980s, John Cairns, of Harvard, and Barry Hall, of the University of Connecticut, were amongst a small number who went further than this and argued on the basis of intriguing circumstantial evidence that bacterial cells have mechanisms for 'adaptive mutation', i.e. for encouraging mutations with advantageous consequences. For example, it was found that mutants of *Escherichia coli* which were unable to utilise lactose or tryptophan as nutrients regained the ability to do so with greater frequency if the nutrient was present than if it was not. This may have been suggestive of Lamarckism, but a Darwinian explanation was far from impossible, although it needed to involve something more complex than fortuitous reverse mutations, as more than one mutation was required on some occasions.[28]

Work on the so-called 'adaptive mutations' in bacteria progressed during the 1990s, and it began to emerge that genetic material can be passed 'horizontally' from one cell to another to produce the observed effects. Thus, the mechanism of 'adaptive mutation' seems to be a complex form of Darwinism, rather than Lamarckism.[29] Moreover, it is uncertain whether the findings have any significance in terms of 'vertical' transmission of genetic material, i.e. inheritance, and hence evolution, particularly in the case of more complex multi-celled organisms.

Self-organisation

To explain how systems might increase in complexity, without any involvement of Lamarckism, some evolutionary biologists drew attention to the non-equilibrium thermodynamics of Nobel laureate Ilya Prigogine, of the Free University of Brussels. The second law of thermodynamics made it clear that a system could only decrease in entropy, i.e. become more ordered, if its surroundings became more disordered, so that there was an overall gain of entropy in the system plus surroundings. Prigogine showed during the 1970s that a spontaneous increase in order (self-organisation) was most likely to occur in a system which could exchange both energy and matter with its surroundings, and which was kept well away from a position of equilibrium.[30] These concepts were then developed by others, particularly enthusiasts attached to the Santa Fe Institute in New Mexico.

One of the members of the Institute, Stuart Kauffman, a biologist with medical qualifications, who was formerly at the University of Pennsylvania, presented detailed arguments for self-organisation in his book *The Origins of Order*, which was published in 1993.[31] In the introduction, Kauffman summarised his case as follows:

> It is not that Darwin is wrong, but that he got hold of only part of the truth. For Darwin's answer to the sources of order we see all around us is overwhelmingly an appeal to a single singular force: natural selection. It is this single-force view which I believe to be inadequate, for it fails to notice, fails to stress, fails to incorporate the possibility that simple and complex systems exhibit order spontaneously.[32]

After that, he claimed that our understanding of evolution could be transformed by linking recent developments in molecular biology with the outcome of studies into self-organisation. Thus, complex features in living cells might have arisen in spontaneous fashion, fully formed, rather than emerging gradually through the operation of natural selection. Kauffman believed that the reductionist approach had been taken about as far as it could go, and only by re-introducing elements of holism could further progress be made.

Many of Kauffman's ideas emerged from a mathematical study of systems of binary variables called NK Boolean networks. In *The Origins of Order*, he explained that each element in a network can be in one of two states, active or inactive, and the activity of each of the N elements is linked to the prior activity of K other elements. Such networks have three main regimes of behaviour, which are termed ordered, chaotic and complex. In the ordered regime, many of the elements are frozen in fixed states of activity, and form a large, inter-connected cluster. This frozen component percolates across the system, leaving behind islands of unfrozen elements, i.e. ones whose activities can fluctuate. The opposite situation applies in the chaotic regime, where an inter-connected cluster of unfrozen elements percolates across the system, leaving behind just a few islands of elements whose activities are frozen. In this regime, the dynamics are extremely sensitive to initial conditions, small varia-tions causing avalanches of changes in the activities of the unfrozen elements. The third regime, the complex one, is located in the border region between order and chaos.[33]

Kauffman's study indicated that the transition from chaos to order could take place very easily, without any need for precise control. Even systems with millions of elements could develop order in spontaneous fashion provided each element could be influenced by a few others. Kauffman applied this approach to a model system for the study of molecular evolution, and he concluded that Boolean networks in the boundary region have an optimised capacity to evolve by accumulating advantageous variations. Extending this further, he formulated the general hypothesis, 'Living systems exist in the solid regime near the edge of chaos, and natural selection achieves and maintains such a poised state'.[34] He pointed out that, in such a complex system, a vast amount of order was available to be used by the processes of selection.

Evidence from the fossil record of population changes during mass extinction episodes and subsequent radiations of new forms was consistent with calculations arising from Kauffman's hypothesis. Generally, movement into the chaotic region would result from an environmental shock, although this was not essential. Order would eventually be restored, but there would be differences from the system which existed previously. In situations where populations of interdependent species were analogous to systems poised at the edge of chaos, then even a relatively small jolt could have a major effect. Kauffman explained that 'at the poised state, coevolutionary avalanches of change propagate through the ecosystem with a power-law distribution of avalanche sizes'.[35] Such avalanches could easily lead to extinction events.

The second section of *The Origins of Order* presented Kauffman's views on the origin of life on Earth. According to him, the first living organisms were complex, self-organised systems. These possessed fully developed metabolic networks in addition to sets of macromolecules, each member of which was capable of catalysing the final step in the formation of another. As Kauffman saw it, life as we know it today developed from just one of the many self-organised systems which could have been formed.

The final section of the book was concerned with ontogeny, i.e. the development of individuals. Not surprisingly, Kauffman again believed that self-organisation was a very important factor. He argued that much of the organisation apparent in ontogeny could be attributed to the 'powerful tendency for order to emerge in massively complex systems', giving natural selection the materials with which it could work.[36]

Many biologists remained sceptical about the relevance of mathematical studies of Boolean networks to their subject area. Nevertheless, Kauffman's conclusions were in line with what developmental biologists had been saying and, indeed, continued to say.

In an article in *The Biologist* in 1990, Dorian Pritchard of the University of Newcastle-upon-Tyne, author of the 1986 textbook *Foundations of Developmental*

Genetics,[37] rejected the idea that we could consider the evolution of the eye in terms of the piecemeal addition of extra components. In his opinion, the evolution of the visual system as a whole, including the role of the eye in relation to the rest of the body, needed to be taken into account. Pritchard pointed out that, in the embryo, the developing system was dynamically self-regulating, so that if part was removed, it could be re-formed from other parts. He eventually concluded, 'What evolution has created within the bodies of animals are integrated, self-organizing systems which are not just defined by their component parts, but actually define those components. In considering animal evolution we need to address primarily the origins of such interactive systems in embryos and secondarily the products of those interactions'.[38]

Similar arguments also featured prominently in *How the Leopard Changed its Spots*, a book by Brian Goodwin, Professor of Biology at the Open University, which was published in 1994.[39] Goodwin, who was Canadian by birth, had been a research student of Conrad Waddington at Edinburgh University (see Chapter 10). In the title of his book, the 'leopard' was biology itself, which Goodwin, like Kauffman and others, believed had gone much too far down the reductionist route, and needed to move back closer to holism. Right at the beginning of his book, Goodwin pointed out that Darwin's belief that evolutionary change was determined by nothing more than the gradual accumulation of small hereditary differences had failed to be substantiated, as further evidence had come to light. Instead, it seemed that some other process must be responsible for the emergent properties of life (see Chapter 10), the characteristic features that distinguished one group of organisms from another. He wrote,

> Clearly something is missing from biology. It appears that Darwin's theory works for the small-scale aspects of evolution: it can explain the variations and the adaptations within species that produce fine- tuning of varieties to different habitats. The large-scale differences of form between types of organism that are the foundation of biological classification systems seem to require a principle other than natural selection operating on small variations, some process that gives rise to distinctly different forms of organism.[40]

Like Kauffman in *The Origins of Order*, Goodwin suggested that D'Arcy Thompson had pointed towards the solution over half a century earlier, when he attempted to re-establish the organism as the dynamic system responsible for biological emergence. Goodwin continued, 'Once this is included in an extended view of the living process, the focus shifts from inheritance and natural selection to creative emergence as the central quality of the evolutionary process'.[41]

Unlike Kauffman, Goodwin based his arguments largely on biological evidence. For example, he drew attention to the Mermaid's Cap (*Acetabularia acetabulum*), a green alga resembling a mushroom, found in shallow water around the Mediterranean coast-line. As it grows, rings of protruding branches, known as whorls, form along the stalk. To the best of anyone's knowledge, these serve no useful purpose and eventually disappear. Hence, Goodwin saw the process of whorl-formation as a consequence of self-organisation rather than natural selection. In his view, the surface features of the stalk were influenced by morphogenetic fields, the mechanism in this case involving changes in the distribution of calcium ions within the organism.[42] However, Goodwin did not restrict himself entirely to biological arguments. He acknowledged that a long working relationship with Kauffman had led to a convergence of ideas, so, despite the very different starting points and methods, they had come to very similar conclusions.[43]

Most biologists were still not convinced, and even some who were sympathetic to a holistic approach were doubtful about the relevance of the computer simulations to living organisms.[44] This was not just a case of biologists being blind to compelling evidence from other disciplines, for such evidence did not exist, in any tangible form. Interviewed in Mitchell Waldrop's 1992 book, *Complexity*, Doyne Farmer, a physicist who was a member of the inner circle of the Santa Fe Institute, admitted that only small, isolated pieces of the puzzle were currently understood. He continued:

> The hope is that we will eventually be able to stand back and assemble all these fragments into a comprehensive theory of evolution and self-organization. This [is] not a field for people who like sharply defined problems. But what makes it exciting is the very fact that things *aren't* laid in stone. It's still happening. I don't see anybody with a clear path to an answer. But there are lots of little hints flying around.[45]

Farmer thought it possible that a detailed theory could be assembled in twenty to thirty years, whereas Kauffman expressed hope of a much shorter time-scale. However, John Horgan, in an article entitled, 'From complexity to perplexity', which appeared in *Scientific American* in 1995, revealed that some members of the Santa Fe Institute were becoming increasingly concerned about exaggerated claims being made for complexity, and were 'beginning to fret over the gap between such rhetoric and reality'.[46] Clearly, although self-organisation offered the promise of eventually filling a gap in evolutionary theory, it was not yet in a position to justify any claim for much more than that.

18   Into the new millennium: evolution today

Regulatory genes and speciation

What, then, can we say about the present state of evolutionary theory? Without doubt, it has been going through a prolonged period of change, and the process is still ongoing. Ernst Mayr (see figure 9.1), the doyen of orthodox evolutionary biologists, admitted in his 1991 book, *One Long Argument*: 'Although rarely completely wrong, the conclusions supported by the followers of the evolutionary synthesis were often incomplete and rather simplistic'. He nevertheless maintained that the current controversies 'all take place within the framework of Darwinism. The basic Darwinian principles are more firmly established than ever'.[1]

Similarly, in 1995, in *Darwin's Dangerous Idea*, the American science historian and philosopher Daniel Dennett commented, 'New discoveries may conceivably lead to dramatic, even "revolutionary" *shifts* in the Darwinian theory, but the hope that it will be "refuted" by some shattering breakthrough is about as reasonable as the hope that we will return to a geocentric vision and discard Copernicus'.[2]

Ernst Mayr again took much the same view in *What Evolution Is*, a work that first appeared in 2001, when the author was aged 97. Towards the end of the book, whilst acknowledging that we still lacked a full understanding of the details of the evolutionary process, Mayr wrote, 'Darwinian evolutionists have every reason to be proud of the paradigm of evolutionary biology that they have constructed. Every attempt in the last 50 years to refute one or the other assumption of Darwinism has been invalidated. Furthermore, no competing evolutionary theory has been proposed, certainly none that was in any way successful'.[3]

Yet, although that may be true, few of the evolutionary issues debated throughout the 1980s and 1990s have been resolved. New discoveries continue to be made in the field of molecular genetics, without a clear enough picture emerging to allow the new data to be incorporated into evolutionary theory. It has been suggested, for example, that the shuffling of exons in eukaryotes could facilitate rapid evolution by joining segments of protein together in a variety of ways; that there could be a non-genetic aspect to inheritance; and that non-coding sequences (often called 'junk DNA') could have a role in evolution. However, those and other ideas remain largely at the level of speculation.[4]

Despite the uncertainties, there is growing acceptance that the mutations of particular importance in evolution are not the ones affecting 'normal' structural genes, but those changing the characteristics of genes with a regulatory function, especially ones involved with development. As Philip Whitfield, of King's College, London, pointed out, in his 1993 book, *The Natural History of Evolution*, 'The development genes can, at a stroke, produce massive potential alterations with enormous adaptive significance. Natural selection can, via these genetic agents, shape the simple buttercup or fashion the dazzling complexities of an orchid.'[5] Similarly, Neil Shubin of the University of Pennsylvania, Cliff Tabin of the Harvard Medical School and Sean Carroll of the University of Wisconsin wrote, in *Nature*, in 1997, 'The origin and diversification of fins, wings and other structures, long a focus of paleontology, can now be approached through developmental genetics. Modifications of appendage number and architecture in each phylum are correlated with regulatory change in specific patterning genes'.[6] Three years later, Carroll, this time in company with Artyom Kopp and Ian Duncan, demonstrated that changes in regulation of the *bric-a-brac* (*bab*) gene have played an important part in the evolution of abdominal segment morphology in *Drosophila*.[7]

Some of the regulatory genes thought to be of evolutionary importance are ones with the so-called homeobox (Hox) sequence. So, for example, Christopher Lowe and Gregory Wray, of the State University of New York, Stony Brook, presented a report of a study of echinoderm evolution in *Nature* in 1997, and concluded that 'the reorganization of body architecture involved extensive changes in the deployment and roles of homeobox genes'. Similarly, Marie Kmita-Cunisse and colleagues, of the Universities of Basel and Reims, wrote in 1998 in the *Proceedings of the National Academy of Sciences, U.S.A.*, 'From our current understanding of the genetic basis of development and pattern formation in *Drosophila* and vertebrates, it is commonly thought that clusters of Hox genes sculpt the morphology of animals in specific body regions'.[8]

Again, in *Nature* in 2002, Matthew Ronshaugen, Nadine McGinnis and William McGinnis of the University of California, San Diego, wrote, 'Mutations in homeotic (Hox) genes have long been suggested as potential causes of morphological evolution, and there is abundant evidence that some changes in Hox expression correlate with transitions in animal axial pattern'. They claimed to have identified the structural changes which had occurred in a Hox protein 400 Myr ago, at the time of the divergence of six-legged insects from multi-limbed arthropod ancestors, and concluded, 'To our knowledge, this is the first experimental evidence that links naturally selected alterations of a specific protein sequence to a major morphological transition in evolution'.[9]

Evidence for punctuated evolution continues to accumulate, as do findings which suggest that evolutionary change can, on occasions, be non-adaptive or influenced by environmental factors. Even the evolution of Darwin's Galápagos finches has been shown to have been a rapid, uneven process, linked to environmental changes.[10] Speciation remains a major problem, for, 150 years after Darwin, there is still no clear understanding of the mechanisms involved in the origin of new species.[11] In

Figure 18.1   Zoologist Richard Dawkins.

marked contrast to what was envisaged by the founders of the Modern Synthesis, macroevolution does not appear to be a straightforward extension of microevolution, with speciation simply being a stage in the even-paced accumulation of small mutations, each producing small effects. Even so, the 'hopeful monster' concept associated with the name of Richard Goldschmidt seems to be quite impossible if, as generally assumed, it involves a new species arising in an instant as a result of a macromutation in a single individual. How could a viable population of the new species, reproductively isolated from the old, ever be established by such a mechanism?

Richard Dawkins (figure 18.1) suggested in the multi-author work *Evolution from Molecules to Men* (a collection of papers marking the centenary of Darwin's death), and again in his 1996 book, *Climbing Mount Improbable* (where he maintained that progress was often achieved by an indirect route), that evolutionary jumps could, in theory, result from two possible kinds of macromutation, which he termed Boeing 747 and Stretched DC8 macromutations.[12] The former was derived from a scenario presented in 1983 by Sir Fred Hoyle (see Chapter 14), in *The Intelligent Universe*, involving a Boeing 747 aircraft being assembled as a result of a tornado blowing through a junkyard (which the author mistakenly regarded as being equivalent to the traditional view of evolution).[13] Needless to say, such an event was extremely unlikely to happen. However, the other possibility, analogous to the conversion of the conventional DC8 aircraft to the stretched-body modification, was perfectly plausible, and could be regarded as a natural extension of gradualism, particularly where the large change in the phenotype had resulted from a small change in the genotype, for example because of a mutation in a regulatory gene.

In many instances of speciation, a Stretched DC8 type of macromutation could be involved. As suggested by Stephen Jay Gould in 1980, developing ideas previously put forward by Goldschmidt (see Chapter 15), a mutation in a regulatory gene, conveying a significant advantage at the phenotype level, could be sufficiently minor at the geno-type level to enable the individual concerned to breed successfully with members of the previously existing species. That would allow the favourable mutation to spread rapidly by natural selection through the population in that locality. Eventually, an unrelated mutation might occur, causing the reproductive isolation of the sub-population pos-sessing the favourable characteristic, and establishing it as a separate species. That, however, is mainly guesswork for, except in a few cases, which may not be typical of the rest, we have no direct information about the formation of new species. Near the end of his 1989 book, *Evolutionary Genetics*, John Maynard Smith admitted:

> This book has been concerned with processes that can be studied in contemporary populations over relatively short time-periods. Our picture of evolution on a larger scale – macroevolution – comes from comparative anatomy and embryology, from taxonomy and geographical distribution, and from palaeontology. The question naturally arises whether the processes of population genetics are sufficient to account for macroevolution. Very different views can be held on this question.[14]

Despite all the subsequent discoveries about aspects of molecular genetics, that still remains the situation. Maynard Smith did not see fit to change a single word of the above passage for the second edition of *Evolutionary Genetics*, which was published in 1998.[15]

Reductionist and holistic approaches

Mae-Wan Ho of the Open University, and her collaborators, Peter Saunders of King's College, London, and Sidney Fox of the University of Miami, were amongst those who argued that genetics was only part of the overall picture of inheritance and mutational change. In a *New Scientist* article in 1986 they wrote:

> The present-day concept of heredity needs to be reformulated. Instead of a linear chain of command from DNA to phenotype, there is a complex of interlocking feed-back processes. Nucleus communicates with cytoplasm and cells communicate with cells. During development interactions between layers of cells induce tissues to form. In relationships between the organism and the external environment the internal processes [are] orchestrated and coordinated. Inheritance is a property of the whole system, not just the genes in the nucleus.[16]

The confusions of modern evolutionary biology were clearly apparent during the sub-sequent arguments about whether this constituted a new paradigm. Without question, evolution covers so many fields of study that nobody can possibly be an expert in them all. As discussed earlier in this book (see, in particular, Chapters 10 and 15), the perspec-tives of palaeontologists, gained as they are from the fossil record, have been difficult to reconcile with those of biologists who investigate living organisms. However, the latter group also covers a very wide range, from naturalists and ecologists to molecular biologists and geneticists, with those working in each of the different specialist areas having a different knowledge base and outlook to the rest. An attempt to provide an overall picture of evolution, accessible to biologists of all backgrounds, was made in 1994 by David Rollo, of McMaster University, Canada, in *Phenotypes: Their Epigenetics, Ecology and Evolution*.[17] In this book, Rollo pointed out that recent developments in molecular genetics had not resolved issues. On the contrary, they had emphasised just how much was still to be explained. Whilst the relationship between individual structural genes and their respective proteins might be straightforward, it was far from clear how the various processes of gene expression combined to give rise to overall characteristics such as arms, legs and eyes. Without question, a proper appreciation of the relationship between genotype and phenotype was still a very long way away. As Rollo saw the situation, we would come to a more complete understanding of evolution only by treating nature holistically. Natural selection might take place at various levels of phenotypic organisation, ranging from molecules to populations. The genome itself might also incorporate a hierarchical regulatory system, influencing processes such as metabolism and development. Taking all this into account, Rollo wrote:

> Rather than stressing the contributions of individual genes mainly in terms of the structural proteins they sometimes make, a regulatory framework stresses the way that genes and networks of genes are interconnected into complex circuits. Such circuitry is subject to selection for precise control of metabolism, differentiation and morphogenesis that may be the main target of phenotypic evolution.[18]

As we discussed in Chapter 15, Niles Eldredge (see figure 15.1) argued during the 1970s that our understanding of evolution was far from complete, and he has continued to do so. In company with those of like mind, such as Rollo (as we have just seen) and Stephen Jay Gould, he has consistently stressed the importance of hierarchies in the evolutionary process. One of these hierarchies is a *genealogical* one, involving reproduction at various levels, whilst another is an *economic* hierarchy, made up of systems concerned with survival. Eldredge explored these ideas in his aptly titled *Unfinished Synthesis*, published in 1985, and again in his 1995 book, *Reinventing Darwin*. In the former, he outlined his thesis in these words:

> Genes, organisms, demes, species and monophyletic taxa form one nested hierarchical system of individuals that is concerned with the development, retention and modification of *information* ensconced, at base, in the genome. But there is at the same time a parallel hierarchy of nested *ecological* individuals – proteins, organisms, populations, communities, and regional biotal systems, that reflects the *economic* organization and integration of living systems. The processes within each of these two process hierarchies, plus the interactions between the two hierarchies, seems to me to produce the events and patterns that we call evolution.[19]

Eldredge went on to argue that the founders of the Modern Synthesis had taken into account only some of these factors, generally ones from the lower levels of the genealogical hierarchy, whilst much the same view was still being propagated by those whom he labelled ultra-Darwinians.[20] He suggested that the reductionist approach was restricting developments, and wrote: 'Epistemologically, hierarchy theory is a formal embodiment of the principle that evolution is probably a more complex affair than the synthesis would have us believe. Hierarchy offers us, I think, a more realistic, hence more useful, framework for the investigation of this complexity'. Similar views continued to be expressed by Eldredge and Gould, and hierarchies were a key feature of 'Towards a revised and expanded evolutionary theory', the second (and final) part of Gould's substantial work *The Structure of Evolutionary Theory*, published in 2002.[21]

Richard Dawkins, on the other hand, is a prominent figure in the alternative, ultra-Darwinian camp. Right at the beginning of his 1995 book, *River out of Eden*, Dawkins expressed an enthusiastic commitment to Darwinism, claiming that in no other theory were so many facts explained by so few assumptions. Before coming to his reductionist message, he explained his main aims in writing the book. The first of these was to give 'due recognition to the inspirational quality of our modern understanding of Darwinism'. It could be seen, claimed Dawkins, that each apparent complication of detail in a living organism fulfilled some useful purpose. He then continued:

> The other feature of earthly life that impresses us is its luxuriant diversity: as measured by estimates of species numbers, there are some tens of millions of different ways of making a living. Another of my purposes is to convince my readers that 'ways of making a living' is synonymous with 'ways of passing DNA-coded texts on to the future'. My 'river' is a river of DNA, flowing and branching through geological time . . . [22]

In *River out of Eden*, Dawkins confronted the issue of the evolution of vision, so often seized upon by critics of Darwinism. He claimed that a frequent question of the creationists, 'What is the use of half an eye?', was easy to answer, for half an eye was better than no eye at all. The value of having a visual system, however basic, was beyond question, for eyes of some type or another had evolved on more than forty separate occasions. Furthermore, Swedish scientists Dan Nilsson and Susanne Pelger had demonstrated by computer simulations that a basic, light-sensitive sheet could evolve into a complex eye in straightforward fashion, guided only by the requirement that any change must be both small and an improvement on what had existed before.[23]

Dawkins was keen to emphasise that this hypothetical process was, in fact, driven entirely by natural selection. However, he made no detailed attempt to explain the origin, in biological systems, of the variants available for selection. As must be apparent, it would be far too simplistic to envisage the evolution of the eye in terms of mutations in a single gene, for the characteristics of a complex visual system depend on the expression of a number of genes. Hence, some rules of organisation could be involved in creating from mutations in different genes a range of variants which could be acted upon by natural selection.

Despite the arguments, there is in fact much common ground between the ultra-Darwinians and the supporters of hierarchies, as Dawkins, Eldredge and others have been careful to point out. There are, of course, some significant differences, with the ultra-Darwinians putting greater emphasis on the genome and on adaptive selection than the other group, yet the two camps are closer together than they may appear to be when they write about the same problem from different perspectives.

Nevertheless, the debates about details of evolution continue, within as well as between the various groups. Writing in 1991 in *One Long Argument*, Ernst Mayr commented: 'Darwinism is not a simple theory that is either true or false but is rather a highly complex research program that is being continuously modified and improved. This was true before the synthesis, and it continues to be true after the synthesis'.[24]

A conference entitled *Creative Evolution?!*, sponsored by the Center for the Study of Evolution and the Origin of Life, was held at the University of California, Los Angeles, in March 1993. At the beginning of the Proceedings, the editors, John Campbell and William Schopf, summarised the current situation in these words:

> One hundred and thirty-four years after Darwin's *The Origin of Species*, the historical fact of evolution has been scientifically established. Natural selection is the central mechanism for evolutionary change, joined by various other processes, such as mutation and random genetic drift. There is also close agreement about the general course of evolution over the past three billion years, as revealed by the geological record and comparative study of contemporary organisms. In contrast, fundamental disagreements remain about some of the *properties* of the evolutionary process.[25]

Later, in bringing the Proceedings to a close, Campbell used the term 'plain vanilla' to describe the traditional view of evolution adopted by the followers of the Modern Synthesis, and he concluded that 'plain vanilla Darwinism is as misleading a paradigm for the totality of the evolutionary process as a calculator is for computers or as baby talk is for human language'.[26]

A few years after this, as the world approached the new millennium, Steve Jones, Professor of Genetics at University College, London, wrote a book entitled *Almost Like a Whale*, which was published in 1999. The sub-title, 'The *Origin of Species* Updated', made it clear that what had been attempted was a re-writing of Darwin's masterpiece in the light of modern knowledge. Jones retained the same structure and chapter titles. In the fifth chapter, concerned with the laws of variation, he wrote about our changing understanding of genetics as follows:

> many apparently eccentric notions – use and disuse, acclimatization, the correlation of different parts of the same animal, the reappearance of characters long lost – turn out, in the light of modern biology, to have a basis in fact. Filled

with complexities and exceptions as it is, genetics remains as the rock upon which the whole edifice of evolution rests.[27]

Later, in the section on morphology in the thirteenth chapter, he described the importance of regulatory genes in development, providing an example by drawing attention to the fact that a single mutation in the fruit fly *Drosophila* can cause the middle section of the body to duplicate itself, resulting in a fly with four wings rather than two. He continued, 'To make a thorax takes thousands of genes, but a simple command can, it seems, set the whole army into motion. The additional organs appear because an order is given in the wrong place and the local cell machinery obeys it'.[28]

From this and many other examples, we now know just how great an influence a mutation in a regulatory gene can have on development. Nevertheless, this has still to be properly incorporated into a comprehensive theory of evolution. As Gabriel Dover pointed out, in 2000, in his book *Dear Mr Darwin*:

> The recent discoveries in biology are challenging how we think about biological evolution . . . To have a theory of evolution we need a theory of development; but to have a theory of development we need a theory of molecular interactions during construction of an organism. We don't have a theory of interactions and so it is difficult to have a comprehensive theory of evolution.[29]

Even if we did have such a theory, it would apply only to evolution taking place at 'normal' times. The issue of the causes and possible evolutionary consequences of mass extinction episodes would still remain. As we discussed in Chapter 14, it may once have seemed perfectly reasonable to imagine evolution proceeding according to its own internal logic, immune from external influences. However, as debates about evolutionary mechanisms were taking place through the 1980s and 1990s, scientists were becoming increasingly aware of factors, both terrestrial and extraterrestrial, which had the potential to interfere in dramatic fashion with the evolutionary process. Furthermore, there were reasons for thinking that such interference was likely to have occurred on a number of occasions. In the early days, evolutionary biologists were far from convinced but, as time passed, the situation started to change. It became increasingly difficult to ignore the physical context within which evolution had taken place, because of the accumulating evidence of potentially disruptive influences in the terrestrial environment, and also in space.

19 Chaos in the Solar System

Planets and satellites: impact craters and anomalous orbits

The Solar System in which we live is generally thought to owe its origin to the condensation under gravity of a large cloud of dust and gas, some 4,600 Myr ago.[1] This model was proposed, independently, by Pierre Simon de Laplace (see Chapter 3) and Immanuel Kant (see Chapter 5).[2] However, it has had to be modified, as a result of a study of lunar craters made during the Apollo missions of the American space agency, NASA, in the 1960s, which showed that the Moon had suffered an intense bombardment when it was very young (figure 19.1). In line with a suggestion made in 1944, prior to the Apollo programme, by the Russian geophysicist Otto Schmidt, and developed by the Russian mathematician Viktor Safronov and the American geologist George Wetherill, it now seems likely that the first product of the gravitational collapse was a spinning disc of dust grains, with a proto-Sun at the centre, very different from the Solar System as we know it today. A process of accretion then began to take place, as collisions between dust particles led, over a long period of time, to the formation of pebbles. Further collisions eventually produced larger stones and then, in turn, huge boulders and planetesimals. During this period, bodies must have been travelling in orbits of all kinds as a result of glancing encounters with other objects, giving rise to a very unstable situation. Head-on collisions could have smashed apart one or both of the objects involved, whilst bodies in similar orbits which came together gently would have tended to merge. Eventually, after hundreds of millions of years, the process gave rise to the present planets, in orbits which no longer brought them into contact with each other.[3]

In the 1980s, supercomputers such as the Digital Orrery at the Massachusetts Institute of Technology and the Cray at the University of London, used in Project LONGSTOP, directed by Archie Roy of the University of Glasgow, assessed gravitational interactions between the various planets and back-calculated planetary orbits for up to one fifth the age of the Solar System. These studies revealed that the orbits, particularly of Pluto and the inner planets, were more chaotic (i.e. less predictable) over time-scales of millions of years than previously thought to be the case. Nevertheless, it seemed clear that, during the past 1,000 Myr, the planets had never been located far from their present orbits, and that view remains the generally accepted one.[4]

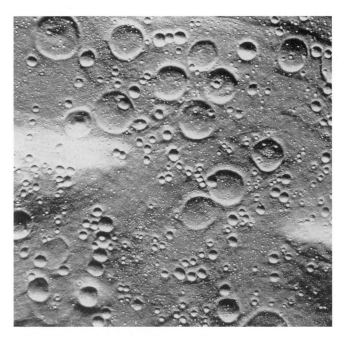

Figure 19.1   Lunar craters (NASA).

On the other hand, there is now plentiful evidence that smaller cosmic bodies can cross these orbits, giving them the potential to interact with the planets and their satellites. Also, Pluto, currently regarded as the outermost planet (although its status as a true planet has recently been brought into question, partly because of its small size) has an orbit with a higher eccentricity, i.e. one which is more elliptical, than that of the next planet, Neptune. This brings Pluto inside the orbit of Neptune on occasions, including the period between 1979 and 1999.[5] Just as Pluto was moving back beyond Neptune in 1999, the Indian-born physicist Renu Malhotra, then at the Lunar and Planetary Institute in Houston, claimed in a *Scientific American* article that interactions with planetesimals during the later stages of planetary formation had increased the average orbital radius of Neptune by about one third, to its present value of 30 A.U. (astronomical units, defined as the average distance from the Earth to the Sun), causing the giant planet to interfere with Pluto. The orbits of the other gas giants, Uranus, Saturn and Jupiter, would have expanded in similar fashion to that of Neptune, although not to the same degree.[6]

It has been known for many years that some satellites of the giant planets orbit in the direction of rotation close to the equatorial plane, and these are thought to be part of natural systems. Others have very different characteristics, and may well be captured intruders. For example, Triton, the largest moon of Neptune, travels in the 'wrong' direction, and calculations indicate that it may spiral in towards the planet and eventually collide with it. Nereid, another moon of Neptune, moves in the more normal direction, but in an orbit which is extremely elliptical. Also, the orbital planes of Triton and Nereid are inclined at unusually large angles to the equatorial plane of Neptune. The similarity of these satellites, particularly Triton, to Pluto suggests that they may originally have been members of the same family of cosmic bodies.[7]

Just like satellites orbiting a planet, most planets travel around the Sun with their orbital planes close to the Sun's equatorial plane, and with their axes of rotation roughly perpendicular to it. In marked contrast to this, Uranus orbits the Sun with its axis of rotation in approximate alignment to its orbital plane, apparently having been knocked over in a collision with a giant comet or other cosmic body. Uranus has five major satellites, and ten minor satellites were detected by the Voyager 2 fly-by in 1986, all in the planet's equatorial plane, at right angles to its orbit. So close together are six of the minor satellites that they could be fragments of a larger moon which has broken up. Voyager 2 also showed that, with the exception of Umbriel, the major satellites have surfaces which appear to be mosaics of rock and ice. After the collision which turned Uranus onto its side, a new satellite system may have formed from the debris. There could then have been solid-state extrusion of ice from the interiors of the satellites, accounting for the surface features. Alternatively, the new satellites might have broken up again as a result of subsequent impacts, and the component parts re-assembled by gravitational attraction, after a certain amount of mixing.[8]

The NASA Voyager missions of the 1980s provided abundant evidence of impact craters on satellites of all the outer planets.[9] It is likely that most, but by no means all, of these impacts occurred in the early days of the Solar System, over 4,000 Myr ago.[10]

Mimas and Tethys, two of the eight major moons of Saturn, have impact craters which cover a significant proportion of the surface, suggesting in each case a collision that must have come close to causing the satellite to disintegrate. Enceladus, another Saturnian moon, shows signs that it might have been broken into fragments and then re-formed. Phoebe, which orbits Saturn in a highly elongated retrograde fashion, is rocky, whereas the other major Saturnian moons have icy surfaces. For these reasons, Phoebe would seem to be an interloper into this family, together with many (if not all) of the 23 minor moons that have now been identified.[11]

Of the various moons of Jupiter, the six inner ones move in circular orbits in the planet's equatorial plane. Hence, despite major differences in size and other characteristics, they presumably constitute a natural satellite system. That is unlikely to be the case with 31 small outer satellites, known to be travelling in unusual orbits, many of them with retrograde motion.[12]

Only two of the four inner planets have satellites, and those of Mars are small, irregularly shaped and in orbits which indicate they may be captured intruders.[13] In contrast, the Earth's Moon is big enough to be considered a planet in its own right, being about the same size as Pluto. It seems likely that the Moon was formed long ago by accretion of material blasted into space from the young Earth, as a result of a

Figure 19.2   The cratered surface of Mercury, photographed by Mariner 10 (NASA).

collision with a body which may have been larger than Mars.[14] The Moon is heavily cratered, as are Mars and Mercury (figure 19.2).[15] Venus, however, as shown by the Magellan probe, between 1990 to 1994, is not cratered to the same extent as some other planets and satellites. A study of the crater densities has indicated that, for reasons which are still uncertain, the present surface of the planet solidified from molten rock only about 500 Myr ago.[16]

Asteroids and comets: the past, present and future threat to planets

The planets and their satellites have revealed a considerable amount about their interactions with other cosmic bodies during the history of the Solar System. This picture of frequent collisions and near-misses will come into even clearer focus as we turn our attention towards these smaller bodies.

It has long been known that a vast asteroid belt lies between the orbits of Mars and Jupiter. The largest members are Ceres (940 km in diameter), Pallas (580 km) and Vesta (575 km), but most are very much smaller than these three. At one time, it was thought that the asteroid belt consisted of material from a shattered planet, but now it is generally believed that no complete planet ever formed in this region of the Solar System, perhaps because of the gravitational influence of Jupiter. Different asteroids consist of a wide range of combinations of rock (predominantly either carbonaceous or silicate minerals) and metal (mainly iron). Many are probably fragments produced from the disintegration of larger bodies which consisted of a metallic core enclosed within a rocky crust.[17]

The spacecraft Galileo transmitted to Earth images of the silica-rich stony asteroids Gaspra, in 1992, and Ida (figure 19.3), two years later, revealing these to be irregular-shaped, cratered bodies, rather like the moons of Mars. Similarly, close-up pictures of the carbonaceous asteroid, Mathilde, were obtained by the Near Earth

Figure 19.3   The main-belt asteroid, Ida, which has a length of about 60 km and a width of about 25 km (NASA).

Asteroid Rendezvous (NEAR) spacecraft in 1997. Mathilde was found to be very dark and approximately spherical, around 50 km in diameter, with five craters larger than 20 km across on the side facing the camera. The NEAR spacecraft was then directed to make a close approach to the stony asteroid, Eros, in 1999, showing this to be heavily cratered and shaped like a peanut, 33 km in length and 13 km in width.[18]

Inside the main asteroid belt, crossing the orbit of Mars, is another group of asteroids, called the Amors. Still further inwards are found the Apollo asteroids (which include Apollo itself, whose discovery in 1952 we noted in Chapter 14), and even closer to the Sun are the Aten asteroids. Each Apollo asteroid has an average orbital radius greater than 1 A.U., whereas each Aten asteroid has an average orbital radius less than that value. However, definitions based on average distances do not tell the whole story, for Apollo and Aten asteroids generally have highly elliptical orbits, causing many of them to cross the Earth's path, with obvious dangers for our planet.[19]

Outside the main asteroid belt, Chiron, an object 200 km in diameter, was first detected in 1977 in an elongated orbit around the Sun that brings it on occasions into the vicinity of Saturn, a situation which could cause it to be deflected towards the inner Solar System, and hence the Earth. Chiron could be either an asteroid or the dormant core of a giant comet, because it sometimes shows signs of possessing a tail. A number of similar bodies, also in orbits crossing those of the outer planets, were discovered in the 1990s. One of these, around the same size as Chiron, and similarly showing some evidence of a tail, is called Pholus, whilst another, perhaps even larger, has been named Chariklo. Slightly smaller ones, with diameters in the range 50–100 km, include the objects named Asbolus, Hylonome and Nessus. All of these objects have been classified as 'Centaurs'. Another body, in a particularly elongated orbit that brings the risk of deflection by every planet from Uranus to Mars, was detected in 1991 and named Damocles. This was because it is a potential threat from above, like the legendary sword of Damocles, which hung by a thread over the head of Dionysius I, tyrant of Syracuse, as he ate his meals.[20]

Gerard Kuiper (1905–1973), a Dutch–American astronomer, suggested in 1951 that there was a belt of comets and planetesimals beyond the orbit of Neptune, up to around 200 A.U. from the Sun. Eight years previously, a similar suggestion had been made, without Kuiper being aware it, by the British astronomer, Kenneth Edgeworth (1880–1972).[21] The first physical signs of this belt came in 1992, when the American astronomers David Jewitt and Jane Luu detected a 200 km object, 1992 QB_1, in the outer regions of the Solar System. Now, several hundred other bodies over 100 km in diameter have been observed between the orbits of Neptune and Pluto, and it has been estimated that there could be more than 100,000 objects of this size in the Edgeworth–Kuiper Belt. Amongst them are 2000 WR_{106} (Varuna) and 2001 KX_{76} (Ixion), both of which have diameters between 500 and 1200 km.[22] Although Pluto itself is much larger than this, with a diameter of around 2300 km, it has been suggested, on the basis of spectroscopic observations, that it should be regarded as being a member of the family of Edgeworth–Kuiper bodies, together with the 'captured' Neptunian satellite, Triton. Perhaps around ten thousand million smaller bodies may be present in the belt, and some evidence of these has been found.[23]

In 1997, Luu and her colleagues reported the presence of a bright object almost 500 km in diameter, labelled 1996 TL_{66}, moving in a highly elliptical orbit which will eventually take it far beyond the supposed limits of the Edgeworth–Kuiper Belt. This could be a member of a 'scattered disc' produced by the gravitational influence of Neptune on bodies originating towards the inner edge of the Edgeworth–Kuiper belt, or the outer edge of another belt postulated to exist between the orbits of Neptune and Saturn.[24]

Even further from the Sun, it is generally believed that a vast number of cometary nuclei (maybe more than one hundred thousand million) lie dormant far beyond the orbit of Pluto, forming a spherical cloud which encloses the entire Solar System. The arguments for this were presented in 1950 by the Dutch astronomer Jan Oort, in support of an earlier suggestion by Ernst Öpik. The evidence came mainly from the characteristics of known long-period comets (those with a periodicity greater than 200 years), whose highly elliptical orbits were found to cover a wide range of angles to the ecliptic. This 'Oort cloud' of cometary nuclei occupies the region of space from around 10,000 A.U. from the Sun to a solar distance which may be greater than 100,000 A.U.[25] Most of these cometary nuclei are likely to be between one and ten kilometres in diameter, typical of those seen travelling past the Earth, but some may be much larger than that.[26]

Although part of our Solar System, the Oort cloud is not isolated from surrounding space. Stars move about relative to each other, and occasionally one will come close enough to the Sun to disturb the Oort cloud (and possibly even to exchange some comets). So, for example, according to observations made in 1997, Gliese 710 will move from its present position in the constellation Ophiuchus and pass within the boundaries of the Oort cloud in about a million years time.[27] In the course of such an event, it is likely that some comet nuclei would be propelled into the inner Solar System. The orbits of these could be changed as a result of encounters with planets and, like Halley's Comet, they might then stay within the orbit of Pluto, regularly travelling past the Earth as short-period comets. Many other short-period comets could originate in a belt believed to be present between Saturn and Neptune, of which the Centaurs may be particularly large representatives, or slightly further out in the Edgeworth–Kuiper Belt.[28]

There are, in fact, two groups of short-period comets. Members of the Halley family, which of course include the comet of that name, complete their circuits in periods between 20 and 200 years, and their orbits show a wide range of inclinations to the Sun's equatorial plane, whereas members of the Jupiter family, comprising about 90% of the total, remain within a relatively flat disc, so are unlikely to have originated within the Oort cloud.[29] Comets in the latter group have periods of less than 20 years and, as their family name implies, frequently pass close to Jupiter. The Russian astronomer, Anders Johan Lexell (1740–1784), calculated that the comet named after him moved into a short-period orbit in 1767 after passing near to Jupiter. After missing the Earth by a mere 2.25 million km (six lunar diameters) in 1770, it then moved back into a long-period orbit after a further brush with the giant planet.[30]

All of this gives a clear picture of bodies, many of them of substantial size, crossing the orbits of planets, including the Earth, and sometimes coming into contact with them. It would seem that some asteroids have been captured by planets, to become the satellites which are additional to natural systems. Many asteroids and comets must also have collided with planets and other large bodies, on the evidence of the number of impact craters observed.[31] As already stated, it is probable that most of the craters were formed early in the history of the Solar System. However, impacts continued to occur after that time, right up to the present day. According to the evidence available, mainly from the Moon, the average rate of impact has changed little over the past 3,000 Myr. Furthermore, there seem to have been episodes of higher-than-normal activity at intervals throughout this time-period.[32]

Meteorites found at the surface of the Earth are generally thought to be remnants of bodies formed in the early days of the Solar System and, in many cases, dating studies have given results consistent with that view. However, members of the Shergottite, Nakhlite and Chassigny (SNC) group of meteorites seem to be much younger, consisting of rock which solidified less than 1,300 but more than 200 Myr ago. Analysis of the SNC meteorites suggests that they may have been hurled to Earth from Mars in an explosion caused by a large impact. Furthermore, a number of older meteorites may also have come from the same planet, or from a different one. Some achondrite meteorites found in Antarctica appear to consist of lunar rock, whilst yet another group may have originally been part of the large asteroid Vesta.[33] Studies on the orbits and surface composition of some minor asteroids have indicated that these might have been blasted from Vesta by an impact.[34] Collisions between main belt asteroids, and close encounters between asteroids and other bodies, may still move material into Earth-threatening orbits, as new members of the Apollo or Aten asteroid belts.[35] Calculations have shown that the asteroid Eros is in an orbit which will be changed by a close encounter with Mars, perhaps eventually bringing it into collision with the Earth. However, we have been reassured that this will not happen within the next 100,000 years.[36]

Until the late 1980s, it was generally believed that the Oort cloud had changed very little since the birth of the Solar System. Then, Victor Clube and some other British astronomers argued that, because of capture during encounters with giant molecular clouds, very few comet nuclei would be found in the Oort cloud at the present time (instead of the many millions apparently there), unless it had been replenished at intervals from other sources, especially from interstellar clouds during a passage of the Solar System through a spiral arm of the Galaxy. Alternatively, many of the comets now in the 'outer' Oort cloud could have been displaced from regions closer to Pluto, where they would have been safe from being stripped by giant molecular clouds. Conversely, at least some of the comets now in the Edgeworth–Kuiper belt might have moved there from original locations further out.[37] The situation has been far from static. Those episodes when impact craters on planets and satellites appear to have been formed at an increased rate may reflect occasions when there has been an influx of comet nuclei into the inner Solar System, whether from the Oort cloud or from beyond.[38]

Cometary nuclei are generally thought to be a blend of ice and dust, forming compacted 'dirty snowballs', a view first put forward by the American astronomer Fred Whipple in 1950. As a comet approaches the Sun, some of this material evaporates, producing the coma around the nucleus as well as the spectacular tail.[39] The 1986 Giotto and VeGa missions to Halley's Comet revealed an icy core, which was largely covered by an insulating crust of dust particles. Through the gaps in the crust, jets of volatile material, mainly water vapour and carbon dioxide, together with dust, surged from the nucleus. Similar findings were also made when Deep Space 1 approached the 8-km-long potato-shaped nucleus of Comet Borrelly, a member of the Jupiter family, in September 2001.[40] In neither case was there evidence of any larger rocks in the ejected material, but it is likely that had some been present within the nucleus, they would have remained there.[41]

It has become increasingly apparent that some asteroids may be the remains of former comets, possibly as a result of an icy nucleus being sealed off completely by the formation of an insulating crust, or by all of the volatile material evaporating away from a rocky interior. Such a scenario might be true both for bodies found in the main asteroid belt and those in the inner Solar System.[42] Some 1–5-km-diameter objects that are regarded as Apollo asteroids, for example Phaethon and Oljato, are associated with streams of meteoroids, which pass through the Earth's atmosphere as meteors (shooting stars). Meteoroids are generally assumed to be material ejected by comets, the Eta–Aquarid and Orionid streams, for example, being linked to Comet Halley. Phaethon's orbit is almost exactly the same as that of the Geminid meteoroid stream, whilst Oljato, which shows signs of having a faint tail, is associated with the Taurids. Like Oljato, other Apollo asteroids in this size range, including 1982 TA and 1984 KB (Jason), have orbits similar to those of the Taurid meteoroids, and so does Comet Encke, suggesting that this entire grouping of dust and larger objects may have arisen from the break-up of a huge comet.[43]

Regardless of that, and whatever the specific details of the processes involved, it is now well established that the inner asteroid belts are not decaying remnants of the early Solar System. They are still supplemented from time to time with new material pushed inwards from the outer regions of the Solar System, some coming from the cometary zones and the rest from the main asteroid belt. The Earth is a dangerous place to live, and will continue to be so.[44]

20   Catastrophes on Earth

Rocks in the skies and holes in the ground

As we saw in the previous chapter, the asteroids which form the Apollo and Aten belts are in Earth-threatening orbits. If the present compositions of these inner asteroid belts are typical of what they have been in the past, it is likely that more than ten asteroids of at least 0.5 km in diameter, including some with diameters in excess of 1 km, will have struck the Earth every million years.[1] These estimates were made during the early 1990s by, in particular, Clark Chapman, of the Planetary Science Institute, Tucson, Arizona, and David Morrison, of the NASA Ames Research Center, California, from a data-base compiled by Eugene Shoemaker (1928–1997) of the US Geological Survey.[2] Further calculations showed that a 0.5 km asteroid should have an impact energy equivalent to about 10,000 megatons of TNT, which is half a million times greater than the energy of the atom bomb dropped on Hiroshima at the end of World War II whilst, for a 1 km asteroid, the impact energy could be in excess of a million megatons. At an even more destructive level, it is likely that the Earth has been struck by between five and ten asteroids with diameters greater than 10 km and impact energies in excess of the equivalent of 100 million megatons of TNT, within the past 500 Myr.[3]

Cometary nuclei may possibly be lighter than asteroids of similar size but, particularly in the case of ones accelerating in from beyond Jupiter, they are likely to be travelling faster relative to the Earth, so could be just as devastating on impact.[4] Although there is less chance of the Earth being struck by a comet than by an asteroid, the possibility is not insignificant. Indeed, it has been argued that around one third of all terrestrial impacts by cosmic bodies 1 km or more in diameter can be attributed to collisions with long-period comets, and the proportion could be even greater in the case of larger impacts .[5]

These various estimates may or may not be accurate, but there is clear evidence from impact craters that extraterrestrial projectiles, whether asteroids or comets, have indeed struck the Earth at intervals over the past 500 Myr. Over 140 impact craters have been identified at the Earth's surface, mostly dating from the past 200 Myr. Craters older than that are difficult to locate, because of erosion and other geological processes, and large craters of any age are hard to identify, because they cannot be easily seen as a whole from the surface of the Earth. Furthermore, parts of the perimeter wall may have collapsed, or much of the crater filled by sediment or by flows of volcanic lava. Many projectiles must have struck in areas of the world where little investigation has taken place, particularly in the oceans, which cover 80% of the Earth's surface, or in areas of great tectonic activity, where craters would be destroyed relatively quickly. Yet, regardless of all these difficulties, three craters of around 100 km diameter have been recognised for some time, each of which must have been caused by the impact of a missile of around 5 km diameter. These are the structures at Puchezh–Katunki, Siberia, and Manicouagan, Canada, both formed, in broad terms, around 200 Myr ago, and another Siberian crater, the 35-Myr-old Popigai structure.[6] Now, several even larger craters are known from the far-distant Precambrian Period, as well as two others from more recent times: a 180-km-diameter crater at Chicxulub, Mexico, now generally regarded as being produced by the impact of a 10-km projectile, and a 120-km crater near Woodleigh, Western Australia. Both of these escaped detection for many years because they are now covered by sedimentary rock to a depth of hundreds of metres.[7]

Moving down in scale, the impact of a 1-km body should produce a crater of 10–20 km in diameter and the geological evidence, given the uncertainties outlined above, indicates an average of about ten 10-km craters being formed every million years, consistent with the astronomical observations.[8]

Impacts by smaller bodies occur on a more frequent basis. One such event produced the most famous impact structure in the world, the Barringer 'Meteor' Crater of Arizona (see Chapter 14), which is around 1.2 km in diameter and 25,000–50,000 years old. Its relatively recent formation means that there has been little chance for significant erosion to take place, which explains its spectacular appearance. A crater of this comparatively small size is likely to have been formed by an asteroid of about 60 metres in diameter with an impact-energy equivalent to some 20 megatons of TNT. The projectile was probably made of iron, because any other type of body less than 150 metres across is unlikely to survive passage through the Earth's atmosphere.[9]

Sometime during the past 10,000 years, a shower of projectiles appears to have struck northern Argentina at a low angle, producing at least nine craters, the largest around 100 metres in diameter, containing metal meteorite fragments. The region in which the event took place is now known as Campo del Cielo, i.e. 'Field of the Sky', in line with a local Indian tradition. That may not have been the only such event to have occurred in Argentina, for there is evidence of another to the northeast of Campo del Cielo, at Rio Cuarto, although the cause of the craters in that location has recently been disputed. It seems that impact glass thought to be associated with them may in fact be from an older tektite field.[10]

A similar group of craters, again just a few thousand years old, was discovered at Ilumetsa, Estonia, in 1938. Yet another cluster of about the same age, or perhaps significantly younger, was found in 1932 at Wabar, in Saudi Arabia. Possibly the famous Black Stone of the Kaaba, in Mecca, is an impactite, a mixture of fused sand and meteoritic material, formed during the production of the Wabar craters. Rocks resembling impactites have also been found in eastern Uruguay, suggesting the possibility of a similar event in that part of the world.[11] Again, an impact in the Egyptian Sahara may have produced some of the precious stones found in the tomb of the pharaoh Tutankhamun.[12]

More recently, three fireballs were reported to have hit the rainforest of western Brazil, near the Curaçà river, in 1930, setting large sections of it ablaze.[13] An 8.5-metre impact crater was formed in Estonia in 1937, and a 5-metre crater was produced by an iron meteorite near Sterlitamak in eastern Russia in 1990.[14] Over 100 small craters, the largest 14 metres across, were formed in 1947 when an iron meteorite exploded in the atmosphere over the Sikhote–Aline region of Siberia.[15]

Also in Siberia, the well-known event which occurred in 1908 in the Tunguska region did not result in the formation of a crater, because the projectile apparently exploded before it hit the ground. Indeed, the fact that four Soviet expeditions, between 1921 and 1938, were unable to find either a crater or traces of meteorites in the area led to many arguments about the possible cause, with wild speculations involving antimatter, black holes and alien spacecraft being put forward.[16] However, it is now generally accepted that the characteristics of the event are consistent with the arrival of a 50–100-metre diameter stony or carbonaceous asteroid, or cometary fragment.[17] The incident happened when the Earth was passing through the ß-Taurid meteoroid stream so, as suggested by the writer Arthur C. Clarke in 1980, the object which caused the damage may have been a member of the system associated with this stream and Comet Encke, mentioned in the previous chapter.[18] It has been estimated that the impact energy was equivalent to about 10–30 megatons of TNT, i.e. at least 500 times more powerful than the Hiroshima atomic bomb.[19] Because of the remoteness of the Tunguska region, it is easy for outsiders to underestimate the full scale of the event, but the first investigator, Leonid Kulik, pointed out that had the object arrived four hours later than it did, it would have wiped out St Petersburg. Similarly, if had it exploded over central Belgium, that entire country would have been devastated, and had it fallen over the sea, it could have produced a lethal tidal wave, or tsunami.[20]

A comparable impact may have occurred about 800 years ago on the South Island of New Zealand, on the evidence of widespread destruction of the forest at the time, and stories in Maori tradition. There could also have been others in the past millennium of which we know nothing, the bolides having exploded over parts of the world which were largely uninhabited. Calculations indicate that Tunguska-scale events could occur, on average, every two or three centuries, and that might even be a conservative estimate.[21]

According to the chronicle of the monk Gervase of Canterbury, a disturbance was observed on the Moon in 1178. Firm evidence is lacking, but this could have been the 100,000-megaton impact which formed the 22-km diameter Giordano Bruno crater.[22]

Hermes, a 1-km diameter Apollo asteroid, passed by the Earth at less than two lunar distances (around 750,000 km) in 1937, as did a similar body, 1989 FC, since named Asclepius, half a century later.[23] Early in 2002, just a month after its discovery, 2001 YB_5, an asteroid slightly less than half a kilometre in diameter, also missed the Earth by around two lunar distances.[24] Toutatis, a 5-km asteroid, will pass within four lunar distances in the year 2004, having missed by a larger margin in 1996, and the 1-km asteroid 1997 XF_{11} will approach to within about two lunar distances in the year 2028.[25] Asteroid 1997 XF_{11} will again pass close to the Earth in 2040, as will asteroid 1998 OX_4 in 2046 and asteroid 1999 AN_{10} in 2044. Similarly, asteroid 1999 RM_{45} will brush past the Earth in both 2042 and 2050, and so will asteroid 2000 BF_{19} in 2022. Asteroid 2001 PM_9 will come close on several occasions, including 2005 and 2007, whilst there is a small risk of a collision with asteroid 2002 CU_{11} in 2049. There is also a possibility of strikes by the 2-km asteroid 2002 NT_7 in 2019, and the 1-km one 1950 DA in 2880.[26]

As for smaller cosmic objects, the Leonid meteoroid stream, consisting of material ejected from Comet Tempel–Tuttle, produces displays of shooting stars when coming into contact with the Earth's atmosphere in mid-November each year, amongst the most spectacular being those in the years 1799, 1833, 1866 and 1966.[27] The display in 1999 was also impressive, although not in Britain, where there was extensive cloud cover, and small flashes were observed in the Moon's surface, apparently as a result of impacts there. There were also reports of flashes on the surface of the Moon in November 2001.[28] In 1975, lunar seismometers left by an Apollo mission recorded that a swarm of large boulders associated with another meteoroid stream, the Taurids, struck the Moon over a five-day period, and the loss of the European Space Agency's Olympus spacecraft has been blamed on damage caused by the Perseid meteoroid stream.[29]

Asteroid 1991 BA, with a diameter of about ten metres, missed the Earth by the narrow margin of 170,000 km in 1991. Over the next decade, several other asteroids of around the same size, including 1993 KA_2, 1994 ES_1, 1994 XM_1 and 2000 LG_6, were seen to miss the Earth by similar distances.[30] Even if one of these small asteroids had hit, it would not have caused much, if any, damage. Indeed, such events are far from uncommon. It has been suggested that house-sized comets bombard the Earth on a daily basis, their nuclei breaking up and melting in the upper atmosphere, but that interpretation of the evidence remains controversial.[31] Whatever the truth about small comets, it is a well-established fact that 10-metre asteroids arrive on a collision course with the Earth at a rate of about one per year and, generally, burn up completely in the atmosphere. An encounter with an asteroid 20 metres in diameter is ten times less frequent than with a 10 metre one, but is potentially more dangerous. However, even so, the projectile rarely reaches the Earth's surface. Even if a small meteorite does penetrate the atmospheric shield, it poses only a very localised threat.[32]

An object of uncertain size was filmed passing almost horizontally over the Rocky Mountains in 1972, appearing to observers in Salt Lake City and the Zion, Grand Teton and Yellowstone National Parks as a spectacular fireball, before returning safely to space. In October 1992, another fireball was seen by thousands of spectators at a football game in Peekskill, New York. This time, a meteorite hit the ground, causing damage to a car as it struck.[33] Sixteen months later, the passage of an object through the atmosphere over the Pacific Ocean, releasing energy equivalent to a 0.1 megaton explosion, was detected by six spy satellites, and apparently led to Bill Clinton, then President of the United States, being awakened by his staff.[34]

In December 1997, a fireball passed eastward before dawn over the southwestern corner of Greenland, and then blew up into at least four fragments. Just five days later, according to reports from Colombia, three fireballs struck Bogotá, one of them causing the death of four children by setting their home ablaze.[35]

In February of 1998, the explosion of a milk churn in Belleek, Northern Ireland, caused a security scare, before it became clear that a meteorite, not a bomb, was responsible. During March of the same year, a fireball exploded over Monahans, Texas, and two stony meteorites, each weighing more than a kilogram, fell to the ground. In the following month, a daytime fireball was seen passing over Queensland, Australia. A few weeks later, a brilliant object passed over New Mexico for about 8 seconds and then exploded. A cluster of meteorites was found, of total weight in excess of 100 kg, the largest weighing around 20 kg. Seven days after that, a dazzling orange light trailing black smoke moved across the daytime sky in Turkmenistan. The bolide broke into several pieces before impact, the largest producing a crater four metres wide, from which a stony meteorite weighing 820 kg was recovered. A month later, in July 1998, a small stony meteorite fell at the feet of a startled golfer in Ontario, Canada. Then, at the very end of 1998, a 30-kg meteorite impacted in southern Portugal, producing a small crater.[36]

In January 1999, a fireball exploded over Alaska. Later, in July of the same year, another exploded over the North Island of New Zealand, showering the Earth with meteoritic fragments. A month after that, a bright object was seen in the evening sky passing over Albuquerque, New Mexico. Then, in October 1999, several fireballs were observed over eastern Canada, after which a number of explosions were heard. Another fireball exploded over County Carlow, Ireland, in late November, after which a number of meteorite fragments, the size of golf-balls, were found lying on a country road.[37]

Early in 2000, an explosion over Yukon, Canada, resulted in the fall of a shower of carbonaceous chondrite meteorites, weighing in total more than 1 kg. Later in the summer of that year, a fireball exploded over Colorado, and another was seen (at a distance) to strike the Earth in north-central Victoria, Australia. A few months afterwards, a bright light passed over Arizona and California, heading towards the Pacific Ocean, and two similar ones were observed travelling over Spain. Then, as 2000 drew to a close, a fireball exploded after passing over Vermont. Similar events also occurred in 2001, over Calgary (Canada), New Hampshire, Colorado, Pennsylvania, Northern Ireland, the North Sea, southern Scotland, northern Germany, Jordan and Western Australia.[38]

In any one of these instances, and many others (excluding, of course, ones where there was an Earth-bound explanation, or where the fireball was subsequently shown to have been caused by the re-entry of an artificial satellite), the small projectile making the hit or near miss could easily have been a much bigger one. The arrival of each came as a complete surprise, and there are many such Apollo asteroids, perhaps 90% of the total, with some (albeit a lesser proportion) one kilometre or more in diameter, whose orbits have still to be determined. Amongst these is Hermes, which was discovered in 1937, but lost again within a few days. Another asteroid of similar size, Albert (named after Baron Albert von Rothschild, a benefactor of the Vienna Observatory), was tracked briefly in 1911, but not re-discovered until 2000. More than 500 near-Earth asteroids with diameters greater than 1 km have now been identified and, as J. Scott Stuart of the Massachusetts Institute of Technology reported in *Science* in 2001, results of the LINEAR (Lincoln Near-Earth Asteroid Research) survey at White Sands, New Mexico, indicated that the actual number in this size range is likely to be around 1,200. The number of smaller ones is certain to be many times greater than that.[39]

Asteroid 1996 JA$_1$, with a diameter of around 0.25 km, was first observed on 15 May 1996 and four days later passed within 450,000 km of the Earth, i.e. just slightly further way than the Moon. On 8 March 2002, an asteroid of about a quarter the size of 1996 JA$_1$, 2002 EM$_7$, travelled past the Earth at a similar distance and was not detected until four days afterwards, because of the glare of the Sun. Three months later, on 14 June, 2002 MN, a similarly sized object (i.e. one in the Tunguska range) came even closer, passing within 120,000 km of the Earth, yet was only detected three days later.[40] Comets with Earth-threatening orbits can also arrive unexpectedly. Comet Hyakutake, with a nucleus over 1 km across, was seen for the first time on 30 January 1996, and passed by the Earth less than two months later. As it happened, it missed by the comfortable margin of fifteen million kilometres but, for all we knew, it could have been on a collision course.[41]

Consequences of impact

A 1-km object striking the Earth from space would cause considerable destruction of living creatures, over a significant proportion of the Earth's surface. What is more, larger-scale events are perfectly possible, and these would be expected to cause even greater devastation. The impact of a missile with a diameter in excess of 10 km could well have effects which were truly worldwide and extremely long-lasting.[42] Comet Iras–Araki–Alcock, with a nucleus believed to be of this size, passed within 4.7 million kilometres of the Earth, i.e. three times closer than Comet Hyakutake, in May 1983.[43] In 1992, concern was expressed that Comet Swift–Tuttle, which was then making a close approach, might hit the Earth on its next visit to the inner Solar System in 130 years time. This comet, which was associated with the Perseid meteor stream, was another with a 10-km diameter nucleus. Swift–Tuttle will indeed cross the Earth's path in the year 2126. However, detailed calculations showed that it will almost certainly miss the planet on that and several subsequent occasions, but it might possibly collide in 3044.[44]

Again in 1992, Comet Shoemaker–Levy 9 passed so close to Jupiter that its nucleus split into several fragments. These spread out and looped around to collide with the giant planet in July 1994, producing spectacular and long-lasting effects. After that, even the most die-hard sceptics had to accept that comets could indeed strike planets.[45]

Shoemaker–Levy 9 was not the first comet known to have fragmented, for one was seen to do so in 1744, and another in 1882. The latter was one of the Kreutz family of 'Sun-grazing comets', named after an early investigator, the German astronomer, Heinrich Kreutz, who lived in the nineteenth century. During the 1960s, Brian Marsden, later to become Director of the Minor Planet Center at the Harvard-Smithsonian Center for Astrophysics, Massachusetts (approved by the International Astronomical Union as the repository for data on asteroids and comets), retro-calculated the orbits of the Kreutz group. He found that all the members may have been derived from a giant comet which entered the inner Solar System about 15,000 years ago, and suggested that the spectacular comets of 371 B.C. and 1106 A.D. were large fragments which subsequently broke down further. Another comet, named LINEAR (since it was discovered during the survey with that acronym, mentioned earlier) travelled into the inner Solar System on a long-period orbit towards the end of 1999 and then unexpectedly began to disintegrate, possibly as a result of a collision with an asteroid, or because of an unstable arrangement of its nuclear components.[46]

Comet Machholz 2, which was discovered soon after the final impact of Shoemaker–Levy 9 on Jupiter in 1994, also turned out to have a fragmented nucleus. It comes close to the Earth every six or seven years, but will not collide with us in the foreseeable future.[47]

Comet Hale–Bopp was discovered in 1995, moving towards the inner Solar System from beyond Jupiter. Its unusual brightness indicated that it had a nucleus which was at least 10 km in diameter (and possibly several times larger than that) but, again, it quickly became apparent that it was not on a collision course with the Earth. Indeed, when it passed by the Earth in spectacular fashion in 1997 (figure 20.1), it was at a safe distance of almost 200 million kilometres. Nevertheless, we cannot guarantee that the next new comet to show up will not be heading straight for us.[48]

Should a 10-km object strike the Earth, projectile and rock would be instantly vaporised, leaving a crater of about 180 km in diameter. For example, an impact centred on Milton Keynes in Buckinghamshire would not only obliterate that much-reviled 'new' town, but would also produce a crater stretching from the centre of London to Nottingham, taking in cities such as Birmingham, Gloucester, Swindon, Oxford, Chelmsford, Cambridge, Peterborough and Derby. An intense fireball would rise from

Figure 20.1   Comet Hale–Bopp, as seen from Nottinghamshire in the Spring of 1997.

the impact site, producing a violent scorching wind, and combining some of the nitrogen and oxygen in the atmosphere to form nitrogen oxides. These would later react with water to form nitrous and nitric acids. Also, sulphuric acid might be produced from the conflagration of plants. Hence, acid rain would fall in the aftermath of the impact, and this might be strong enough to dissolve shells of creatures living in surface waters. Whilst some molten rock and hot ash would drop back to Earth, a large amount of dust would be produced from the rest, which would spread throughout the upper atmosphere, obscuring the light of the Sun for several months. This would interfere with the process of photosynthesis in plants, with obvious consequences for the nutrition of land animals, because herbivores would have nothing to eat. Similarly, in the oceans, where photosynthesising phytoplankton are the base of the food chain, the effect of a prolonged absence of sunlight would have a profound effect on all marine life.[49]

Another consequence of the dust cloud would be a significant cooling at the surface of the Earth, similar to the 'nuclear winter' thought likely to follow a nuclear war. If, during this 'impact winter', much of the Earth was covered with snow, then sunlight would continue to be reflected back before it could be converted into infrared heat, so the low temperatures might last for many years after the dust cloud had dispersed.[50]

On the other hand, if the impact produced 'greenhouse-effect' conditions in the upper atmosphere, the 'winter' might be of short duration, being followed by a lengthy period of global warming, after the dust had fallen back to Earth. This could occur if the impact had released a great deal of carbon dioxide from rocks, or if the projectile had splashed into an ocean, ejecting large amounts of water vapour. Under these circumstances, average surface temperatures could, for a considerable period of time, be well above what they had been before the missile struck.[51]

So, the impact of an asteroid or comet with a diameter greater than 10 km would inevitably cause widespread destruction of animal and plant life, because of prolonged darkness, as well as significant and long-lasting temperature changes, which could be in either direction. In addition, the Earth's ozone layer, which gives protection against damaging ultra-violet solar radiation, would be damaged for many years. Further destruction might result from earthquakes and volcanic activity arising as a direct consequence of the shock of a large impact. Also, if the projectile landed in the ocean, there would be extensive flooding of the continents because of the formation of tsunamis.[52] Death could come in a variety of ways in the aftermath of an impact.

Stochastic catastrophism, coherent catastrophism and other potential dangers from space

Without question, the impact of a large object from space would have devastating consequences, regardless of whether it occurred on land or in an ocean. Knowledge accumulated during the 1980s and 1990s about the likelihood of impacts and their catastrophic consequences thus added substance to the speculations made earlier by Allan Kelly, Max de Laubenfels, René Gallant and others (see Chapter 14).

Moreover, a major catastrophe could also arise on Earth from a totally different cause, for example a nearby supernova explosion, as proposed in the 1950s by Otto Schindewolf (see Chapter 14). It became well established that a star undergoing such an event would increase in brightness to more than a thousand million times the luminosity of the Sun in a period of just a few days, and then become dimmer over a much longer time-scale, emitting great bursts of radiation. Supernovae explosions of this nature, involving distant stars in our Galaxy, had been observed from Earth on three occasions, in the years 1054, 1572 and 1604. More recently, a supernova explosion was seen in the large Magellanic Cloud, a close neighbour of our Galaxy, in 1987. Had such an explosion occurred even closer, within 100 light years of the Earth, it would have caused severe damage to animal and plant life, because of the direct effects of the radiation, and also as a result of associated destruction of the ozone layer.[53] A supernova less than 33 light years away could destroy almost the entire ozone layer, and it has been estimated that this is likely to happen about once every 240 Myr.[54]

Furthermore, similar radiation effects could arise from other causes. For example, an abnormally energetic flare in our own Sun might cause severe radiation damage to terrestrial life, particularly if it occurred during a geomagnetic reversal episode, when the Earth's magnetic field would be weak, offering less protection than usual.[55] A flare in a far-off magnetar, a type of neutron star, sent a burst of energy past the Earth in 1998, and the close approach of such a body could create havoc because of the intensity of its radiation and magnetic effects. In 1996, Arnon Dar and other Israeli astrophysicists suggested that a collision between two neutron stars some distance away could devastate life on Earth, releasing cosmic rays with an output energy around one hundred times greater than that resulting from a supernova explosion. On a different track, in the following year, the Indian physicists Samar Abbas and Asfar Abbas speculated that the Earth might encounter a dense clump of 'dark matter' in space. This would cause the interior of the planet to heat up, resulting in outpourings of molten lava on a large scale.[56] Such catastrophes, of whatever nature, might already have affected the Earth in the distant past.

However, even though scientists are now becoming increasingly aware of a variety of threats to the Earth from the cosmos, Immanuel Velikovsky's theories concerning catastrophic planetary interactions, whilst retaining a group of devoted supporters, remain just as far from mainstream acceptance as previously (see Chapter 12). Earth scientist Richard Huggett, in his 1989 book, *Cataclysms and Earth History*, wrote that the Velikovsky theory 'is regarded by most astronomers as ridiculous'.[57] Indeed, at around the same time, astronomers and space scientists such as Clark Chapman and David Morrison (in *Cosmic Catastrophes*), Victor Clube and Bill Napier (in *The Cosmic Serpent* and *The Cosmic Winter*) and Gerrit Verschuur (in *Impact!*) used words like 'wrong',[58] 'impossible',[59] 'absurd'[60] and 'laughable'[61] to describe the Velikovsky scenario. Similarly, Duncan Steel (in *Rogue Asteroids and Doomsday Comets*, published in 1995) argued that 'Velikovsky's idea is in breach of various laws of physics and hence is untenable'.[62]

Even Velikovsky's enthusiastic supporters had to accept that there was a problem. So, for example, the American engineer James Strickling wrote in the 1996 edition of his book *Origins* that Velikovsky's overall scenario, incorporating the Saturnian hypothesis, had 'not yet been supported by known principles of physics'. Nevertheless, because of the mythological evidence, he went on to suggest that 'there may be astrophysical principles with which we are not yet acquainted'.[63] Similarly, Wallace Thornhill, who studied physics at Melbourne University, admitted in 1999 in *Chronology and Catastrophism Review* that the proposals of Velikovsky and the Saturnists were incompatible with our present understanding of the nature of gravity, before continuing, 'It is time to re-examine those "laws" or long-held beliefs that have diverted scientific curiosity away from uncomfortable questions about the safety of our spaceship Earth. We can no longer afford to deny the possibility that global myths and images of the planetary gods may refer to a frighteningly close-up view of the planets within the memory of the human race'.[64] The interpretation of myth is, however, a subjective process. Hence, it is hardly surprising that most scientists, when confronted with a situation in which the choices are either that ancient myths have been misinterpreted, or that there is something fundamentally wrong with the laws of physics, have unhesitatingly supposed the former to be the case.

Nevertheless, whilst rejecting almost every aspect of Velikovsky's hypothesis, astronomers accept that a large cosmic body could pass close to the Earth and cause catastrophes. In particular, the Earth may be under threat from giant comets which, although small in comparison to planets, are very much bigger than the comets that

Figure 20.2   Left to right: Bill Napier, Victor Clube and Mark Bailey, at Fitzwilliam College, Cambridge, in July 1997.

are normally seen. That view has been developed mainly by British astronomers, including Victor Clube (see Chapter 12), together with Bill Napier and Mark Bailey of the Armagh Observatory (figure 20.2), Duncan Steel of Salford University (and formerly of Spaceguard Australia), and David Asher of Oxford University and the Communications Research Laboratory, Japan.[65] On the basis of the likely range of diameters of cometary nuclei in the regions beyond Jupiter, and in particular in the Oort cloud, they have estimated that, several times every million years, giant comets may become trapped in the inner Solar System. Even if there was no direct collision with the huge cometary nucleus, dust and boulders released from it could cause serious problems for life on Earth, especially if the nucleus started to break up under the gravitational influence of the Sun, which would be a distinct possibility.[66] Such a scenario, involving devastation on Earth because of a cluster of small impacts over a short period of time, coupled with global cooling caused by the dusting of the upper atmosphere, all resulting from the break-up of a giant comet, has been termed *coherent catastrophism*. This distinguishes it from the concept of larger impacts occurring in isolation at longer intervals, which might be called *stochastic catastrophism*.[67] The long-term adverse effects on the environment at the Earth's surface could be very similar with either scenario.

There is some astronomical evidence to support the concept of coherent catastrophism, such as the meteoroids and cluster of asteroids associated with Comet Encke, which could be the remnants of a giant comet (see Chapter 19). Also, Chiron, a 200-km-diameter object with some cometary characteristics, is currently in an unstable orbit between Saturn and Uranus (see Chapter 19), and a deflection from one of the giant planets could bring it into the inner Solar System within 100,000 years. This we know, but, if the assumptions mentioned above are correct, there are many other bodies of similar size in the outer Solar System, with the potential to move in and threaten the Earth with catastrophe.[68]

Taking seriously the threat of catastrophe, from above and below

Setting aside speculative scenarios, even ones derived by mathematical calculations from astronomical observations, it remains an established fact that there are currently many asteroids and comets in Earth-crossing orbits. Although these are small by cosmic standards, they are, nevertheless, capable of causing great damage in the event of a collision. Furthermore, the craters found at the Earth's surface demonstrate that a considerable number of impacts of significant size have already occurred. In 1990, concerned about the risk of a collision with a near-Earth object (NEO), of whatever nature, the U.S. House of Representatives directed the space agency, NASA, to set up two committees, one to suggest how the detection of potential impacting bodies could be improved, and the other to advise on what could be done to avert a catastrophe, should a large object be found to be heading straight for the Earth.[69]

In its 1992 report, the first of these committees, chaired by David Morrison, recommended the setting up of a Spaceguard Survey, taking its name from a project featured in Arthur C. Clarke's science-fiction novel, *Rendezvous with Rama*.[70] The intention was that Spaceguard should build upon the previous Spacewatch programme, which had led to the detection of near-Earth asteroids such as 1991 BA, 1993 KA_2 and 1994 XM_1, but insufficient money was made available to accelerate the detection process to any significant extent. In 1996, concerned astronomers from around the world formed the Spaceguard Foundation to draw attention to the issues, and to try to persuade governments to do more. The setting up of NASA's Near-Earth Asteroid Tracking (NEAT) system in Hawaii at around this time improved matters, and NASA also supported other programmes, including the LINEAR survey in New Mexico, a Spacewatch programme at the University of Arizona, and the Lowell Observatory Near-Earth Object Survey (LONEOS).[71]

An effective programme for intercepting and deflecting a threatening object would require an even larger budget than one for detection.[72] Even if one had been made available (which it wasn't), concerns still remained about the possible misuse of such a programme, since the procedures developed for the protection of the Earth could be employed to divert a projectile towards an unfriendly country.[73] Regardless of that, it seemed clear that deflection of an approaching asteroid or comet would be most effective if carried out a long way from the Earth. Also, making such an object miss the Earth by accelerating or retarding its motion appeared to be a more practical proposition than attempting to force it to one side. According to the calculations of Thomas Ahrens and Alan Harris, of the California Institute of Technology, significant changes could be made to the orbital characteristics of a 1-km projectile by a 0.1 megaton nuclear explosion 400 metres above its surface. Similarly, a 10-km object could be deflected to a significant extent by a 10-megaton explosion at an altitude of four kilometres.[74]

Discussions continued, and various alternative ideas were put forward.[75] So, for example, in 1998, NASA funded the Applied Physics Laboratory of Johns Hopkins University, which controls the NEAR spacecraft, to develop ideas about a comprehensive Earth protection system called SHIELD. This concept, devised by Robert E. Gold, involves a space-based detection system. If an Earth-threatening asteroid is discovered, a spacecraft would be directed to attach itself to the surface, and fire its booster rockets to deflect the hazard. Calculations indicated that a spacecraft of the NEAR type would be able to divert a 300-metre asteroid from a collision course with the Earth, provided it was placed in position ten years before the estimated time of impact, and a more powerful one would of course be even more effective.[76]

Without international agreement and a very large budget, all of this remains hypothetical at the present time, but it indicates that the threat of an extraterrestrial impact is starting to be taken seriously. There have been some backward steps, such as the Australian government leaving the southern skies with very limited coverage by withdrawing the funding for an NEO search-and-tracking programme that had operated under the direction of Duncan Steel between 1990 and 1996. Nevertheless, the situation continues to develop, particularly in the northern hemisphere. Previously unknown NEOs are being identified at a steady rate, mainly in NASA-supported surveys, for NASA is committed to identifying 90% of near-Earth asteroids with a diameter greater than 1 km by 2008. Around one hundred objects of this size were detected for the first time in 2001. Another development has been the introduction of the Torino scale, first suggested at a conference in Turin, Italy, in 1999, which enables the general public to be given a simple measure of the threat to the Earth from a newly discovered asteroid or comet. That should help to avoid the situation which occurred a few years ago, when there was a hysterical over-reaction in some sections of the media to preliminary calculations of the orbit of asteroid 1997 XF_{11} suggesting a possible collision with the Earth in 2028. Shortly afterwards, this was followed by a backlash, in which scientists were accused of having caused unnecessary alarm, when more detailed calculations showed that the Earth would emerge unscathed on this particular occasion. The Torino scale combines collision probability with energy of impact, should one occur. An event classified as zero on the scale would have no likely consequences, either because there was no possibility of a collision within the next few decades, or because the object would burn up in the atmosphere anyway, whereas an event classified as 10 on the scale would be a certain collision with an object large enough to cause a global catastrophe. As to other developments, the British government set up a task force in January 2000 to consider the issues of detection and protection. Seven months later, it recommended the setting up of a programme to identify and monitor all near-Earth asteroids with diameters larger than 300 metres and, as part of that programme, to find partners to build an advanced three-metre telescope in the southern hemisphere. In the meantime, it was announced early in 2002 that the Isaac Newton Telescope at La Palma in the Canaries was to be used in a pilot study, and that an NEO Information Centre was to be set up within a few months in Leicester, which duly occurred. Also, a non-governmental Comet and Asteroid Information Network was established, under the management of the Spaceguard Centre in Wales.[77]

Looking back rather than forward, the estimated rate of impacts of asteroids and comets is information which cannot be ignored when assessing the likely extent to which events on planet Earth have been influenced by previous catastrophes. Other types of extraterrestrial event, such as a supernova explosion or a collision between neutron stars, may have caused devastation on Earth, or we may have been spared them. However, we can be certain that some catastrophes have been caused by encounters with asteroids or comets, even if these were just localised events resulting from relatively small impacts. Moreover, according to the astronomical evidence, several projectiles larger than 10 km in diameter are likely to have struck the Earth since animals with skeletons first appeared around 550 Myr ago. If so, then the destruction from the impact event and its aftermath would have been on a global scale on each occasion, and the fossil record should show evidence of it.

Even disregarding the hazards from space, we know that earthquakes and volcanic eruptions, principally related to the movements of tectonic plates, have the potential to cause considerable devastation. Much effort is being put into trying to predict where and when these will next occur. Earthquakes are quite common, around 100 being recorded every year with a magnitude in the order of 6.3 on the Richter Scale. Although each such event releases energy equivalent to the explosion of a 1 megaton bomb, this usually occurs well below the Earth's surface and too far from any major centre of population to do a great deal of damage. There are, however, exceptions. For example, the focus of the Great Hanshin Earthquake of 1995, magnitude 6.9 on the Richter Scale, was only 14 km below the surface and only 20 km in terms of horizontal distance from the Japanese city of Kobe, with the consequence that 6,000 people died and another 35,000 were injured. Similarly, in March 2002, a shallow earthquake of around magnitude 6.0 struck northern Afghanistan, killing a thousand people or more as houses with walls of dried mud collapsed throughout the Nahrin district.[78]

Larger earthquakes are less frequent, with one of magnitude 8.3 or more occurring, on average, about every 10 years. The Richter Scale is an exponential rather than a linear one, so an 8.3-magnitude earthquake is in fact 2,500 times more powerful than one of magnitude 6.3 and accordingly, all other factors being equal, will be much more destructive. The great Lisbon earthquake of 1755 apparently had a magnitude of around 8.75 and its effects, together with those of the tsunami it created, caused devastation in the Portuguese capital. Some 30,000 people were killed, with a similar number dying from starvation and disease in the aftermath. A tsunami was also responsible for much of the devastation caused to relatively low-populated coastal regions of Alaska, following an 8.5-magnitude event in 1964. The earthquake which struck Messina, Italy, in 1908 was smaller, being around magnitude 7.5 on the Richter Scale, but 120,000 inhabitants perished as buildings collapsed in catastrophic fashion. Similarly, around 20,000 people died in collapsed buildings after the 7.9-magnitude earthquake in western India in January 2001. Earlier in the same month, landslides caused by a slightly smaller earthquake in El Salvador killed almost a thousand people. Fire is also a great danger in the aftermath of an earthquake, this being responsible for many of the 200,000 deaths recorded in Tokyo and Yokohama in 1923, as a consequence of the 8.3-magnitude Great Kanto Earthquake, whilst tsunamis can pose a threat to far-distant places, as well as to ones close to the epicentre. The 8.6-magnitude earthquake which occurred off the coast of southern Chile in 1960 killed about 5,000 people on the mainland as buildings collapsed, but thousands more died as tsunamis generated by this event and its aftershocks struck coastal regions in South America, Hawaii and Japan.[79]

An earthquake detection system was set up in northeastern China in 1966, after two tremors of around magnitude 7 were experienced within a few months of each other. This enabled the city of Haicheng to be evacuated just prior to a 7.3-magnitude earthquake in 1975, saving many lives, but it failed to predict the 7.8-magnitude event which killed 250,000 people in Tangshan in the following year. In the U.S.A., the first moves towards what is now a vast system of earthquake monitoring and (hopefully) prediction began after a 7.6-magnitude event in California in 1906. Although this killed 'only' 700 people, it caused fires which devastated the city of San Francisco. As is well known, densely populated California, where the Pacific Plate is slipping northwards past the North American Continental Plate along the line of the San Andreas Fault, is a high-risk earthquake zone. The Los Angeles area has suffered eight major earthquakes in the past 1,500 years, the last in 1857, so another can be expected in the not-too-distant future.[80]

As with 'small' earthquakes, 'small' volcanic eruptions are quite common, yet can be deadly to people unfortunate enough to be close to the centre of activity. In 1902, a 'pyroclastic flow' of superheated volcanic gases swept down the slopes of Mt Pelée into the town of St Pierre, on the island of Martinique, in the West Indies, causing the deaths of 29,000 people. In Colombia in 1985, a flow of mud and debris from the Nevado del Ruiz volcano hit the town of Armero, leaving 23,000 people dead. There might have been similar consequences in the Philippines in 1991, when Mt Pinatubo exploded, producing a crater 2 km in diameter, had not vulcanologists managed to predict that an eruption was about to happen, enabling many thousands of people living on the slopes and in the valley beneath to be evacuated. Even more recently, the inhabitants of Montserrat, in the West Indies, were moved from the island to save them from volcanic outpourings. Then, in January 2002, lava from Mount Nyiragongo in eastern Congo engulfed the town of Goma, causing hundreds of thousands of people to flee from their

Figure 20.3   The civic forum at Pompeii, overlooked by Vesuvius.

homes. The streams of lava from Nyiragongo also poured into Lake Kivu, threatening to cause 'lake overturn', when inflammable methane and suffocating carbon dioxide would be released from the depths. In all, it has been estimated that something like 500 million people around the world are at risk from volcanic eruptions and, in the hope of being able to provide warning of an impending eruption, many active or potentially active volcanoes are being carefully monitored. These include Vesuvius, whose eruption in 79 A.D. destroyed the Roman cities of Pompeii and Herculaneum, and now threatens millions of people in and around the city of Naples (figure 20.3).[81]

Even people who live a considerable distance from an erupting 'small' volcano can be affected, in certain circumstances. Whilst a volcano which releases copious amounts of molten lava, such as Kilauea on Hawaii, may be dangerous on a local scale, one which explodes and blasts magma and gases into the upper atmosphere, leading to reduced penetration of solar radiation and falls of ash over a large area, is generally a greater hazard. An example is Mount St Helens, in Washington State, U.S.A., which exploded in 1980, producing a crater 3.2 km in diameter, and pyroclastic flows which killed 62 people, mainly members of investigating teams. The death count was low because very few people lived in the area, but the devastation was clear for all to see: a wasteland with an area greater than 500 square kilometres was created in a few minutes, and ash fell all over Washington State and into Oregon, disrupting traffic and damaging crops.[82]

Yet, the Mount St Helens eruption was only given a Volcanic Explosivity Index (VEI) of 5 out of a possible 8. What, then, of larger events, with a VEI of 6? We have already considered a likely one, the catastrophic explosion of Thera, during the Bronze

Age, which deposited ash over a wide area and produced tsunamis that smashed into the islands of the Eastern Mediterranean (see Chapter 13). When the explosion was over, the centre of the volcano collapsed into the empty magma chamber, leaving the outermost parts (which became the present-day islands of Thera and Therasia) as a roughly circular caldera, over 10 km in diameter. The eruption of Krakatoa, in the East Indies, in 1883 was on a slightly smaller scale, giving rise to a caldera which was about half the area of that of Thera. Nevertheless, it has been estimated that the Krakatoa explosion released energy equivalent to 50–100 megatons of TNT, i.e. more than in the Tunguska event of 1908. What is known for certain is that tsunamis produced by the blast killed almost 40,000 people.[83]

The eruption of nearby Tambora in 1815 had been even more powerful, with a VEI of 7. After the explosion, a dust cloud blocked out the Sun's light for two days up to a distance of 600 km from the volcano. The following year, 1816, was remembered as the 'year without a summer' in the northern hemisphere, abnormally low temperatures being recorded in both Europe and North America. Similarly, low temperatures and dense fogs caused by acid rain were recorded in North America and Europe in 1783 and 1784, following the opening of the 27-km-long Laki system of fissures in the Lakagigar region of Iceland, which spewed out vast quantities of gas, dust and lava. Another very bad summer had been reported in 1801, the poor weather apparently resulting from a massive eruption of Volcán Huaynaputina in the Peruvian Andes in 1800.[84]

It seems that volcanic explosions on the Tambora scale can be expected to recur on Earth at intervals of between a few hundred to a few thousand years. Furthermore, even the Tambora eruption was relatively mild compared with some more distant events, which are classified as having a VEI of 8: for example, the super-eruption of Toba, also in the East Indies, around 75,000 years ago, was perhaps 100 times more energetic than the Krakatoa blast, leaving a caldera over 50 km in diameter. Events on this scale may recur every 100,000 years or so, on average. Three have taken place in the Yellowstone Park region of Wyoming in the past 2 Myr, the last around 600,000 years ago. Another supereruption, perhaps involving the same 'hotspot', in the Bruneau Jarbridge region of southwestern Idaho, occurred around 10 Myr ago. The scale of this can be appreciated from the fact that hundreds of rhinos, horses, camels and smaller animals were killed and preserved by ash at a waterhole 1,600 km to the east of the supervolcano, forming what are now known as the Ashfall Fossil Beds, near Orchard, in northern Nebraska.[85]

Moreover, it is known that there were several episodes during the past 250 Myr where volcanic activity occurred on a massive scale, over large geographical regions. In particular, there were huge outpourings of lava in Siberia around 250 Myr ago, in the Pacific and Indian Oceans between 120 Myr and 80 Myr ago, and in India around 65 Myr ago. Such vulcanism, even if of a non-explosive kind, must have caused considerable destruction of living organisms.[86] Like the 'impact winter' (also called a 'cosmic winter') that could arise from a large impact or a cluster of smaller impacts, it has been argued that an episode of vulcanism on this scale could lead to a 'volcanic winter' lasting many years.[87]

Large-scale outbreaks of volcanic activity, when they occurred, may have been phenomena of purely terrestrial origin, as generally supposed,[88] or they may have been triggered by shocks resulting from huge extraterrestrial impacts.[89] Other links between impacts and phenomena previously thought to have an entirely Earth-centred explanation are also possible. For example, some geomagnetic reversals are co-incident with fields of microtektites, which are generally believed to result from impacts. Michael Rampino of New York University and NASA suggested that shock waves from a large impact could cause a geomagnetic reversal by disrupting currents in the liquid core of the Earth.[90]

As all this evidence became available, it was apparent to many that the evolution of life on Earth had to be considered in the context of a Solar System where major collisions between bodies are relatively common, on a time-scale of millions of years, and an Earth where huge volcanic explosions or outpourings of lava occur with a similar regularity. Some events causing terrestrial catastrophes in the past, and perhaps mass extinction episodes, may have arisen on the Earth itself, whereas others might have been of extraterrestrial origin.

On the other hand, the fact that a particular mass extinction episode *might* have been the result of such a catastrophe did not necessarily mean that it *must* have occurred in this way, in whole or even in part. As we have seen, tectonic factors, acting over a long time-scale, could cause mass extinctions. Also, studies on Boolean networks (see Chapter 17) have shown that avalanches of change could take place, even in the absence of a major environmental crisis. At the present time, we are entering a new mass extinction episode, not because of any physical catastrophe, but because of the overwhelming dominance of a single species, *Homo sapiens*.[91] For each event, the evidence must be considered as a whole, and in all its detailed aspects. Nevertheless, when this evidence is assessed, the possibility of the involvement of a catastrophist mechanism can no longer be dismissed out of hand, as was often the case during the past century and a half. Attitudes have changed considerably, not only because of what we now know about the dangers of our environment, but also as a consequence of attempts to find out precisely what happened at the end of the Cretaceous Period. That is the topic we shall now go on to discuss.

21 The death of the dinosaurs: iridium and the K–T extinctions

Alvarez and Alvarez

Of all the transitions between geological periods, the Cretaceous–Tertiary boundary is the most clearly defined in rocks throughout the world. Dating from around 65 Myr ago, it is the youngest of the major mass extinction horizons known from the fossil record (see figure 5.2), so has had less chance than the others of being disrupted by tectonic changes. Large areas of continental land masses were covered by shallow seas during the Late Cretaceous Period, so rocks formed at that time often consist of chalk, a limestone composed of the fossilised remains and secretions of marine plankton. Examples include the cliffs of southeastern England and northwestern France, and the rocks which Cuvier described as the underlying ones of the Paris basin (see Chaper 3). The term 'Cretaceous' is derived from the Latin word for chalk, which is kreide in German and krete in Greek. Hence, the Cretaceous–Tertiary boundary is generally written for convenience as the K–T boundary, the use of the letter 'K' avoiding possible confusion with the Cambrian or Carboniferous Periods.

In limestone rocks, the K–T boundary is generally marked by a clay layer, which is no more than a few centimetres in thickness. Immediately below this, the next few millimetres of rock show that populations of the lime-containing plankton were severely depleted as the Cretaceous drew to a close, with some groups, especially the planktonic foraminifers and nanoplankton, almost disappearing completely. From an estimate of the likely sedimentation rate, it was suggested during the 1970s that these marine extinctions could have taken place over a period of 200 years or less, although others continued to argue for a much longer time-scale. The ammonites, with their characteristic spiral shells (see figure 24.2), died out completely at the end of the Cretaceous Period, together with some other forms of marine life, including the huge sea-reptiles. On land at around the same time, the dominant group of large vertebrates, the dinosaurs, also became extinct, with no species of terrestrial animal weighing more than 25 kg (the weight of an average-sized dog) surviving the transition to the Tertiary.[1]

The debate about the nature of these events took a surprising turn in 1977, when geologist Walter Alvarez, of the University of California at Berkeley, enlisted the help of his father, Luis (1911–1988), an eminent physicist who was also based at Berkeley, to try to get an accurate estimate of the time-scale of the K–T transition. This simple move led

simple, for 'what good is a theory, even a correct theory, that can generate no confirming evidence?' Now, such evidence was available, in the form of an iridium abundance
anomaly.[8]

Soon afterwards, other investigators confirmed the finding of high concentrations of iridium in the K–T boundary clay at Stevns Klint, together with similar
elements, such as osmium.[9] At around the same time, iridium abundance anomalies
at the K–T boundary were found in limestone rocks at Caravaca in Spain,[10] in a core
from the Deep Sea Drilling Project in the Central Pacific[11] and in sedimentary rocks
laid down under freshwater swamp conditions in the Raton Basin of New Mexico.[12] In
addition to the high iridium concentrations in the clay layer at Caravaca, the presence
of sanidine spherules of around 1 mm in diameter was taken as evidence of an impact,
for they could have solidified from molten rock.[13] Again, at the Raton Basin site, and
elsewhere, there was less angiosperm (flowering plant) pollen immediately above the
K–T boundary layer than below it, whereas, in the case of fern spores, it was the other
way round. Such a discontinuity, known either as a 'pollen break' or a 'fern spike', was
seen to be consistent with the impact hypothesis, because ferns were known to recover
from crises such as forest fires much faster than angiosperms.[14]

Iridium, tektites, quartz and soot

The iridium abundance anomaly which Charles Orth, of Los Alamos National
Laboratory, New Mexico, found at the Raton Basin site was the first to have been
identified in rocks which had not been formed under an ocean. This knocked out of
court a suggestion by Michael Rampino that the abundance of iridium at the other
sites might have been caused by living organisms building up concentrations of the
metal by extracting it from sea-water.[15] A second anomaly at a non-marine site soon
followed, when the Alvarez group, investigating rocks collected in the region of Hell
Creek, Montana, by the Berkeley palaeontologist William Clemens, found elevated
iridium levels in a thin coal band which marked the K–T boundary. By April 1982, less
than two years after the publication of the *Science* paper, iridium abundance anomalies
had been detected at the K–T boundary at thirty-six different locations scattered around
the world, and within two more years, the number of such sites had reached almost
twice that figure.[16]

Nevertheless, arguments about the nature of the K–T boundary layer continued.
It was possible that the iridium had been released from the Earth's core by vulcanism,
but geochemists Jean-Mark Luck and Karl Turekian of Yale University showed that the
osmium-187/osmium-186 ratios in K–T boundary samples were more characteristic
of ones in material which had come from space than those generally found in terrestrial rocks. Also, the total amount of osmium present seemed too great to have been
deposited by volcanic activity, and so indicated an extraterrestrial impact.[17]

When, back in 1973, Harold Urey had put forward an impact hypothesis to
account for the end of certain geological periods, he predicted that microtektites

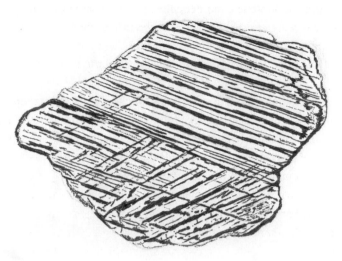

Figure 21.2   Shocked quartz grains from the Cretaceous–Tertiary boundary
(drawing based on a U.S. Geological Survey photograph).

would eventually be discovered at the K–T boundary (see Chapter 14). The sanidine
spherules in K–T boundary samples from Caravaca (found by the Dutch geologist
Jan Smit and his American colleague Miriam Kastner) were not glassy enough to be
classified as microtektites, but they had many characteristics in common. Sanidine
spherules were also present in the K–T boundary clay of Italy, as were similar spherules
of glauconite and magnetite. This was demonstrated by the Italian geologist Alessandro
(Sandro) Montanari, then a graduate student at Berkeley. Soon, spherules were found
co-incident with the K–T iridium abundance anomaly at many other sites. Supporters
of the impact hypothesis claimed that these spherules must once have been droplets
of molten rock, which supported their case. Furthermore, they argued that they had
been formed as a result of an impact into basaltic rock, like that forming the oceanic
crust, because an impact into a more silica-rich rock would have produced glassy
microtektites.[18]

More evidence for a major impact at the end of the Cretaceous Period came
in 1984 when Bruce Bohor and his colleagues of the U.S. Geological Survey found
shocked quartz grains at the K–T boundary (figure 21.2). These contained several sets
of lamellae, each set intersecting the others, which provided strong support for the
impact hypothesis. The only other possible cause involved vulcanism, but this had
never been shown to produce sufficient shock pressure to give rise to more than one
set of lamellae in quartz grains.[19]

In 1985, a team which included Wendy Wolbach, a geochemist from the University of Chicago, discovered aggregates of carbon at the K–T boundary at several sites in
Europe and New Zealand. They concluded that this was evidence of a worldwide layer
of soot, produced by conflagrations of plant material, and was therefore consistent

with the impact hypothesis. Wolbach and her colleagues calculated that the amount of soot in this layer was equivalent to about 10% of the total carbon present in living organisms today. Hence, if it had been thrown up into the atmosphere during the widespread fires following an impact, it would have blocked the Sun's rays to a very significant extent.[20]

Another relevant finding was made in carbonate rocks, which reflect the composition of the sea water from which they were precipitated. In rocks of this type, it was discovered that the carbon-13/carbon-12 ratio dropped sharply during the K–T transition. A carbon-isotope shift of this nature indicates depleted populations of phytoplankton, because these normally remove much of the carbon-12 from surface waters, utilising it during photosynthesis in preference to the heavier isotope, carbon-13.[21] Hence, this finding, like that of the layer of carbon, was seen to be consistent with the impact hypothesis, even if not much more than that.

What killed the dinosaurs (and lots of other animals)?

As we have seen, there was, by the middle of the 1980s, much direct and indirect evidence supporting the hypothesis that a projectile from space had struck the Earth at the end of the Cretaceous Period. However, no impact crater of appropriate size or age was known at the time. Fred Whipple suggested that the missile might have hit the mid-Atlantic ridge, triggering the volcanic activity which led to the formation of Iceland. That was an attractive theory, but it had to be rejected. Greenland and Norway had not separated by the end of the Cretaceous Period (see figure 14.2), so there could not have been a mid-Atlantic ridge when the impact took place.[22] Supporters of the impact hypothesis therefore believed that the crater had been destroyed by subduction during tectonic movements, or that it lay undiscovered, perhaps covered by sediment at the bottom of an ocean.[23]

As the search for a possible Cretaceous–Tertiary impact site continued, detailed assessments were made as to the likely consequences of a strike by a 10-km cosmic body at this time. It was clear that, regardless of location, the blast from an impact of this size would have devastated animal populations over a wide area. Later, decreased temperatures associated with an 'impact winter', perhaps reinforced by the radiation resulting from a breakdown of the ozone layer and the collapse of food chains when photosynthesis ceased, would have led to continuing mayhem. Yet more problems for animal life may have occurred if temperatures then rose significantly because of a greenhouse effect (see Chapter 20). As for the survivors, small animals require less food than large ones and could have found shelter in holes and under vegetation or mud, whereas large animals could not. Hence, it was seen to be entirely consistent with the impact hypothesis that, on land, only small animals made it through to the Tertiary Period. The simple factor of size could have been instrumental in determining which animal species became extinct and which survived.[24]

Adult dinosaurs varied considerably in size and other characteristics, but none was particularly small. Of the many dinosaur genera now known and characterised, the smallest appears to be the bird-like *Compsognathus*, which had a length within the range 60–140 cm, about the size of a domestic cat.[25] Some larger reptiles, including turtles and crocodiles, lived through into the Tertiary Period, but such animals are aquatic and practise hibernation, which might have helped their survival. It is also possible that only their eggs remained alive, buried in the ground, hatching out later when conditions had improved. Even so, a number of species of each became extinct as the Cretaceous Period ended.[26]

Animals living in fresh water generally survived, perhaps because the base of their food chain was decaying vegetation, which could have been in plentiful supply during a time of environmental crisis. The situation was more complex in the oceans. The lime-containing plankton largely died out, possibly because of poisoning, whilst other plankton, including dinoflagellates, survived. As a result, the food chain was broken in some places but remained intact elsewhere, producing an irregular pattern of extinctions. In general, marine organisms living close to the equator suffered more than those at higher latitudes.[27]

Taken as a whole, the evidence available in the mid-1980s neither proved nor ruled out the impact hypothesis. This continued to be developed by its supporters, whilst many others looked elsewhere for an explanation of the events around the time of the K–T transition.

In addition to extraterrestrial hypotheses involving an impact or a supernova explosion, over fifty different terrestrial explanations had been put forward at various times to account for the K–T extinctions. Amongst these were suggestions that the dinosaurs as a group had become senile, and that animals had evolved which used dinosaur eggs as a source of food. Other theories included the poisoning of dinosaurs by alkaloids in flowering plants, which first appeared during the Cretaceous Period, (but some 50 Myr before its end); the dinosaurs going blind as a result of increased solar radiation; and the disruption of their gender balance by climatic changes. A number of hypotheses referred only to the death of the dinosaurs, ignoring the other extinctions taking place at around the same time, so none of these was likely to provide a satisfactory explanation for events at the end of the Cretaceous Period.[28]

Nevertheless, some Earth-centred extinction mechanisms deserved serious consideration. Evidence was produced to indicate that regression of shallow epicontinental seas, a process which had been linked to the Permian–Triassic and other mass extinctions, also occurred during the K–T transition, suggesting that the explanation for the mass extinctions at this time could involve plate tectonics. However, in contrast to the situation in the Late Permian (see Chapter 14), there was no obvious reason why sea-floor spreading should have stopped, leading to a fall in sea-level, at the end of the Cretaceous Period, when the break-up of Pangaea was only partially complete (see figure 14.2). In any case, whilst the regression of shallow seas indicated a cessation of tectonic activity, there were also abundant signs of volcanic activity at the close of the Cretaceous, which suggested that sea-floor spreading was taking place with full vigour.[29]

Regardless of any association with sea-floor spreading, vulcanism itself, operating on a large scale, came to be regarded by some as a possible cause of the K–T mass extinctions, as well as releasing the iridium found in the boundary clay. Evidence was produced which indicated that there had been extensive volcanic activity throughout the last 10–15 Myr of the Late Cretaceous, including the outpouring of a million cubic kilometres of lava in India to form the Deccan traps. Daniel Axelrod, a botanist from the University of California, Davis, had claimed in 1968 (in company with Harry Bailey) that the dinosaur extinctions were due to a gradually deteriorating climate and later, in 1981, he argued that massive, sustained volcanic activity provided a mechanism for this, by reducing the amount of solar heat reaching the Earth's surface, as well as contributing more directly to the killing. For example, falls of ash and acid rain could have poisoned organisms both on land and in the seas, disrupting food chains.[30] Amongst the other early supporters of the vulcanism hypothesis were Peter Vogt (who later converted to the impact theory), Dewey McLean, a geologist at the Virginia Polytechnic Institute, and Vincent Courtillot, a geophysicist attached to the Institut de Physique du Globe, in Paris.[31] Without question, this scenario provided a genuine challenge to the impact hypothesis.

Conflicts during the early 1980s

As we have seen, the *Science* paper written by the Alvarez group in 1980 stimulated much interest in the possible evolutionary effects of impacts of asteroids and comets. To discuss the issues, a major conference, entitled 'Geological Implications of Impacts of Large Asteroids and Comets on the Earth', was sponsored by the Lunar and Planetary Institute and the National Academy of Sciences, and held at Snowbird, Utah, in October 1981, attracting over one hundred delegates. The proceedings were published as a special paper of the Geological Society of America in the following year.[32]

However, most of those who gave sympathetic attention to the Alvarez impact hypothesis were non-biologists, being mainly geologists, chemists, physicists and astronomers.[33] At first, just a few palaeontologists were prepared to accept the impact hypothesis as a possibility. These included Digby McLaren, whose arguments for catastrophes at some mass extinction horizons had first been presented in 1970, as we saw in Chapter 14. Prominent amongst the others were David Raup, of the University of Chicago, whose research was mainly computer-based, but who had been President of the Paleontological Society. There was also Dale Russell, a palaeobiologist from the National Museum of Natural Sciences, Ottawa, Canada, who had been convinced for many years that mass extinctions required an extraterrestrial explanation and, in collaboration with physicist Wallace Tucker, had in 1971 given sympathetic consideration to the supernova hypothesis. However, most palaeontologists remained sceptical about claims that impacts of projectiles from space could be responsible for any of the mass extinction episodes, including that at the K–T boundary.[34]

One of these sceptical palaeontologists was William Clemens, who made much of the fact that the iridium-rich layer at Hell Creek, Montana, was about three metres above the highest dinosaur fossil, a *Tyrannosaurus rex* thigh-bone. In contrast, at the marine sites, the iridium anomaly was exactly co-incident with the extinction of the foraminifers which were characteristic of the Cretaceous Period.[35] However, given the incomplete nature of the fossil record, it was pointed out by others that these findings did not necessarily mean that the extinction of the dinosaurs occurred before that of the marine plankton. Foraminifers, as well as being much smaller than dinosaurs, appear in the fossil record in much larger numbers. In limestone rocks, foraminifer fossils are found with an average vertical spacing of less than a millimetre, whereas at Hell Creek, a continental formation, there is an average of less than one dinosaur fossil per metre of the vertical column. The extinction of the dinosaurs was therefore likely to have occurred some considerable time after the death of the most recent T. rex specimen found at Hell Creek, and quite possibly at the time marked by iridium anomaly.[36]

Phil Signor and Jere Lipps, two palaeontologists from the University of California, argued out that such a sampling problem, which was particularly serious with large (and hence rare) fossils, would make even a genuinely instantaneous mass extinction event appear gradual or step-wise. The last representatives of the various species involved would be found at different vertical distances from the level which marked the actual time of extinction.[37] This phenomenon has become known as the Signor–Lipps effect.

However, whilst accepting the logic of the arguments of Signor and Lipps, most palaeontologists continued to believe that the K–T extinctions had occurred at a variety of different times and, indeed, that the extinctions in many groups of organisms had started well before the end of the Cretaceous Period. The Alvarez group came to accept that, at least in the case of marine invertebrates, the extinction process appeared to have begun before the K–T boundary, although they emphasised that, in their view, it had been greatly accelerated by the subsequent impact event. As for the dinosaurs, some palaeontologists claimed that they had started to decrease in taxonomic diversity (i.e. in the number of species in existence) well before their eventual extinction,[38] whereas Dale Russell concluded from his studies that there had been no such decline in diversity over the last 10 Myr of the Cretaceous Period.[39]

In Britain, the dinosaur specialists Alan Charig and Beverley Halstead attacked the impact hypothesis at a meeting of the British Association for the Advancement of Science in Brighton, 1983. They believed that there was no single K–T event for, as they saw it, the final dinosaur extinctions on land had taken place hundreds of thousands of years *after* the extinctions of marine plankton associated with the high iridium levels.

They also thought that this iridium was more likely to be of volcanic origin rather than coming from space.[40] Two years earlier, Halstead had made these same claims in *Dinosaurs* (written in collaboration with his wife, Jenny, a biological illustrator).[41] In this book, Halstead expressed vehement opposition to extinction hypotheses involving impacts of asteroids and comets, writing, 'Such theories are certainly an advance on invoking the wrath of a Deity but not very much. The real trouble with this category of theory, in which dinosaurs left the stage with a bang, is that it is almost impossible with such a scenario to explain why so many other groups seemed not to have been aware of the great global commotion that was supposedly enveloping them'.[42]

However, Halstead's view that 'even among the reptiles lizards and snakes, turtles and crocodiles seem to have been singularly unaffected' was not supported by the evidence presented by Russell, who pointed out in a 1982 *Scientific American* article that almost half of the families of reptiles (other than dinosaurs) living in North America disappeared at the end of the Cretaceous Period, although a few of these survived elsewhere. Amongst the families which died out completely were one of marine turtles, one of crocodiles, two of ichthyosaurs, three of plesiosaurs and two of lizards (one terrestrial and one marine, the mosasaurs).[43] Steven Stanley, in his 1987 book, *Extinction*, argued in similar fashion that the extinctions of dinosaurs on the land were paralleled by extinctions in the oceans, with all the mosasaurs and plesiosaurs, as well as the largest marine turtles, being unable to survive through into the Tertiary.[44]

Yet, as Halstead saw the situation, there was nothing to suggest there had been a 'great global commotion' at the time. In the style of Lyell and Geikie (see Chapters 2 and 5), he maintained that catastrophist arguments were subjective and over-simplistic, whilst gradualistic ones were objective and based firmly on scientific observations. He wrote, 'The other brand of theory involves a careful consideration of all the evidence that can be accumulated, drawing both from biology and geology. These more synthetic theories tend to be less exciting but are more likely to approximate to what actually occurred. They are not as popular as the cosmic cataclysms but are able to encompass more evidence'.[45]

Halstead argued that the disappearance of the dinosaurs could have lasted 'anything up to 500,000 years'. He concluded that the most acceptable theory to account for their extinction was the 'combination of climatic changes and mammalian competition in the changing conditions'. In his view, the crucial environmental change was a general rise in temperature towards the end of the Cretaceous Period, possibly because of a greenhouse effect resulting from reduced photosynthesis by marine organisms as shallow epicontinental seas were drained.[46]

Although it soon emerged, on the basis of oxygen-isotope studies in marine microfossils (see Chapter 27), that average temperatures may indeed have risen, by 6 °C or so, during the early Tertiary Period, they fell by a similar amount over the last 15 Myr of the Cretaceous. This was different to what Halstead had suggested, but, nevertheless, it was a point seized upon by most of those who opposed the Alvarez impact hypothesis, including Steven Stanley, writing in 1987 in *Extinction*. Indeed, it was generally agreed that, regardless of causal mechanisms, this drop in temperature during the Late Cretaceous could have contributed to a general decline in populations of marine organisms, particularly in tropical areas, before the main wave of extinctions. As for the reversal of this downward trend in temperatures, carbon-isotope studies have indicated that the productivity of phytoplankton was low for perhaps hundreds of thousands of years after the K–T transition. Since marine phytoplankton normally release the dimethyl sulphide which acts as a precursor for cloud formation over the seas, it has been argued that the extinction of many of them at the end of the Cretaceous Period would have resulted in fewer clouds to reflect the Sun's rays back into space, causing global warming. This in turn could have played a role in maintaining low productivity during the early part of the Tertiary.[47]

However, regardless of ongoing trends, had there been a particularly large extinction episode, caused by an impact, at the very end of the Cretaceous Period? That question was addressed in 1983 by the geologists Charles Officer and Charles Drake, both of Dartmouth College, New Hampshire, in a paper published in *Science*. As we shall see, Officer (generally known as 'Chuck') was to become one of the most vociferous critics of the impact hypothesis.[48]

According to Officer and Drake's *Science* paper, results of studies of stratigraphic layers near the K–T boundary, which paid particular attention to the direction of geo-magnetic polarity, indicated that the dinosaurs became extinct in the San Juan Basin, New Mexico, 400,000 years after the foraminifer extinctions at Gubbio. Moreover, it seemed that not even the marine extinctions at different sites had been simultaneous, or had taken place over the same intervals of time. From this, it seemed clear to the Dartmouth College geologists that, if there had been an asteroid impact at the end of the Cretaceous Period, 'then the faunal transitions did not occur instantaneously in response to it'. Whilst they acknowledged that difficulties of interpretation could result from the mixing of stratigraphic layers by burrowing animals, the fact that the K–T iridium band varied significantly in thickness and intensity at different sites seemed inconsistent with it having an impact origin. Further, there was nothing about the mineralogy of the K–T boundary clay which required an extraterrestrial explanation, so they concluded that it had probably been produced by volcanic activity. To Officer and Drake, the evidence, taken as a whole, did not support the hypothesis that a large extraterrestrial impact was the cause of the K–T iridium anomaly. Their conclusion was that a series of extinctions had taken place over a period of hundreds of thousands of years against a backdrop of large-scale volcanic eruptions and a draining of epicontinental seas. They continued:

> Investigation of the environmental effects of terrestrial events including extensive volcanism, major regression and transgression of sea level, paleoclimatic, and paleoceanographic changes, and the significance of geochemical changes such as

oxidation–reduction conditions and oxygen and carbon isotopic variations may be useful to an understanding of the K–T faunal transitions. The effects of a single environmental change may not be applicable to the entire range of faunal transitions from nannofossils to dinosaurs.[49]

Despite a counter-offensive, by the Alvarez group and others, who produced evidence of the extreme sharpness of the K–T boundary, and challenged Officer and Drake's assertions to the contrary,[50] the Dartmouth College geologists continued to maintain that the extinctions which were claimed to have been instantaneous and synchronous with an impact had in fact occurred over a time interval of 100,000 years, and the iridium could have been deposited over a similar period.[51] In a paper in *Paleobiology* in 1984, the Berkeley palaeontologist Lowell Dingus pointed out that, because of the nature of the evidence, neither side was being objective. He wrote that 'it seems unlikely that we can distinguish episodes of extinction lasting 100 years or less from episodes lasting as long as 100,000 years'.[52]

In 1983, an iridium abundance anomaly was demonstrated in tiny spheres, less than 1 millimetre in diameter, which had been found in soil in the Tunguska region of Central Siberia and also in a South Pole ice-core. These apparently resulted from the Tunguska impact event of 1908. However, particles thrown into the air during the 1983 eruption of the Kilauea Volcano, Hawaii, were also found to be rich in iridium. Taken together, these findings supported the view that an iridium abundance anomaly could equally well indicate vulcanism or an impact. Some supported one view and some the other.[53]

In April 1984, in an editorial in *Nature*, John Maddox, wrote how 'it is proper to acknowledge that the intellectual climate has changed in favour of catastrophism, especially now that Lui[s] and Walter Alvarez appear to have proved their original case that the massive extinction at the end of the Cretaceous Period was caused by the impact of some extraterrestrial object'.[54]

In contrast, reflecting a significant body of scientific opinion, an editorial in the *New York Times* a year later stated categorically that the extinctions of dinosaurs and other species were far from instantaneous, for some were in decline before the end of the Cretaceous Period. It went on to conclude, 'Terrestrial events, like volcanic activity or changes in climate or sea level, are the most immediate possible causes of mass extinctions. Astronomers should leave to astrologers the task of seeking the cause of earthly events in the stars'.[55]

Around this time, Earth-scientists Antoni Hoffman and Matthew Nitecki carried out a survey of views on the impact hypothesis, and the results were published in *Geology* in December 1985. Opinions had been sampled in five groups of subjects: American geophysicists, British and German palaeontologists, and Soviet and Polish geoscientists. Results showed that 31% of the American geophysicists and 9% of the British palaeontologists thought that an impact had caused the K–T extinctions, percentages for the other groups falling between these extremes. The figures for those who believed that an extraterrestrial impact had occurred at the end of the Cretaceous Period, but had *not* caused the K–T extinctions, ranged from 40% (Polish geoscientists) to 15% (American geophysicists). Some still denied that any mass extinctions had taken place at the end of the Cretaceous Period, the numbers in this category ranging from 25% of Soviet geoscientists to 3% of American geophysicists.[56]

This survey was criticised by some for presenting too limited a choice of positions. It did not, for example, cater for those who accepted the reality of the K–T mass extinctions, and of an impact at the end of the Cretaceous, but believed that the impact was just one of several factors responsible for the extinctions.[57] Stephen Jay Gould was one who would have fitted into that category, for he believed that the dinosaurs suffered a decline as a result of a gradually deteriorating climate, making them susceptible to the *coup de grâce* of an extraterrestrial impact. As he wrote in an essay included in *Hen's Teeth and Horse's Toes*, 'Without the great Cretaceous extinction, dinosaurs might have rallied and still dominate the Earth'.[58] Others were of the opinion that, rather than a single major impact, there might have been several impacts from a cluster of comets, or from the breakdown products of a giant comet, as the Cretaceous drew to a close.[59] The situation remained confused.

Whatever simplistic views may have been taken by some who adopted the impact hypothesis in the early 1980s (or were assumed by others to have been taken by them), it soon became apparent to all that the Late Cretaceous and other mass extinctions were complex events.[60] As long ago as 1986, Digby McLaren wrote, in the multi-author book *Dynamics of Extinction*, 'In general, no claim is being made or has been made that impacts cause all extinctions or that any one extinction has been caused only by an impact. Like many other phenomena in geology, such causes may be multiple and resolved only with difficulty'. Nevertheless, he continued:

> But surely the burden of proof rests, now, firmly on the shoulders of those who would deny the existence of this violent phenomenon, and for them the proof must be in two phases: (1) evidence and argument leading to the denial of the demonstration of Earth-crossing asteroids and their periodic arrival on Earth and, if that cannot be achieved, (2) the demonstration that such energetic events have had no effect, detectable by geological means, on crustal evolution, including the development of life.[61]

22 The continuing K–T debate

Ammonites and dinosaurs: rapid extinction or gradual decline?

Whilst there were indications during the first half of the 1980s that the intellectual climate was becoming more receptive to catastrophist ideas, there was still no general agreement about the precise nature and cause of the mass extinction episode at the end of the Cretaceous Period. In particular, the issue of whether an extraterrestrial impact had been involved in the K–T extinctions remained unresolved.[1] The evidence was capable of being interpreted in different ways, which led to vigorous debates, especially since some evolutionary biologists reacted with hostility every time a scientist with an expertise in a completely different subject area expressed views about evolution. So, Steven Stanley wrote, at the beginning of *Extinction*:

> In part, the fervent new efforts of paleontologists to unravel the puzzle of mass extinctions have resulted from our chauvinistic impulse to convince the world that astronomers do not have simple answers to complex geological problems. It might be argued that we are reacting with bias, but there is a logical rejoinder. Paleontologists have data that raise serious doubts about the idea that impacts of extraterrestrial objects have caused most episodes of mass extinction. Moreover, we have data suggesting that many of the mass extinctions resulted from certain other, more mundane, causes.[2]

Stanley saw changes in the Earth's climate as being of particular importance, these often resulting from new configurations of land masses and modifications to ocean currents, as a consequence of continental drift.

Arguments about the Late Cretaceous extinctions continued throughout the second half of the decade and into the next. So, for example, Charles Officer and Charles Drake cited the existence of spherules found above and below the K–T boundary at Gubbio as evidence against the impact hypothesis,[3] but Alessandro Montanari was able to show that, although they may have looked similar to the sanidine spherules present at the boundary itself, they were in fact modern seeds and insect eggs.[4] At El Kef, Tunisia, on the other hand, contrary to what might have been expected from the impact hypothesis, no sanidine spherules were found at the K–T boundary in marine rocks similar to those at Gubbio.[5]

A team from Princeton University, led by Gerta Keller, claimed to have found evidence that the extinctions across the K–T boundary at El Kef had taken place in an episodic fashion over millions of years, with a series of minor extinctions occurring in step-wise fashion prior to a much larger event.[6] Similarly, observations at other sites suggested that there had been extinctions throughout the closing stages of the Cretaceous Period. Some groups of organisms, including all the ammonites, apparently disappeared completely before the main K–T extinction event.[7] However, subsequent investigations by the University of Washington palaeontologist Peter Ward, on the cliffs of the Biscay coast, in both France and Spain, showed that ammonites could be found right up to the K–T boundary.[8]

Officer and Drake had assigned the foraminifer extinctions at Gubbio and the final dinosaur extinctions in New Mexico and Montana to magnetic polarity zones 29R, 29N and 28R, respectively, which meant they could not have taken place simultaneously.[9] However, William Clemens came to different conclusions, placing them all within polarity zone 29R. Nevertheless, because of the length of this zone, Clemens found it impossible to be sure whether these extinctions had occurred at precisely the same time as each other or whether they were separated by an interval of up to 500,000 years.[10]

Findings in the San Juan Basin of New Mexico indicated that the dinosaurs may have survived beyond the K–T boundary, defined by a pollen break.[11] In contrast, further north in Alberta, Montana and Wyoming, it was possible that they died out before the end of the Cretaceous Period, as already mentioned in connection with the Hell Creek formation in Montana (see Chapter 21). Furthermore, Leigh Van Valen of the University of Chicago and Robert Sloan of the University of Minnesota claimed to have found evidence in the Bug Creek area of Montana that mammals characteristic of the Palaeocene Epoch (the first part of the Tertiary Period) were present during the Late Cretaceous Period[12] That indicated a gradual transition from the Cretaceous to the Tertiary.

Van Valen was a persistent critic of catastrophism, especially of some of its more assertive advocates. He frequently pointed out that selective use of the evidence could appear to prove the impact hypothesis, whereas in reality there were many signs pointing in the opposite direction.[13] It seemed likely to him, and to Sloan, that the mammals in Montana had simply out-competed the dinosaurs as climatic conditions gradually changed, with the result that the dinosaurs were forced into extinction. On the other hand, Jan Smit and Sander van der Kaars concluded in 1984, after an investigation of the area, that water channels which cut through the K–T boundary had confused the stratigraphy. They saw no reason to doubt that all the dinosaur fossils dated from before, and all the mammals from after, the K–T boundary.[14]

Sloan, Van Valen and colleagues re-examined the evidence but, in a report published in *Science* in 1986, they again came to different conclusions from Smit and van der Kaars, believing that competition between dinosaurs and mammals had taken place both before and after the K–T boundary, which was marked by a pollen break and an iridium abundance anomaly. Dinosaur teeth found in Tertiary deposits were too sharp to have been moved there by disturbances of the Earth, which would have blunted them. Furthermore, they were not accompanied by traces of other species which were common during the Late Cretaceous. The paper concluded that there had been thirty dinosaur genera present in the Bug Creek area 8 Myr before the end of the Cretaceous, but only twelve were still there when the K–T boundary event began, and more than half of these survived for tens of thousands of years into the Tertiary. The authors were prepared to accept that an extraterrestrial impact occurred at the K–T boundary, which would not have helped the dinosaurs' attempts at survival, but neither could it have been the sole reason for their extinction.[15]

The claim that dinosaurs survived well into the Tertiary Period in Montana continued to be disputed.[16] In the late 1980s, a detailed investigation in the Hell Creek Formation carried out by a Milwaukee Public Museum team led by Peter Sheehan did not find any Tertiary dinosaurs, but it reduced the dinosaur fossil gap beneath the K–T boundary from three metres to sixty centimetres.[17]

Whilst these arguments were taking place, a controversial, although not entirely new, thesis was put forward by palaeontologist Robert Bakker, of the University of Colorado, Boulder, in his 1986 book, *The Dinosaur Heresies*. This was that dinosaurs, despite being descended from reptiles, were warm-blooded, and had lifestyles similar to those of modern mammals. Indeed, if Bakker had correctly interpreted the evidence of dinosaur footprints, *Tyrannosaurus rex* could run as fast as the large predatory animals of today. In *The Dinosaur Heresies* he presented other arguments to support his belief, such as the fact that bone microtexture, rate of growth and predator–prey ratios of dinosaurs were typical of those of warm-blooded animals. Modern mammals, because of their warm-blooded characteristics, are able to suppress the evolutionary potential of today's reptiles, whereas the dinosaurs dominated the mammals of their time, even though the proto-mammals (therapsids) had become established before the proto-dinosaurs (thecodonts). Whilst, for a time after the disappearance of the Sun below the horizon, a large body mass can help to maintain temperatures suitable for activity in cold-blooded animals, that factor could not account for the success of dinosaurs at the smaller end of the size range, nor explain how some dinosaurs could live in very cold environments, close to the poles.[18] Despite Bakker's arguments, the issues remained largely unresolved. However, for whatever reason, dinosaurs had unquestionably been the dominant land animal for around 150 Myr, throughout the Jurassic and Cretaceous Periods.[19]

Why, then, did these highly successful creatures become extinct at the end of the Cretaceous? Throughout the book, Bakker acknowledged the importance of extinction events. The rise of the proto-dinosaurs was facilitated by the extinctions of proto-mammals in the Middle and Late Permian, and their disappearance at the end of the Triassic cleared the way for the emergence of the dinosaurs themselves. Later, in North

America and elsewhere, the lumbering dinosaurs of the Jurassic, whose long necks gave them easy access to food growing on trees, were replaced by the more nimble ground-feeders of the Cretaceous, following the appearance of flowering plants. Bakker might have been expected to suppose that the complete extinction of these extremely well-established (and, in his view, warm-blooded) animals at the end of the Cretaceous Period required some sensational event, yet he was content to go along with the theory advanced by Norman Newell over twenty years earlier, that the main cause was the draining of the shallow epicontinental seas (see Chapter 14). This would have changed habitats and allowed the mixing of species which had previously been kept separate from each other, encouraging competition and the spread of disease. Without question, shallow seas covered large areas of the continents during the Late Cretaceous. This meant, for example, that the Hell Creek area of Montana, now an arid, inland region, was then a fertile plain linking the Rocky Mountains to an epicontinental sea, which retreated towards the south as the Period drew to a close. Bakker was prepared to accept that a large extraterrestrial impact occurred at the end of the Cretaceous but, if so, this would have had little bearing on extinction processes which were already nearing completion. The main problem, as he saw it, was that a few dinosaur species had become too successful, so the group as a whole had lost the range and diversity to cope with changing conditions.[20]

However, a comprehensive survey of dinosaur diversity in each of the stratigraphic stages of the Jurassic and the Cretaceous Periods, carried out by palaeontologist Peter Dodson of the University of Pennsylvania, found no evidence of a reduction in species numbers as the Cretaceous drew to a close. Indeed, in his report, published in *Proceedings of the National Academy of Sciences, U.S.A.* in 1990, Dodson concluded that the dinosaurs had been 'near the peak of their historic diversity' during the Maastrichtian, the final stratigraphic stage of the Period, and of their existence.[21] Nevertheless, he thought it likely that they had suffered a gradual decline as the stage progressed.[22] On the other hand, Peter Sheehan and the Milwaukee Public Museum group, focusing on the final few million years of the Maastrichtian, could find no indications in the Hell Creek region of any reduction in dinosaur diversity before their complete disappearance.[23] This study, however, failed to convince William Clemens and some other palaeontologists, including David Archibald of San Diego State University, that the extinction of the dinosaurs had been a sudden rather than a gradual event.[24]

Arguments for and against a catastrophic K–T impact

Whilst these investigations of dinosaur diversity were taking place, new botanical evidence was seen to be consistent with the hypothesis that a major impact occurred at the end of the Cretaceous Period.[25] In 1986, Jack Wolfe and Garland Upchurch, of the U.S. Geological Survey, deduced from a study of fossil leaves that there had been a distinct but relatively brief temperature drop at the K–T boundary in southern U.S.A., with similar but smaller effects in the northern states and also in South America.[26]

Subsequently, a more detailed investigation around Teapot Dome, Wyoming, led Wolfe to believe that the sudden drop in temperature occurred in the middle of what had been the summer season. What is more, the overall patterns of change in leaf fossils suggested to him that there had been two impacts at the end of the Cretaceous Period, a small one close to Wyoming and a larger one in a more distant location.[27]

Further support for a K–T impact event came in 1987 from Ronald Prinn and Bruce Fegley, of Massachussetts Institute of Technology, who concluded from their studies that the patterns of extinction, particularly the sudden depletion of limestone-forming organisms, were consistent with the large-scale atmospheric production of nitrous and nitric acid, and the subsequent fall of acid rain. Prinn and Fegley considered it most likely that the projectile involved was a long-period comet, since an impact by a lower velocity short-period comet or an asteroid would have produced much less acid.[28]

On the other hand, Charles Officer continued to argue with vigour the case for a volcanic explanation for the K–T extinctions. In 1986, he was amongst a group who announced in *Geology* the finding of shocked quartz grains in material which had been ejected from the Indonesian supervolcano, Toba, during the Pleistocene Epoch.[29] However, the lamellae were not as complex as those in grains found at known impact sites, and at the K–T boundary, an issue which remained a problem for the vulcanism hypothesis.[30] As Richard Kerr explained in *Science* early in 1987, 'Try as they might, advocates of a volcanic end to the Cretaceous have failed to find the same kind of so-called shocked quartz grains in any volcanic rock. Because shocked quartz continues to maintain its exclusive link to impacts, the impact hypothesis would seem to be opening its lead over the sputtering volcano alternative'.[31]

Nevertheless, Officer stood firm and, accompanied once again by Charles Drake and reinforced by others, including Anthony Hallam, he presented an updated version of his arguments in *Nature* in 1987. Although, in this paper, it was acknowledged that many types of plankton died out in less than 10,000 years, Officer and his colleagues continued to maintain that the K–T extinctions as a whole took place over a much longer time-scale, and were due to the effects of extensive volcanic activity, coupled with the disappearance of the epicontinental seas.[32]

It would be easy to see these various exchanges as a renewal of the long-discontinued debate between gradualists and catastrophists. However, that would be a mistake, for large-scale vulcanism is itself a catastrophist process, so neither side was supporting the kind of gradualism proposed by Lyell and Darwin. Indeed, in an article in *Science* in 1987, Hallam emphasised the catastrophist nature of the scenario he was supporting when he pointed out that the draining of the shallow seas from over the land in the Late Cretaceous Period occurred too rapidly to be caused by an increase in the volume of ocean basins, and was more likely to be related to vertical tectonic movements on the continents, linked to the vulcanism.[33] Vincent Courtillot, one of the first to argue that volcanic eruptions had been a causal mechanism in the K–T extinctions, wrote in *Scientific American* in 1990, 'Both the asteroid impact and volcanic hypotheses imply that short-term catastrophes are of great importance in shaping the evolution of life'.[34]

By this time, it was becoming increasingly clear that the massive volcanic activity of the Late Cretaceous was concentrated much more towards the end of the Period than had been supposed. Previously it had been thought that the lava forming the Deccan traps in India had been laid down over a period of anything up to 50 Myr but, in the mid-1980s, Courtillot and colleagues reduced that figure to less than 4 Myr. It was even possible that the vulcanism was restricted to within magnetic polarity zone 29R, during which the Late Cretaceous extinctions occurred.[35] Nevertheless, the average rate of lava release over this half-million-year period would only have been about the same as that observed in the Hawaiian islands in the present century.[36] However, the indications were that the formation of the Deccan traps was far from being a gradual process. Instead, it involved a series of discrete, and hence more intense, events, which would have affected the environment in catastrophic fashion.

Earth-scientists generally believed that the probable cause of the Deccan eruptions was the rise of a plume of hot, light, free-flowing material from the Earth's mantle.[37] They also drew attention to the fact that increased amounts of iridium had been found in particles emitted by the Piton de la Fournaise volcano on the island of Réunion, linked to the Deccan traps by a chain of underwater volcanoes.[38] Many of the features of the impact hypothesis for the K–T extinctions, including iridium enhancement, acid rain and the blocking of the Sun's rays, leading to lower temperatures and reduced photosynthesis, also formed part of the theory that volcanic activity was the primary cause. Therefore, much of the evidence held to be consistent with the former was also consistent with the latter.[39]

However, there were points of difference, and some pieces of new evidence were tending to strengthen the view that, in addition to the large-scale episodes of vulcanism which had clearly occurred at the end of the Cretaceous Period, there had also been a massive extraterrestrial impact.[40] For example, as reported in *Science* in 1988, a team led by Joanne (Jody) Bourgeois of the University of Washington, Seattle, discovered an unusual sandstone bed showing signs of ripples caused by waves, immediately under the K–T boundary layer at sites along the Brazos River, near Waco, Texas. To Bourgeois and her colleagues, the most likely explanation was a tsunami 50 to 100 metres high, produced by an impact into water. Certainly, it seemed unlikely that volcanic activity in far-off India could be implicated.[41] At around the same time, the powerful argument that the shocked quartz grains at the K–T boundary were produced by the arrival of a projectile from space was strengthened even further when Michael Owen of St. Lawrence University, New York, and Mark Anders of Berkeley showed that the cathodoluminescence colour of the grains was characteristic of impact-derived ones, but not of ones produced by vulcanism.[42] Again, a Russian team concluded that the rhodium/iridium ratios at the K–T boundary indicated an extraterrestrial event rather than release of metals from the Earth's core by volcanic activity.[43]

But, if there had been a major impact, where could the missile have struck? The evidence was somewhat contradictory. The minerology of the shocked quartz suggested an impact on land, possibly on the American continent, whereas the presence of sanidine spherules and the indications of a tsunami were more consistent with an ocean impact. Five impact craters were known which had been provisionally dated to around the end of the Cretaceous Period, but all of these seemed much too small to mark an impact which could have caused the K–T extinctions, having diameters considerably less than the 100–200 km which would have resulted from the impact of the 10-km asteroid featured in the Alvarez hypothesis. The largest, located in the Russian Arctic, was the 65-km diameter Kara crater, which had apparently been formed at the same time as the nearby 25-km Ust-Kara impact structure. Two of the others were also seemingly related to each other. These were the Kamensk and Gusev craters of southwestern Russia, which had diameters of 25 km and 4 km, respectively. The fifth, located near Manson, Iowa, was a 35-km diameter structure lying under debris deposited by glaciers of the subsequent Pleistocene Ice Ages.[44] Despite the fact that it was only of moderate size, much interest was shown in the Manson structure, because of its accessibility and the indications in the geological record that there had been an impact in the North American region at the end of the Cretaceous Period.[45]

By 1990, no-one could seriously question the fact that catastrophism had begun to feature once more in the mainstream of scientific debate, however controversial the particular issues and details might still be. Charles Officer, in his 1993 book, *Tales of the Earth*, written in collaboration with journalist Jake Page, commented, 'Whether Luis Alvarez's idea proves to be right or wrong – and the recent geologic and paleontological findings suggest that it has serious problems [a comment which was only to be expected] – he deserves a great deal of credit for reviving interest in one of the fundamental geologic problems: the causes of mass extinctions that have occurred over the past 600 million years'.[46]

When Anthony Hallam wrote *Great Geological Controversies*, published in 1983, he included just a single sentence about modern theories of catastrophism, to note that 'mass extinction episodes such as those at the close of the Palaeozoic and Mesozoic eras are widely considered to be the result of some global catastrophe, though the precise nature of such catastrophes remains a matter of dispute'.[47] The second edition of this book, which appeared in 1989, devoted a whole chapter to the subject, whilst retaining much the same conclusion. Thus, Hallam commented, 'In marked contrast to the so-called earth sciences revolution involving the acceptance of plate tectonics, where magnetic anomalies provided the key to the problem and the consensus of geologists was converted within a few years, over eight years have elapsed since the Alvarez hypothesis was put forward and no sign of a consensus about the causes of extinction is yet in sight. The jury is still out'.[48]

The search for conclusive evidence of an impact

Whilst, in the late 1980s, there may have been no general consensus about the causes of mass extinctions, the balance continued to shift towards an acceptance of the hypothesis that an extraterrestrial impact had played a major role in the K–T extinction episode. As mentioned in Chapter 21, a conference had been held at Snowbird, Utah, in October 1981, to discuss reactions to the original Alvarez paper. Seven years later, in October 1988, another conference was held at Snowbird to consider the many developments which had taken place since the first one.[49] Of this second Snowbird conference, which was entitled, 'Global Catastrophes in Earth History', Richard Kerr wrote in *Science*, 'No one asked for a show of hands, but a vote among those attending the conference . . . would have given a clear-cut victory to an asteroid or comet impact as the most likely explanation of the mass extinction 66 million years ago'.[50]

Whilst the conference participants may not have constituted a fully representative sample of scientific opinion, it was evident that an increasing number of scientists were beginning to warm to the impact hypothesis as more evidence came to light.[51] Another significant development came in 1989 when John McHone and colleagues found stishovite, a dense form of silica, at the K–T boundary in New Mexico. Stishovite is generally regarded as an indicator of impacts, and it has not been found in the outpourings of volcanoes.[52] Similarly, in the same year, geochemists Meixun Zhao and Jeffrey Bada discovered α-amino-isobutyric acid and isovaline in K–T boundary sediments at Stevns Klint, Denmark. These particular amino acids, unlike others, are rarely found in terrestrial rocks, but are known to be components of carbonaceous meteorites.[53] Again, in 1991, David Brez Carlisle and Dennis Braman found minute diamonds at the K–T boundary in Alberta, Canada. These nanometre-sized precious stones were similar to ones found in carbonaceous meteorites, and could not have been formed by volcanic activity, although it was within the bounds of possibility that they could have been produced from Earth rocks by the shock of an impact. However, further investigations demonstrated that these particular diamonds must have arrived, fully formed, from space.[54]

Carlisle, of Environment Canada, came to the conclusion that the supposed K–T projectile was a comet, or something derived from a comet, rather than an object displaced from the main belt of asteroids. He explained his reasoning in his 1995 book, *Dinosaurs, Diamonds and Things from Outer Space*. Since comets are formed much further out in the Solar System than asteroids, they should consist of material which is in a less oxidised form, together with ice. Thus, unlike asteroids, comets should contain free carbon (including diamonds), silicon carbides and minerals such as spinels and chromite. Such substances are found in carbonaceous meteorites, which themselves may be derived from cometary material, but are not present in those meteorites which consist predominantly of iron or stone. These same materials are also found in the K–T boundary clay, together with abnormally high amounts of iridium. Naturally, a missile made of pure ice could not produce an iridium abundance anomaly, except indirectly by stimulating volcanic activity. However, this metal is likely to be present in the dust and rock which form a significant part of a comet. Thus, although iridium would also be found in metal-containing asteroids, the findings as a whole convinced Carlisle that the K–T projectile must have originated in the cometary zones.[55]

However, that was only part of Carlisle's scenario. He suggested that the Late Cretaceous mass extinctions could have been caused by the joint effects of a nearby supernova explosion, involving a star more than twenty times the mass of the Sun, about five light years away, and one or more subsequent cometary impacts. The energy released from the supernova would have caused widespread fires on Earth and killed many land animals outright. It would also have disrupted the comets of the Oort cloud, causing at least one to strike the Earth centuries later. That was consistent with the fact that the K–T soot layer was found beneath the boundary clay. Furthermore, this clay contained anomalously high concentrations of silver-107, which could have been produced in a supernova explosion.[56]

Regardless of this, the search had been continuing for the site of a major impact which could be linked unequivocally to the events at the end of the Cretaceous Period. As we have already noted, an impact was known to have occurred at Manson, Iowa, but this event by itself seemed far too insignificant to account for the K–T mass extinction episode and the layer of iridium-enriched clay found throughout the world. On the other hand, an impact in the western hemisphere appeared to have played a part in the events at the end of the Cretaceous. Even if not the main one, an impact site in or near the U.S.A. was indicated, because only in that region were there two K–T boundary layers. In the U.S.A., the K–T layer familiar from around the world, containing the iridium, sanidine spherules, shocked quartz grains and soot, was found overlying a second layer, which appeared to consist of more-localised products of an impact. Furthermore, the discovery of larger and more abundant shocked quartz grains in North America than elsewhere suggested that even the main event might not have been too distant from the U.S.A.[57] The Manson event could possibly have been responsible for the layer of localised ejecta, whilst the Texan tsunami could have been the consequence of a larger impact somewhere off the American coast.[58]

Alternatively, the ejecta layer itself could have been produced by an off-shore impact. The Caribbean area began to seem a likely location for such an event when, in 1990, Alan Hildebrand and William Boynton, of the University of Arizona, together with colleagues from Florida International University, showed that the ejecta layer near Beloc, Haiti, was 50 cm thick, some twenty times thicker than at sites in the U.S.A. This layer, as shown by Haraldur Sigurdsson and colleagues of the University of Rhode Island, contained abundant glassy tektites, in addition to microspherules. Evidence of a tsunami was particularly strong around the Caribbean, with possible impact-wave deposits being found in many places, including sites in the Colombian

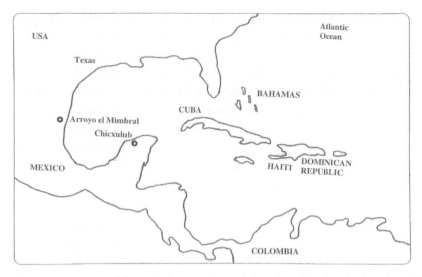

Figure 22.1   Map of the Caribbean region as it is today, showing the location of the impact site at Chicxulub, and other places mentioned in the text. Note that North and Central America were not connected to South America at the time of the Cretaceous–Tertiary transition (see figure 14.2).

Basin, the 'Big Boulder Bed' of Cuba, and at Arroyo el Mimbral in northeastern Mexico. The Colombian Basin and a region southwest of Cuba were suggested as possible sites for the impact.[58]

However, an area around Progreso, on the northern coast of the Yucatán Peninsula of Mexico, soon became the focus of interest. The specific site centred on the village of Chicxulub (figure 22.1) which, in the local Mayan language, means 'Tail of the Devil'. In the Late Cretaceous Period, and for millions of years afterwards, this area was covered by sea-water to a depth of several hundred metres, so the suspected impact crater had long-since been filled in and buried under layers of sedimentary rock.[59] It was first discovered in 1981 by geologists Glen Penfield and Antonio Camargo, who were working for Pemex, the national oil company of Mexico. News of the finding was released into the public domain, but all the survey details were kept secret, for commercial reasons. Therefore, even though the Chicxulub structure seemed to be the largest impact crater known from the past 1,500 Myr, with a diameter of 180 km, within the range required for the Alvarez K–T scenario, very few people were aware of the fact. When Alan Hildebrand came to hear of it in 1990, he contacted Penfield, and investigations began again shortly afterwards. Shocked quartz grains were discovered at the site, and although the age could not at first be determined with certainty, it was at least approximately right for association with the K–T mass extinctions.[60]

However, Charles Officer remained as opposed as ever to the K–T impact hypothesis. In *Tales of the Earth*, he and his co-author, Jake Page, suggested that the origin of the Chicxulub structure was 'still under debate', and continued, 'The advocates for terrestrial causes for the extinctions may seem to be party poopers, but the facts seem to lie on their side'.[61] In their view, and that of some others, all of the supposed evidence for a major impact in the Late Cretaceous, including the features around the Caribbean, could be explained by volcanic activity.[62] Three years later, in their 1996 book, *The Great Dinosaur Extinction Controversy*, Officer and Page claimed that, even though the impact hypothesis had become accepted faster than any other they could recall, this acceptance had come largely from journalists and physicists.[63] They went on to suggest that Earth-scientists were more aware than others of the complex issues involved, and they could see that the impact scenario had 'collapsed under the weight of accumulated geologic and other evidence to the contrary'.[64] The Alvarez hypothesis was dismissed as a 'degenerative' research programme, as it had changed from the idea of a single impact causing all the K–T extinctions to a broader one which allowed for the involvement of other factors over an extended time-scale. Officer and Page derided as 'pathological science' all continuing attempts to argue that the Chicxulub structure was an impact crater, explaining that 'pathological science arises from self-delusion'.[65]

However, such out-and-out antagonism to the possibility that a major impact had occurred at the end of the Cretaceous Period was becoming very uncommon, even amongst those Earth-scientists who remained convinced that the majority of the K–T extinctions were caused by terrestrial mechanisms. David Archibald, for example, who had always maintained that the dinosaurs declined throughout the Late Cretaceous, acknowledged in 1996 that the process had probably been accelerated by an impact, writing in his book, *Dinosaur Extinction and the End of an Era*, 'Suddenly, a literally earth-shattering event magnified the differences between species doing well and species doing not so well. An asteroid or comet struck the area that today we call the Yucatán'.[66] Similarly, Vincent Courtillot, whilst maintaining his belief that volcanic activity was the main cause of the K–T and other mass extinctions, admitted in his 1999 book, *Evolutionary Catastrophes*, 'I now think that quite an exceptional extraterrestrial object must indeed have struck the Earth 65 million years ago'. Although he was convinced from detailed geological studies that the Deccan vulcanism had already started by this time, the consequence was that 'species that were already hard pressed [suffered] the additional catastrophe of an impact by an asteroid or comet'.[67]

Inevitably, some discoveries only served to confuse the various issues, by demonstrating the complexity of the situation. That was particularly so at high latitudes, where extinctions were less severe than in tropical areas.[68] Nevertheless, evidence continued to be found which strengthened the argument that there had been a major impact in the area of Chicxulub at the end of the Cretaceous Period.[69] According to Walter Alvarez, writing in his 1997 book, *T. rex and the Crater of Doom*, the winter of 1991–92 'seemed like the turning point' towards general acceptance of that view.[70] The strength of support for it was clear at the third 'Snowbird' conference in 1994, entitled, 'New developments regarding the K/T event and other catastrophes in Earth history', when not a single speaker questioned the age or impact origin of the Chicxulub crater (note that, somewhat confusingly, this conference was held not in Snowbird, Utah, like the first two, but in Houston, Texas).[71]

The geology of the area around the impact site was such that many features of the K–T boundary could now, for the first time, be properly explained, and some apparently irreconcilable facts could be fitted into a coherent picture. The missile (apparently a comet or a carbonaceous asteroid) must have passed through half a kilometre of water and a layer of sediment before striking the continental rock beneath, this giving rise to a tsumami and creating sanidine spherules, as well as shocked quartz grains and localised tektites. Perhaps significantly, an impact in such a sulphur-rich region would also have given rise to severe acid rain in the aftermath. Furthermore, large amounts of carbon dioxide would have been released from carbonate rocks, producing a greenhouse effect.[72]

More speculatively, some suggested that shock waves generated by the collision could have liberated vast amounts of methane, formed from rotting vegetation and trapped underground at sites throughout the world. The released methane would have been ignited by electrical storms in the disturbed atmosphere, with devastating consequences.[73] Again, those who sought to establish an impact-related origin for the extensive vulcanism of the time drew attention to the fact that the Deccan eruptions of the Late Cretaceous occurred close to the point on the Earth's surface which was furthest away from Chicxulub. Michael Rampino and David Brez Carlisle both argued that shock waves from a large impact might travel around the world and produce volcanic activity at the opposite side.[74]

And what of the possibility, mentioned above, that there had in fact been *two* impacts in quick succession at the end of the Cretaceous Period?[75] An investigation at Beloc (Haiti) in 1995 concluded that most of the tektites and other impact remains were from the Chicxulub event, but this may have been closely followed by a second, more distant, impact. Similarly, two years later, Gerta Keller and colleagues suggested that the Chicxulub impact, giving rise to the tektites around the Caribbean, may have occurred a short period of time (in geological terms) before the world-wide K–T event, marked by mass extinctions, the iridium abundance anomaly and the presence of spinels. Then, in 1998, evidence of foraminifer extinctions and a tsunami at Poty, northeastern Brazil, was taken to indicate the possibility of an impact in the South Atlantic less than 100,000 years after the Chicxulub event.[76]

In the early 1990s, argon-40/argon-39 dating of the Manson structure, the Chicxulub structure and of tektites from Haiti had all given values indistinguishable from the age of the K–T boundary.[77] However, concern was expressed about the nature of the rock used for dating the Manson structure, and when the process was repeated using sanidine linked to the formation of this crater, a date of 74 Myr was obtained, ruling the Manson impact out of consideration for involvement with the K–T boundary events.[78] Very similar results were obtained when the same technique was applied to the Kara and Ust-Kara structures, whilst a 23-km diameter crater at Lappajärvi, Finland, was found to be about three million years older, showing that there had been a series of moderately sized impacts in the Late Cretaceous, prior to the huge Chicxulub event. On the other hand, argon-40/argon-39 studies showed that the Kamensk crater (and, by implication, the Gusev one) was much younger, being about 49 Myr old. That leaves no known crater with a date very close to that at Chicxulub, so, if the Poty tsunami was indeed caused by an impact, then the location of this has still to be identified.[79]

The end of the ultra-gradualistic paradigm

Without question, a Kuhnian paradigm shift (see Chapter 14) occurred during the 1980s and early 1990s, concerning the history of life on Earth. Whilst it may not have been as complete or as dramatic as that of the 1960s, when the reality of continental drift became accepted over a very short time-scale, after many years on the periphery of orthodox thought, it was nevertheless of great significance. Whereas catastrophist mechanisms had previously been dismissed from any consideration by establishment figures upholding the ultra-gradualistic paradigm, it was now accepted that such mechanisms were indeed possible. However, that is not to say it was agreed that they must have played a significant role in any particular event, including the extinctions at the end of the Cretaceous Period.

In 1994, William Glen, who had made a study of the previous paradigm shift concerning continental drift (see Chapter 14), summarised the state of current thinking about the K–T extinctions at the Snowbird/Houston conference and in the multi-author book, *The Mass-Extinction Debates*.[80] As he saw the situation:

> Definitive closure has not been reached on any of the major issues entrained in these debates, but the vast majority of earth scientists are now convinced of at least one impact at the Cretaceous/Tertiary boundary, and a substantial number are inclined to think in terms of multiple impacts, either instantaneous or spread over 1 to 3 million years; however, far from all who subscribe to impact(s) – especially among the paleontologists – view impacts as the chief cause of the mass extinction(s).[81]

One who saw no need to change his long-standing gradualistic views was John Briggs, a semi-retired marine scientist from the University of Florida. In *Mass-Extinction Debates* he claimed that species turnover at certain times had been greatly exaggerated, so, in consequence, there was 'no evidence that global mass extinctions (defined as short-term catastrophic events) ever took place.'[82] Without question, it was clear that, even at the end of the Cretaceous Period, extinctions close to the equator were more marked than those elsewhere, and some extinctions occurred over a longer time interval than others. However, extensive volcanic activity of a catastrophic nature undoubtedly occurred at the time, regardless of whether or not it was linked to an impact. Hence, it

Figure 22.2   The end of the Cretaceous Period, when there was a major extraterrestrial impact off the coast of central America, and widespread volcanic activity. All dinosaurs, including herbivores such as *Triceratops* (left) and meat-eaters such as *Tyrannosaurus* (right) became extinct during the Cretaceous–Tertiary transition.

seemed difficult to dispute the claim that one or more catastrophic events, terrestrial or extraterrestrial, or both, must have played a significant role in the disappearance of the dinosaurs and of other Cretaceous life-forms (figure 22.2).

However, that is not so say that gradualistic mechanisms, especially those involving tectonics, were insignificant. So, for example, David Archibald still continued to believe that marine regression was the main reason for the extinction of the dinosaurs in North America, regardless of the possible effects of impacts and vulcanism. In *Dinosaur Extinction and the End of an Era*, he wrote that the dinosaur extinctions started 'well back before the K/T boundary, maybe as much as five million years before, as the intercontinental seaway began to slip away from the interior of North America'.[83] Also, as we have noted, average temperatures fell towards the end of the Cretaceous, whether because of the consequences of continental drift, atmospheric pollution resulting from volcanic activity, a series of impacts, cometary dust, or a mixture of causes. A recent study at Hell Creek, led by Jack Horner, the curator of paleontology at the Museum of the Rockies, in Bozeman, Montana, supported the view that significant climate changes had taken place in that region prior to the Chicxulub impact.[84]

Back in 1993, in the conclusion to a chapter entitled 'Dinosaur summer' in the multi-author work *The Book of Life*, the Bristol University palaeontologist Michael Benton commented, 'The balance of opinion concerning K–T seems to be about equal between the impact-catastrophe model and the gradual-global-cooling model. There is evidence for both, and both may be significant'.[85]

Anthony Hallam took a similar view in his 1997 book, *Mass Extinctions and their Aftermath*, written in collaboration with Paul Wignall of the University of Leeds. As we saw earlier in this chapter, Hallam dealt with mass extinctions in a single sentence in the 1983 edition of his *Great Geological Controversies*, but expanded the treatment to a full chapter in the 1989 edition, whilst remaining non-committal about the Alvarez impact hypothesis. Now he was devoting a complete book to the subject and in it, he and Wignall wrote:

> Even if one accepts the likelihood of end-Cretaceous impact, it cannot be ignored that the latest Cretaceous was a time of considerable environmental change involving both climate and sea level . . . A compound scenario involving both gradual extinctions followed by a catastrophic *coup de grâce* seems to be the one best fitted to the facts as we know them at present. One suspects, furthermore, that even without a bolide impact there might well have been a mass extinction recorded at the end of the Mesozoic era.[86]

It is now apparent that the detailed effects of impacts on the environment are difficult to assess, particularly since it is possible that they may influence geological processes such as sea-floor spreading, mountain building and vulcanism.[87] Some environmental changes, including ones occurring over a long time-scale, could thus be linked to impacts, whereas others might be purely Earth-centred phenomena. The K–T event, as well as some other mass extinction episodes, could have been the result of an extraterrestrial impact coming at a time when previously dominant species were already suffering, because of previous impacts, stresses of terrestrial origin, or a combination of factors. The issues are complex, to an extreme degree. What cannot be denied, however, is that evolutionary gradualism, previously unchallengeable, and catastrophism, contemptuously brushed aside for most of the twentieth century, now have to participate in the debates on an equal footing. Evidence, not presumption, must determine which theory is correct in any given situation.

This change in attitude can be clearly seen if we compare what was said about the K–T extinctions in four books by authors associated with the Natural History Museum, London, published over a period of 18 years, starting in 1979 with *A New Look at Dinosaurs*. The author of this book, Alan Charig, who at the time was head of the section in the Museum responsible for fossil reptiles and birds, began by stressing that the extinctions 'were not as immediate and as sudden as some people like to believe'. He then considered a list of possible explanations for the death of the dinosaurs, starting with the ones he thought most sensible, such as changes in climate due to the effects of continental drift. After dismissively mentioning others which seemed to him more implausible, such as species senility, disease, a tendency for slipped discs and a trend towards smaller brains, Charig continued:

> Among the even less likely causes suggested for the death of the dinosaurs are poison gases, volcanic dust, meteorites, comets, sunspots, God's will, mass suicide (like lemmings!) and wars.[88]

Writing along similar lines in his 1982 book, *Hunting the Past*, Beverley Halstead (who had worked at the Museum before moving to Reading University) concluded:

> There is still no satisfactory theory to account for the extinction of the dinosaurs, but at least one can discount the extra-terrestrial catastrophic ones. It seems reasonable to interpret the disappearance of the dinosaurs as a process which occurred gradually over millions of years.[89]

Eleven years later, Tim Gardom, writing in collaboration with Angela Milner (head of the Museum's fossil vertebrate division) in *The Natural History Museum Book of Dinosaurs*, took a rather different perspective, starting the chapter on the death of the dinosaurs with the words:

> Some catastrophic event wiped out the dinosaurs. Was it an asteroid collision? Or a climate change brought about by massive volcanic eruptions? Buried under tonnes of rock, the evidence is elusive and controversial. Yet it all points to the dinosaurs' involvement in one of the most disastrous mass extinctions in the Earth's history.[90]

Similarly, in *Life – An Unauthorised Biography*, published in 1997, Richard Fortey (a senior palaeontologist at the Museum) drew attention to a 1994 letter in the journal *Geology*, which claimed, 'The reality of a large impact event marking the K–T boundary cannot be denied'. Fortey continued:

> Well, it obviously can be denied, since that is exactly what the critics are doing. Even if an impact is accepted, a direct link to extinctions is a separate step. This discussion of the K–T crater is at that critical stage in a historical argument where evidence and counter-evidence are paraded through the pages of the journals. Things have changed since the early days, because it now seems that it is the critics who are in the more defensive position.[91]

In this and the previous book, the issues were not brought to any definite conclusion. However, in marked contrast to the first two books, catastrophist explanations were clearly considered to be perfectly possible. Without any question, by the mid 1990s, catastrophism was back in business.

23   Mass extinctions and the course of evolution

The K–T extinctions and the rise of the mammals

It has now become clear that explosive radiations of new species do not generally cause or even accompany mass extinctions, but follow them, as previously existing ecological niches are vacated or new ones created. That is typified by the K–T extinctions, and their consequences. During the Cretaceous, whilst a few small, furry mammals scurried around in the undergrowth, the dinosaurs were dominant over all the land environments. The break-up of Pangaea into the northern supercontinent of Laurasia and the southern one of Gondwana (see figure 14.2) resulted in different patterns of dinosaur evolution with, in general, existing characteristics such as large size being developed further in the south, whilst a rich variety of novel forms were brought into being in the north.[1] Yet all of these died out in the K–T extinctions.

After the disappearance of the dinosaurs, the mammals were left largely without rivals on the continents, and new species, with a range of novel characteristics, emerged to fill most of the vacant ecological niches. This occurred rapidly, in geological terms, but it still involved a time-scale of many millions of years. Of the fifteen known families of Late Cretaceous mammals, ten survived the K–T extinctions and entered the Palaeocene Epoch, but the number of mammalian families did not exceed seventy until after the start of the Eocene Epoch, 10 Myr later. At a lower level of classification, 40 mammalian genera in the Late Cretaceous became 200 in the Early Eocene, rising to over 500 later in the same epoch. The first mammals were all tiny and, although a few moderately large ones appeared during the Palaeocene, it was not until the Oligocene, after the Eocene had come and gone, that the diversity of large mammals started to become comparable with that of the dinosaurs of the Cretaceous Period. Whilst there were some mammalian radiations during the Late Cretaceous, there was nothing to suggest that the mammals would have gone on to challenge the dominance of the dinosaurs had it not been for the K–T extinctions. The notion that the replacement of the dinosaurs by the mammals was an inevitable consequence of evolutionary progress has long been discredited. It seems, instead, that the dominance of mammals was accidental. The early mammals may just have been smaller, better insulated and generally more capable of coping with harsh conditions than the dinosaurs at a time when these characteristics made the difference between survival and extinction.[2]

After the removal of the large reptilian animals that had dominated the continents for the previous two geological periods, the mammals had an unexpected opportunity to diversify and dominate in their turn. As Michael Benton wrote in *The Rise of The Mammals*, published in 1991, 'Whatever its cause, the K–T event was the trigger for the beginning of the Age of Mammals – our present age. It cleared the stocks and left the way open for mammals to show their true potential, after 160 million years of living in the shadow of the dinosaurs'.[3] Similarly, in the same year, Niles Eldredge commented, in his book, *The Miner's Canary*, 'Dinosaurs did just fine, filling ecological space the world over for 145 million years, yielding just enough space for us mammals to cling to. Mammals arose at just about the same time that dinosaurs got going ... It is now abundantly obvious that it took the eradication of dinosaurs to give mammals a chance to take over, which we did, but not until after the great K–T extinction'.[4]

In general, not only has the significance of mass extinction events in helping to determine the overall course of evolution become well recognised, but evolutionary theory must now accommodate both gradualist and catastrophist processes, and mechanisms of extraterrestrial as well as Earth-centred origin. Thus, Charlotte Avers of Rutgers University wrote in *Process and Pattern in Evolution*, published in 1989, 'Mass extinctions occurred at ... the end of the Cretaceous ... leading to the disappearance of the dinosaurs and many other animals. The Cretaceous mass extinctions may have been triggered by the impact of an extraterrestrial body on the Earth, raising clouds of dust that could have blotted out the sun and produced a "nuclear winter", lasting for months or, perhaps, years. Mammals superseded the reptiles as the predominant animal life in the Cenozoic Era'. She went on to add that, together with similar events at other times, 'these extinctions and adaptive radiations indicate that evolution is unpredictable and irreversible'.[5] Similarly, in the 1993 standard textbook *Evolution*, Mark Ridley, of Emory University, Atlanta, Georgia, wrote, 'Mass extinctions are real events in the history of life ... The best studied mass extinction happened at the Cretaceous–Tertiary boundary, and for this there is a growing body of evidence that it was caused by a collision of the Earth with an asteroid. This idea has clear implications for the pattern of fossil extinctions.'[6]

Back in 1982, Dale Russell speculated how life on Earth might have developed had the dinosaurs not disappeared at the end of the Cretaceous Period. Without the K–T extinction event, would the Earth now be dominated by an animal with some human features, but with more reptilian characteristics, perhaps descended from a relatively large-brained dinosaur such as *Stenonychosaurus*? That remains just speculation.[7] The dinosaurs *did* become extinct, and humankind as we know it eventually emerged.

Yet, even that line of evolutionary descent might have been cut off during the K–T transition, had circumstances been slightly different. As science journalist Tim Haines wrote in *Walking with Dinosaurs*, the book of the 1999 BBC television series, 'Research has revealed that there were at least three forces at work at the end of the Cretaceous, which would have made life pretty miserable for most organisms'. These were the Deccan

Figure 23.1   Palaeontologist David Raup.

volcanic eruptions, the Chicxulub impact, and the fall in sea-level. He concluded, 'No single doomsday theory fits all the evidence, but since we know that all these events were taking place at the same time, it is possible that they all played a role in the mass destruction. Indeed, taken altogether, it is a wonder that anything at all survived'.[8]

Nevertheless, whilst the dinosaurs disappeared completely, some other groups of organisms, including mammals, managed to cling on to life. As a consequence, these, and only these, could provide the genetic material for future developments. The course of evolution changed in very significant fashion at the end of the Cretaceous Period.

The evolutionary importance of extinctions

The view of evolution that emerged as the world approached and entered the twenty-first century, although still Darwinian, is actually closer to the process envisaged by Patrick Matthew than that of Darwin himself. As we noted earlier, in Chapter 7, Matthew saw natural selection operating in a more catastrophist context than his contemporary, which seems nearer the truth as we now know it. So, whatever developments take place, no new evolutionary synthesis can possibly disregard mass extinctions, as the so-called Modern Synthesis did, or maintain that, even if they occurred, they were caused in every instance by a combination of ordinary gradualistic processes. There is no justification for taking a view of evolution based solely on biological and palaeontological data, whilst totally ignoring all that geologists now know about the occurrence of massive volcanic eruptions, or the astronomical evidence which shows that extraterrestrial impacts are a major feature of the history of the Solar System. Concerning the relevance of impacts to biological evolution, David Raup (figure 23.1) wrote in *Paleobiology* in 1992, 'Following a decade of controversy and research on the impact hypothesis for the terminal Cretaceous mass extinction (K–T event), two things have become clear, independent of whether the K–T event was actually triggered by comet or asteroid impact: (1) large-body impacts are common on geological time scales, and (2) impacts are capable of producing devastating biological consequences'.[9]

According to Raup's calculations, more than half of all species extinctions during the history of life on Earth may have resulted from extraterrestrial impacts. As he saw the situation, small impacts could have been responsible for a significant proportion of 'background' extinctions, with larger impacts being the principal cause of at least some of the well-recognised mass extinction events.

Whilst there continue to be major disputes about the precise causes of mass extinctions, it has become increasingly accepted that such events, especially but not only those at the end of the Permian and Cretaceous Periods, have been of great evolutionary importance. So, for example, writing in 1990 about reptile evolution in the multi-author book *Evolutionary Trends*, Michael Benton concluded, 'Most large-scale trends seem to relate to major extrinsic causes, such as mass extinction events . . . '.[10] Similarly, in an article about foraminifers which appeared in *Paleobiology* in 1991, Richard Norris of Woods Hole Oceanographic Institution wrote, 'Trends in body size and skeletal shape may be dictated more by variations in survivorship . . . than by long-term directional changes in the environment. Hence, mass extinctions can help drive evolutionary trends by selectively eliminating some morphologies and permitting the survivors to found the next radiation'.[11] Again, though in less scientific language, Richard Fortey commented in his 2000 book, *Trilobite!*, 'Drama is often used as a metaphor in the history of life. Animals have been described as actors on the ecological "stage". Mass extinction events that have interrupted more routine oscillations of fortune, the stuff of everyday evolution and decline, are "dramatic" interruptions in the "narrative"'.[12]

The picture presented in these three passages is far removed from Darwin's view of evolution as an ultra-gradualistic, progressive, continuous and essentially even-paced process, as expressed, for example, in the fourth chapter of *On the Origin of Species*:

> It may be said that natural selection is daily and hourly scrutinising, throughout
> the world, every variation, even the slightest; rejecting that which is bad,
> preserving and adding up all that is good; silently and insensibly working,
> whenever and wherever opportunity offers, at the improvement of each being in
> relation to its organic and inorganic conditions of life. We see nothing of these
> slow changes in progress, until the hand of time has marked the long lapses of
> ages, and then so imperfect is our view into long past geological ages, that we only
> see that the forms of life are now different from what they formerly were.[13]

Towards the end of the same chapter, Darwin summarised his vision of unending competition for expansion and survival, with extinct forms being those which had lost out in this struggle. He gave no hint that an occasional pruning of the 'tree of life' (not, of course, by a supernatural gardener, but by natural processes) might stimulate fresh growths. Indeed, Darwin concluded: 'As buds give rise by growth to fresh buds, and these, if vigorous, branch out and overtop on all sides many a feebler branch, so by generation I believe it has been with the great Tree of Life, which fills with its dead and broken branches the crust of the earth, and covers the surface with its ever branching and beautiful ramifications'.[14]

For Darwin, not surprisingly, the process of extinction was usually a gradual one. So, in the tenth chapter of the *Origin*, he wrote that 'species and groups of species gradually disappear, one after another, first from one spot, then from another, and finally from the world', and that 'the complete extinction of the species of a group is generally a slower process than their production'.[15] Where the situation appeared to be different, as at the end of the Cretaceous Period, this was probably a consequence of the 'wide intervals of time between our consecutive formations' for, 'in these intervals there may have been much slow extermination'.[16] Nevertheless, he allowed for one situation where there could be an exception to this general rule, because 'when by sudden immigration or by unusually rapid development, many species of a new group have taken possession of a new area, they will have exterminated in a correspondingly rapid manner many of the old inhabitants'.[17] His theory did not, however, admit the possibility that it might have been the other way round, with the extinctions coming before the appearance of new species.

In *Evolutionary Trends*, Stephen Jay Gould argued that Darwin was wrong not only in seeing macroevolution as a continuation of microevolution, but also in failing to recognise the evolutionary importance of extinctions. He went on to write, 'Some evolutionists have trouble envisioning an extinction-driven trend in the speciational mode. Extinction makes nothing. How can the mere elimination of part of a spectrum of variation among species achieve anything new in evolution? People impressed with this claim should adopt a historical perspective to grasp its fallacy'. Indeed, the nineteenth and early twentieth century arguments against the evolutionary importance of natural selection were framed along very similar lines and, as Gould pointed out, the Darwinian response 'correctly held that selective elimination can drive evolutionary change so long as any new modal state retains the capacity for generating a random spectrum of variation about itself'.[18]

Chance and necessity, the creative duet

Of course, as the final quotation in the previous paragraph makes clear, natural selection cannot be regarded as the creative force in evolution, without also acknowledging the importance of the production of the variants on which it can act. So, Francisco Ayala, of the University of California, Irvine, writing in 1994 in *Creative Evolution?!*, referred to 'the creative duet: chance and necessity'. Chance produces a range of variants, and necessity determines which of them survives.[19]

In fact, it is likely that when environmental conditions are stable and population density is high, the main effect of natural selection is to eliminate novel forms, thus limiting evolutionary change. David Raup wrote, in his 1991 book, Extinction – Bad Genes or Bad Luck?:

> without species extinction, biodiversity would increase until some saturation level was reached, after which speciation would be forced to stop. At saturation, natural selection would continue to operate and improved adaptations would continue to develop. But many of the innovations in evolution . . . would probably not appear. The result would be a slowing of evolution and an approach to some sort of steady-state condition. According to this view, the principal role of extinction in evolution is to eliminate species and thereby to reduce biodiversity so that space – ecological and geographic – is available for innovation.[20]

In that context, it can be seen why mass extinctions, particularly if linked to environmental change, have an important bearing on the course of evolution. After such an episode has occurred, the tight control normally exerted by natural selection would be relaxed, with many existing ecological niches lying vacant, and possibly some new ones having been created. New variants arising from those individuals fortunate enough to survive might find enough time and evolutionary space to proliferate, and eventually give rise to further variants. Should environmental conditions return to what they had been before the mass extinction took place, a vacated ecological niche would, in the course of time, become occupied by something which resembled the previous occupant, even if unrelated to it. Similarly, if a new ecological niche became available, variant forms would eventually arise which could take advantage of the opportunity.

In the same year that Raup's book appeared, Niles Eldredge (see figure 15.1) presented a similar view of evolution in *The Miner's Canary*. To start with, he pointed out that specialist species (stenotopes) are more numerous than generalists (eurytopes) because each of the former occupies a narrow ecological niche, whereas the latter are broad-niched, allowing a single species to occupy what could be regarded as a series of niches. However, should conditions change, generalists would be more able to find a niche which could sustain their existence. Furthermore, the species would be likely to remain relatively stable, since fledgling species arising from a generalist one would be in direct competition with their ancestors, and so unlikely to gain an ecological foothold. Specialists, on the other hand, would be making use of only part of the habitat available, providing opportunity for some fledgling species to gain a foothold in a similar, but distinct, sub-habitat. So Eldredge concluded that 'the fledgling specialist has a greater probability of survival than the fledgling ecological generalist, even though it is the generalist species, once established, that has the greater chance for truly long-term survival.'[21]

Whilst speciation and extinction must, of course, have taken place throughout the long, relatively quiet periods of the Earth's history, Eldredge believed that the major evolutionary events were linked to very exceptional times. As he saw it, 'massive across-the-board extinctions free up ecological space, nurturing and perhaps even goading new species to appear and survive'.[22] During the 1990s, that view, markedly different from anything envisaged over the previous half-century by the founders of the Modern Synthesis and their followers (see Chapter 14), became widely accepted. Ernst Mayr, one of two founders to remain alive and active throughout that period (Ledyard Stebbins, who died in 2000 at the age of 94, being the other), acknowledged the new outlook in his 2001 book, *What Evolution Is*, writing, 'Background extinction and mass extinction are drastically different in most aspects. Biological causes and natural selection are dominant in background extinction, whereas physical factors and chance are dominant in mass extinction'.[23]

There seem to be good reasons, therefore, for thinking that mass extinction events had a crucial bearing on the course of evolution. Mass extinctions disrupt systems which would normally maintain the *status quo*, providing increased opportunity for new species, arising through normal mechanisms, to become established. It has even been suggested that additional mechanisms of variant production may be introduced at such times, and there are reasons for taking that supposition seriously. So, for example, Suzanne Rutherford and Susan Lindquist of the University of Chicago produced evidence in *Nature* in 1998 to show that the effects of the heat-shock protein Hsp90 on developmental processes can be modified by a temperature change, possibly giving rise to new evolutionary traits.[24] However, the radiations of new species after a mass extinction can be explained perfectly well without the involvement of a mechanism such as this, and there is some evidence against it, particularly the time lag between an extinction episode and the subsequent radiation, which we discussed at the beginning of this chapter.

As we have seen, many now believe that the mass extinctions could have been caused by catastrophes, and even by cosmic catastrophes. So, in 1993, in *The Book of Life*, Michael Benton wrote, 'Some of the boundaries between geological periods seem to represent mass extinction events whose causes have become a major target of research. Asteroid impacts are a strong possibility, and other suggestions implicate changes in climate, sea level and evolutionary rates'.[25] Concerning one of these events, Peter Whitfield commented in the same year, in *The Natural History of Evolution*, 'The extraterrestrial impact theory, put forward by the US scientists Luis Alvarez, Walter Alvarez, Frank Asaro and Helen Michel in 1980, is becoming recognized as the accepted view of the event at the so-called K/T boundary.'[26]

In the foreword to Whitfield's book, Roger Lewin explained how the current view of mass extinctions differs fundamentally from Darwin's concept of evolutionary change: 'The mechanism of natural selection implies that a species' success is determined by how well it is fitted to prevailing circumstances, including its interaction with other species – the struggle for existence, as Darwin put it. A species that fails to compete may become extinct. When mass extinctions occur, however, these rules change. Whatever their cause – whether through global climate change or asteroid impact – mass extinctions elect as their victims species with characteristics having nothing to do with everyday success or failure.'[27]

Similarly, Simon Lamb and David Sington wrote, in their 1998 book, *Earth Story*, 'Each mass extinction has proved to be a turning point in the development of life as the extinct species are replaced by an even greater variety of new creatures, often characterized by the sudden appearance of novel features. Thus the slow and stately progress of evolution, envisaged by Darwin, seems to have been interrupted by a number of abrupt events that have had a decisive impact on the direction life has taken'.[28]

Without question, our view of evolution, whilst still remaining consistent with many of Darwin's ideas, has changed considerably over the past twenty years, not least because of increased knowledge about mass extinction events. In 1999, Steve Jones, in his book *Almost Like a Whale* (which, as we saw in Chapter 18, was an attempt to show how Darwin might have written *On the Origin of Species*, had he known all that we know today), began his discussion of extinctions in traditional fashion, writing, 'Many creatures disappear because they are replaced by something better. That process, busy all over the world, is itself evidence of natural selection'. However, the emphasis quickly changed, when he admitted there were other ways to become extinct, and pointed out that 'lots of life has gone out with a bang'. After summarising the threat to the Earth from asteroids and comets, he turned his attention to the specific issue of the K–T extinctions. Although emphasising that these were far from instantaneous, Jones acknowledged that 'an event out of the ordinary' occurred at this time. He continued, 'It took two million years to get back to normal; and the new normality was noticeably different from the old, with many novel kinds in the place of those who had gone'.[29]

Darwin would surely have struggled to recognise that as a straightforward extension of his own vision. As we have seen, his world-view could not accommodate the possibility of major catastrophes, at least partly because he was totally unaware of the variety of potential threats to life on Earth. We in the twenty-first century, however, cannot use the same excuse.

Part II

Catastrophes and the history of life on Earth

24 Extinctions large and small

Life and death in the Precambrian

As we saw in Part I, it is now abundantly clear that a number of mass extinction episodes have occurred during the course of the Earth's history. That does not, of course, imply there must have been major catastrophic events at each of those times. Indeed, we noted some alternative explanations for mass extinctions towards the end of Chapter 20. Nevertheless, because of the interest in the possibility that an extraterrestrial impact and large-scale volcanic activity both played a significant role in the K–T extinctions, much effort has been put into looking for the involvement of similar mechanisms on other occasions. In general, it has not been possible to reach definite conclusions, particularly about the involvement of impacts.

To start with, the absence of a well-defined iridium abundance anomaly at a particular site cannot be taken to exclude the possibility of a major impact. In the 1989 multi-author book *Mass Extinctions – Processes and Evidence*, Charles Orth pointed out that 'thin fallout layers are extremely sensitive to erosion and mixing processes, and many sections from widely separated localities should be examined before any firm conclusions can be drawn'.[1] Regardless of that, a projectile made largely of ice or stone, without the inclusion of any significant amount of metal, would bring very little iridium to the Earth.[2] We know that a large extraterrestrial object must have struck the Earth around 214 Myr ago, because of the evidence of the 70–100-km-diameter crater produced by the impact at Manicouagan, Quebec, but melt rock from the site is not enriched with iridium.[3] Conversely, a finding of an iridium abundance anomaly at a particular site would not necessarily constitute proof of an extraterrestrial impact, because there could be more earthbound explanations, such as volcanic activity, or the accumulation of the metal in that locality by living organisms. The evidence has to be interpreted with caution.

With that in mind, let us now consider some of the mass extinction episodes which seem to have been of particular evolutionary significance. We shall try to place these within the context of the development of life on Earth, starting at the very beginning.

As we discussed in Chapter 19, it is now well-established that the primaeval Earth was bombarded by comets and asteroids for hundreds of millions of years. Some of the important biochemicals necessary for the development of living organisms may have been synthesised using energy derived from these impacts. Possibly some were even brought to Earth by the impactors, or in falls of cosmic dust. Regardless of what precisely happened, it seems that life did not become established until after the massive, sustained bombardment eased off, around 3,800 Myr ago.[4]

For most of the time between the origin of life and the present day, all living organisms were unicellular and, to start with, they were all simple prokaryotes. Nevertheless, some of them were capable of carrying out photosynthesis, thus producing oxygen which gradually built up to significant concentrations in the atmosphere. Larger, more complex eukaryotic cells eventually appeared about 2,100 Myr ago, perhaps as a result of symbiosis between different types of prokaryotic cells, as suggested by Lynn Margulis, a biologist from the University of Massachusetts, Amherst. Micro-organisms unable to tolerate the new oxygen-rich atmospheres on their own may have been able to do so in association with oxygen-utilising ones, leading eventually to the creation of larger single-celled organisms, with sub-cellular compartments having specialised functions. The first multicellular organisms (metazoans) appeared during the Late Precambrian Period, around 700 Myr ago, in the form of the soft-bodied Ediacara fauna, named after the Ediacara Hills of South Australia, where traces were found.[5]

Very little is yet known about the Late Precambrian and the extinctions which preceded the explosive radiation of types of hard-shelled invertebrates early in the Cambrian. Significant continental movements appear to have taken place during the Late Precambrian, following the break-up of the supercontinent of Rodinia, which comprised the entire land-mass of the world, and these may have been quite rapid. There could also have been a very significant change in tilt of the Earth's axis. A major Ice Age occurred around this time, possibly covering the whole of the Earth's surface with ice for several million years. As it finished, there was a steady rise in sea-levels, probably accompanied by anoxic (i.e. oxygen-poor) conditions, towards the end of the Period. This process may have been interrupted briefly by further episodes of glaciation and regression of the seas.[6]

During the early 1980s, at sites in both China and Russia, increased levels of iridium were found at the Precambrian–Cambrian boundary, together with a sharp carbon-isotope shift. This latter finding indicated a sudden decrease in the number of organisms carrying out photosynthesis in the seas (see Chapter 21), shortly before the Cambrian explosion of new invertebrate species. However, it is far from clear whether this depletion of marine organisms occurred because of the preceding glacial conditions, an extraterrestrial impact or some other cause. The iridium anomalies could not be confirmed when samples from the same sites were analysed in the U.S.A., except in the case of one from the Guizhou Province of China, where the range of elements found in excess amounts suggested that metals had been precipitated from seawater.[7]

The impact of a huge nickel-rich asteroid during the Precambrian produced the Sudbury Basin in Canada, a multi-ringed impact crater around 200 km in diameter.

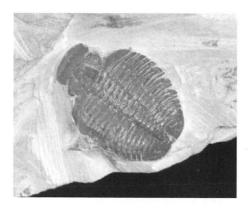

Figure 24.1   A Cambrian trilobite of the genus *Elrathia*, from North America, around 500 Myr ago.

This, however, has been dated at 1,850 Myr before the present time, over 1,000 Myr before the end of the Period. Another large crater, that at Vredefort in South Africa, is of similar age to that at Sunbury so, again, its formation was far too early to have had any effect on events in the Late Precambrian.[8] The 60-km-diameter Beaverhead impact structure of Montana is much younger than these, having been dated at around 600 Myr ago. Similarly, the Acraman impact structure in Australia, which has an intermediate ring 90 km in diameter and a possible outer ring 160 km in diameter, dates from around 570 Myr ago. Yet, even the Acraman impact came too early for it to have caused the extinctions preceding the appearance of the first forms of Cambrian life, around 550 Myr ago (see figure 5.2).[9]

Possibly the environment suffered a further jolt at the very end of the Precambrian, but as yet there is no clear evidence for this. Indeed, the extinctions around the Precambrian–Cambrian boundary may have occurred as a consequence of ongoing processes, whatever the cause or nature of these. Gradualist explanations include sea-level changes and anoxia, as well as the possible emergence of eukaryotic herbivores which could feed on the dominant algal community, creating the ecological space for the first major radiation of new forms early in the Cambrian.[10] However, catastrophist explanations cannot be excluded.

The Cambrian explosion and its consequences

Regardless of possible causes, the 'Cambrian explosion' of new animal forms was a very significant event. Within a few million years, the seas contained a wide range of soft-bodied creatures and others with external skeletons, including the trilobites (figure 24.1). Much of our early knowledge of this came from the Burgess Shale in Canada, which was formed around 530 Myr ago as a result of a mudslide. Burgess Shale fossils which had been collected by the American palaeontologist Charles Doolittle Walcott around 1910 were re-examined in the 1970s by Harry Whittington, of Cambridge University, and his research students Derek Briggs (who later moved to Bristol University) and Simon Conway Morris (who remained at Cambridge). This investigation revealed an extremely diverse fauna, with creatures such as *Wiwaxia*, a headless organism covered in scales and spines, and the weird *Hallucigenia*, possessing two rows of spines along its back and a row of tentacles underneath, which were difficult to classify, since they apparently fitted into no known phyla.[11]

Moreover, similar findings were later made in other locations, including the Buen formation of North Greenland, and Chengjiang in Southern China. Amazingly, it appeared that not only did the Cambrian explosion produce most of the animal phyla in existence at the present time (which number between twenty and thirty) but, according to some initial estimates, perhaps as many as one hundred altogether, the majority not enduring for any great length of time.[12] That view owed much to *Wonderful Life*, Stephen Jay Gould's 1989 book devoted to the Burgess Shale fossils.[13]

However, the classification was controversial, some arguing that the individual species were capable of being grouped into far fewer phyla than had been supposed, including ones still existing today.[14] As for the Burgess Shale fossils, Simon Conway Morris suggested in 1998, in his book *The Crucible of Creation*, that the claims made by Gould were somewhat exaggerated. He concluded that some of the strange fossils might simply have represented early stages in the development of known phyla. *Wiwaxia*, for example, could have been on the route which led to the annelids (worms).[15] Nevertheless, the Cambrian explosion clearly had a unique place in evolution. As the Swedish palaeontologist Stefan Bengtson wrote in *Nature* in 1991, 'The abundance of problematic fossils in the Cambrian is now widely seen to reflect unpruned diversity after the first major metazoan radiation'.[16]

Subsequent pruning occurred through a series of extinction events. Allison (Pete) Palmer, of the Institute of Cambrian Studies, Boulder, Colorado, together with others, demonstrated that at least four major episodes of trilobite extinctions had taken place during the Cambrian. Some argued that the first of these, dated around 520 Myr ago, was the biggest mass extinction of all in terms of the percentage extinction of genera. However, background extinction was high throughout the Cambrian, so the significance of this particular event remained uncertain.[17]

As to events later in the Cambrian, it seems that, on several occasions, deep-water species of trilobites invaded vacant areas of shallower water over continental rocks, and radiated into a variety of ecological niches. Then they suddenly disappeared, to be replaced, after a slight but significant delay, by further types invading from the depths.[18]

There is still little indication of the cause or causes of the Cambrian extinctions, including those at the end of the Period, when almost half of American trilobite families became extinct, apparently in step-wise fashion. A 1980s investigation of the Late Cambrian extinction boundary, dated at about 500 Myr ago, in uplifted marine limestone in western Utah, did not reveal any iridium abundance anomaly. It was suggested that the extinctions might have been caused by a drop in water temperature or a reduced availability of oxygen but, however plausible, that was little more than speculation.[19]

Whatever mechanisms might have been involved, the Cambrian extinctions cannot be explained simply as culminations of trends set up by Darwinian competition for limited resources. For example, Stephen Westrop of Brock University studied a Late Cambrian mass extinction event in North America and found that the survival pattern of trilobite families could not be predicted from patterns of turnover amongst their component species. His report, which appeared in *Paleobiology* in 1989, concluded: 'The results affirm the importance of a hierarchical approach to the interpretation of macroevolutionary patterns and provide some support to the suggestion that sorting processes operating during mass extinctions differ from those of background times'.[20]

Similarly, in *Wonderful Life*, Gould argued that there could have been no way of predicting which of the forms represented in the Burgess Shale would disappear quickly and which would endure. He acknowledged that the fossils included all four of the long-lasting arthropod groups, i.e. trilobites, crustaceans, uniramians and chelicerates, represented by *Naraoia, Canadaspis, Aysheaia* and *Sanctacaris*, respectively. However, these were accompanied in the Burgess Shale by at least thirteen other arthropod lineages, one of which, *Marrella*, was amongst the most abundant of the fossils, whilst another, *Leanchoilia*, was one of the most specialised. Hence, Gould went on to write:

> I challenge any paleontologist to argue that he could have gone back to the
> Burgess seas and, without the benefit of hindsight, picked out *Naraoia, Canadaspis,*
> *Aysheaia*, and *Sanctacaris* for success, while identifying *Marrella, Odaraia, Sydneyia*,
> and *Leanchoilia* as ripe for the grim reaper. Wind back the tape of life, and let it play
> again. Would the replay ever yield anything like the history that we know?[21]

Only a small number of the variety of forms produced during the Cambrian explosion survived to the end of the period and beyond. That had implications for all subsequent evolution for, as we have noted previously, each new radiation could only be derived from the genetic material of creatures which happened to have survived preceding extinction events. Although our knowledge of convergent evolution suggests that, had different groups survived, superficially similar organisms might still have emerged to fill the various ecological niches, these would have been based on body plans with significant differences from ones with which we are familiar.[22]

Times of crisis in the Ordovician and Devonian Periods

The animals which lived in the Cambrian seas were all, to a greater or lesser extent, generalists. They could occupy a range of niches, and they had legs which could serve more than one function. However, those, such as the trilobites, which survived into the Ordovician Period were joined by an increasing number of more specialised creatures, including molluscs, brachiopods, corals and graptolites. These were adapted to particular habitats, and had different sets of limbs to carry out different functions. This process was curtailed by the Late Ordovician mass extinction episode, about 438 Myr ago, when around 80% of all species died out.[23] As Richard Fortey wrote, in *Life – An Unauthorised Biography*:

> The end of the Ordovician was a punctuation mark in the history of life, providing a natural end to the first great phase of diversification and reorganization of marine life. Those animals and plants that survived went on to make the modern world; had the list of survivors been one jot different, then so would the world today.[24]

The initial phase of the Late Ordovician mass extinction event appears to have been linked to the onset of glaciation, perhaps facilitated by the crossing of the South Pole by Africa, which was then part of the southern supercontinent of Gondwana (see figure 14.2). Although conditions may have fluctuated, temperatures at the time were generally lower than what they had been previously and, because of water trapped in the ice-caps, there was a fall in sea-level. That situation persisted for hundreds of thousands of years. Then, another major wave of extinctions took place as the ice-caps melted, temperatures increased, sea-levels rose rapidly and the environment became anoxic. There could, in fact, have been a connection between anoxia and sea-level changes. A fall in sea-level would have exposed much organic matter to oxidation by the atmosphere, leading to an increase in carbon dioxide concentration and a reduction in free oxygen. On this and other occasions, it seems that anoxic oceans, resembling large, stagnant ponds, apparently occurred when sea-levels started to rise again following a period when they had been low.[25]

As yet, there is little evidence to suggest that extraterrestrial impacts played a part in these events. In the early 1980s, increased iridium was found in Kentucky in possible association with a crater that might have dated from the time of the Late Ordovician extinctions but the details were far from certain. A more reliable finding of raised iridium at the end of the Ordovician Period came from a location in Quebec, Canada, as well as at Dob's Linn, Scotland, and the Yangtze Basin, China. However, no shocked quartz, microtektites or other signs of an impact were present at any of these sites, and it seemed likely that the high iridium concentrations were due to erosion of upper mantle rocks which had become exposed. Although three impact craters have been dated to around the time of Ordovician–Silurian boundary, these are all relatively small ones, the largest being the 24-km-diameter Strangways crater in Australia.[26]

After the Ordovician extinctions, the number of species in the seas began to increase once again. A few isolated fish types had appeared during the Ordovician, but fish remained relatively rare throughout the remainder of this period and the next, the Silurian. Eventually, early in the Devonian Period, they started to proliferate.[27] However, around 367 Myr ago, in the Late Devonian, between the Frasnian and Famennian stratigraphic stages, another major mass extinction episode took place, with the loss of around 60% of existing species. This occurred on land as well as in the seas. The first invasion of the continents by amphibians took place during the Frasnian but, according to the fossil evidence, the invaders had all disappeared by the start of the Famennian.[28]

This transition seems to have taken place during a cold period, possibly linked to the crossing of the South Pole by the part of Gondwana which is now South America. Extinctions occurred over several million years, but there was an intensification around the Frasnian–Famennian boundary. Although the sea-level was generally high throughout, there may have been a fluctuation at the end of the Frasnian, and anoxia was a consistent feature. Also, extensive vulcanism took place on the East European platform in the Late Frasnian, and again at the end of the Famennian.[29]

In the early 1980s, normal iridium concentrations were found at the Frasnian–Famennian boundary in sedimentary rocks in New York State and Belgium, and investigations elsewhere revealed nothing suggestive of an impact. Although a slight abundance anomaly was detected in 1984 in Late Devonian rocks in the Canning Basin, Western Australia, this coincided with a stromatolite bed, produced by microorganisms, which could have absorbed iridium from sea water, producing a relatively high concentration of the metal in that location. Similarly, a moderate increase in iridium concentration was detected in Late Devonian rocks in Germany, but the context suggested a terrestrial rather than an impact-related cause.[30]

However, evidence for one or more impacts at this time then began to emerge. Microtektites were found in association with high iridium levels in Late Devonian rocks in South China, and a possible impact crater linked to them was discovered under the nearby Taihu Lake. Furthermore, an appropriately dated microtektite field was located in Belgium, and shocked quartz grains were found in Late Devonian rocks in the U.S.A. Six impact craters of suitable age were identified in North America and Europe, including the 54-km Charlevoix crater in Quebec and the 55-km Siljan crater in Sweden. In addition, there were indications of a significantly larger one in the Alamo region of Nevada, but this could not be investigated properly because it was located on military land, including the much-discussed Area 51, which features strongly in UFO legends.[31]

In 1996, after carrying out a detailed study of the Frasnian–Famennian boundary in Alberta, Canada, Kun Wang and colleagues from the local University concluded that the extinctions had probably been caused by a large extraterrestrial impact at a time when living organisms were already under considerable strain.[32] In the same year, the Rutgers University geologist George McGhee suggested in his book *The Late Devonian Mass Extinction* that multiple impacts might have been involved. He wrote, 'The Late Devonian is a time of at least three known continental impacts, possibly as many as six, and who knows how many more oceanic impacts of which no evidence remains . . . Did several impacts trigger the Frasnian-Famennian? That scenario would indeed match the biological evidence'.[33]

Kun Wang and his group also demonstrated a widespread iridium abundance anomaly in rocks from the very end of the Devonian Period, at the boundary with the Carboniferous Period, around 360 Myr ago, when there was an anoxic episode and extinctions took place which were only slightly smaller in scale than the Frasnian–Famennian ones. A line of impact craters in northern Chad, the largest 17 km in diameter, might date from this time, although some estimates place it much later. A huge 120-km-diameter crater discovered under sediment near Woodleigh, Western Australia, in 2000 has been found to have an age of around 360 Myr, but this may turn out to be another impact site associated with the Frasnian–Famennian extinctions. The involvement of a catastrophic event in the Devonian–Carboniferous extinctions remains far from certain. Investigations in at least some of the sites suggested that the iridium excess might not have resulted from an impact. Similarly, it was generally thought that normal terrestrial processes had given rise to an iridium abundance anomaly around 320 Myr ago, between the Mississipian and Pennsylvanian Periods, which are subdivisions of the Carboniferous Period in North America. However, it then emerged that eight circular structures, which form a line from Kansas to Illinois, could be impact craters dating from this time.[34] Even so, these could only be evidence of a relatively localised event, very different from the Frasnian–Famennian transition, or even the Devonian–Carboniferous one, when extinctions occurred throughout the world.

The great Permian extinction

After the Late Devonian crises, new forms of life once more started to appear. Amphibians again invaded the land, this time more successfully than in the Frasnian, and diversified rapidly during the Early Carboniferous Period. Not long afterwards, reptiles emerged from amphibian stock and became the dominant large land animal in the Late Carboniferous and throughout the Permian Period.[35]

Then came the transition between the Palaeozoic and Mesozoic Eras, around 250 Myr ago (see figure 5.2), marked by the Late Permian mass extinction event which is generally regarded as the largest of them all. Around 90% of species became extinct, including the last representatives of some long-surviving groups, such as the trilobites. Many of the details are still unclear because the stratigraphic record is even more incomplete than usual in Late Permian rocks, but it is believed that the extinctions occurred at a time when all the major continental plates locked together to form the supercontinent Pangaea (see figure 14.2). We discussed tectonic explanations for the extinctions earlier, in Chapter 14. There was a very significant reduction in the habitat area available to organisms which lived in shallow seas, but that factor alone does not seem sufficient to account for the events that took place.[36]

However, the locking of the plates and the formation of Pangaea must have led to climatic changes which also played a part in the extinctions. There seems to have been a significant decrease in salinity of the oceans, and the seas are likely to have become anoxic as the waters started to rise rapidly at the beginning of the Triassic Period, following their fall in the Late Permian. The concentration of oxygen in the atmosphere may have dropped to about half its present level during the transition, because of increased oxidation of organic matter exposed by the retreating seas, and also because of increased aridity (and hence reduced oxygen production by plants) on the immense continental mass. Another consequence of the oxidation of exposed organic matter was an increased atmospheric carbon dioxide concentration, possibly leading in turn to global warming. Furthermore, the northwards drift of Pangaea led to the onset of glaciation in the northern hemisphere, whereas previously there had been glaciation in the south.[37] Clearly, tectonic factors must have caused far-reaching environmental changes towards the end of the Permian Period. As Digby McLaren wrote, in *New Scientist* in 1983, 'To demand, in addition to the extreme harsh conditions at the time, a catastrophic event, would possibly be putting gilt on the ginger bread'.[38]

During the early 1980s, it was claimed that an iridium abundance anomaly had been found at the same level as Late Permian extinctions at several locations in China, but this could not be confirmed when specimens from these sites were analysed at Berkeley, Los Alamos or Los Angeles. There seemed to most people, therefore, no compelling reason to think that an extraterrestrial impact had been involved in the events at the end of the Permian Period.[39]

However, in 1990, McLaren, together with Wayne Goodfellow, argued in an article in *Annual Review of Earth and Planetary Sciences* that, although environmental changes linked to tectonic factors may well have caused the disappearance of many species during the Late Permian Period, the final extinction event at the Permian–Triassic boundary was on too large a scale and took place over too short a time-scale to be regarded simply as a continuation of the same process. McLaren and Goodfellow concluded:

> if one sums disappearances observed at each of the relatively few well-studied sections, it is legitimate to invoke a single cause that is independent of local facies, faunal province, or tectonic event. It evidently acted on land and in the sea, and it affected terrestrial animal and plant life. The coup de grace at the end of the Paleozoic was an unusually energetic event superimposed on a biota already deeply stressed environmentally. An extraterrestrial impact is very probably the ultimate cause of the Permian–Triassic killing event.[40]

Later work supported the view that the Permian–Triassic boundary extinction episode was an abrupt one. Furthermore, it appeared that there had been another sharp extinction event 5 Myr earlier. Catastrophic events undoubtedly took place near the end of the Permian Period, but the cause could have been purely terrestrial. As in the Late Cretaceous, extensive vulcanism occurred, with the laying down of the flood basalts of the Siberian Platform over an area greater than that of the Deccan traps, and there was further volcanic activity in South China. Possibly the vulcanism in Siberia, like the formation of the Ural Mountains, was a consequence of the collision between the Siberian and European plates, the final act in the creation of Pangaea.[41]

Douglas Erwin, a palaeontologist from the National Museum of Natural History, Washington, reviewed the geological and palaeontological evidence in *Nature* in 1994 and could find nothing to suggest there had been a major impact in the Late Permian. Instead, he concluded that the extinctions were due to a deterioration of the environment caused by a 'tangled web' of factors, with sea-level changes and volcanic activity playing significant roles.[42]

Before Erwin's review appeared, however, increased levels of iridium were reported in Permian–Triassic boundary rocks in the Meishan section, China, coincident with microspherules and a carbon-isotope shift. Soon afterwards, a finding of shocked quartz grains was also claimed, together with other evidence of an impact. Opinions were beginning to change. In *Science*, in 2000, Erwin, together with a group from the Nanjing Institute of Geology and Paleontology, reported that a detailed study of the Meishan section had confirmed that the extinctions at the same level had been sudden, the findings as a whole being 'consistent with the involvement of a bolide impact'. Furthermore, in the same journal a year later, a team led by Luann Becker, of the University of Washington, and including Michael Rampino, reported the discovery of fullerenes ('Buckyballs') at the Permian–Triassic boundary at sites in China, Hungary and Japan. These contained trapped helium and argon with isotope ratios suggesting an extraterrestrial origin, which indicated a major extraterrestrial impact at this time. In the opinion of Becker and her colleagues, this impact (apparently of an asteroid or comet with a lower iridium content than the Chicxulub impactor) set into motion a complex series of events, including vulcanism, anoxia and sea-level and climate changes, which together were responsible for the greatest of all mass extinction episodes.[43]

On the assumption that the formation of the Siberian traps might have been stimulated by an impact at the opposite side of the world (as with the suggested association between the Deccan traps and the Chicxulub event), the search for a possible crater has centred around South America. The 40-km-diameter Araguinha Dome structure in Brazil dates from around the time of the Permian–Triassic transition, but the impact which produced it was unlikely to have been large enough to cause such a massive effect in distant Asia. Also, the Araguinha Dome is too far way from the theoretical location, which is in the region of Tierra del Fuego, diametrically opposite from Siberia.[44] Michael Rampino, concentrating his search in the area around Tierra del Fuego, has located two large circular basins in the Falkland Plateau, under the waters of the South Atlantic, where the underlying rocks are Late Permian, from around 248 Myr ago. However, the two basins (which are of 200 and 355 km diameter) have yet to be confirmed as impact structures. Another possible Late Permian impact crater is a 200-km-diameter structure called the Bedout, which lies under sediment off the coast of northwestern Australia but, again, firm evidence is lacking.[45] Although opinion is

Figure 24.2 *Asteroceras obtusum*, a member of the ammonite group of ammonoids, from 180-Myr-old Jurassic rocks near Lyme Regis, Dorset.

moving towards acceptance of the occurrence of an impact, most of the specific issues concerning events at the end of the Permian Period remain unresolved.

Radiations and extinctions during the Mesozoic Era

After the Permian–Triassic extinctions, which marked the transition to the Mesozoic Era, the early Triassic Period saw one of the biggest ever increases of biological diversity, second in scale only to the Cambrian explosion. In the seas, the ammonoids (figure 24.2) were one of the major groups involved. Ammonoids originated in the Early Devonian and quickly established themselves, but were severely depleted in the Late Devonian mass extinction episode. After staging a recovery and thriving once more during Carboniferous and Permian times, they almost disappeared as the Permian Period came to an end.[46] Of the consequences of that for subsequent evolution, Niles Eldredge wrote, in his 1987 book, *Life Pulse*:

> The extinction greatly biased which particular set of ammonoid genetic instructions was to form the basis for future ammonoid evolution . . . But we are likely to see a repetition of old adaptive themes – ways of being an ammonoid that worked in the past and are refashioned when life starts bouncing back. And that really is the theme for all of early Triassic life. Few kinds of organisms are evident in the rock layers immediately above the extinction. But after a few million years, there is always a sudden proliferation, very much as if life has exploded in all directions.[47]

The situation on land was similar to that in the oceans. As the continents came together to form Pangaea during the Late Permian Period, the mammal-like therapsids replaced the pelycosaurs, their fellow synapsid reptiles, as the dominant large animals. The therapsids included the carnivorous gorgonopsians, the herbivorous dicynodonts and the smaller cynodonts, the probable ancestors of present-day mammals. At the end of the Permian, the gorgonopsians became extinct, as did many species of dicynodonts, and those therapsids which managed to survive the transition to the Triassic then had to struggle for dominance with two groups of large diapsid reptiles. These were the rhynchosaurs, with their beak-like jaws, and the thecodonts, the fore-runners of the dinosaurs.[48]

One of the first palaeontologists to appreciate the key role played by mass extinctions in therapsid evolution was Tom Kemp, the curator of vertebrates at the Oxford University Museum. In 1982, Kemp argued in *New Scientist* that mass extinctions were important 'both as a filter that allows the differential survival of some lineages and not others, and also because of the events which follow a mass extinction.' He continued, 'Conditions at such times seem to have been so different from normal that certain evolutionary events could occur which would not be predictable from a knowledge of the ecology of modern animals alone. These times seem to have been the periods of maximum evolutionary change and were therefore of the greatest consequence in determining the whole course of evolution'. Hence, at least partly because of the occurrence of extinctions, as Kemp maintained in his 1999 book, *Fossils and Evolution*, mammals did not evolve from their ancestors along strictly neo-Darwinian lines, but by means of 'a true species-selection process superimposed upon organism-level selection'. [49]

For much of the Triassic, the plant-eating rhynchosaurs were amongst the most common large land animals in various parts of the world but, towards the end of the Period, their habitats were taken over by the proto-dinosaurs and, later, by dinosaurs themselves. It seems likely that this take-over did not occur until *after* the rhynchosaurs became extinct, the dinosaurs and their immediate ancestors triumphing not through competition, but simply by radiating into vacant ecological space. Similarly, the therapsids retained dominance over large areas of the supercontinent of Gondwana, including what is now South Africa, throughout much of the Triassic Period, before they too disappeared, to be replaced by dinosaurs.[50] Michael Benton, in his 1990 book, *The Reign of the Reptiles*, summarised the main evolutionary events leading to the dominance of the dinosaurs, as he saw them, in the following passage:

> There were two mass extinctions during the 20–25 million years of the
> Late Triassic – one near the beginning, the other at the end. The rhynchosaurs,
> dicynodonts and many cynodonts died out in the first event and the dinosaurs
> were able to exploit the niches left vacant in a way analogous to the situation at the
> beginning of the Triassic. At the end of the Triassic, the remaining thecodontians
> and most therapsids disappeared and the dinosaurs and other reptile groups were
> able to diversify further on land, in the air and in the sea.[50]

Niles Eldredge commented, in *Life Pulse*, 'That the therapsids gave way to the dinosaurs shows that there is no inherent superiority in the synapsid line (which includes mammals) over the line that includes most of the familiar reptiles and birds'.[51] Indeed, as is well known, dinosaurs went on to dominate the land for another 150 Myr or so. During that time, flying reptiles dominated the skies, but the first birds appeared during the Jurassic Period, perhaps evolving from early carnivorous dinosaurs. Whilst flying reptiles and dinosaurs disappeared at the end of the Cretaceous Period, birds are still very much with us today.[52] However, let us not rush ahead of ourselves but stay, for now, with the subject of the Late Triassic extinctions.

There is still uncertainty about the time-scale and characteristics of the events taking place at the Triassic–Jurassic boundary, around 208 Myr ago, when over 60% of existing species disappeared, for this is another instance where the fossil evidence is patchy.[53] Some palaeontologists believe that there was a wave of extinctions in the Carnian, the last-but-one stratigraphical stage of the Late Triassic, during a period of exceptionally heavy rainfall in Europe and North America, but that is disputed by others. However, there is general agreement that an intense extinction event occurred at the end of the final stage, the Norian (now generally taken to include the formerly separate Rhaetian stage), when sea-levels appear to have been rising. Evidence from Europe had previously indicated that the sea surface fell relative to the level of land at the time of the Triassic–Jurassic transition, but that could have been due to uplift rather than an actual drop in sea-level. Alternatively, an initial fall in sea-level may have been followed by a rise, bringing about anoxic conditions and a decrease in salinity of surface waters. More certainly, basalts in eastern North America, Brazil, Europe and West Africa, together constituting the Central Atlantic Magmatic Province (CAMP), were laid down around this time, during a period of massive volcanic activity that heralded the break-up of Pangaea and the opening of the Central Atlantic Ocean. According to some interpretations of fossil plant evidence, the atmospheric carbon dioxide content increased fourfold at the Triassic–Jurassic boundary, which could have increased average temperatures by several degrees.[54]

One study carried out during the 1990s concluded that the vertebrate extinctions of the Late Triassic were either gradual or stepwise. However, another found that selective extinction among European bivalve molluscs was inconsistent with either anoxia or sea-level changes being responsible, a better explanation being a reduction in primary productivity (i.e. a depletion of phytoplankton populations), as would have resulted from an extraterrestrial impact. Consistent with that, a team led by Peter Ward reported in *Science* in 2001 that there had been a sudden productivity collapse at the very end of the Triassic Period. Evidence for an impact (or possibly multiple impacts), came in 1990 with the discovery near Corfino, in the Italian Apennines, of shocked quartz grains within three closely spaced shale beds at the Triassic–Jurassic boundary. Furthermore, these quartz grains possessed several sets of lamellae. As we have seen, only impacts have been shown to produce grains with more than one set of lamellae and, although none of the grains at Corfino had as many sets as some of those found at the K–T boundary, they suggested that there may have been one or more impacts of significant size at the end of the Triassic. Pollen evidence at the Triassic–Jurassic boundary, showing a sudden increase in the proportion of opportunistic ferns to other plants, such as occurred at the K–T boundary, provided additional support for an impact-related scenario for the extinctions. In 2002, in a report in *Science*, following a detailed investigation at 70 sites in North America, an international team led by Paul Olsen of Columbia University, New York, confirmed that the final Triassic extinction event, which occurred less than 30,000 years before the first appearance of large dinosaurs, was marked by a pollen break, and they also found it to be associated with a small iridium abundance anomaly.[55]

The huge Manicouagan crater of Quebec, mentioned earlier, dates from the Late Triassic Period, and its formation might have been part of a multiple-impact event, together with a series of smaller craters, all less than 25 km diameter, at Saint Martin (Canada), Red Wing (U.S.A.), Rochechouart (France) and Obolon' (Ukraine). However, they all seem to have been produced a few million years before the actual end of the Triassic so, whatever may have happened at the time, there is unlikely to be a direct link between these craters (including Manicouagan) and the final extinctions of the Period. Similarly, another possible candidate, the 80-km-diameter Puchezh–Katunki crater of Russia, does not appear to have been formed at the precise time of the Triassic–Jurassic extinctions. If these extinctions were caused, in whole or in part, by a major impact, then its location has still to be identified.[56]

Twenty million years or so after the Triassic–Jurassic event, relatively minor extinctions took place at the end of the Pliensbachian stage of the Early Jurassic, when the Karroo traps were laid down in South Africa and the Farrar lavas in Antarctica. An anoxic episode occurred around this time, but most of the extinctions may have taken place after this was over, as indicated by studies in northern Europe.[57] In rocks from a little later in the Jurassic Period, the Callovian–Oxfordian stage boundary (dated at about 160 Myr ago), iridium abundance anomalies have been demonstrated in Spain and Poland. These were associated with evidence of volcanic eruptions, suggesting that the immediate cause involved tectonic activity.[58]

There was another episode of extinctions at the end of the Jurassic Period, around 144 Myr ago, particularly in shallow seas around Europe. Some evidence of an iridium abundance anomaly at this time has been found. The 70-km-diameter Morokweng crater in southern Africa has been dated to the Late Jurassic, and so has the 40-km Mjølnir impact structure under the Barents Sea, which is associated with shocked quartz grains. The extensive Parana traps in Brazil were also laid down around this time.[59]

Two relatively small iridium anomalies, one of them coincident with an extinction episode, were found in 1988 near the boundary between the Cenomanian and Turonian stratigraphical stages of the Cretaceous Period, dated to 91 Myr ago, at a site near Pueblo, Colorado. However, it was soon discovered that iridium concentrations were much lower at sites outside the immediate region, indicating a localised and possibly terrestrial cause. On the other hand, there appears to have been a worldwide carbon-isotope shift at around the same time, together with fluctuations in sea-level and extensive volcanic activity in Madagascar, suggesting the possibility of an event on a more global scale. Whilst it is generally believed that almost half the existing species of marine organisms disappeared at the time of the Cenomanian–Turonian transition, it has recently been argued that this apparent extinction episode is actually a sampling artifact, linked to sea-level changes. The situation remains uncertain.[60]

There is no such uncertainty about the significance of the next mass extinction episode, as we discussed in detail in Chapters 21 and 22. This was the great K–T event which brought the Mesozoic Era to an end around 65 Myr ago (see figure 5.2). As Roger Osborne and Michael Benton wrote in 1996, in *The Viking Atlas of Evolution*, 'The Chicxulub crater, located in southeastern Mexico, is almost certainly the site where a vast asteroid hit the Earth 65 million years ago'.[61] Whether because of that or other causes, including massive outpourings of lava in India, around 70% of existing species became extinct as the Cretaceous came to a close. In the seas, the victims included the ammonites, the final representatives of the ammonoid group. Major extinctions also occurred on land, as is well known, the dinosaurs being amongst those who departed the scene forever, vacating a large number of ecological niches.[62] Concerning the consequences of that, the University of Cambridge palaeontologist David Norman wrote in his 1994 book, *Prehistoric Life*, 'From an evolutionary point of view, any survivors of the Cretaceous extinctions were handed a wonderful opportunity to evolve into the ecological niches that had been vacated by dinosaurs. The story of the Early Tertiary is just that – a time of rapid evolution and adjustment among the various surviving groups'.[63] Similarly, in 2001, Tim Haines wrote, in *Walking with Beasts*, the book of the BBC television series which followed on from *Walking with Dinosaurs*, 'The Cenozoic Era started 65 million years ago with the catastrophic extinction that obliterated the giant dinosaurs and left the birds as the only legacy of their 170 million-year-old reign. But mammals didn't just take over. It took the world a long time to recover from the KT extinctions which saw the end of the dinosaurs'.[64]

The Tertiary Period and its environmental crises

Temperatures had fallen during the Cretaceous–Tertiary transition, but then started to rise again (figure 24.3). One possible cause was the release of vast amounts of carbon-rich gas from under the oceans, as a result of volcanic activity 55 Myr ago at the Palaeocene–Eocene boundary, when there was a mass extinction episode largely restricted to organisms living on the sea-floor. During the early part of the Eocene Epoch, temperatures were generally much higher than they are now, particularly at high latitudes, for semi-tropical forests extended northwards to well inside the Arctic circle.[65] At the other end of the world, Antarctica, still attached to both Australia and South America as the last remnant of Gondwana, was moving over the South Pole, where it lies today, but its shores were kept relatively ice-free by a warm current of water originating in the South Pacific. However, by the end of the Eocene, about 35 Myr ago, the three continents had separated, allowing a cold current to flow around Antarctica and start to deflect the flow of warm water. As the gaps between the continents widened

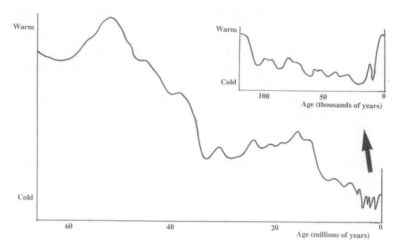

Figure 24.3   Trends in average temperature over the past 65 million years, according to oxygen isotope evidence. The major trends over the last 120,000 years, omitting short-term fluctuations, are shown in greater detail in the smaller figure to the right.

and the cold circumpolar current strengthened during the Oligocene Epoch, which followed the Eocene, ice-sheets formed on Antarctica and the surrounding seas, with cold water also sinking to the depths. Cool conditions were then transmitted from this region to other parts of the world by ocean currents and winds.[66]

But, could this have been the complete, or even the main, explanation for the general switch to harsher climates, which undoubtedly occurred? Even such a staunch gradualist as Beverley Halstead (who died in 1991) thought differently, writing in his 1979 book, *The Evolution of the Mammals*, 'During the Oligocene the Earth's axis of rotation tilted slightly and an ice-cap formed at the South Pole'.[67] Although he was presumably thinking about a slow change linked to natural cycles, rather than a sudden one associated with a catastrophic event, he was clearly unconvinced that continental drift could have led to the formation of the Antarctic ice-sheet.

Another factor which may have contributed to the deteriorating climate was the extensive vulcanism in the South Pacific, Colorado and northeastern Africa (including the laying down of the Ethiopian traps during the Early Oligocene). Whatever the cause or causes, the general lowering of temperatures around the world was accompanied by a marked fall in sea-level. There was also a series of extinction episodes which started in the mid-Eocene (before the formation of the Antarctic ice-cap and the onset of volcanic activity in Ethiopia) and continued right through to the mid-Oligocene.[68] Marine organisms, such as foraminifers and silicate-shelled radiolarians, were affected to the greatest extent, but land animals also suffered. After the K–T extinctions, mammals had diversified to the extent that around 575 genera were in existence by the middle of the Eocene, 45 Myr ago. However, that figure was reduced to around 380 by the middle of the Oligocene. The mammals may have replaced the dinosaurs as the dominant land animals, but they were having to face set-backs on their way to inheriting the Earth.[69]

Extraterrestrial objects of significant size struck the Earth several times during the Eocene Epoch, producing, for example, the 25-km-diameter Kamensk crater in Russia and the 45-km Montagnais crater in Canada around 50 Myr ago. Such events were particularly common during the final few million years of the Eocene, according to the evidence of tektite and microtektite fields, as well as craters. Amongst these impacts were the ones which produced the 17-km crater at Logoisk, Belarus, and the 28-km crater at Mistastin, Labrador, both considered to be between 37 and 40 Myr old. However, the largest of the Late Eocene impact craters is the 100-km-diameter Popigai structure, in Siberia, which was formed about 36 Myr ago. An iridium abundance anomaly and spinels have been found in rocks of the same age near Ancona, Italy, together with shocked quartz grains which were linked by mineralogical evidence to the Popigai crater. There is still some uncertainty about the precise dating of the Eocene–Oligocene transition, but the impact in the Popigai region undoubtedly occurred very close to the boundary.[70]

The so-called North American tektite field was laid down over the southern U.S.A. around this time, as was a field of smaller microtektites, distributed over the Gulf of Mexico, Caribbean Sea and western North Atlantic. The impact which produced these tektite and microtektite fields apparently occurred in Chesapeake Bay, where a 90-km-diameter crater now lies buried under sediment. Shocked quartz grains have been found associated with both tektites and crater, dated between 35 and 36 Myr before the present time. It seems that the Popigai and Chesapeake Bay impacts occurred in quick succession, separated by an interval of no more than a few hundred thousand years. About a million years earlier, a field of microspherules was laid down over the equatorial Pacific and the Indian Ocean and, shortly before the North American event, another one was scattered across the equatorial Pacific, Caribbean Sea and Gulf of Mexico, both of these fields being associated with iridium abundance anomalies. A further microspherule layer was laid down across the Indian Ocean, equatorial Pacific and the Gulf of Mexico during the Middle Oligocene, around 31 Myr ago.[71]

The first of these microspherule layers coincided with the extinction of some foraminifers, and the second with the disappearance of several radiolarian species. However, there was not always a clear association between extinctions and impacts during the Eocene and Oligocene Epochs. Two foraminifer extinction events occurred towards the end of the Middle Eocene, but this was before the impact which produced the earliest of the microspherule layers. An extinction of forest-dwelling animals also took place in the Middle Eocene, apparently linked to a change to cooler climates. More extinctions occurred, in fairly even fashion, as temperatures continued to fall throughout the Late Eocene and Early Oligocene, and yet another extinction episode took place in the Middle Oligocene. Hence, many Earth-scientists remained convinced that terrestrial factors were the main cause of the extinctions of the Eocene and Oligocene

Epochs.[72] So, for example, Donald Prothero, a geologist from Occidental College, Los Angeles, wrote in 1989, 'At best, it appears that extraterrestrial materials influenced only the weakest and least dramatic of the extinction events. All the rest are clearly a result of terrestrial causes'.[73]

Without question, the detailed evidence gives no indication that the Late Eocene extinctions occurred as a direct consequence of a single major impact. That had seemed a possibility in the early 1980s, when high iridium levels and microtektites were found in a layer of appropriate age, marking the extinction of several radiolarian species, in cores from the Venezuelan and the Caribbean Sea.[74] However, subsequent investigations only served to confuse the situation, by showing that other extinctions took place at different times. Nevertheless, it remained possible that extraterrestrial events could have contributed in significant fashion to the environmental changes which occurred in Eocene and Oligocene times. In 1987, in an article in *Nature*, the Princeton astronomer Piet Hut, together with Walter Alvarez, Eugene Shoemaker and others, argued that the stepwise extinctions and deterioration of the climate could have been due to a series of impacts over a period of several million years, because of the intersection of the Earth's orbit with a comet shower. More recently, Kenneth Farley and colleagues from the California Institute of Technology have also suggested that the Earth was subjected to multiple impacts from a comet shower during the Late Eocene.[75] Similarly, Victor Clube and Bill Napier, in their 1990 book, *The Cosmic Winter*, proposed that a giant comet broke up in the vicinity of the Earth during the Eocene Epoch, leading to a variety of catastrophic effects. They wrote:

> the overall extinction, lasting some millions of years, is best explained by a bombardment episode of a few million years duration and dominated by the disintegration products of great comets – kilometre-sized asteroids and dust infall from dense meteor streams, the dust especially plunging the Earth into long, cold epochs. The dust intermittently kicked up by the multiple impacts would of course add to the chaos. Overall, the effect is that of a prolonged battering of the environment.[76]

Whatever the cause, average temperatures fell by between 5 and 10 °C between the middle of the Eocene and the middle of the Oligocene (see figure 24.3), resulting in habitats which were significantly different from what they had been ten million years previously, with winters in particular becoming much colder.[77] It is hardly surprising that many animal species became extinct during the long Eocene–Oligocene transition. In *Walking with Beasts*, Tim Haines envisaged the scene as follows:

> It is a disaster for wildlife, much of which cannot cope with the swings of climatic change. After 30 million years of prosperity, the mammals will suffer their first ever mass extinction. In only a few million years, an estimated 20 per cent of all life on Earth will become extinct. This is not nearly as dramatic as the extinctions at the end of the Cretaceous, but it is enough to change the flow of evolution.[78]

Over the same period, continental drift was bringing Africa closer to Eurasia. This would have implications for climate in the future. However, even by the Middle Oligocene, these land masses were still completely separated by the shallow Tethys Sea along the line of the Mediterranean Sea and the Persian Gulf (see figure 14.2).[79] To the south, thick forests and woodland covered most of the African continent, where developments of significance for the evolution of apes, and hence humans, were taking place. We shall take up that particular story later, in Chapter 26, after first considering another possible aspect of mass extinction events.

25 Cyclic processes and mass extinctions

Is there a periodicity in mass extinction episodes?

In the previous chapter, we discussed current thinking about the various mass extinction episodes which had occurred during the history of the Earth. Although the nature and cause of these events remains controversial, their existence has been recognised for some time, particularly following the work of Norman Newell in the 1960s (see Chapter 14). In 1977, Alfred Fischer and Michael Arthur of Princeton University proposed that there was a 32 Myr cycle in extinctions, linked to changes in sea-level and sea temperature, which in turn were associated with tectonic processes. That hypothesis produced little interest at the time, because no precise mechanism could be suggested. However, in the mid-1980s, with arguments taking place about all aspects of mass extinctions, the issue of a possible periodicity, and its cause, became a focus of attention.[1]

Because of the incomplete nature of the fossil record, there was too much uncertainty about when individual species and genera became extinct for meaningful statistical analysis to be carried out on the times of their disappearances. However, it was found that the situation improved if investigation was carried out at the next level of classification, the family. A compilation of the disappearance of 2,400 families of marine animals was produced by John Sepkoski (1948–1999), a palaeontologist from the University of Chicago, who assigned each extinction event to one of the stratigraphical stages in the record, those at a boundary being regarded as belonging to the preceding stage. So, for example, all the extinctions which took place during the 8 Myr duration of the Frasnian stratigraphical stage of the Devonian Period, including those at the Frasnian–Famennian boundary, were grouped together.[2] On the basis of these data, Sepkoski and his Chicago colleague, David Raup (see figure 23.1), saw that five extinction events, including that associated with the Frasnian, stood out from the background. The other four were those associated with the final stratigraphic stages of the Ordovician, Permian, Triassic and Cretaceous Periods (see figure 5.2). Over 10% of all marine families and well over half of all marine species disappeared on each of these occasions, which have been termed the 'big five' extinctions.[3]

Raup and Sepkoski found that if they restricted their analysis to the most fully documented of the families which became extinct during the most recent 250 Myr, when the fossil record was considered more reliable than for earlier times, then twelve of the extinction peaks stood out from the background to varying degrees. The first of these was the great Permian–Triassic event, whilst the last, a relatively minor affair, took place during the Miocene Epoch about 11.3 Myr ago. The Chicago pair dated all these extinction peaks in line with the Harland time-scale (which was almost identical to the one in the process of being introduced by the Geological Society of America) and then tested for any regular features by applying Fourier analysis. This attempted to fit a combination of mathematical curves to the extinction data plotted against time. Although a periodicity of about 30 Myr was indicated, it was realised that this could have been an artifact arising from the irregular lengths of the stratigraphical stages, so the best fit of regular cycles to the real extinction data was compared to 8,000 randomised versions, termed Monte Carlo simulations, after the random nature of the events which determine whether fortunes are won or lost in the Monte Carlo casinos. It was found that the cycle providing the best fit to the actual extinction data had a periodicity of about 26 Myr, suggesting the occurrence of ten main extinction events during this period, starting with that at the end of the Permian. This cycle gave a much better fit than any obtained with the random simulations.[4] Assuming that the periodicity of extinctions was genuine, which seemed probable, according to their statistical analysis, Raup and Sepkoski asked whether it was likely to reflect a purely biological phenomenon or the effect of cycles in the physical environment. If the latter, were these cycles likely to be shaped by terrestrial processes or ones out in space? They concluded, 'Although none of these alternatives can be ruled out now, we favor extraterrestrial causes for the reason that purely biological or earthbound physical cycles seem incredible, where the cycles are of fixed length and measured on a time scale of tens of millions of years'.[5]

Possible causes of periodic extinctions

The claims by Raup and Sepkoski about a possible 26 Myr periodicity of extinctions were published in February 1984 in the *Proceedings of the National Academy of Sciences, U.S.A.* Pre-prints had been distributed four months earlier, stimulating several groups to begin working on possible explanations. A series of short papers on the subject appeared in *Nature* in April 1984.[6]

Generally, the explanations suggested for the supposed periodicity centred on the possibility that the comets in the Oort cloud might be disturbed on a regular basis. Two such scenarios, based on the movements of stars in the Galaxy, had been proposed a few years previously. In 1979, and again in their 1982 book, *The Cosmic Serpent*, Victor Clube and Bill Napier (see figure 20.2) had argued that comets would be captured by the Solar System as it passed through a spiral arm of the galaxy, which happens at intervals of somewhere between 50 and 400 Myr. Such an event would lead to a period of increased activity, during which several comets, or asteroids derived from them, might collide with the Earth.[7] Similarly, Jack Hills, an astronomer from Los Alamos, had suggested in 1981 that the Sun might come within 3,000 A.U.

(450,000 million kilometres) of another star of similar size roughly once every 500 Myr. This would disrupt the comets of the Oort cloud, causing some to strike the Earth over the next few million years.[8] However, with both of these mechanisms, the interval between mass extinction episodes would be much greater than 26 Myr.

When the conclusions of Raup and Sepkoski became known, two groups of American physicists pointed out independently that effects similar to those postulated by Hills, but with a shorter periodicity, would occur if the Sun had a small companion star going round it in a highly elliptical orbit. According to Marc Davis and Richard Muller of Berkeley, working in collaboration with Piet Hut, the companion, which they named Nemesis, travelled around the Sun with an orbit whose eccentricity was around 0.7, bringing the two stars to within 30,000 A.U. of each other at closest approach. A companion at this distance would have much the same effect as a star of similar size passing by ten times closer (as in the Hills model), since it would be moving at a slower relative speed. However, if Nemesis was small, with a mass less than one tenth that of the Sun, the disruption of the Oort cloud would be weak, and only a few comets would reach the Inner Solar System. An orbit which brought the companion star closer to the Sun would have more effect, but would be less stable, so that possibility was dismissed by Davis and his colleagues.[9]

A different view was taken by the other group, consisting of Daniel Whitmire, of the University of Southwestern Louisiana, and computer scientist Albert Jackson. They considered that the orbit must be extremely elliptical, with an eccentricity above 0.9, to bring the two stars to within 20,000 A.U. of each other on occasions. This close approach was necessary to the theory because the companion was likely to be very small, having given astronomers no previous reason to suspect its existence. Whitmire and Jackson considered that the companion was a 'black dwarf' with a mass below the limit for hydrogen ignition, which is 0.07 times the mass of the Sun. However, although it would not be visible to observers on Earth, it ought to be detectable at infra-red wavelengths. According to the calculations of Whitmire and Jackson and, indeed, of those of the other group, the companion should at present be far from its closest approach to Earth, perhaps between 130,000 and 150,000 A.U. away.[10]

Stephen Jay Gould questioned whether Nemesis was the best name for the 'death star', suggesting that a more appropriate link would be forged between mythology and evolution by calling it Siva. Nemesis, a Greek goddess, brought down those who deserved it, whereas there was a more random element to the destructive activities of the Hindu god, Siva (also called Shiva), just as there were distinctly random aspects to the consequences of an extraterrestrial impact. If one occurred, the existence of species which had hitherto appeared to be evolutionary winners would be threatened, as well as those which were already struggling.[11]

However, it was pointed out that if the mass extinctions had a precise periodicity, then a solar companion of whatever name was unlikely to be the cause, for its orbital period was likely to increase by about 15% in a 250 Myr period, as a result of the gravitational effects of other stars.[12] Nevertheless, the concept of a 'death star' proved an attractive one, and so did the name 'Nemesis', despite the arguments of Gould. A detailed search for Nemesis began, but no solar companion was ever found.[13]

An alternative explanation for the supposed periodicity of mass extinction events was suggested by Michael Rampino and Richard Stothers of NASA. Their assessment of the data indicated a slightly different extinction periodicity of 30 Myr, i.e. about the time taken by the Solar System to complete a half-cycle as it moves up and down through the plane of the Galaxy. Rampino and Stothers argued that when the Solar System crossed the galactic plane, the clouds of gas and dust concentrated in that region would disrupt the Oort cloud comets, resulting in collisions with the Earth. The fact that the interstellar clouds were not all located precisely on the galactic plane would explain why not all the extinctions events occurred exactly on schedule.[14] An extinction periodicity of around 30 (in fact, 31) Myr was supported by a study carried out by Jennifer Kitchell and Daniel Pena, of the University of Wisconsin. However, Kitchell and Pena were far from convinced about the involvement of an astronomical mechanism.[15] Whilst the University of Missouri astronomers Richard Schwartz and Philip James had no such reservations, they pointed out that the Sun currently lies close to the mid-plane of the Galaxy whereas, according to the data of Sepkoski and Raup, we are about halfway between extinction episodes. To resolve this problem, they suggested that the extinctions might have occurred when the Solar System was at the extreme positions above and below the galactic plane. Thus, they could have been caused by increased bombardment of the Earth in these exposed locations by X-rays or cosmic rays from within the Galaxy, rather than being linked to cometary impacts.[16]

Clube and Napier developed their earlier theory in line with the Rampino/Stothers hypothesis, arguing that two separate processes were involved. Some extinction episodes were due to the passage of the Solar System through a spiral arm of the Galaxy, which they now thought occurred every 250 Myr or so, whilst further extinctions took place at roughly 30 Myr intervals, as the Solar System crossed the galactic plane.[17] Rampino and Stothers went along with this in general terms, calculating the two periodicities to be 260 Myr and 33 Myr.[18]

However, some claimed that molecular clouds were not sufficiently concentrated near the galactic plane, nor was the Sun's vertical trajectory large enough, to account for the apparent periodicity in extinctions. In response, Stothers presented additional evidence to support the view that the up-and-down motion of the Solar System could result in periodic cometary showers.[19] Support came from Clube and Napier, who argued that the Sun's vertical trajectory may currently be anomalously flat, as a result of encounters during the recent passage of the Solar System through Gould's Belt, which is linked to the Orion arm of the Galaxy. As Clube and Napier saw matters by the time they came to write their 1990 book, *The Cosmic Winter*, encounters with molecular clouds would be an irregular feature in cometary disruption, whereas a more regular periodicity would result from tidal effects caused by the motion of the Solar System relative to the rest of the Galaxy. In fact, they now claimed that this periodicity was 15, not 30 Myr, there being a weak surge when the Solar System was at the extreme positions of its vertical oscillation, as well as the strong surge when it crossed the mid-plane of the Galaxy.[20]

Yet another possible mechanism for a periodicity in extinctions involved planet X, a tenth planet of the Solar System whose existence had previously been proposed to account for discrepancies in the motions of the outer planets. Daniel Whitmire (one of the original proponents of the companion star hypothesis) and his fellow University of Southwestern Louisiana astrophysicist John Matese acknowledged that such a planet beyond the orbit of Pluto would long ago have cleared out a path through the Oort cloud of comets. However, they argued that if it had an elongated orbit that itself precessed around the Sun every 52 Myr, due to the effects of the outer planets, then twice during that period planet X would be in an orbit taking it close to the inner edge of the gap in the Oort cloud. On such occasions, it would disturb nearby comets, perhaps propelling some towards the inner Solar System.[21] The question was whether such a mechanism could produce sufficiently intense episodes of cometary disruption to cause periodic extinctions on Earth, which seemed doubtful. Furthermore, several years of searching for planet X, including an infra-red survey of the Outer Solar System by the IRAS satellite in the mid-1980s, failed to provide any evidence of its existence. For these reasons, Matese and Whitmire eventually went on to support the hypothesis that movement of the Solar System through the Galaxy was responsible for the periodicity of mass extinction events. Nevertheless, in 1999, Matese argued that a new analysis of cometary orbits indicated the presence within the Oort cloud of a planet larger than Jupiter.[22]

Regardless of cause, confirmatory evidence for a periodicity of impacts consistent with the extinction data of Raup and Sepkoski was claimed by Walter Alvarez and Richard Muller, using a list of craters produced by Richard Grieve, of the Geological Society of Canada. Thirteen craters with diameters in excess of 10 km were known from the past 250 Myr, each of which had been dated with reasonable certainty. According to Alvarez and Muller, Fourier analysis showed that the cycle of crater formation with the best fit to the data had a periodicity of 28 Myr (although Rampino and Stothers claimed that the best fit was 31 Myr), and Monte Carlo simulations led to the conclusion that this periodicity was likely to be a genuine phenomenon. Although Grieve's data base covered only a small proportion of the Earth's surface, most of the extinction episodes could be linked to at least one of the craters. Taken together with the evidence of multiple microtektite levels near the Eocene–Oligocene boundary (see Chapter 24), this led Alvarez and Muller to conclude that there had been periodic encounters with cometary showers, interspersed with more random impacts involving asteroids.[23] On the other hand, as pointed out by Paul Weissman, an astronomer from the Jet Propulsion Laboratory, Pasadena, analysis of the impact melts of the craters seemed to show that in most cases the impactor had a high stone or iron content, indicating an asteroid rather than a comet consisting principally of ice and dust.[24] However, the possibility of large rocks being present in cometary nuclei could not be excluded.[25]

Whilst all these discussions were taking place about the possibility of periodic cometary showers, Alfred Fischer, who, together with Michael Arthur, had introduced the idea of a periodicity of extinctions in the 1970s, continued to argue that it was related to tectonic activity. According to Fischer, convection currents within the Earth's mantle caused supercycles in the climate, and mass extinctions occurred when prevailing conditions changed from a 'greenhouse climate' to an 'icehouse climate', or vice versa. It appeared that a switch of this nature occurred approximately every 150 Myr. In Fischer's view, the change from icehouse to greenhouse conditions, characterised by increases in temperature and sea-level, was caused by increased atmospheric carbon dioxide concentrations, probably produced by volcanic activity. Such an event occurred in the Late Cambrian, and then icehouse conditions returned in the Late Devonian. There was a switch back to greenhouse conditions at the end of the Triassic, before a return to icehouse conditions in the Late Eocene.[26]

However, there were no such long-term changes in climate at the times of some of the most important mass extinction episodes, including those at the ends of the Permian and Cretaceous Periods. Furthermore, Fischer's theory could not explain the occurrence of an Ice Age at the end of the Ordovician Period, which should have been in the middle of a greenhouse phase. Hence, even if these climatic supercycles had occurred, and played a part (perhaps a major one) in some mass extinction episodes, they could have made little or no contribution on other occasions. It was becoming increasingly apparent that a range of possible factors, including ones of extraterrestrial origin, had to be considered as possible causes of mass extinction events.

Reactions to claims of periodic catastrophes

Many Earth scientists remained cautious about accepting the claims put forward in favour of extraterrestrial catastrophism. One of these was the British geologist Anthony Hallam, whose views we discussed in Chapter 22. Hallam felt no compulsion to support gradualism for reasons of dogma, as he made clear in his book *Great Geological Controversies*.[27] Again, in an article in *Nature* in 1984, he acknowledged that 'catastrophists such as Cuvier were the true empiricists of the day, interpreting the stratigraphic record as it appeared', whereas 'Lyell introduced confusion into the argument'.[28] As we saw earlier, Hallam eventually came to support a scenario for the K–T extinctions which included some catastrophist elements. Nevertheless, in the *Nature* article, he argued that:

> serious doubts and difficulties persist about extraterrestrially-induced catastrophes, especially as a *general* explanation for mass extinctions. The issue is likely to remain unresolved for a considerable time yet, and much cooperative analysis is called for between teams of specialists working on the Phanerozoic as a

whole. Lyell accused his catastrophist opponents of trying to cut, rather than patiently unravel, the Gordian Knot. He would no doubt shake his head sadly at those modern catastrophists who favour a quick fix for a long-standing problem.[28]

Hallam saw several reasons for being sceptical about the claims of periodic catastrophes associated with cometary showers. For example, the impact cycle proposed by Alvarez and Muller was based on evidence from only thirteen craters. Also, the extinction cycle of Raup and Sepkoski depended on the validity of the Harland time-scale (there being significant differences between this and other published time-scales, with variations in stage boundaries of 5 Myr or more) and, if minor extinction episodes and those different from the predicted time by more than 1 Myr were ignored, only five of the twelve extinction peaks remained.[29] Richard Grieve and colleagues similarly warned of the dangers of basing claims for a periodicity of crater formation on analyses of an incomplete record.[30]

As well as Hallam, several others, including the Columbia University palaeontologist Antoni Hoffman, questioned the statistical basis of the speculations about a periodicity of extinctions, pointing out that the conclusions of Raup and Sepkoski depended upon arbitrary decisions concerning the absolute dating of stratigraphical boundaries and the elimination of all but the best-known extinct families of the last 250 Myr. Some suggested that, despite the use of Monte Carlo simulations, it remained possible that the apparent periodicity was an artifact of the procedures that had been employed.[31] However, Raup and Sepkoski continued to maintain that their conclusions about periodic extinctions, and those of Alvarez and Muller concerning an impact periodicity, were fully justified.[32]

Raup initially believed he had also detected a 30 Myr periodicity in geomagnetic reversals, but this time the periodicity really did prove to be an artifact of the statistical method, so he retracted his claim in 1985.[33] Nevertheless, arguments suggesting a link between catastrophic episodes and geomagnetic reversal events continued to be put forward.[34] In addition, Rampino and Stothers argued in *Science* in 1988 that there was evidence for a 32 Myr periodicity in major volcanic episodes.[35]

Two years earlier, in the same journal, Raup and Sepkoski extended their analysis of marine extinction data from the family to the genus level. Although they found nothing to invalidate their earlier work, they acknowledged that 'completely satisfactory conclusions will be reached only with higher resolution data'.[36] The debate continued but, as Raup and Sepkoski could not provide any clear proof of the case for periodicity, many remained sceptical about their ideas.[37]

In 1997, in *Mass Extinctions and their Aftermath*, Hallam and Paul Wignall suggested that the main argument against a 26 Myr periodicity for mass extinctions was the number of events in the scheme of Raup and Sepkoski which were hard to justify. Thus, the extinctions in the mid-Miocene and the Aptian stage of the Cretaceous were negligible, those across the Eocene–Oligocene boundary took place over a long time-period, the Late Jurassic extinctions were known mainly from evidence relating to Europe, and there was no evidence at all for any Bathonian–Callovian event in the Jurassic. Hallam and Wignall concluded that 'out of a total of ten mass extinctions predicted by the periodicity hypothesis, only five or six occurred with certainty'.[38]

Nevertheless, some significant mass extinction events had undoubtedly occurred and, regardless of the issue of periodicity, debates continued to take place about their nature. David Jablonski, of the University of Chicago, together with a number of other palaeontologists, claimed that the characteristics of marine mass extinctions were different from those of normal marine extinctions, incorporating a more random element. Even Ernst Mayr, one of the founders of the Modern Synthesis, eventually came to accept that this was the case, as we noted in Chapter 23.[39] Not everyone agreed with Jablonski but, if the differences in characteristics were genuine, they seemed to be most marked for the K–T boundary event. A detailed statistical analysis reported in *Paleobiology* in 1992 confirmed that this and at least two other mass extinctions episodes, those at the ends of the Ordovician and Permian periods, were real events, regardless of the actual cause or causes.[40] In 1999, in an article in *Science* entitled, 'The future of the fossil record', Jablonski commented:

> An increasingly interdisciplinary paleontology has begun to formulate the next generation of questions, drawing on a wealth of new data . . . Key issues related to evolutionary biology include the biotic and physical factors that govern biodiversity dynamics, the developmental and ecological basis for the non-random introduction of evolutionary innovations in time and space, rules of biotic response to environmental perturbations, and the dynamic feedbacks between life and the Earth's surface processes. The sensitivity of evolutionary processes to rates, magnitudes, and special scales of change in the physical and biotic environment will be important all these areas.[41]

During the 1980s and early 1990s, the arguments about periodicity and the nature of mass extinctions were based largely on evidence from the oceans, because the fossil record for land animals was not considered complete enough for definite conclusions to be drawn about precise trends and patterns. Michael Benton was amongst those who pointed out that mass extinctions on the continents were not statistically distinguishable from background extinctions.[42] However, after more information became available, Benton reported in *Science* in 1995 that it was now clear that mass extinction episodes affecting both marine and non-marine life had on several occasions interrupted the exponential growth in diversity which had being going on since the end of the Precambrian Period.[43] Vincent Courtillot and colleagues, writing in *Nature* a year later, disagreed with Benton about matters of detail, but accepted that mass extinctions had indeed produced a significant effect on changing patterns of diversity.[44] Courtillot expressed this view again in his 1999 book, *Evolutionary Catastrophes*, whilst Benton argued in *Nature* in the following year that, contrary to what Raup and Sepkoski had supposed, the fossil record over the past 540 Myr as a whole was just as good as over the second half of that period, and showed the same characteristics.[45]

In his 1995 *Science* article, Benton concluded that there had been no periodicity of extinctions over the past 250 Myr, because the seven largest mass extinction peaks were spaced anything between 20 and 60 Myr apart.[46] Similarly, in the same year, David Brez Carlisle wrote in *Dinosaurs, Diamonds, and Things from Outer Space*, 'Many believe that mass extinctions have occurred on a strictly periodic timetable during the Phanerozoic, while others, including myself, believe that mass extinctions occur *on average* every 30 million years or so, with no measure of a regular period'.[47] Again, Steven Stanley, in an article in *Paleobiology* in 1990, argued that the apparent periodicity might be explained by delayed recovery following each mass extinction episode. The consequences of this would be that the fossil record would be unable to show evidence of a second crisis following too closely behind the first.[48] To add to the scepticism of Earth-scientists, there was no obvious upsurge of support within the community of astronomers for any of the mechanisms proposed to account for a possible periodic disruption of the Oort cloud.[49] Writing in 1991, in *Extinction – Bad Genes or Bad Luck?*, David Raup concluded:

> My own view is that periodicity is alive and well as a description of extinction history during the past 250 million years – despite the lack of a viable mechanism. The debate has pretty much died down because most scientists involved have decided that the extinction data do not show periodicity. But the proposal is still on the table, awaiting new data or new ways of looking at old data.[50]

If there was indeed a periodicity in mass extinctions, linked to extraterrestrial events, then the implications for evolutionary theory were clear. In an article in *Nature* in 1985, Stephen Jay Gould commented:

> Obviously, if the Alvarez impact theory holds (and it now seems virtually established as a major factor in the Cretaceous debacle, with intriguing hints for its general validity as a primary trigger of disaster), then mass extinction is not competition by natural selection extended (by Darwin's metaphor of the wedge), but a separate process with rules for differential termination that cannot reduce to simple extensions or intensifications of ordinary adaptive struggles. Organisms cannot track or prepare for periods of cosmic disturbances (including major impacts) that occur once every 26 million years or so.[51]

Although support for the periodicity hypothesis subsequently dwindled, it remained an established fact that five major mass extinction events and several lesser ones had occurred between the Late Ordovician and the Early Tertiary, a period of less than 400 Myr. If these extinction episodes were linked to abrupt changes in the environment, then, regardless of any arguments about a precise periodicity, the principles stated by Gould must have had some cogency. Raup made the same point in *Extinction – Bad Genes or Bad Luck?*, going on to prove a 'kill curve' for marine species, based on data provided by Sepkoski. This showed the average waiting time between extinction events of a particular magnitude, no matter what causal mechanisms might have been involved. So, for example, Raup estimated that an event killing about 5% of species came along roughly every million years, whilst extinction episodes of the 'big five' type, resulting in the disappearance of around 65% of all species, occurred at approximately 100 Myr intervals.[52]

Five years later, in 1996, Alessandro Montanari and colleagues concluded, after investigating the thirty-three most reliably dated impact craters greater than 5 km in diameter formed during the past 150 Myr, that there was no convincing evidence of a periodicity of impacts. However, the production of craters did not appear to be an entirely random process, for impacts often occurred in clusters. The formation of a crater with a diameter in excess of 50 km, which would have involved the impact of a projectile at least 3 km across, generally seemed to be associated with a large-scale extinction episode.[53]

As will be apparent, a time-scale of hundreds of thousands of years for extinction episodes, which some had sought to establish in the belief that it was evidence against the Alvarez impact hypothesis, was now accepted by many who supported an extraterrestrial scenario.[54] In 1987, in a review article in *Nature*, which we noted in Chapter 24, Piet Hut, Walter Alvarez and others, including Gerta Keller (who had consistently argued against a single extinction event at the end of the Cretaceous Period), proposed a causal link between comet showers, clusters of impact events and stepwise mass extinctions. In support of their case, they pointed to evidence from around the Cretaceous–Tertiary, Eocene–Oligocene and Cenomanian–Turonian boundaries. As they saw it, a cometary shower could threaten the Earth for several million years, although the majority of the impacts should occur over a shorter period of time.[55] Another catastrophist model which operated over a lengthy time-scale was that of Clube and Napier. As we have seen, they, together with some other British astronomers, were advocates of the model of *coherent catastrophism*, in which the break-up of a giant comet in the vicinity of the Earth led to a prolonged dusting of the upper atmosphere and a cluster of impacts.[56]

Anthony Hallam, Leigh Van Valen, David Archibald and others had, of course, consistently argued that, even though impacts may have occurred, the major causes of mass extinction events were likely to be terrestrial rather than extraterrestrial. Following the lead of Norman Newell, they saw a significant correlation between several mass extinction episodes and regressions of the sea.[57] Significant changes in climate could have occurred as part of the same processes. Thus, in his article in *Nature* in 1984, Hallam pointed out that, although the falls in sea-level at the end of the Permian, Triassic, Cretaceous and Eocene would have extended the habitat area of large terrestrial vertebrates, they suffered mass extinctions, possibly because of the change to a more continental climate, with greater seasonal extremes of temperature.[58]

Hallam also drew attention to the fact that there was evidence of anoxia at the times of the Late Ordovician, Late Devonian and Late Triassic mass extinction events, as well as those at the Pliensbachian–Toarcian and Cenomanian–Turonian boundaries. This anoxia was likely to have contributed to the critical deterioration of the environment, and might have been linked to the sea-level changes, in the way outlined in Chapter 24. All this, together with the known occurrence of large-scale vulcanism at relevant times, meant that there was simply no need to invoke impacts of asteroids or comets as causes of mass extinctions. Even when there had been a large impact at the time of an extinction episode, as at the K–T boundary, there was no reason to think that it did more than accelerate events which were already in progress.[59]

In 1997 in *Mass Extinctions and their Aftermath*, Hallam and his co-author, Paul Wignall, maintained, 'It must be concluded that, contrary to Raup's (1991) opinion, bolide impact cannot plausibly be invoked as a general cause of extinctions'.[60] Instead, they argued that 'for the majority of extinction episodes regressive–transgressive couplets are the norm . . . The rapidity and magnitude of these spectacular sea-level oscillations must have caused rapid habitat tracking by both the marine and terrestrial biota and extinction for those that failed to keep up'. Thus, Hallam continued to believe that major fluctuations in sea-level were the most important cause of mass extinction episodes.[61]

Cosmic causes for terrestrial events?

Without question, Hallam was correct to insist that the possible involvement of Earth-centred mechanisms should be given very serious consideration when trying to explain mass extinctions. It would be poor scholarship to consider the effect of an impact without placing it within the context of terrestrial processes taking place at the time, or to assume that no major extinction episode could have occurred without the arrival of a large object from space. However, it would be equally wrong to ignore the possibility that extraterrestrial factors could have caused or made a significant contributions to mass extinction events, given what we now know about our cosmic environment, and the likely consequence of a large impact or a cluster of smaller impacts.

Some even maintained that there could have been linkage between impacts and tectonic activity, the main driver of terrestrial processes. Other than the one instance in the Late Permian Period, when the continents locked together, there was no convincing terrestrial explanation as to why sea-floor spreading should have stopped on occasions during the course of the Earth's history, to cause regressions of the seas. Clube and Napier argued throughout the 1980s and 1990s that tectonic theories were incomplete without an extraterrestrial dimension.[62] In 1982, in *The Cosmic Serpent*, they assessed the likely effect of a large impact on the Earth's tectonic system and concluded:

> The picture then is one of *episodes* of rapid continental drift, with all the associated worldwide sea-floor spreading, mountain building, vulcanism and so on,

immediately after an impact, followed by a gradual decline of activity as the disturbed Earth settles into a new pattern of movement with more gradual splitting or colliding of continents. If these ideas are correct, large impacts will thus set in motion a complex chain of interacting phenomena – sea-level changes, climatic excursions, violent tectonic episodes, magnetic-field reversals and, of course, mass extinctions.[63]

In contrast, the Cambridge University geologists Robert White and Dan McKenzie maintained that the Deccan traps and similar large-scale outpourings of lava resulted from a plume of magma rising above an anomalously hot part of the Earth's mantle. In *Scientific American*, in 1989, they dismissed the possible involvement of extraterrestrial impacts, writing:

> we do not think such catastrophes are required. Thick marine sequences of igneous rock, flood basalts on land and perhaps even mass extinctions – all can be explained by the interaction of familiar, ongoing earth processes.[64]

Vincent Courtillot pointed out in the same journal in the following year that no evidence of a large impact had been found in India, where the Deccan traps were located. Furthermore, even in the event of the Earth's crust being punctured by a projectile from space, Courtillot considered that extensive volcanic activity was unlikely to result, because there would not normally be large reservoirs of molten rock underneath.[65]

On the other hand, in 1993, in *Earth and Planetary Science Letters*, Michael Rampino and Ken Caldeira claimed to have to have identified a 27 Myr periodicity linking extinction events, regressions of the sea, flood-basalt eruptions, mountain-building events, changes in sea-floor spreading and anoxic events in the oceans.[66] Similar findings were reported in the same year by Ioannis Liritzis of Athens.[67] Because there was evidence of a major impact on some of the occasions when these periodic events took place, Rampino and Caldeira suggested that the unifying cause was more likely to be cosmic than terrestrial.

Following the example of Clube and Napier, and of earlier writers such as Kelly, Dachille and Gallant (see Chapter 14), Rampino and others maintained that tectonic effects, including an episode of extensive vulcanism, could be triggered by the arrival of an extraterrestrial object. Although the impact would have to be a huge one to penetrate the continental crust, a much smaller one, involving a missile not much more than 1 km in diameter, might be sufficient if it occurred on the thinner crust of the ocean bed or at a boundary between tectonic plates. However, the outpourings of lava would probably not be a direct consequence of this, but something which happened in indirect fashion. As previously noted (see Chapter 22), it was suggested that a large impact might cause waves to pass through or round the Earth, resulting in volcanic activity at the opposite side from where the projectile struck. In addition, an impact in an ocean could disrupt systems of layers that had been established for a long time, bringing oxygen-poor deep-water to the surface. The transmission of this by ocean currents might then make shallow seas anoxic.[68]

Further evidence for a periodicity in terrestrial phenomena, linked to impact events, was given by Napier at a prestigious Fermor Lecture Meeting at the Geological Society in London in 1997, and also by Rampino and colleagues in *Earth, Moon and Planets* in 1996 and again in the *Annals of the New York Academy of Sciences* in 1997. Taking his lead from Stephen Jay Gould, and his references to Siva (or Shiva), the Hindu god of destruction, Rampino gave the name 'Shiva hypothesis' to the notion that cyclical mass extinctions of life on Earth result from extraterrestrial impacts.[69]

In response, Courtillot argued, in *Evolutionary Catastrophes*, that the link between extinctions, sea-level regressions, anoxia, geomagnetic reversals and flood basalt vulcanism was real, but had nothing to do with extraterrestrial impacts, and did not have a strict periodicity. Instead, he suggested that these events were caused, through mechanisms still to be worked out in detail, by the intermittent rise of a hot plume from the Earth's mantle. This happened every 30 Myr or so, giving the impression of a periodic cycle, albeit one without precise characteristics. Similar views were expressed in 2001 by a team of geologists from Ottawa, after a detailed examination of the fossil record of planktonic foraminifers. Courtillot, as evidence for his belief in recurrent catastrophes of terrestrial origin, pointed out that seven of the ten major extinctions episodes in the last 300 Myr had coincided with huge outpourings of lava, and usually falls in sea-level, whereas only at the K–T boundary was there clear evidence of an impact.[70]

As we saw in Chapter 24, evidence for the involvement of one or more impacts in several of these mass extinction episodes is increasing. However, regardless of whether these were the primary cause of the events which took place or just a contributory factor, and of whether a precise periodicity was involved, it now seems to be an undeniable fact that the Earth suffers major environmental crises at an average rate of around one every 30 Myr.

26 The uncertain origins of humankind

Apes and environmental changes in the Miocene

Let us now resume our chronological survey of important events in the history of life on Earth, which we left at the end of Chapter 24, having reached the Oligocene Epoch of the Tertiary Period (see figure 5.2). In particular, we must now look at modern ideas about human evolution, having discussed earlier views on the subject in Chapters 8, 11 and 16. As we shall see, it is now generally thought that human evolution did not take place by gradual and continuous development, as envisaged by Darwin and the founders of the Modern Synthesis, but by a more punctuational process, responding in unpredictable fashion to changes in the environment.

During the first half of the Oligocene, a diverse group of primates, which included the ancestors of Old World monkeys and apes, lived in the trees of northeastern Africa. Although average temperatures had fallen since the Middle Eocene, they were still significantly higher than those of today and, indeed, they then started to drift upwards until the Early Miocene, around 20 Myr ago (see figure 24.3), when the first animals to show distinctly ape-like features appeared. The climate of Africa then remained fairly consistent for about 5 Myr, after which time it started to become generally cooler, drier, and more seasonal, the immense tropical forest covering the continent giving way in many places to woodland and occasional open patches, in which a new plant, grass, appeared. These open areas eventually expanded, albeit in erratic fashion, to become plains of grassland, i.e. savannah.[1]

There are several possible reasons for the downturn in temperature. Extraterrestrial impacts undoubtedly occurred both before and after the Oligocene–Miocene boundary, for a 20-km-diameter crater was produced at Logancha, Russia, 25 Myr ago, and a 24-km crater at Haughton, Canada, 23 Myr ago, followed eight million years later by the formation of a similar-sized crater at Ries, Germany, and the laying down of the Moldavite tektites.[2] However, none of these impacts was large enough to have caused the long-term changes in climate. Of much more likely significance was the coming together of Africa and Eurasia around 17 Myr ago. This led to the disappearance of the Tethys Sea, which had previously separated them, changing the circulation patterns of ocean currents. The collision between the continents also caused the uplift which produced the Taurus Mountains of Turkey, the Zagros Mountains of Iran and the Alps of Europe whilst, to the east, the Himalayas were still rising as a result of India having come into contact with the main mass of Asia. The exposure of fresh rock by mountain uplift would have led to increased absorption of carbon dioxide from the atmosphere, and hence to global cooling.[3] Further adverse climatic effects may have resulted from the volcanic activity which produced the Columbia River basalts, covering 200,000 square kilometres of the northwestern USA around 17 Myr ago.[4]

By this time, the separation of Greenland from Europe, apparently initiated by the major outbreak of vulcanism which resulted in the laying down of the Greenland traps near the end of the Palaeocene Epoch, as well as the birth of Iceland, had advanced sufficiently to allow cold currents from the Arctic Ocean to flow into the Atlantic. In the southern hemisphere, Australia, Antarctica and South America continued to move apart, affecting ocean currents in that part of the world, and creating circumstances favourable for the expansion of the polar ice-cap.[5]

Although there were fluctuations, average temperatures continued to drift generally downwards, and environments changed in consequence. There was an extinction peak, albeit a small one, in the marine fossil record during the Middle Miocene. An iridium abundance anomaly has been found in 11.7-Myr-old Middle Miocene sediments in a deep-sea core collected between Antarctica and New Zealand, and an impact crater from around the same time has been located at Karla, Russia, although this is only 12 km in diameter. Around 10 Myr ago, when the climate of North America was still semi-tropical, and herds of rhinoceros roamed the great plains, a supervolcano exploded in southwestern Idaho, depositing a layer of ash 2 metres deep which stretched halfway across America to the eastern seaboard.[6]

In Africa, tectonic movements in the east of the continent had produced uplift of land and, at a later stage, rift valleys which were orientated roughly north–south, whilst vulcanism was an ongoing feature. These factors, together with generally reduced rainfall and falling temperatures, caused the tropical forest to recede from eastern areas, to be replaced by savannah and woodland.[7]

From the Middle Miocene onwards, populations of apes of various types lived throughout Africa and Eurasia, amongst them the well-known genera *Sivapithecus* and *Dryopithecus*. However, there is still considerable uncertainty as to which of the Miocene apes were of particular significance for human evolution.[8] As we saw in Chapter 11, there is a lack of fossil evidence from Africa for the several million years prior to the appearance there of the first unequivocal hominids, the bipedal australopithecines, in the Middle Pliocene, around 4 Myr ago. In the 1960s, Robert Ardrey argued that the australopithecines were 'mighty hunters', citing evidence of their remains having been found in association with collections of animal bones at several sites. However, it now seems that all of the bones, including the australopithecine ones, had been brought there by predators such as leopards and birds of prey, and that the early hominids were timid creatures, foraging rather than hunting for food, and eating mainly fruit, seeds and nuts.[9]

The common ancestor of australopithecines, chimpanzees and gorillas must have lived in tropical Africa around 10 Myr ago, when there was still a thick band of forest from west to east. During the next five million years, as climates changed, the developing mosaic of forest, woodland and savannah across the continent, with volcanic eruptions in the eastern mountains, would have resulted in disruption and isolation of populations. This presented ideal conditions for rapid evolution. Although a small minority believe that our ancestors adopted an aquatic way of life for a short period, which could have facilitated the development of bipedalism, it is generally thought that the ability of hominids to walk upright arose as an adaptation to life on the savannah or, as now seems more likely, in areas of broken woodland.[10] The ancestors of the gorilla and chimpanzee presumably stayed within the forests as their limits receded westward (and towards the equator in the north–south orientation), leaving behind the ancestors of the hominids in more open, drier habitats.[11]

Early bipedal hominids

Amongst the earliest species capable of bipedal behaviour, albeit whilst retaining many more primitive charactcristics, were *Ororin tugenensis*, *Ardipithecus ramidus* and *Australopithecus anamensis*, dating from between 6 and 4 Myr before the present. Traces of these have been discovered only within the past ten years, in Kenya and Ethiopia. Even more recently, it was reported in *Nature* in July 2002 that a skull dating from almost 7 Myr ago, showing a mixture of ape and hominid features, has been found in the Saharan Sahel region of Chad, and classified as *Sahelanthropus tchadensis*. However, the precise evolutionary significance of each of these species is far from ccrtain.[12]

Although details are extremely sketchy, *Australopithecus anamensis*, whose fossils are just slightly more than 4 Myr old, may have been the ancestor of *Australopithecus afarensis* (named after the Afar region of Ethiopia). The latter hominid species is much more clearly defined than the former, existing throughout the period between 4 and 3 Myr ago. As with the apes of today, males were larger than females, but even the male *Australopithecus afarensis* would only have been about four feet (1.2 m) tall. The skull was small, with typically ape-like characteristics, and the brain would have been similar in size to that of a modern chimpanzee. However, from the shape of the pelvis and other bones, *Australopithecus afarensis* must have been able to move around on two legs. Fossils have been found at Laetoli near the Olduwai Gorge in Tanzania and near Hadar in the Awash Valley, part of the Afar region. The Hadar fossils included the well-known 'Lucy', whose skeleton was almost 50% complete. At Laetoli, about 3.7 Myr ago, three hominids, apparently a man, woman and child, walked across freshly fallen volcanic ash. A shower of rain then hardened the footprints before they were covered by another fall of ash. These footprints, discovered in 1976, provided clear evidence of an upright stance.[13]

Perhaps because of the influence of the 'mighty hunter' hypothesis, there had been an initial assumption that australopithecines walked around much as we do.

They were thought to have 'the head of an ape and the body of a man', as the American science writer Ruth Moore put it in her 1962 book, *Man, Time and Fossils*.[14] However, doubts soon began to emerge about the *degree* to which bipedalism was a feature of *A. afarensis*.[15]

Hominid foot bones from Sterkfontein, South Africa, investigated in 1995 and dated to be around 3.5 Myr old, raised further questions. Their anatomy suggested to Phillip Tobias and one of his colleagues from Witwatersrand University, Ron Clarke, that the hominid in question (generally called 'Little Foot', and apparently a member of a previously unknown australopithecine species) could climb trees effectively but, in addition, was well equipped for walking upright. That raised the possibility that members of this species, not *Australopithecus afarensis*, could have been responsible for the Laetoli footprints, although Laetoli is far to the north of Sterkfontein. More parts of the skeleton have now been located and, when examined in detail, may provide useful information about the early stages of human evolution. However, to complicate the picture still further, other distinct hominid species from the same period have been claimed on the basis of finds made in Chad and the Lake Turkana region of Kenya, being named *Australopithecus bahrelghazali* and *Kenyanthropus platyops*, respectively.[16]

Whatever the identities of the species involved, the key advantages gained by early hominids in becoming bipedal are still unclear. Certainly, it seems unlikely that they had anything to do with enabling weapons to be carried and used effectively, as in the discredited 'mighty hunter' hypothesis. Perhaps the ability to walk upright would have enabled males to collect and bring back food for females and infants, thus possibly giving hominid species an evolutionary advantage over contemporary apes in more open habitats. Alternatively, the main benefit may have been in keeping the brain cool, by minimising the effects of the Sun's rays. For whatever reason, bipedalism was well established by the time of *Australopithecus afarensis*. However, early australopithecines probably spent much of their time in and around trees.[17]

That situation continued long after *Australopithecus afarensis* disappeared from the fossil record. A well-known group of hominid fossils dating from around 3 to 2.5 Myr before the present time, found at various sites in South and East Africa, show small but distinct differences from those of *Australopithecus afarensis*, and are assigned to the species *Australopithecus africanus*. Individuals were light-boned (i.e. 'gracile'), but marginally bigger than the hominids from the earlier period, and had slightly larger brains, with a volume up to about 500 cubic centimetres. However, at the same time, their limbs were apparently more ape-like than their predecessors, their arms being longer and legs shorter. A fossil of a similar type, but showing some significant differences (e.g. larger back teeth), was discovered in Ethiopia in 1999, and classified as a separate species, *Australopithecus garhi*.[18]

It appears that a supernova explosion may have taken place within 130 light years of the Earth about 2 Myr ago, causing extinctions amongst marine molluscs. As evidence for this supernova, the isotope iron-60 has been found in cores from the ocean floor.[19] If the event occurred as suggested, it may have influenced the course of hominid evolution, for some important developments took place around this time. However, some of these may already have been in progress before the supernova explosion took place.

Around 2.5 Myr ago, towards the end of the Pliocene Epoch, average temperatures, which had resumed their downwards drift several hundred thousand years previously, following a period of stability, dropped by several degrees throughout Africa, accompanied by a reduction in rainfall. Shortly afterwards (and presumably not without some linkage), several new species of hominids emerged. In contrast to the gracile australopithecines, with their relatively humanlike bone structures, some larger, more 'robust' australopithecines, five feet tall with thicker bones, huge molar teeth, prominent skull ridges and brains comparable in size with those of modern gorillas, came onto the scene, and representatives of this type survived until about 1 Myr ago. The best-known species were *Australopithecus robustus* and *Australopithecus boisei* (some palaeoanthropologists preferring to classify these as a separate genus, *Paranthropus*). Fossils of *Australopithecus robustus* are known from South Africa, and of *Australopithecus boisei* from Ethiopia, Kenya and Tanzania. Both may have been descendants of *Australopithecus aethiopicus*, whose remains have been found in East Africa. It seems that, whereas the gracile forms kept close to woodland, eating mainly soft fruit, leaves and berries, the robust types ventured more into the open, their strong teeth and jaws enabling them to eat nuts and seeds, as well as tough varieties of fruit and leaves.[20]

Another species which emerged between 2.5 and 2 Myr ago was *Homo habilis*. Although resembling the gracile australopithecines in many respects, these had more human characteristics, including flatter faces and larger brains, some with a volume which crossed the 'cerebral Rubicon' of 750 cubic centimetres, justifying the inclusion of the species within the genus *Homo*. Fossils have been found in East and South Africa, occasionally in close proximity to what are taken to be crude stone tools. From a partial skeleton discovered at Olduvai in 1986, it could be seen that, despite the more humanlike skull, *Homo habilis* had a small body and long arms, much like *Australopithecus africanus*.[21] Besides *Homo habilis*, other species of *Homo* may also have been present in the period around 2 million years ago, one of which has been named *Homo rudolfensis*. However, the details remain controversial. There is no consensus about how many species were present, and which fossils should be assigned to which.[22]

It is generally thought that at least one species of *Homo* from this period must be a direct ancestor of *Homo sapiens*, but the precise details remain far from clear (figure 26.1). Similarly, there is still no agreement as to where the various australopithecine species fit into the picture. Many palaeoanthropologists consider *Australopithecus afarensis* to be ancestral to *Homo*, but some believe *Australopithecus africanus* to have been involved as an intermediate, whereas others take a different view. To complicate matters further, *Kenyanthropus platyops*, discovered in 1999 by a

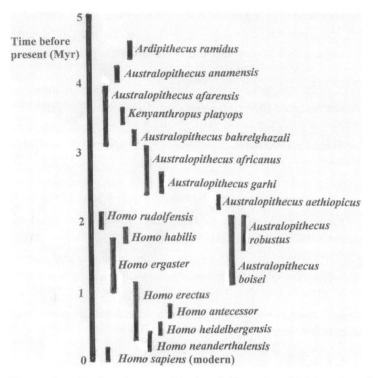

Figure 26.1   The best-known of the hominid species which have been identified from the fossil record, and a number of others, with their approximate time-spans. Some classifications are controversial, and a number of palaeoanthropologists regard the robust forms (on the right) as belonging to the genus *Paranthropus* rather than *Australopithecus*. It is still far from certain which of these various hominid species were on the line that led to modern *Homo sapiens*.

team led by Maeve Leakey, had facial characteristics (*platyops* means 'flat-faced') such that some palaeoanthropologists now think it a more likely ancestor of *Homo* than its contemporary *Australopithecus afarensis*. Again, whilst Lee Berger, American-born successor to Phillip Tobias as director of the palaeoanthropology research unit at Witwatersrand University, has argued that *Homo* emerged from *Australopithecus africanus*, he relegated *Australopithecus afarensis* to a side pathway. Furthermore, as with *Homo*, it is still uncertain whether the robust australopithecines were descendents of *Australopithecus afarensis* and, if so, whether this was direct or via *Australopithecis africanus*. And where do *Australopithecus bahrelghazali* and *Australopithecus garhi* fit into the picture? The situation is not getting any easier to interpret.[23]

Paths to *Homo sapiens* and modern apes

Today, *Homo* is represented by our own species, whilst it is generally thought that the australopithecines have become extinct. However, the British science writers

John Gribbin and Jeremy Cherfas suggested, in their 1983 book, *The Monkey Puzzle*, that the gorilla and chimpanzee might be their descendants. No gorilla or chimpanzee fossils of any great age have ever been found, but it is not known whether this is due to the inaccessibility of appropriate beds or because the African apes have emerged very recently from the australopithecine line. The suggestion of Gribbin and Cherfas made very little impact at the time, but a similar theory was put forward in 1997 by Simon Easteal of the Australian National University. This prompted Gribbin and Cherfas to return to the subject, developing their arguments in *The First Chimpanzee*, a book published in 2001.[24]

Whilst the origin of the African apes remains obscure, the fossil record has provided much information about the possible line of descent of *Homo sapiens*, particularly over the last million years. During the 1980s, the situation looked relatively straightforward, suggesting that *Homo sapiens* arose directly from *Homo erectus*, which itself was a descendant of *Homo habilis*.[25] More recently, the picture has become much more cloudy, as a result of the disputes concerning the classification of early *Homo*, and also because of similar controversies about *Homo erectus*.

Homo habilis apparently became extinct between 1.8 and 1.6 Myr ago, the precise date depending on which specimens are regarded as belonging to that species. From about the same time, which was soon after the beginning of the Pleistocene Epoch, *Homo erectus* spread in dramatic fashion all over the Old World, apparently from an original home in Africa, settling in Java and China more than 1 Myr before the present time. Despite their successful radiation, they had disappeared from most of these areas by around 400,000 years ago, but may have survived in southeastern Asia until less than 100,000 years ago. They may even have reached Australia towards the end of that period, which, even though the present islands of the East Indies were then part of Sunda, a land-mass attached to Asia, would have involved crossing wide stretches of ocean. Their arrival in Australia, however, remains uncertain.[26]

Homo erectus was similar in height and skeletal structure to modern humans. However, it also had a thickish skull, and a characteristically long, flat brain with a volume no more than about 1000 cubic centimetres. This gave it a brain to body size ratio very similar to that of *Homo habilis*, but its overall body structure was far better suited to a bipedal way of life. Because of its narrow hips and pelvis, it is even possible that *Homo erectus* could move around on two legs more efficiently than modern humans. One of the best-known *Homo erectus* finds was made in 1984 at the Nariokotome site west of Lake Turkana, and assembled into an almost complete skeleton by Maeve Leakey and Alan Walker (the team who had previously put together a *Homo habilis* skull from fragments found at Koobi Fora, as we noted in Chapter 11). The specimen (known by some as 'Turkana Boy' and by others as 'Nariokotome Boy') was a youth over five feet tall, who died around 1.6 Myr ago.[27]

By the standards of modern humans, the brain of *Homo erectus* was small in relation to body size. Also, there was no indication in the detailed skeletal features of any capability for vocal communication, and the only associated artifacts were crude stone tools. After making a detailed examination of the Nariokotome Boy over a period of many years, Alan Walker, now at Pennsylvania State University, summarised his findings in his 1996 book, *The Wisdom of Bones*, written in collaboration with his wife, Pat Shipman, another Pennsylvania State palaeoanthropologist. Walker and Shipman concluded:

> the long history of brain-first theories in paleoanthropology implied that the missing link lying between apes and humans would be behaviorally and intellectually human while trapped in an animal body. In that case, the fossil and archaeological record would yield evidence of human capabilities and behaviors while the bones themselves would speak of apelike proportions and imperfect physical adaptations to human behaviors. But the Nariokotome boy proved the inverse. He was in many ways an animal in a human body.[28]

Homo erectus came into existence at around the time when the first of a series of episodic expansions of the northern ice-sheet took place. During this period, the tropics and temperate zones experienced environmental fluctuations as dramatic as those in the polar regions. When glaciation at high latitudes was at its greatest extent, reduced rainfall closer to the equator caused the African rainforest to shrink to an isolated pocket, surrounded by savannah and sandy desert. In contrast, during the wet conditions of a major interglacial period, many areas which were deserts during glacial periods (including some which remain so at the present time, such as the Sahara) were covered by grass.[29]

As the Pleistocene Epoch progressed, hominids with characteristics somewhere between those of *Homo erectus* and modern *Homo sapiens* began to replace the more traditional *Homo erectus* forms throughout Europe, Asia and Africa. Amongst the oldest specimens are the ones discovered at Mauer (Heidelberg) and Boxgrove (Sussex), which date from a relatively warm period around 500,000 years ago. Fossils of a similar type from an even earlier time were found in 1995 in association with crude stone tools at Atapuerca, northern Spain, and shown to be around 800,000 years old. All these types were generally known as 'archaic *Homo sapiens*'. They were hunter-gatherers, perhaps carrying out hunting to an increasing extent as time passed. Wooden throwing-spears 400,000 years old have been found near Schöningen, in Germany.[30]

One group of archaic *Homo sapiens*, the Neanderthals, lived in Europe from about 160,000 years before the present. Neanderthals had brains at least the size of modern humans, with a volume around 1,400 cubic centimetres, encased in long, thick-walled crania, and they left behind artifacts which, for all their limitations, were clearly an advance on those of *Homo erectus*. It is still uncertain whether Neanderthals should be regarded as a sub-species of *Homo sapiens* (in which case they may be classified as *Homo sapiens neanderthalensis*), or a separate species, *Homo neanderthalensis*. Regardless of classification, the Neanderthals, apparently the last of the archaic *Homo sapiens* forms,

disappeared about 35,000 years ago, their areas of occupation being taken over by
Homo sapiens types indistinguishable from those of the present day.[31]

Did we all come from Africa?

One of the most controversial issues in human evolution over the past twenty
years has been whether modern *Homo sapiens* had a single or a multiple birth.[32] In the
1980s, Milford Wolpoff and Australian palaeoanthropologist Alan Thorne revived the
multi-regional hypothesis of Franz Weidenreich, which maintained that *Homo sapiens*
had arisen from *Homo erectus* by parallel development over the same period of time in
several areas of the world, in progressive, Darwinian fashion (see Chapter 11). This
hypothesis was termed the 'candelabra model' by William Howells, in his *Mankind in
the Making*, because it suggested that the line of human evolution diverged after the
emergence and spread of *Homo erectus*, each separate strand continuing to develop in
an almost identical way, resulting in an overall pattern which resembled the shape of
a candelabra. As with Weidenreich's original theory, Wolpoff and Thorne's updated
version of the multi-regional hypothesis was based largely on the interpretation of
fossil evidence.[33]

However, a very different scenario was indicated by evidence from human genes.
Not only do modern humans, for all their variety of life-styles, skin colours and other
physical characteristics, constitute a single species, but they also have a narrower
range of genetic variability than many other species. That was difficult to reconcile
with the separate development of the various human races over a long period of time,
even though Weidenreich had suggested that they would not have been in complete
genetic isolation from each other. Furthermore, molecular clock evidence indicated
that, although *Homo erectus* was spread throughout the Old World a million years ago,
the ancestors of modern *Homo sapiens* may have emerged from Africa within the last
200,000 years. Consistent with this model, which has been termed both the 'Noah's
Ark' and the 'out of Africa' hypothesis, the earliest fossils of modern *Homo sapiens* have
been found in Africa and the Levant, whilst Neanderthals, regarded by many as the last
representatives of archaic *Homo sapiens*, vacated the Middle East well before their final
disappearance from western Europe.[34]

In 1987, in *Nature*, Berkeley geneticists Rebecca Cann, Allan Wilson and Mark
Stoneking (who later moved to Pennsylvania State University) reported the results of
population studies of mitochondrial DNA, which, unlike nuclear DNA, is inherited
solely from the mother. The mitochondrial DNA data confirmed previous indications
that there is a greater diversity of gene structure amongst Africans than amongst
Europeans or Asians. Furthermore, the construction of a family tree on the basis of
this genetic information suggested a common ancestor with a gene structure closer
to those of some present-day Africans than of anyone else, suggesting an African
origin for all modern humans. Using molecular clock calibrations, the Berkeley group
concluded that this common ancestor (who has been called 'mitochondrial Eve') must

have lived around 200,000 years ago. They were not, of course, suggesting that this particular individual was the first modern human, but simply that she was a member of the population in which the transition from archaic *Homo sapiens* to modern humans took place. 'Mitochondrial Eve' just happened to give rise to the one maternal lineage that remained unbroken between her time and the present. Nevertheless, the hypothesis was attacked by those who supported the multi-regional model, and further research did not settle the issue one way or the other. However, similar conclusions were reached from studies of inheritance down the male line, through investigation of the human Y chromosome, initially by Robert Dorit and colleagues of Yale in 1995.[35]

The implication of the 'out of Africa' model is that a population of modern *Homo sapiens*, carrying genes inherited from a long-dead ancestor, began to spread around the world and eventually replace all the well-established archaic *Homo sapiens* populations, without apparently interbreeding with any of them. This is certainly consistent with the prolonged co-existence of different populations at various locations. In the course of time, the archaic populations may eventually have been exterminated by the modern ones, or simply out-competed by them. According to that view, the Neanderthals, for instance, were not our ancestors, but the end-product of a separate branch of hominid evolution, which became extinct as a result of a struggle for existence with *Homo sapiens*.[36]

In contrast, supporters of the multi-regional hypothesis believe that modern *Homo sapiens* might have evolved separately from local populations of *Homo erectus*, via archaic forms of *Homo sapiens*, in different parts of the world. In 1992, Li Tianyuan, of the Hubei Institute of Archaeology, China, and Dennis Etler, a palaeoanthropologist from Berkeley, described two fossil skulls from Yunxian, central China, which, in their view, showed a mixture of 'archaic' and 'modern' features, the latter including a flat face. These were dated at 350,000 years before the present, so could not be attributed to the arrival of any descendants of an African 'mitochondrial Eve'. It was concluded that the Yunxian fossils represented an intermediate stage in the process of continuous development of *Homo erectus* into modern *Homo sapiens* in the region.[37] That would have been readily accepted twenty years earlier, when the gradualistic paradigm was still all powerful, but, in the changed climate of the 1990s, both the interpretation and the dates were challenged.

Less controversially, other archaic *Homo sapiens* fossils have been found in China, dating from around 200,000 years ago and, as we have seen, archaic *Homo sapiens* was widely distributed elsewhere by this time. One particularly intriguing specimen (known as SM3) is the top of the skull of a female from Java, showing characteristics of both *Homo erectus* and *Homo sapiens*, which could date from anywhere between 100,000 and 800,000 years ago.[38] However, the evolutionary significance of these fossils is far from clear. Whatever the origin and history of archaic *Homo sapiens* in a particular region, it might not have gone on to develop into modern *Homo sapiens*. Instead, it could have been replaced or assimilated when more modern forms migrated into the area.

Studies of the characteristics of fossil hominids have yet to provide conclusive evidence for or against the 'out of Africa' and multi-regional hypotheses. Even with the help of statistical analyses, some investigators have taken the fossil evidence to suggest a common origin for all humans, whereas others see indications of separate, regional trends.[39]

There were undoubtedly major fluctuations in the environment over this period (see figure 24.3), which, whilst not making regional continuity impossible, would at least make it seem unlikely, as populations would be expected to migrate to try to find conditions similar to those they had enjoyed previously. In contrast, the details of the 'out of Africa' scenario fit in well with the known features of the climates of the Pleistocene Epoch. Around 200,000 years ago, northern Europe was covered by the glaciers of the Riss Ice Age (as it has traditionally been called, although, as we shall see in the next chapter, the use of such names can over-simplify a complex situation). At the same time, arid conditions affected much of Africa, which would have made life very difficult for the population supposedly in the process of giving rise to modern *Homo sapiens*. The formation of the 1.2-km-diameter Tswaing (or Saltpan) impact crater in South Africa during this period, whilst a minor event by global standards, would not have improved the situation. Because of the long-enduring period of hardship, no significant radiation of the new human form would have been likely until conditions eased with the coming of an interglacial period, about 130,000 years ago. This is consistent with the evidence for, after previously leaving very few traces, early modern *Homo sapiens* appears to have spread throughout Africa and into the Middle East around 100,000 years ago, as indicated by finds in Ethiopia, Zaire, South Africa and Israel. Further expansion would then have been stopped by the onset of the next (and so far, last) Ice Age, the Würm, which was possibly linked to the huge eruption of Toba in Sumatra, larger than the 1980 Mount St Helens eruption by a factor of a thousand.[40]

As Bill McGuire of University College London wrote in his 1999 book, *Apocalypse*, 'Barring impacts by asteroids and comets, gigantic eruptions on the scale of the great Toba blast represent the most devastating of all natural catastrophes'.[41] According to the University of Utah geneticists Lynn Jorde and Henry Harpending, the remarkable lack of diversity in human mitochondrial DNA indicates an evolutionary bottleneck around the time of the Toba explosion, with the total population of modern *Homo sapiens* falling to less than 10,000 individuals. In his 2001 book, *The Seven Daughters of Eve*, the Oxford University geneticist Bryan Sykes argued on the basis of mitochondrial DNA investigations that 95% of modern Europeans are descended from just seven women, who lived between 10,000 and 45,000 years ago.[42]

If the 'out of Africa' hypothesis is correct (and, although still facing considerable opposition, it seems to be the model most consistent with the fossil, genetic and environmental evidence, taken as a whole), then the thread which connected the African emigrants to ourselves must have been very close to being severed as the Würm Ice Age began. By this time, the Neanderthals were already well-established in Europe and parts of the Middle East, and their shorter, more thick-set bodies would have given them a significant advantage in coping with the cold weather.[43] Nevertheless, as conditions eased, modern *Homo sapiens*, having already started to spread into Asia, eventually moved into Europe. Only a few tens of thousand years later, our species dominated the entire Earth.[44]

Processes in human evolution

Regardless of the 'out of Africa' debate, which concerns events only within the last million years, there is no way that hominid evolution as a whole can be regarded as demonstrating straightforward progressive development. For example, *Australopithecus afarensis* appeared earlier than *Australopithecus africanus*, a species which, despite its larger brain, had more ape-like body proportions. As we have seen, the ancestry of the hominid species living around 2 Myr ago is far from clear, and there is no agreement about which of these were directly ancestral to present-day humans. The robust australopithecines came after the more human-like gracile australopithecines, and appeared at much the same time as the first representatives of the genus *Homo*. Brain size may have increased along what, during the past thirty years, has generally seemed the most likely sequence, from *Australopithecus afarensis* to *Homo habilis* to *Homo erectus* to *Homo sapiens*, but the transition from *Homo habilis* to *Homo sapiens*, both of which had rounded, thin-walled, high-vaulted crania, was through *Homo erectus*, with its long, thick-walled crania and jutting brow-ridges. Furthermore, there was a significant difference in size between the tiny *Homo habilis*, which was only slightly larger than *Australopithecus afarensis* (and still spent much of its time in trees), and the much bigger *Homo erectus*, which was as large as a modern human (and completely bipedal). Even so, only with the appearance of *Homo sapiens* was there a major increase in brain to overall body size.[45]

During the past ten years, many palaeoanthropologists have begun to re-allocate some fossils formerly classified as *Homo erectus* to a separate species, *Homo ergaster*. They believe that representatives of *Homo ergaster* (such as the Nariokotome Boy) remained in Africa for over a million years, eventually giving rise to modern *Homo sapiens*, probably via at least one form of archaic *Homo sapiens*. In contrast, although originating from African stock, *Homo erectus* lived mainly in Asia, and had characteristics slightly more advanced than those of *Homo ergaster*. Fossils which, in the revised scheme, continue to be regarded as representatives of *Homo erectus* include Java Man and Peking Man, which we discussed in Chapter 11. However, to complicate matters, a 1-Myr-old skull very similar to these Asian ones has recently been found in Ethiopia.[46]

Regardless of whether they really are separate species, it is clear is that, from the point of view of evolutionary progression, the substitution of *Homo ergaster* for *Homo erectus* as an intermediate in the sequence linking *Homo habilis* to *Homo sapiens* would not result in it appearing to show any more straightforward trends. Similarly, whilst the suggested substitution of *Kenyanthropus platyops* for *Australopithecus afarensis* might support a more progressive development of a flat face, it would not otherwise change the overall picture of seemingly erratic transitions.

Indeed, as pointed out by Alan Walker and Pat Shipman in *The Wisdom of Bones*, it is a mistake 'to cast the course of hominid evolution as a trajectory *toward* humanness'. Had that been the case, the Nariokotome Boy would have possessed more of the characteristics of modern humans, including a much larger brain. Walker and Shipman emphasised that extinct hominid species were not 'failed attempts to become humans', but creatures 'adapted to niches and life-styles that go beyond those of the few species who still survive'.[47]

Whatever the precise details, extreme and unstable climatic conditions coincided with the emergence and spread of hominids, and so must have influenced events. Indeed, there is growing acceptance that speciation and extinction events, linked to changes in the environment, may have been of significance in human evolution. So, Niles Eldredge wrote in 1992, in *The Miner's Canary*, that 'the hominid story is much more than a geometrically mushrooming cloud of population expansion. It is a story, as well, of speciation and extinction, and of exploitation of a variety of different ecological niches, as various of our progenitors and collateral kin have found themselves as parts of a large array of different ecosystems'.[48]

To Australian palaeoanthropologist Colin Groves, writing in *A Theory of Human and Primate Evolution*, published in 1989, speciation not only appears to have played an important role in hominid evolution but, furthermore, it could involve mutations arising according to certain laws, not by chance. He accepted the view of the punctuationalists (see Chapters 15 and 16), writing that 'natural selection is primarily an agent of conservation, not of change'[49] and that 'new species appear suddenly, and begin by being very variable', as well as suggesting that heterochrony could be important in explaining abrupt changes between species. Groves concluded that modern *Homo sapiens* could be 'the product of at least ten quasi-speciation events'.[50]

Similarly, Jeffrey Schwartz, an anthropologist from the University of Pittsburgh, wrote in his 1999 book, *Sudden Origins*, 'Since hominid evolution is as much a history of regulatory changes as is the evolution of any other group of animals, we should not expect to find a trail of intermediates proceeding slowly from one morphological state to another'.[51]

As with *Homo erectus*, there has been an increasing tendency to split archaic *Homo sapiens* into a number of separate species, such as *Homo neanderthalensis*, *Homo rhodesiensis* and *Homo heidelbergensis*.[52] The Atapuerca specimens from around 800,000 years ago have been classified as *Homo antecessor*, a species which, according to some palaeoanthropologists, may have evolved into *Homo heidelbergensis*, and then perhaps into *Homo neanderthalensis*.[53]

In an article in *Scientific American* that appeared in 2000, Ian Tattersall, of the American Museum of Natural History, admitted that the twenty or so hominid species he currently recognised were unlikely to be the only ones which had actually existed. Nevertheless, it was abundantly clear that human evolution had not been a linear process. Tattersall continued,

> Our biological history has been one of sporadic events rather than gradual accretions. Over the past five million years, new hominid species have regularly emerged, competed, coexisted, colonized new environments and succeeded – or failed. We have only the dimmest of perceptions of how this dramatic history of innovation and interaction unfolded, but it is already evident that our species, far from being the pinnacle of the hominid evolutionary tree, is simply one more of its many terminal twigs.[54]

Whilst, as we noted in Chapter 18, the mechanism of speciation is still far from clear, divergence must be facilitated if a population splits into groups with different environments. Palaeoanthropologist Chris Stringer, of the Natural History Museum, London, together with science journalist, Robin McKie, suggested in their 1996 book, *African Exodus*, that such fragmentation might have been of great significance when groups of our ancestors moving out of Africa were trapped by the onset of the Würm Ice Age 75,000 years ago. Stringer and McKie wrote:

> Having evolved in the warm savannah sun, we nearly perished, huddled in cold dismal misery as volcanic plumes straddled the earth. The links between modern human pioneers in Asia and their motherland were severed, allowing cold-adapted Neanderthals to become the sole occupants of the Middle East for the next 30,000 years. Ironically all this fragmentation and environmental pressure may have been the stimulus for the final crucial changes that transformed these hominid bit players into masters of the planet.[55]

As we have seen, that view of the emergence of modern *Homo sapiens*, which is now accepted by many, although certainly not all, palaeoanthropologists, was based initially on findings in the genes of living people. Unfortunately, the same procedures cannot generally be applied to fossil bones, which would not retain any of the DNA of the original animal. However, if fossilisation is incomplete and even a tiny amount of DNA can be recovered, it is possible to amplify this by means of the polymerase chain reaction (PCR), to enable its base sequence to be determined. Thus, DNA recovered from Neanderthal bones was amplified by the PCR technique, allowing the samples to be analysed. Results reported in the journal *Cell*, in July 1997, by a multi-national team including Mark Stoneking and Svante Pääbo, from the University of Munich, showed, apparently conclusively, that modern humans are *not* descended from Neanderthals.[56]

That may have settled the argument as far as one particular aspect of our recent evolutionary history is concerned, or it may not, for fossil bones have since been found near Leiria, Portugal, which seem to represent a form intermediate between Neanderthals and modern humans.[57] In any case, there are many other issues still unresolved. The older the fossil, the less likely it is that any intact DNA would remain, but should any be found in fossils of the earlier hominids, it might at last clarify the relationship between the various *Homo*, *Australopithecus* and other species.

Regardless of that, however, it is now clear that human evolution was not gradual or progressive, as Darwin had believed it to be. It was, instead, an erratic process, influenced in no small measure by dramatic environmental fluctuations.

27   Ice Ages in the Pleistocene Epoch

Possible causes of Ice Ages

Environmental changes must have played a major role in human evolution, although many of the details are still unclear. Some of the suggested causes remain speculative, but the involvement of continental drift is now well established. Without question, the linking of Africa to Eurasia in the Middle Miocene allowed the Miocene apes and, later, early *Homo*, to radiate from Africa, regardless of the arguments about whether modern *Homo sapiens* also emerged in this way. Furthermore, as discussed in the previous chapter, continental drift played a significant role in the deterioration of the climate from around 15 Myr onwards, which led eventually to the occurrence of Ice Ages in the Pleistocene Epoch (see figure 24.3). Amongst the final contributions of continental drift to this process were changes in the East Indies which prevented warm Pacific currents entering the Indian Ocean, and the coming together of North and South America, leading to the formation of the Isthmus of Panama, around 3 Myr ago.[1] Whether or not volcanic eruptions, impacts or the close passage of a large cosmic body played a part in reducing temperatures remains uncertain, but there are reasons for taking such possibilities seriously.

As noted in Chapter 25, Victor Clube and Bill Napier have drawn attention to the fact that the Solar System passed through part of Gould's Belt, a disintegrating molecular cloud associated with the Orion arm of the galaxy, relatively recently, perhaps between 9 and 6 Myr ago. They argued, in their 1990 book, *The Cosmic Winter*, that this would have disturbed the Oort cloud and pushed many comets into the inner Solar System during the Pliocene Epoch, leading up to the start of the Pleistocene. Clube, Napier and others then went on to suggest that, as in the Eocene Epoch (see Chapter 24), the break-up of a giant comet in an Earth-crossing orbit might have resulted in a series of impacts and an atmospheric dust-cloud which persisted for a long period of time, causing a significant deterioration of the climate. Amongst the impacts known to have occurred in the Pliocene Epoch is the one which produced the 18-km-diameter El'gygytgyn crater in eastern Siberia, which is thought to be about 3.5 Myr old.[2]

In 1988 and on subsequent occasions, Frank Kyte and others from the University of California, Los Angeles, produced evidence to show that the impact of an asteroid of between 0.5 km and 4 km in diameter in a deep basin in the southeast Pacific occurred at the same time as a sharp drop in temperature 2.3 Myr ago. This heralded the approach of the first of the Pleistocene Ice Ages, termed the Danube glaciation in Europe and the Nebraskan in North America. Analysis of the debris from the impact, which was spread over a large area of the ocean floor, indicated that the missile was a meteorite with a low metal content. Iridium abundance anomalies were demonstrated in several deep-sea cores from the region, magnetic stratigraphy giving an age coincident with the Late Pliocene temperature drop. The projectile responsible is generally refered to as the Eltanin asteroid, after the name of the ship used to collect the cores. As pointed out at a conference in April 2001 by Liverpool John Moores University social anthropologist Benny Peiser and Australian astronomer Michael Paine, the Earth is likely to have experienced something like twenty devastating impacts on this scale during the past 5 Myr. Had one of these made a direct hit on the area in which the ancestors of modern humans were living, the course of subsequent history would have been very different. Even as it was, impacts might well have influenced events.[3]

An impact of an object approximately 1 km in diameter is known to have occurred just over one million years ago, producing the 10.5-km-diameter Lake Bosumtwi crater in Ghana and the Ivory Coast tektites. This was followed by a similar one about 900,000 years ago, which gave rise to the 13.5-km-diameter Zhamanshin crater in Kazakhstan and more tektites, known as the Irgizites. Then, shortly after the Zhamanshin event, came an impact which resulted in the laying down of the Australasian field of tektites. Evidence of these and related fields has been found in various parts of southeast Asia, and they may be spread over something like 30% of the Earth's surface, suggesting an impactor of up to 5 km in diameter.[4]

The event which produced the Australasian tektites came around the time of onset of the Günz glaciation, known in America as the Kansan. However, despite the impression given by the use of four or five such names for individual Ice Ages stretching throughout the two million years of the Pleistocene, the climate rarely remained stable for any period of 10,000 years, even if short-term fluctuations are disregarded. There were over ten significant glacial–interglacial cycles during the Epoch, and many others of a more minor nature. Average temperatures could change by 10 °C or more over a time-scale of a few decades, or possibly just a few years.[5]

A variety of mechanisms have been put forward to explain the occurrence of Ice Ages. The most enduring gradualistic explanation was proposed over 60 years ago by Milutin Milankovitch, who developed ideas suggested in the previous century by James Croll (see Chapter 6). Milankovitch argued that the changes in climate could be linked to the combined effects of three astronomical cycles affecting the intensity of sunlight at the surface of the Earth, particularly at middle and high latitudes. The first cycle was a variation in the tilt of the Earth's spin axis from 22° to 24° and back over a period of 41,000 years, the second involved changes in the eccentricity of the Earth's orbit, with a periodicity of 100,000 years, and the third concerned the precession, or wobble, of the Earth's spin axis over a period of 23,000 years.[6] However, a detailed study of Pleistocene temperatures at Devil's Hole, Nevada, carried out in the years around 1990, showed that the Milankovitch theory did not provide a particularly good explanation for the pattern of change at this location. More generally, astronomical calculations have indicated that the 23,000-year periodicity should be the most significant one, whereas if there are any genuine underlying patterns in the Pleistocene climate, they appear to indicate the involvement of something with a periodicity of around 100,000 years. Also, cutting across these stately trends are a number of 'interstadial events', when temperatures rose extremely rapidly, but then took thousands of years to fall to their previous values. Even if the Milankovitch theory can explain some of the temperature fluctuations of the Pleistocene, it cannot account for them all, and it is certainly unable to explain why there were no Ice Ages in the 250 Myr period between that time and the Late Permian.[7]

Hence, it has become increasingly common for other possible causes for Ice Ages to be considered. In a 1978 paper and in his 1981 book, *Ice*, Sir Fred Hoyle (see figure 14.3) suggested that Ice Ages are initiated by the creation of an atmospheric dust layer which would prevent heat from the Sun reaching the Earth, caused either by passage through cometary dust or by the impact on the Earth's surface of a stony asteroid around 1 km in diameter. Conversely, Ice Ages could be brought to an end by the impact of a metal-rich asteroid, scattering metallic particles which could absorb heat from the Sun and warm the atmosphere. Asteroids of a similar size also featured in an argument outlined in 1990 by Emilio Spedicato, a mathematician from the University of Bergamo. In an internal publication of his University, Spedicato suggested, like Hoyle, that a glaciation can be brought about by a dust-producing impact on land, but believed that a deglaciation was probably caused by a greenhouse effect following an impact in an ocean. Hoyle subsequently accepted this as the most likely scenario for the ending of an Ice Age. Frank Kyte and his colleagues, however, thought that the consequence of a projectile of this size striking the ocean, as occurred in the Late Pliocene, would be the formation of high-altitude clouds, leading to global cooling.[8]

Without question, the impact of an asteroid of about 1 km in diameter would have significant effects. However, it seems unlikely that an event on this scale could, by itself, and regardless of other circumstances, bring about the start or finish of an Ice Age, particularly since astronomical evidence suggests that several such impacts would be expected every million years (see Chapter 20). Even mechanisms involving larger impacts have problems, because, as we have seen, the best evidence for a major impact event is at the end of the Cretaceous Period, 65 Myr ago, when no Ice Age occurred.

Although still speculative, a situation which could lead to an Ice Age is an intense period of impact activity and atmospheric dusting, possibly linked to the break-up of a giant comet. Victor Clube and Bill Napier believe that Comet Encke, the asteroid Oljato and the Taurid meteoroids (see Chapters 19 and 20) are the remnants of such a comet, the successor to the ones which threatened the Earth in Eocene and Pliocene times. Retro-calculations from current orbits indicate that the parent body may have disintegrated about 9,500 years ago. Although that is after the end of the Würm, the most recent Ice Age, Clube and Napier, together with David Asher and others, argued that the giant comet had probably been in the vicinity of the Earth for thousands of years prior to this, causing an atmospheric dust-cloud which had largely cleared by the time the nucleus broke up. In support of that hypothesis, ice-core studies have indicated that a large amount of dust was deposited on the Earth's surface during the last 10,000 years of the Würm glaciation. Furthermore, this dust has the same chemical composition, including a high iridium content, as that recovered from peat moss in the Tunguska region, where another fragment of the same cometary system may have struck early in the twentieth century.[9]

Volcanic activity, which occurred extensively during the Pleistocene, with out-breaks in Antarctica, Iceland, the North Pacific and elsewhere, is another factor which may have caused the cold climates of the time. Perhaps of particular significance, as already mentioned, the Toba super-eruption took place sometime around the start of the Würm glaciation, 75,000 years ago.[10] However, from a more general perspective, it should be noted that the extensive volcanic activity which occurred during the Late Triassic, the Late Cretaceous, the Early Oligocene and on other occasions between the Permian and the Pleistocene, did not give rise to Ice Ages.

As we saw earlier (in Chapters 14 and 24), plate tectonics seems to provide at least a partial explanation for the Permian, Carboniferous and Ordovician glaciations, for they occurred when a continent approached or covered one of the poles. Land at or near a pole would provide a platform for snow to settle on, reducing temperatures by reflecting the Sun's rays back into space. Also, ice-sheets form and spread more easily over land than over sea. Nevertheless, even with such a base, the further spread of ice-sheets occurs only when there is an appropriate arrangement of ocean currents. Antarctica moved into a position over the South Pole during the Eocene Epoch, long before the start of the first Pleistocene Ice Age, and it is still there today, after the termination of the last of them.[11]

Another relevant factor is the concentration of 'greenhouse' gases such as car-bon dioxide, methane and water vapour in the atmosphere. It is believed that an increase can lead to global warming and a decrease to global cooling. The main source of methane is bacteria in oxygen-poor environments, particularly swamps and bogs. Carbon dioxide is removed from the atmosphere by some living organisms (e.g. plants and phytoplankton, during photosynthesis) and released into the atmosphere by oth-ers (e.g. mammals, as they breathe). It may also be taken up or emitted by rocks and oceans, according to the circumstances. Furthermore, carbon dioxide may by re-leased into the atmosphere during an episode of volcanic activity, or if an asteroid or a comet strikes carbonate rocks, whilst the atmospheric water vapour content would be increased following an oceanic impact.[12]

From these various considerations, it seems likely that Ice Ages result from the cumulative effect of a variety of factors from amongst a group including asteroid impacts, vulcanism, continental drift, changes in solar radiation and Milankovitch cycles, as well as others which appear more mundane. These include the dust, carbon dioxide, methane and moisture content of the atmosphere, as well as deep-water flux (which we shall discuss later) and phytoplankton concentrations in the oceans.[13]

Ice Ages were clearly complex events. So, for example, in *Ice*, Hoyle accepted that an asteroid impact on a continent would only bring about an episode of glaciation if it occurred at a time when at least one of the poles was covered by land. Even then, however, other factors were likely to be involved. Since the publication of *Ice*, it has become increasingly clear that terrestrial climates are far from stable. In 1999, Hoyle, in collaboration with Chandra Wickramasinghe, argued that the Earth could easily slip into an Ice Age at the present time, even without the impetus of an impact, as a consequence of Antarctica covering the South Pole and Greenland being close to the North Pole. Today's climates were relatively warm only because of high atmospheric concentrations of greenhouse gases, a situation that could easily change, particularly if the efforts of environmental campaigners prove successful.[14]

Unexplained findings and heretical theories

As we have seen, the causes of Ice Ages remain uncertain, and may be complex. What is more, even some of the characteristics of glaciations are less than straight-forward. The Pleistocene Ice Ages in the northern hemisphere were not simply times when the polar ice-cap expanded evenly, for Siberia and Alaska (whose names now invoke images of extremely long, cold, winters) remained largely free of ice when glaciers more than a kilometre thick covered much of northern Europe, Greenland and North America.[15] Intact carcasses of mammoths (figure 27.1), dating from the Late Pleistocene, have been found buried in frozen ground in regions of Siberia north of the Arctic circle where no large animals can survive at the present time. In the Pleistocene, in contrast, sufficient vegetation must have been available, at least during the summer months, to provide food for herds of grazing animals. Investigations suggested that some of the mammoths had died by suffocation, so they might have been buried alive, before being frozen. There was certainly no reason to think that they froze to death as temperatures suddenly plummeted, as some writers of sensational accounts have imagined. Possibly some were the victims of landslides, whereas others may have fallen into mud-traps, or been drowned and then covered by silt[16] Nevertheless, even if the mammoths died during the final cooling episode of the Pleistocene, it still seemed strange to some that, as temperatures subsequently rose very significantly in other parts of the world, the land which had previously given sustenance to herds of large animals remained completely frozen, apart from the thawing of a thin surface layer during the brief summer season.

For this and other reasons, a number of writers have challenged the generally held view of Ice Ages, including the notion that the Pleistocene glaciations were long lasting, and a continuation of climatic trends which began much earlier. So, for

Figure 27.1   A mammoth of the species *Mammuthus primigenius*, based on a reconstruction in the Jardin des Plantes, Paris.

example, it has been suggested that the freezing and thawing episodes which occurred between 12,000 and 10,000 years ago were not part of an ongoing process, but the result of a catastrophic event. In their 1995 book, *When the Earth Nearly Died*, the retired British geologists Derek Allan (1917–2000) and Bernard Delair argued that the axial tilt of the Earth increased around 11,500 years ago, as a result of the close passage of a large cosmic body, which gave rise to the Phaeton legend.[17] They suggested that this was a product of the Vela supernova explosion, thought to have occurred in a part of the Galaxy close to our Solar System between 14,300 and 11,000 years ago.[18]

In addition, according to Allan and Delair, smaller companions of the main body passed through the atmosphere and struck the Earth whilst travelling in a northeast to southwesterly direction from Alaska to South America. They pointed to the presence along the supposed path of a large number of oval lakes with an appropriate orientation, including the Carolina Bays, following the example of Otto Muck (see Chapter 13) and also Allan Kelly and Frank Dachille (see Chapter 14) in claiming that these were impact craters. Although it is accepted that the Carolina Bays probably originated during the Pleistocene, it has generally been thought that they were created by a purely terrestrial mechanism, such as the effects of winds on what was then desert. However, it remained possible that they could have been produced by small, non-explosive impacts, although no direct evidence has been produced to support that hypothesis.[19]

Allan and Delair went on to claim that this encounter was the primary cause of the extensive volcanic activity of the Late Pleistocene, together with hurricanes, massive floods (the source of the various flood myths from around the world) and other catastrophic effects. The increased tilt of the Earth would have led to colder climates at high latitudes, made worse by the dust thrown up into the higher atmosphere as a consequence of impacts and vulcanism. In this scenario, as with that of the catastrophists of the early nineteenth century, the 'erratic' boulders and undifferentiated gravel deposits of northern Europe and the U.S.A. owed more to transport by flood-water than by glaciers. However, at high latitudes, flood-water which could not quickly drain back to the sea would have been trapped on the continents as ice-sheets, as temperatures plunged. Allan and Delair believed that this hypothesis provided a plausible explanation for the frozen mammoths of Siberia, as well as the extraordinary 'muck' deposits of Alaska (frozen, tangled masses of animal remains, molluscs, vegetation, volcanic ash and ice) and the similar mixed deposits packed into caves and gorges in more southerly regions. Over thirty years previously, René Gallant (see Chapter 14) had similarly suggested a major impact event as the cause of these phenomena.[20]

A completely different, but equally unorthodox, viewpoint on what happened during the Late Pleistocene was provided by the American science historian Charles Hapgood, who died in 1982. Hapgood's ideas were popularised during the 1990s by the British journalist Graham Hancock and the Canadian librarians Rose and Rand Flem-Ath, as we shall discuss later, in Chapter 28. We have already noted how Hapgood argued in a series of books, from 1958 onwards, that the entire crust of the Earth must, on occasions, have suffered slippage relative to the core, some areas moving into polar regions, whilst others did the opposite (see Chapter 12). The effects of this could have been more far-reaching than a simple exchange of climates. For example, the arrival of land over a pole, where previously there had been just frozen water, would cause the ice-cap to expand. More generally, a dislocation would stretch the crust in some regions and cause it to wrinkle in others, because the Earth is not a perfect sphere. This would result in the levelling of some old mountain ranges and the creation of new ones, as well as causing earthquakes and floods.[21]

Geologists were firmly of the opinion that the forces required to bring about a crustal dislocation would be so great as to be far beyond anything which could be achieved by processes known to be operating within the Earth–Moon system. Hapgood countered that there were, nevertheless, good reasons for thinking that such slippages had actually occurred. Making the assumption that the magnetic poles always remain close to the Earth's axis of rotation, and so give a good indication of the positions of the geographic poles at any particular time, he argued that palaeomagnetic evidence showed that the location of the polar regions had changed in significant fashion during the course of the Earth's history. Thus, it appeared that, around 80,000 years ago, an area of the Yukon district of Canada lay over the North Pole, to be replaced a few thousand years later by a region between Greenland and Scandinavia. Around 50,000 years ago, the pole was located in the vicinity of Hudson Bay, Canada, and then, between 17,000 and 12,000 years ago, another movement resulted in the situation as we know it today. On the same basis, an area off the coast of Queen Maud Land, Antarctica, apparently covered the South Pole about 80,000 years ago. The next region to be over the pole was the Ross Sea, followed by an area between Antarctica and Western Australia, and finally the present location on the main Antarctic continent.[22]

In Hapgood's scenario, Canada moved away from the North Pole at the end of the Pleistocene Epoch, whilst Siberia moved closer to the polar region, which could account for the frozen mammoths. Also, the fact that the pole was no longer covered by land could explain why the northern ice-cap became very much smaller at this time.

However, results of recent studies of ancient climates have provided strong support for the orthodox view of the Pleistocene Ice Ages. Indeed, these findings, based on oxygen-isotope determinations, are contrary to what would be expected from either the crustal displacement theory of Hapgood or the hypothesis of Allan and Delair, that the last Ice Age was an event of short duration, caused by a single, recent catastrophe. Water molecules contain one of two possible isotopes of oxygen, those incorporating the lighter one (oxygen-16) evaporating more easily than those incorporating the heavier one (oxygen-18). When temperatures are low and ice-sheets are spreading, trapping, as frozen snow, water taken by evaporation from the oceans, the oxygen-16/oxygen-18 ratio in the water remaining in the oceans will be relatively low. Conversely, when temperatures are high, and more water is being returned to the oceans from ice-sheets than is being removed by evaporation, the oxygen-16/oxygen-18 ratio in the oceans will be relatively high. The same ratios would be found in the shells of creatures living in the seas at the time, so the measurement of oxygen-isotope ratios in marine fossils gives an indication of the ocean temperature when they were living. Similar conclusions can also be drawn from oxygen-isotope ratios in the individual layers of the northern and southern ice-sheets, for it is generally believed that each layer was formed from the compressed snows of a single year.[23]

Oxygen-isotope ratios in the shells of microfossils in deep-sea cores have demonstrated temperature fluctuations throughout the Pleistocene and Late Pliocene, with even the highest average temperatures of these times being far lower than the typical temperature of the Miocene. A correlation in climatic events has been demonstrated between cores from the North Atlantic and ones from China.[24]

Similarly, ice-core data have demonstrated fluctuating but generally low temperatures throughout the last 100,000 years of the Pleistocene (see figure 24.3).[25] That is at odds with the belief of Allan and Delair that conditions had remained mild until the temperature suddenly dropped around 11,500 years ago. Comparisons of Antarctic and Greenland climates over this entire period suggest that the same glacial–interglacial sequences took place in both polar regions, and these were consistent with temperature changes in the oceans and climate changes in Africa. Although there were some variations in timing between events in the north and south polar regions and, indeed, between different locations in Antarctica, which may have been due to the effects of ocean currents, there were no times when climatic trends in Greenland and Antarctica were moving consistently in opposite directions, as might have been expected if the crustal displacement theory was correct. Furthermore, the Taylor Dome, Byrd and Vostok ice-cores, from different parts of Antarctica, all show, despite short-term fluctuations, a generally upward drift in temperatures between 20,000 and 10,000 years ago.[26] This was the period during which, according to Hapgood, the continent moved over the south pole so, according to his theory, the ice-cap should have been expanding, not shrinking, and Antarctica should have become significantly colder, not warmer.

In any case, modern geologists, with the benefit of greatly increased knowledge, see matters very differently from Hapgood. As it now appears, palaeomagnetic data from rocks of all ages (not just the Pleistocene) can be explained by the effects of two separate processes, neither of which was established at the time Hapgood originally formulated his theories (although he was aware of speculations about them). The first of these processes is an apparent change in position of a magnetic pole, resulting not from a crustal displacement but the much more plausible mechanism of continental drift. The tectonic plates carrying the continents move relative to each other at an average speed of around ten centimetres per year, which would give the appearance of a large-scale polar wander over a time-scale of millions of years. Over the past 80,000 years, the period on which Hapgood concentrated, continental drift could account for a movement of only a few kilometres, but the palaeomagnetic changes to which he drew attention can easily be explained by the second process, a genuine if relatively small-scale wander in the position of a magnetic pole within the confines of an area centred on the geographic pole. Since 1831, when the position of the magnetic north pole was first located, in northwestern Canada, it has travelled almost 800 km in the approximate direction of the geographic pole, whilst in the same time the continents have moved no more than a few metres. Thus, Hapgood's assumption that a change in position of a magnetic pole relative to the continents must be evidence of a crustal displacement can be seen to be false.[27]

For all these various reasons, it must be concluded that neither the single-catastrophe theory nor the crustal displacement one can be sustained. The evidence continues to indicate that a number of major cooling episodes, affecting climates in all parts of the world in similar fashion, took place at intervals throughout the two million years of the Pleistocene Epoch.

The Late Pleistocene extinctions
Extinctions of animal species had occurred throughout the Pleistocene, but were particularly marked at or near its conclusion, when temperatures rose higher than they

had been for the previous 100,000 years and, after a period of continuing fluctuations, eventually became much more stable.[28] The first marked rise in temperature occurred around 14,000 years ago, but then conditions started to become colder again during the final stage of the Pleistocene, known as the Younger Dryas, before a more sustained increase in temperature started around 11,500 years ago, heralding the start of the Holocene Epoch.[29]

As a whole, the Late Pleistocene extinctions were on a smaller scale than some earlier mass extinctions, such as those at the ends of the Permian and Cretaceous Periods, but large land animals were profoundly affected. In North America, thirty-three genera (three quarters of the total, including all the mammoths, mastodons, horses, tapirs and camels) disappeared between 12,000 and 10,000 years ago, and possibly in a much shorter time period than that. In South America over the same period, forty-six genera of large animals became extinct.[30] The extinctions in the Old World were more modest, but the long-horned buffalo, giant hartebeest and several other species of large animal disappeared completely from Africa at the end of the Pleistocene, and the European mammoth, woolly rhinoceros and giant deer became extinct at this time. Nine other genera disappeared from Europe but survived on other continents. The record in Asia is less well-documented but, as already noted, mammoths were prominent victims of the transition to the Holocene.[31]

All of this is generally agreed, the major ongoing argument being about reasons for the extinctions. Whatever causal mechanisms may have been involved, major environmental changes undoubtedly took place over the period in question, linked to increases in average temperature and changing patterns of rainfall. The Late Pleistocene extinctions in northern U.S.A. and Canada were coincident with the retreat of the ice-sheet of the Wisconsin glaciation (the American equivalent of the Riss/Würm) north of the Great Lakes, and with the replacement of spruce woodland and tundra by pine and deciduous species. In the southern part of North America, they were synchronous with the greatest change in climate and vegetation since the previous interglacial, which occurred 100,000 years earlier. In both Europe and Asia at middle latitudes, mammoths disappeared rapidly from the huge region that had previously been grassy steppe but was in the process of being converted to forested taiga. Further north, similar events took place as the relatively fertile steppe–tundra of Arctic Siberia changed to the boggy, vegetation-poor tundra we know today. Neither taiga nor tundra were suitable habitats for the mammoths. Even in the far north, contrary to what Hapgood, Allan and Delair had supposed, the immediate cause of this environmental transition may have been nothing more dramatic than a change in patterns of precipitation. In the Late Pleistocene, northern Siberia had a relatively dry climate (consistent with the absence of an ice-sheet), and the clear skies during the summer months allowed sunlight to penetrate. This facilitated the growth of a rich variety of plants, which could not be sustained when conditions became wetter at the end of the Epoch. Similar associations of extinctions with climatic changes are found throughout the world. Even in Australia, where the megafaunal extinctions occurred earlier than elsewhere, over 20,000 years ago, the death of the giant marsupials was synchronous with a long period of heat and drought. The end of the Pleistocene also saw a change to increased seasonality. This would have led to bigger climate changes at higher latitudes rather than near the equator, and the extinction pattern is consistent with this.[32]

However, another factor that cannot be ignored is the emergence of humankind, and its spread into new areas. Although there are hints that there may have been isolated settlements in the New World at an earlier time, it now seems clear, from both fossil and molecular clock evidence, that the main wave of settlers crossed from Asia into Alaska by means of a land-bridge less than 30,000 years ago, when sea-levels were low as a result of water being trapped as ice. They then spread over the northern and southern continents, reaching the southern tip of Chile around 10,000 years before the present time. The Clovis Stone Age culture of New Mexico was well-established around 11,000 years ago, and clear evidence of the systematic butchering of large animals has been found at some Clovis sites. Humans may also have reached Australia shortly before the time of the extinctions of giant marsupials and flightless birds there, although that remains uncertain.[33]

Arguments have been going on for many years about the causes of the Late Pleistocene extinctions. In particular, the possible contributions of climate change and human contact (humans killing animals directly by hunting and indirectly by introducing new diseases) have been much debated.[34] Regardless of which was the most significant, it is reasonable to conclude that both climatic and human factors must have played a part. That was the view of Peter Ward, in his 1997 book, *Call of Distant Mammoths*, after taking into account the latest computer modelling studies. Furthermore, the extinctions continued after the end of the Pleistocene. Ward wrote:

> Of one thing I am sure. There must be a connection between the cessation of mad temperature swings, 10,000 years ago, and the rise of human agriculture and civilization. And as we learned to sow and reap, surely our numbers rose as never before. With these changes our wandering and exploring may have even increased, enabling us to find the last regions of mammoth steppe, to sail the seas, to conquer the last great islands . . . to seek out the remaining refuges of the great mammals or flightless birds, to find these lost worlds, and to destroy them.[35]

Catastrophes during the Pleistocene–Holocene transition

What brought the Pleistocene Ice Ages to an end? According to the theory of Clube and Napier, the clearing of a cometary dust cloud from the atmosphere was a significant factor. Others, including Otto Muck (see Chapter 13), Fred Hoyle and Emilio Spedicato (mentioned earlier in the present chapter), suggested the involvement of an extraterrestrial impact in an ocean.[36] In 1997, the Italian mathematician Flavio Barbiero calculated that, given an Earth acting like a gyroscope with a liquid core, the impact of a 1-km-diameter asteroid, if striking from an appropriate direction, could change the positions of the points on the Earth's surface marking the axis of rotation, in rapid fashion. He believed this had happened 70,000, 50,000 and 11,600 years ago, explaining the palaeomagnetic changes noted by Hapgood although, as we have seen, an association between the change in position of the magnetic and geographic pole is unnecessary. Nevertheless, Barbiero maintained that the supposed change in the axis of rotation 11,600 years ago brought about the end of the Pleistocene Epoch.[37]

This was purely theoretical and speculative, but evidence for an impact at this time was claimed by the Russian geologist E.P.Izokh, in the form of tektites at the Pleistocene–Holocene boundary off the coasts of Australia and Vietnam. The Australasian tektites had actually been given much earlier radiometric dates, but Izokh argued that these, like the Vietnamese ones, had been formed in the outer regions of the Solar System around 770,000 years ago, and travelled to the Earth with the comet which he believed had terminated the Pleistocene.[38]

Similar arguments were put forward by the University of Vienna geologist Alexander Tollman, writing in collaboration with his wife, Edith Kristan-Tollman. The Tollmans believed that the Australasian and Vietnamese tektite fields were just two of seven caused by fragments of a comet which exploded in the Earth's atmosphere, landing in an ocean. They associated this event not so much with the Pleistocene–Holocene transition as with the legend of Noah's Flood, dating it at around 9,600 years ago on the basis of an acidity peak in a Greenland ice-core and a carbon-14 maximum (possibly indicating damage to the ozone layer) in tree-rings.[39]

However, subsequent examination of the Australasian and Vietnamese tektite fields by Eugene Shoemaker (just before his tragic death in a car accident) and others showed that the apparent location of tektites in Pleistocene–Holocene sediments was due to re-working, whereas they had actually arrived on Earth at a date consistent with their radiometric age.[40] Thus, there is still no convincing evidence of an impact which could have caused the end of the Pleistocene Ice Ages.

Indeed, many consider that the Pleistocene–Holocene transition was simply a continuation of processes which began thousands of years earlier, involving changes to currents in the Atlantic Ocean. At the present time, the Gulf Stream brings warm surface water from equatorial regions into the North Atlantic. However, in the region of Iceland, the density of surface water increases as a result of evaporation and the formation of ice, setting into motion what is known as the 'thermohaline circulation'. The cold, salt-rich water sinks to the bottom of the ocean, being replaced at the surface by warmer water from intermediate depths, which keeps air temperatures higher than would otherwise be the case. Meanwhile, the cold, dense water flows south along the ocean bottom until it meets a similar current coming from the Antarctic, when both are diverted to the east, past the southern tip of Africa. According to some, including the Columbia University oceanographer Wallace Broecker, this thermohaline circulation, although a long-established feature, may have been disrupted as temperatures rose at the end of the last glacial maximum, around 14,000 years ago. At that time, the surface water of the North Atlantic might have been so diluted by fresh water from the melting ice-sheets that it could no longer sink to the bottom. In consequence, water from the intermediate depths could not rise to the surface, so its warming influence was lost. That could have been the factor which sent the rise of temperatures into reverse, bringing about the start of the Younger Dryas around 13,000 years ago. Eventually, as less water flowed into the Atlantic from melting ice, the salt concentration of the surface water increased once more, to the point where the thermohaline circulation was re-instated, and the great final thaw of the Pleistocene Epoch began.[41]

This time, the melting of the ice did not stop the thermohaline circulation, so why it should have done so on the previous occasion, if indeed it did, is far from clear. That just reinforces the point made earlier in this chapter, that the beginnings and ends of Ice Ages, and the fluctuations in between, are likely to be complex affairs, involving a variety of factors.

Whatever events may have conspired to terminate the final Pleistocene Ice Age, and whatever role humans played in the Late Pleistocene mass extinction of animal species, it is clear that natural catastrophes occurred as the ice-sheets melted at the end of the epoch. So, for example, the retreat of the glaciers removed the barrier which held back the waters of Lake Missoula in western Montana, causing devastating floods over a wide area of the Columbia Plateau beneath the lake, and gouging out deep channels in the scablands of eastern Washington. This happened not once, but several times, as conditions fluctuated. When, in the 1920s, the Chicago geologist J. Harlen Bretz first suggested that the channels of the Washington scablands had been created by catastrophic floods, he was attacked by his colleagues, particularly those of the U.S. Geological Survey, for challenging the assumptions of the gradualist orthodoxy. Unless gradualist processes could be shown to be incapable of explaining a particular feature, which, they believed, Bretz had not done, then there was no reason to consider catastrophist ones. Thus, James Gilluly maintained that the channels could have been by produced by floods of the same magnitude as ones which still occur in the region of the Columbia river or, by floods 'at most a few times as large'. That, however, is no longer seen to be the case, given the short time-scale, and also (a fact not known at the time), a source for the catastrophic flood-water in glacial Lake Missoula. It is now believed that channelled scablands were also produced by waters released in similar catastrophic fashion from other glacial lakes in the western United States, such as Lake Bonneville, Utah.[42] As the American ice continued to melt, a super-lake, Algonquin, was formed in the northeast. This consisted of the present Lakes Superior, Michigan and Huron, but occupied a much greater area.[43]

On a more global scale, the release of water which had previously been stored as ice eventually led to a rise in sea-level of over 100 metres. For many years, it was generally assumed that this had been a gradual, even-paced process. However, it now seems that the deglaciation, and associated changes in the oceans, took place in rapid fashion. A team led by Jeffrey Severinghaus of the University of Rhode Island announced in 1998, on the basis of Greenland ice-core evidence, that, around 11,000 years ago, average temperatures rose by almost 10 °C in a short period of time, probably less than a decade.[44]

Low-lying regions throughout the world were flooded as sea-levels rose. Sometimes there was a long delay between cause and effect, increasing the catastrophic nature of the latter. So, for example, although it had generally been assumed that the Black Sea expanded in area and volume in a gradual fashion after the end of the Pleistocene, with excess water flowing in from the Atlantic Ocean via the Mediterranean Sea and the Bosporus as the ice melted, it now seems that the Black Sea was sealed off from the Mediterranean by a natural dam in the Bosporus region which eventually burst around 5600 B.C. Water then rushed into the Black Sea, flooding over 150,000 square kilometres of its low-lying coastal regions within a period of a year or so. Evidence for this was presented by geologists William Ryan and Walter Pitman, of Columbia University, in their 1999 book, *Noah's Flood – The New Scientific Discoveries about the Event that Changed History*. Previously, the Black Sea had been an oxygen-rich, freshwater lake, but the incoming salt-water sank to the bottom, causing anoxic conditions in the depths, a situation which still exists today. Radiocarbon dating studies on cores taken from the bed of the Black Sea at various locations have shown that oxygen-dependent shellfish living in deep water all became extinct around 5600 B.C., whilst saltwater molluscs made their first appearance in the Black Sea at exactly the same time. Ryan and Pitman argued that recollections of this catastrophic flooding, passed on by people who managed to escape and migrate towards Mesopotamia, gave rise to the Sumerian Epic of Gilgamesh and, in turn, to the *Genesis* story of Noah and his family. That remains controversial, and the evidence for the event itself has been challenged.[45] What is beyond challenge, though, is that, despite some continuing uncertainty about timescales, sea-levels throughout the world rose in significant fashion during the early Holocene.

28 Modern views of Atlantis

Berlitz and Kukal: conflicting arguments

Because Plato, in his *Timaeus*, wrote that the destruction of the island civilisation of Atlantis took place 9,000 years before the visit of Solon to Egypt, i.e. around 11,500 years before the present time (or 9500 B.C.), some writers have associated the supposed event with the Pleistocene–Holocene transition. As we noted in Chapter 13, there has been a long history of speculative theories concerning Atlantis, and that tradition has continued right up to the present day. Several best-selling books have claimed to have identified evidence of the destruction of an advanced civilisation at the end of the Pleistocene Epoch so, although most of the arguments put forward can be easily dismissed, the present work would not be complete without some brief consideration of them.

During the 1970s, the writer and linguist Charles Berlitz began to publicise the 'mysteries' of the Bermuda Triangle, in which he detected the influence of Atlantis.[1] However, most of the so-called mysteries turned out to have a very mundane explanation, as demonstrated by Lawrence David Kusche, in his 1975 book, *The Bermuda Triangle – Mystery Solved*.[2] Undaunted, Berlitz claimed, in 1984, in *Atlantis – The Lost Continent Revealed*, that the scenario presented by Plato was essentially correct, and the legend of Atlantis was 'now becoming a recognizable reality'. Much of what Berlitz wrote was intriguing, but always the hard evidence was lacking. A diver had apparently located a huge underwater pyramid near the Bahamas (see figure 22.1), but would not reveal its location. An island supposedly emerged from the Atlantic Ocean in 1882, complete with bronze swords and other artifacts, but was never found again, and the log of the ship which discovered it was destroyed in the London blitz in 1940. Buildings and a pyramid were said to have been found under an 8,000-year-old lava flow in Mexico, but the source of the information was not given.[3] Berlitz was presumably referring to the Cuicuilco pyramid, discovered by Byron Cummings of the University of Arizona in 1922. If so, the lava field in question, from the Xitle Volcano, is now generally regarded as dating from no earlier than 600 B.C., and probably from the first century of the Christian Era.[4]

Several pages of the book were devoted to a discussion of the so-called 'Bimini Road'. This structure, consisting of stone blocks stretching over several hundred metres, with a sharp curve near one end, was discovered under five metres of water off the coast of Bimini, in the Bahamas, in 1968. Much was made of the fact that the date, and the location, fitted in with Edgar Cayce's predictions about the rediscovery of Atlantis (see Chapter 13).[5] However, there was an element of self-fulfilment in this as the structure was found, albeit indirectly, as a consequence of intensive searches made by Cayce's followers in the Bimini region during the first of the years in question. In any case, the University of Iowa anthropologist Marshall McKusick, together with Eugene Shinn, of the U.S. Geological Survey, had presented evidence in *Nature* in 1980 to show that the 'Bimini Road' was a natural structure, formed long after the supposed time of Atlantis. Grains within the stones ran from each 'block' directly through to the next, and shells contained within them were dated by the radiocarbon technique to a period between 2,500 to 3,500 years ago. According to McKusick and Shinn, all the evidence pointed to the 'road' being beach-rock, originally laid down along the coastline of the time, and subsequently fractured and eroded to give the appearance of a series of stone blocks.[6]

Berlitz failed to mention this. Instead, he threw in a reference to various classics from the literature of 'unexplained phenomena', such as the lines and patterns marked out centuries ago over a large area of desert north of Nazca, Peru, to demonstrate that our knowledge of the ancient world is far from complete.[7] That proposition in itself can hardly be contested, but it does not prove the existence of Plato's Atlantis. Despite Berlitz's skilful attempts to create the impression that there must be some truth to the story, no single piece of evidence for it came near to being established. His style of argument was generally to present an association of loosely connected facts, many unproven, bringing them to a resounding but unjustified conclusion. That may be judged from a claim that, when the first Portuguese arrived in the Azores, they found that someone with knowledge of the Americas had been there before them. According to Berlitz, the Portuguese discovered the statue of a horseman, pointing west, on the island of Corvo. They dismantled it and sent the pieces back to Portugal, where they were lost. Berlitz then added, 'An unusual legend associated with the statue recounts that it was called Cate or Cates. This word resembles a word in a language not of Europe but of the New World, Quechua, the language of the Inca empire of South America. In Quechua cati means 'that way' – in other words towards the American continents.'[8]

No source was given for the story of the finding of the statue, but it was presumably based on an account by the Portuguese historian Manuel de Faria y Sousa. This was written in 1628, two centuries after the supposed event so, in the absence of other evidence, cannot be considered reliable. It is even less clear where the 'legend' of Cates came from. It could not have been transmitted via an oral tradition, because the Azores were uninhabited when the Portuguese arrived, and Sousa's account referred only to some indecipherable markings. Furthermore, Berlitz's statement that Cates 'resembles a word in a language not of Europe but of the New World' ignores the obvious similarity to Gades (Cadiz), the Spanish port founded by the Phoenicians. If anyone had visited the Azores before the Portuguese, the sea-faring Phoenicians must be considered amongst the most likely candidates.[9] So, even if the initial statement was correct, which is unsupported by evidence, the arguments which Berlitz tried to build upon it were far from convincing.

In the same year, 1984, Zdenek Kukal, of the Central Geological Survey, Prague, came to quite different conclusions from Berlitz in a 200-page article, entitled 'Atlantis in the light of modern research', which was published as a special issue of *Earth Science Reviews*. Kukal wrote:

> We have neither found a sunken island nor an ancient advanced civilization. Not a single trace! It appears that there was no Atlantis 11,500 years ago and if there was no Atlantis there was no destruction of Atlantis either. No trace has been found by anybody, including geologists and archaeologists, on either islands or the sea floor.[10]

The only land above or below the Atlantic Ocean which comes near to fitting the general description of Atlantis given by Plato is the extensive submarine plateau on which stands the nine islands of the Azores. It has an area of 135,000 square kilometres and, as we have seen, was considered to be the location of Atlantis by Muck and others. It is not on continental rock but welded to the Mid-Atlantic Ridge, so subject to frequent volcanic and seismic activity. However, the geological evidence suggests that the area has been rising rather than falling for millions of years.[11]

Berlitz was vague about the location of Atlantis and the cause of its destruction, despite giving sympathetic attention to Muck's asteroid impact hypothesis. However, others argued that Atlantis had not so much sunk at the end of the Pleistocene as been flooded by the rising ocean as the ice-sheets melted. The University of Miami geochemist Cesare Emiliani was one who, in an article in *Science* in 1975, linked the rapid rise in sea-level 11,500 years ago with the widespread flood myths, and with the destruction of Plato's island civilisation.[12] However, that scenario too cannot support the hypothesis that the Azores region was the location of Atlantis: the Azores plateau is now 2,000 metres below water, whereas the overall rise in sea-level since the last glacial maximum 18,000 years ago has been little more than a hundred metres.[13] Kukal suggested that the rate of elevation, despite showing variations, had never been greater than about 8 millimetres per year and, although a very different view is now taken about the rate of deglaciation around 11,500 years ago (see Chapter 27), it does not appear that the Azores plateau was ever in a position to have become submerged, whether 'in a single day or night' or over a longer time-scale. Even a tidal wave resulting from an asteroid impact in the ocean, as suggested by Muck, would have no relevance if the Azores plateau was already under more than a thousand metres of water at the time.

Archaeological evidence and 'alternative research'

Plato described Atlantis as an advanced, literate, metal-using civilisation capable of subjugating half the countries bordering on the Mediterranean. However, archaeologists have no knowledge of the existence of any such civilisation until 6,000 years *after* the supposed destruction of Atlantis. Apart from the dubious evidence of the equestrian statue mentioned above, there is no indication that the Azores were inhabited before the Portuguese arrived in 1427. The nearby Canary islands were settled earlier, but archaeological evidence, backed up by radiocarbon dating, limits human activity to the last few thousand years. Throughout the world, all of the human artifacts found at the level of the Pleistocene–Holocene transition are characteristic of Stone Age cultures.[14]

On the other hand, Berlitz and other 'alternative researchers' (i.e. ones without university posts, who challenge orthodox views of science and history) have pointed out that there are known instances of an artifact such as a screw being found at a level thousands of years before its time. However, this could be the result of a recent disturbance of the ground, or a fissure in a rock. If Plato's account had been correct, there should be much more evidence than an occasional, isolated finding of a small metal object in the 'wrong' stratum. Atlantis itself may have sunk beneath the waves, taking its inhabitants and its artifacts with it, but traces should have remained throughout its empire. Moreover, if Athens was able to defeat Atlantis in battle, it must have had a comparable culture. Plato was certainly correct when he said that the Athens of 11,500 years ago had higher and more fertile hills than in his own day, when floods had washed much of the soil into the sea. However, could it have had, as he claimed, a highly organised society with a permanent garrison of 20,000 armed men and women? If so, absolutely no traces have been found, either in Athens or elsewhere in Greece. Plato also claimed that these very early Athenians tilled the soil. It seems that the first deliberate attempt to cultivate crops, by the Natufian communities of southwestern Asia, did indeed occur towards the end of the Pleistocene, apparently as a consequence of the climate changes of the Younger Dryas, which made it impossible for them to continue with their hunter–gatherer way of life. However, there is no evidence of farming reaching Greece until several thousand years afterwards.[15]

Yet, throughout the rest of the 1980s and the 1990s, there was still much popular interest in the notion that, despite the lack of archaeological evidence, Plato's story was essentially true. Nor was this interest confined to less-educated members of the general public: Ken Feder, an archaeologist from the Central Connecticut State University, found in a survey carried out in 1984 that almost 30% of the students at his University believed there was good evidence for the existence of the lost continent of Atlantis, whereas less than 15% took the opposite view, the rest being undecided. Similar results were obtained when the survey was repeated in 1994.[16]

Feder, in his book *Frauds, Myths and Mysteries*, demonstrated that alternative researchers could (and often did) make even the most unlikely argument seem plausible to an undiscerning readership, by distorting some of the facts, and by failing to mention any of the evidence (sometimes of a substantial nature) which pointed to quite different conclusions. Similar points were made by the British historian and freelance writer

Peter James, in collaboration with Nick Thorpe, an archaeologist from King Alfred's College, Winchester, in *Ancient Mysteries*, and by American geologist Robert Schoch, in *Voices of the Rocks*, as part of balanced and open-minded investigations of Atlantis and other controversial issues.[17]

Regardless of this, books claiming to have identified the original source of human civilisation continued to appear. Even by the standards of alternative research, some of these were full of wild speculation, giving the credit to beings from other worlds.[18] More rationally, although not necessarily being much more convincing in their arguments, others claimed to have identified the location of the historical Atlantis, and/or the cause of its destruction. So, for example, writer Herbie Brennan, in his 1999 book, *The Atlantis Enigma*, accepted the arguments of Allan and Delair (see Chapter 27) that the Pleistocene had been brought to a catastrophic end by the arrival of fragments from a supernova explosion, and went on to argue that Atlantis had been destroyed in the process.[19] Emilio Spedicato, in his *Apollo Objects, Atlantis and the Deluge*, followed Cesare Emiliani's line and suggested that Atlantis was flooded during the rapid rise in sea-level at the end of the Pleistocene, which he believed to have been caused by an asteroid impact in the Atlantic Ocean. In Spedicato's scenario, Atlantis was Hispaniola, the large, mountainous island now divided into the territories of Haiti and the Dominican Republic (see figure 22.1). Hispaniola, of course, still stands above the waves, but Spedicato argued that Plato's story referred to the submerging of the capital city, on the former coastal plane to the south of the Dominican Republic.[20] Even before the rise in sea-level, however, Hispaniola covered a much smaller area than Plato's Atlantis.

In *Gateway to Atlantis*, published in 2000, writer Andrew Collins accepted much of Spedicato's thesis, but placed the capital city to the west of Hispaniola, on the island of Cuba. In his view, the most plausible location was the southwestern coastal plain, which was largely flooded as the Pleistocene ice-sheets melted, creating the Gulf of Batabano. This separated the citadel of Atlantis, on what is now the mountainous Isle of Youth, from the Cuban mainland. Like Otto Muck, as well as Allan and Delair, Collins associated the end of the Pleistocene with an extraterrestrial bombardment arriving from the northwest. Unlike them, however, Collins believed that the intruder was the disintegrating nucleus of a comet, fragments of which exploded in the air (as in Tunguska in 1908) to produce the Carolina Bays, whilst the main body impacted into the Atlantic. He accepted that there was no metal-using civilisation in the Caribbean (or elsewhere) at the time, but maintained, on the basis of myths, artifacts and cave-drawings of uncertain age, that there was a significant Neolithic culture, comparable to ones in the Near East, on Cuba and surrounding islands around 9000 B.C., which later transferred to North and Central America. Nevertheless, despite the claims from Bimini and elsewhere, Collins acknowledged that no-one had yet provided convincing evidence of submerged 11,000-year-old buildings in the region, let alone demonstrated the existence of an Atlantean city.[21]

Yet another variation on the theme that Plato's account told of the sinking of the capital city of Atlantis, rather than the entire country, was put forward by the cartographer Jim Allen, in his 1998 book, *Atlantis – The Andes Solution*. As the title of the book implies, Allen located Atlantis in South America. To be precise, he argued that the plain of Atlantis corresponded to the Bolivian Altiplano, with the capital city built around the lava rings of a volcano. According to Allen, it was this city that sank, in 'a single day and night', beneath the waters of a large inland sea, of which the present-day Lake Poopo is a remnant. Again, although with only a passing mention of Atlantis, the medical geneticist Stephen Oppenheimer suggested in his 1998 book, *Eden in the East*, that an influential early civilisation in southeast Asia was destroyed as sea-levels rose following the last Ice Age. This civilisation, which had a Neolithic culture, was located in the low-lying region of Sunda (see Chapter 26), of which the promontories and islands of the East Indies are the only parts now remaining above the water.[22]

Closer to Plato's account, Jacques Collina-Girard, a reputable scientist from the University of the Mediterranean in Aix-en-Provence, suggested in the journal *Comptes Rendus de l'Academie des Sciences*, in 2001, that Atlantis had been an island now referred to as Spartel, in the Atlantic west of Gibraltar, which disappeared as sea-levels rose at the end of the Pleistocene. However, even at its greatest extent, Spartel was no more than 14 km long and 5 km wide, far smaller than Plato's Atlantis.[23]

In 1995, in *When the Sky Fell*, Canadian librarians Rand and Rose Flem-Ath (originally Rand Flemming and Rose De'Ath) combined Plato's Atlantis story with Charles Hapgood's theories of catastrophic crustal displacement (see Chapter 27).[24] Hapgood was impressed by ancient maps, said to be copies of even older maps, which appeared to depict the Antarctic continent free of ice. This, he argued in 1966, in his best-known book, *Maps of the Ancient Sea Kings*, showed that Antarctica had moved into the polar regions, as a result of a crustal displacement, within the period of human civilisation.[25] However, Hapgood's interpretations of the ancient sea maps were challenged by orthodox historians, who suggested that what he had taken to be Antarctica was, in some cases, a part of South America, and, in others, an imaginary continent inserted into an unexplored region to fill up space. Also, as we noted in the previous chapter, modern scientific investigations have revealed the existence of hundreds of thousands of what are generally accepted to be annual layers of compressed snow in ice-cores from sites in various parts of Antarctica, indicating that the continent had never been ice-free during this period. Disregarding this scientific evidence, the Flem-Aths maintained that Antarctica had indeed moved into the south polar region less than 20,000 years ago, as Hapgood had suggested. Their main thesis was that Atlantis was Lesser Antarctica, the smaller of the two main land masses making up the Antarctic continent. Hence, the ancient island civilisation disappeared around 9500 B.C., as suggested by Plato, not because it sank beneath the waves, but because it became covered by a thick layer of ice.[26]

If this theory was true, all traces of the Atlantean civilisation on Lesser Antarctica would now be well hidden from our eyes. Hence, direct supporting evidence would be impossible to find. However, according to the Flem-Aths and Colin Wilson, who joined with Rand Flem-Ath in writing a follow-up book, *The Atlantis Blueprint*, published in 2000, there may be indirect evidence of it at other locations, including South America and Egypt.[27] Some other alternative researchers took the same view. For example, Graham Hancock, writing in support of Hapgood's thesis in 1995, in *Fingerprints of the Gods*, considered the temples constructed from massive stones at Tiahuanaco, near Lake Titicaca, in the mountainous Bolivia–Peru border region, to be extremely old, citing the arguments of the Austrian engineer Arthur Posnansky, who made a detailed investigation of the site over four decades, starting in 1904. Posnansky's conclusions are believed by conventional scholars to have been discredited by further work carried out at the site, but Hancock failed to mention this in his book (as did Wilson and the Flem-Aths in theirs). Hancock simply repeated Posnansky's view that astronomical alignments suggested that construction took place more than 10,000 years ago. Similarly, he claimed that Tiahuanaco was intended to function as a port, whereas it now stands twenty kilometres from Lake Titicaca, and over thirty metres higher than the present water-level. If the claim was true (which is considered unlikely by orthodox scholars), it indicates that Tiahuanaco was built a long time ago, but whether that was around 500 A.D., as generally supposed (a belief supported by radiocarbon dating evidence), or many thousands of years earlier, remains a matter for conjecture.[28]

It is clear, therefore, that the evidence for a very ancient civilisation at Tiahuanaco does not stand up to close scrutiny, but what about the possibilities elsewhere? The Flem-Aths, in *When the Sky Fell*, drew attention to the many catastrophe myths originating in different parts of the world, and the traditions of a 'golden age' preceding these catastrophes. Similarly, Hancock pointed out that many traditions from the Andes of South America refer to a character called Viracocha or Kon Tiki, who appeared after the deluge and taught the local populations, then little more than savages, the essentials of medicine, metallurgy, farming and the art of writing, setting them on the road to civilisation. According to some of these stories, Viracocha was tall, fair-skinned and bearded, very different from the indigenous Indians. Furthermore, there was a very similar figure in the legends of Central America, who was known as Quetzalcoatl or Kukulkan, and yet another, called Osiris, in Egyptian mythology. These characters, reputed to be founders of civilisations on both sides of the Atlantic Ocean, could have been refugees from some lost civilisation, perhaps Atlantis, following its destruction in a catastrophe.[29]

However, arguments of this nature, based on myths and legends of uncertain age and origin, could only really be of value as a support to more direct evidence for a particular theory, if such was available. It is far from established, for example, that South American legends of fair-skinned, bearded gods were formulated before the arrival of the Spaniards. Nevertheless, the possibility, however remote, that a great civilisation was brought to an end by a natural disaster 11,500 years ago cannot be ignored in a consideration of how catastrophes might have influenced the development of life on Earth.

Hancock and Bauval: evidence for an extremely ancient civilisation?

An apparent attempt to provide detailed evidence for a very early lost civilisation was made in 1996 in *Keeper of Genesis* (known in the U.S.A. as *The Message of the Sphinx*). *Keeper of Genesis* was written as a collaborative effort by Hancock and the Egyptian-born Belgian construction engineer Robert Bauval. The latter had previously been co-author (with writer Adrian Gilbert) of *The Orion Mystery*, which was published in 1994. *Keeper of Genesis* developed the arguments of *The Orion Mystery*, the starting point being the claim that the arrangement of the three main pyramids at Giza, Egypt, with the smallest (that of Menkaura) slightly off the line of the other two (those of Khufu and Khafra), was a representation of the three stars in Orion's Belt, constructed in relation to the Nile as a representation of the Milky Way.[30] Although both books were highly readable, they had a tendency to stretch points beyond reasonable limits, and to resort to circular reasoning. They were also inclined to interpret evidence in a cavalier fashion, and to use numbers in an unscientific way to bolster weak arguments. To some extent, the style in each case resembled that of the books of Berlitz, where each speculation became, by implication, the 'fact' on which the next speculation could be based. The problems with the arguments put forward have been well documented by various authors (including the present one).[31] Here we shall consider just the main points relating to our theme.

As Bauval and Hancock made clear in *Keeper of Genesis*, they went along with the orthodox view that much of the Giza pyramid complex in Egypt (figure 28.1) was built during the fourth dynasty, by its second pharaoh, Khufu, his son, Khafra, and grandson, Menkaura, around 2500 B.C. However, they claimed that some of the features, including the Sphinx, dated from earlier times.[32]

Following American writer John Anthony West, who in turn was influenced by the writings of René Schwaller de Lubicz (1887–1961), a French mystic, Nazi guru and amateur Egyptologist, Bauval and Hancock believe that the Sphinx is much older than many of the surrounding structures because it shows more signs of erosion.[33] Most geologists (including Lal Gauri and James Harrell, who have extensive experience of fieldwork in Egypt) are not impressed by this argument, because wind and sand erosion is an erratic process.[34] Although the British civil engineer and geologist Colin Reader has controversially argued that the very earliest phase of the Sphinx Temple and, by implication, the Sphinx itself, predates the fourth dynasty, he was suggesting re-dating the construction of the Sphinx by no more than a few hundred years, not the many thousands of years proposed by Bauval and Hancock.[35]

Nevertheless, one academic geologist, Robert Schoch of Boston University, has come out in favour of a more radical scenario, as a consequence of his belief that

Figure 28.1   The author, Trevor Palmer, on camel at Giza, with the pyramid of Menkaura on the left, that of Khafra in the centre and part of that of Khufu on the right.

the erosion of the Sphinx was caused by prolonged rainfall. Since the climate of Giza is generally thought to have been exceedingly dry for the past 4,000 years, and only moderately wet for 3,000 years before that, he concluded that the Sphinx must have been constructed between 7000 B.C. and 5000 B.C. This was the period of the Nabtian pluvial, when the Sahara was fertile savannah, and a megalithic Stone Age culture existed on the shores of a lake at Nabta, in southern Egypt.[36]

However, as West acknowledged, there are no indications that a civilisation capable of producing the Sphinx existed in Egypt between 7000 and 5000 B.C. (although the first towns, e.g. Çatal Hüyük and Jericho, were starting to appear elsewhere, as centres for Stone Age farming communities[37]). West tried to overcome this problem in a strange way, suggesting that the Sphinx was the creation of an even earlier civilisation, whose traces are buried where no-one has yet looked.[38] Bauval and Hancock used similar logic to argue that, because we cannot easily explain how the megalithic pillars and architraves in the Valley Temple of Khafre next to the Sphinx were put into place with the technology of 2500 B.C., this too must have been erected much earlier, with both it and the Sphinx dating from 10,500 B.C.[39]

The reason for choosing this particular date apparently followed from the belief, discussed in *The Orion Mystery* and *Fingerprints of the Gods*, that the Giza complex and surrounding features formed a vast map of the sky. In *Keeper of Genesis*, Bauval and Hancock investigated the possible significance of alignments in the sky and on the ground, extrapolating current processes back in time. They concluded that there was a perfect meridian-to-meridian alignment of the dominant axis of the Orion's Belt/Milky

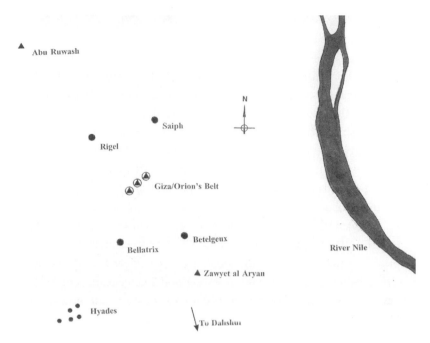

Figure 28.2   Stars in the Orion region of the sky superimposed on a map of
pyramids in the Memphis region of ancient Egypt, on a scale which gives the best
'fit' between the three stars of Orion's belt and the main pyramids at Giza (stars
are indicated by circles, pyramids by triangles).

Way region with that of the corresponding features on the ground in 10,500 B.C. but not
in 2500 B.C.[40] However, the reality is that, regardless of what they meant by 'dominant
axis', which was not made clear, the correspondence between the alignment of the
Giza pyramids and Orion's Belt (in 10,500 B.C.) to their respective meridians is far
from exact.[41]

Bauval and Hancock also claimed that, as part of the ancient Egyptians' supposed map of the heavens, the two large pyramids built at Dahshur by Sneferu, the first
pharaoh of the fourth dynasty, were representations of stars of the Hyades (the head of
Taurus). They wrote, 'If we regard the Giza Pyramids (in relation to the Nile) as part of
a scaled-down 'map' of the right bank of the Milky Way, then we would need to extend
that 'map' some 20 miles to the south in order to arrive at the point on the ground where
the Hyades–Taurus region should be represented. How likely is it to be by accident that
two enormous Pyramids – the so-called 'Bent' and 'Red' Pyramids of Dahshur – are
found at this spot?'[42] The reality, however, was that, instead of being in the correct
location, Dahshur was much too far away from Giza, and to the southeast rather than
the southwest (figure 28.2)[43]. A further claim that the 'pattern on the ground' formed
by the Bent and Red pyramids 'correlates very precisely with the pattern in the sky of
the two most prominent stars in the Hyades' is not, in itself, very much to get excited

about, regardless of whether it is correct.[44] In fact, as the Canadian Egyptologist, Robert Chadwick, has pointed out, 'Taurus only matches the layout of the Dashur pyramids when it is rising in the east, and Orion only matches the Giza pyramids when it is on the meridian, ninety degrees away. What kind of correlation is that?'[45]

In *The Orion Mystery* it was suggested that other components of the same scheme included the unfinished pyramid on a hill-top near Abu Ruwash, built by Djedefra (the pharaoh between Khufu and Khafra) and another incomplete fourth dynasty pyramid at Zawyet al Aryan. Bauval wrote, 'I laid out a map of the Memphis area and compared it with a picture of the region of the sky containing Orion. Carefully aligning the Giza group pyramids with the stars of Orion's Belt, I saw that the pyramid . . . at Abu Ruwash corresponded with the star Saiph . . . and that at Zawyat al Aryan represented Bellatrix'.[46] Again, though, that is not the case, for neither pyramid is even close to the precise location required. The Abu Ruwash one, for example, is around twice the appropriate distance from Giza, and too far west (see figure 28.2).[47] As for pyramids which could have been representatives of Rigel and Betelgeux, the two brightest stars of Orion, Bauval acknowledged that he had no candidates to put forward, suggesting that all traces might have been buried by sand or taken away.

So, there seems little reason for thinking that the pyramids around Giza were intended as components of a map of the sky in 10,500 B.C., or at any other time. All that remains is the possible correspondence between the three stars of Orion's Belt and the three main pyramids of Giza, and even this is far from certain. Some discrepancy in the relative positions is found, and there is little correlation between the sizes of the pyramids and the relative brightnesses of the stars. If there was any particular design in the lay-out of the pyramids, that could have been simply to build them such that the southeast corner of each fell on the same straight line. Such an alignment is indeed found, even though the centres of the three pyramids are offset.[48]

In general, it seems far more plausible to think that the ancient Egyptians built their pyramids where they did because those locations provided firm ground and prominent positions, rather than because they corresponded to certain stars in a sophisticated scale-map.

Apart from the orientation of the site, the arguments for Giza being under the control of civilised people from as early as 10,500 B.C. centred on the Sphinx itself. This, with its leonine body, faces due east and at sunrise on the spring equinox would have gazed directly towards the constellation of Taurus the bull in 2500 B.C., whereas in 10,500 B.C. it would have faced the more appropriate Leo.[49] However, to see significance in that assumes that the ancient Egyptians associated the constellation Leo with a lion, which is far from certain. Our present zodiac derives from one developed in Greece around 400 B.C. and, although there were earlier versions in Babylon and Egypt, their details are unclear. The best-known Egyptian zodiac (now in the Louvre Museum, Paris) is from the temple of Hathor at Dendera, but this dates from no earlier than the first century B.C.[50]

In a discussion of Orion, Bauval and Hancock quoted the American astronomer Virginia Trimble, saying, 'What constellations the Egyptians saw in the sky is still something of a mystery . . . but they had one constellation that was an erect standing man, Osiris the god. And the one candidate that looks like a standing man to everyone is Orion.' They did not challenge the general point made in that statement yet, on the very next page, they wrote about Taurus and Leo as if it was established that the ancient Egyptians associated these groups of stars with the same figures we do.[51]

Bauval and Hancock maintained that a group of educated astronomer priests (perhaps the 'Followers of Horus' of Egyptian mythology) based close to Giza at Heliopolis, constructed the Sphinx and Valley Temple in 10,500 B.C. to mark the establishment of the kingdom of Osiris. It was unclear whether this was meant literally or whether it alluded to the fact that the celestial Osiris, Orion, reached the lowest point in its precessional cycle around this time.[52]

As for the major construction work around 2500 B.C., Bauval and Hancock suggested that this followed the realisation that the spring equinox sunrise had moved into the Hyades–Taurus region of the sky.[53] The pyramids were built with remarkable precision, each of the four sides of the Great Pyramid of Khufu, for example, being the same length, to within a few centimetres. Nevertheless, we now know enough about its construction to remove any need for fanciful explanations.[54] Bauval and Hancock tried to suggest otherwise, by greatly exaggerating the precision involved. Even though the top of the Great Pyramid is now missing, its limestone facing-stones long-since removed and the underlying building blocks badly worn and damaged, they claimed to know its original height to within a tenth of a millimetre, and argued that the dimensions had been carefully specified to encode aspects of advanced knowledge. In this and other ways, they tried to maintain that the construction was supervised by the long-enduring group of astronomer priests, who had access to seemingly miraculous technologies.[55] However, if so, what were they, why were they kept so secret, why were they used so sparingly over immense time-intervals, and what subsequently happened to them? Bauval and Hancock did not appear to be prepared to hazard opinions on these questions.

They were, however, happy to speculate about other matters. For example, they claimed that just before dawn at the spring equinox in 10,500 B.C., Leo would have risen to sit on the eastern horizon. At that precise moment, the Sun would have been 12° below the horizon, directly under the back paws of the lion. They therefore suggested that there is a very important chamber, which they called the Genesis Chamber, still to be discovered thirty metres beneath the rear paws of the Sphinx.[56]

Indeed, there was much talk in *Keeper of Genesis* of undiscovered passages and chambers. It was said that Zahi Hawass, Director General of the Giza Pyramids, was currently investigating newly discovered tunnels immediately southeast of the Sphinx. Edgar Cayce had claimed a collection of records from the lost civilisation of Atlantis was buried at Giza in 10,500 B.C., and the American visionary predicted that a passageway leading from the right forepaw of the Sphinx to the Record Chamber would be discovered before the end of the twentieth century. A researcher once funded by the Edgar Cayce Foundation, Mark Lehner, is now an eminent Egyptologist with conventional views, happy to date the whole Giza complex, including the Sphinx, to around 2500 B.C. The reasons for his conversions to orthodoxy, detailed in correspondence published in *Keeper of Genesis*, seemed to puzzle Bauval and Hancock. Lehner claimed simply that he became convinced after several years of investigation that the entire complex was planned and built during the fourth dynasty.[57]

Apart from references to the visions of Cayce, *Keeper of Genesis* avoided mention of the central thesis of *Fingerprints of the Gods*, that the development of civilisation in Egypt (and other places) owed much to the arrival of refugees from Atlantis, following its destruction in a natural catastrophe at the end of the Pleistocene Epoch, when climates changed in dramatic fashion throughout the world.[58] Indeed, the date of 10,500 B.C., although the one envisioned by Cayce for the sinking of Atlantis, was a thousand years earlier than that given by Plato for the disappearance of the island kingdom.

However, in 1998, in *Heaven's Mirror*, Hancock (writing this time in collaboration with his wife, the photographer Santha Faiia) returned to the *Fingerprints of the Gods* theme of ancient sites being constructed on the basis of information supplied by the scattered remnants of a destroyed civilisation, perhaps with the intention of conveying information to future generations.[59] He even tried to extend it to the Angkor complex in Cambodia, referring misleadingly to the 'pyramid-temples' there, despite the fact that their high, pointed towers are nothing like the squat pyramids of Egypt and Central America. Following the example of Ignatius Donnelly (see Chapter 13), Hancock suggested that the building of pyramids on both sides of the Atlantic indicated a common culture, brought by refugees from a destroyed civilisation.[60] However, although the pyramids of Egypt and Central America had much more in common with each other than with the temples of Cambodia, there were still significant differences in style and, apparently, function between them. For example, many of the Central American pyramids, unlike the Egyptian ones, had external stairways and platforms at the top for human sacrifice. Furthermore, by the time Hancock wrote his book, it was well established that the pyramid-building period in Egypt began no less than 7,000 years after Plato's date for the supposed destruction of Atlantis (and 8,000 years after 10,500 B.C.), whilst the corresponding period in Central America was more than a thousand years later still. Even that was a very long time ago compared with the construction of the Angkor complex, which started as recently as the twelfth century of the Christian Era.[61]

Disregarding these problems for his theory, Hancock selected ten from amongst the many 'pyramid-temples' at Angkor (which number more than sixty) and argued that they were representations of the constellation Draco, aligned as it would have been in 10,500 B.C. However, even with such a large number of structures to choose from to get a good 'fit', the correspondence between the positions of the temples selected and the stars in the constellation was far from exact. Claims were also made in *Heaven's Mirror* that the stone statues of Easter Island in the Central Pacific were products of the same ancient civilisation, yet these were still being constructed three centuries ago, and their alignment involved nothing more astronomically sophisticated than, for example, being placed in rows to face the setting sun.[62]

In an attempt to give some scientific credibility to his speculations, Hancock drew attention to a paper by the Russian astronomer Alexander Gurshtein, which appeared in the May 1997 issue of *American Scientist* (not the more widely circulated *Scientific American*, as stated by Hancock). In this paper, Gurshtein had put forward what he termed 'provocative hypotheses' that some of the present signs of the zodiac may have been recognised earlier than usually thought, linking the time of their naming to dates when the Sun lay within their boundaries at equinoxes and solstices. On this basis, Sagittarius, Virgo, Gemini and Pisces could have been identified between 6500 B.C. and 4400 B.C., and Scorpio, Leo, Taurus and Aquarius between 4400 B.C. and 2200 B.C., leaving the remaining four to be named at around the time generally supposed. Hancock did not appear to have read Gurshtein's actual paper, for the only reference he quoted was to a newspaper article about it. Perhaps for this reason, he stretched the Russian's arguments somewhat, claiming, 'The full implication of Gurshtein's argument is that these constellations could have been recognized – very much as we see them today – as early as 6000 B.C.'. That was much more than Gurshtein was suggesting but, even so, it still left Hancock 4,500 years short of a 10,500 B.C. zodiac.[63]

Nevertheless, Hancock thought that the site of a 12,000-year-old civilisation had been found in shallow waters off the Japanese island of Yonaguni, between Okinawa and Taiwan. He arranged for the geologist Robert Schoch, who had challenged the orthodox dating of the Sphinx at Giza, to visit Yonaguni to examine the apparent artifacts there. Schoch, however, concluded that they were natural geological formations, much to Hancock's disappointment.[64]

Writing in 1999 in *Voices of the Rocks*, Schoch concluded that Plato probably used aspects of several unrelated events to compile his story of Atlantis. If one of his sources really was a report of human conflicts many thousands of years before his time, as he claimed, it could only have referred to a situation such as that suggested in 1986 by prehistorian Mary Settegast, involving battles between different groups of Stone Age people. She suggested that Plato's Atlanteans were in fact the Magdalenians, who lived in southern France and northern Spain between 18,000 and 9,000 years ago, and are renowned for their cave paintings, bone and antler carvings, range of stone-tipped weapons and the invention of a device to allow a spear to be thrown with greater force.[65] No matter how cultured the Magdalenians might have been for their time, or how ferocious and extensive the fighting between them and their neighbours (which can only be a matter for conjecture), this scenario contains few of those features of

Plato's story which have caught the imagination of many people over the years. There was no city of Atlantis, no island civilisation, no stepping stones to other continents, no use of metals and no wars between Atlantis and Athens – just conflicts between different Stone Age tribes.

Despite the many thousands of pages devoted to the subject by Bauval and Hancock, separately and together, and by other authors such as Colin Wilson and the Flem-Aths, in their various books, there is still no convincing evidence of a civilisation flourishing around 10,500 B.C. or, indeed, for thousands of years afterwards. The various arguments and claims, powerful though they may seem when first encountered, collapse and disappear under closer examination. For example, the claims that new passageways and chambers had been found at Giza were investigated by Ian Lawton and Chris Ogilvie-Herald, who, as reported in their 1999 book, *Giza: The Truth*, concluded that 'much of the rumour mill surrounding the Plateau is at best misleading and at worst complete garbage'.[66] Bauval's 1999 offering, *Secret Chamber*, did not provide any significant evidence to the contrary. It was suggested that something dramatic would happen as the world entered the new millennium, but nothing did. Similarly, although Hancock's 2002 book, *Underworld – Flooded Kingdoms of the Ice Age*, avoided the bizarre speculative excesses of some of his previous works, concentrating on searching for underwater structures which might have been man-made, he failed by some distance to substantiate his thesis, i.e. to establish that any of the structures were traces of cities which had been covered by rapidly rising seas when the last Ice Age came to an end.[67] So if, as seems likely, only scattered Stone Age communities existed at the time, the fall of an advanced Atlantean civilisation cannot be counted amongst the natural catastrophes which undoubtedly occurred during the Pleistocene–Holocene transition.

Could Atlantis have been a Bronze Age civilisation?

As we saw in Chapter 13, some have suggested that the destruction of Atlantis occurred in the Bronze Age, only a few hundred years before the time of Plato, possibly caused by the eruption of Thera. However, there were few points of similarity between Plato's Atlantis and either Thera or Crete. Angelos Galanopoulos claimed that many of the details of Thera matched Plato's account of Atlantis, but only on the assumption that Plato had exaggerated his measurements of size, like his time-scale, by a factor of ten.[68] Even then, there was no mention of any volcanic eruption in Plato's story. Furthermore, the supposed linkage between the eruption of Thera and the fall of the Minoan civilisation was far from secure. Whereas, at the First International Conference on Thera in 1971, the prevailing view was that there was a direct link between these two events, by the time of the Third International Congress on Thera, held in 1989, it was clear from archaeological evidence that the Minoan civilisation on Crete was *not* destroyed by the eruption of Thera. Indeed, it survived for another fifty years or more, albeit in a weakened state, and perhaps ruled by a new dynasty.[69]

From Aristotle onwards, some have argued that everything Plato wrote about Atlantis was a complete fiction, designed to allow him an opportunity to express his views on how an ideal society might operate. So, for example, Kukal, in his 1984 review, concluded that the Atlantis story was largely a product of Plato's imagination, incorporating for dramatic effect incidents from the war between Greece and Persia in 490 B.C., and from the ongoing struggle between Carthage and Greek colonists for the control of Sicily. Aspects of the Atlantis story might also have been suggested by the destruction of the Greek city, Helike, by earthquake and flood during Plato's lifetime, in 373 B.C.[70]

In *The Flood from Heaven*, published in 1992, the German geoarchaeologist Eberhard Zangger refused to accept the 'fiction' hypothesis, largely because the Atlantis civilisation described by Plato was far from ideal, being one with corrupt and tyrannical rulers. Moreover, Zangger's previous archaeological investigations at Bronze Age Tiryns, in the Greek Argolid, had revealed a devastating flood and possible earthquakes near the end of the Late Helladic IIIB period, which had distinct echoes of the Atlantis story. The submerging of Atlantis is well known, but Plato's account also referred to the disastrous consequences of earthquakes and floods in Attica, which is not far from the Argolid. Accordingly, Zangger began to look for a Late Bronze Age source for the Atlantis legend, and he gave his solution in *The Flood from Heaven*. In doing so, he linked 'the most famous legendary place in the western world' with 'the most famous archaeological site'. In other words, to Zangger, Atlantis was Troy.[71]

Zangger was not, of course, arguing that Troy was located on a large island in what is now the Atlantic Ocean. He was happy to go along with the general, if not universal, view that it centred on the mound of Hisarlik, on the coastal plain to the south of the Dardanelles (the Greek Hellespont). How, then, was this to be reconciled with Plato's account? Zangger had four main answers. The first was that the 9,000 years mentioned in the Egyptian records referred to lunar rather than solar years (an idea first suggested by Pedro Sarmiento de Gamboa in 1572), which would have placed the conflict with Atlantis at around the time of the Trojan war. Secondly, the ancient Egyptians used the term 'island' in a much vaguer way than we do, so did not necessarily mean a land completely surrounded by water. Thirdly, it is possible from the Greek words used that, instead of Atlantis being 'larger' than Libya and Asia Minor, it was simply 'of greater significance' than them. Fourthly, Solon had translated some place names from those originally given into what he thought was the Greek equivalent, thus probably confusing the issue for Plato and for us. This might be the reason, for example, why Athens seemingly played a much more prominent role in the conflict with Atlantis than in the Trojan war.[72]

According to Servius, in a commentary on Virgil's *Aeneid* written around 400 A.D., there were at least two separate locations for the 'pillars of Heracles', the well-known set, the Rock of Gibraltar and the Monte del Hacko, bordering the straits leading to the Atlantic Ocean, and another lining the entrance to the Black Sea (i.e. past

Troy). Solon and Plato would have been aware of both of these, whereas there is no evidence that Mycenaean Greeks (Achaeans) of the Late Bronze Age knew of the former. Nevertheless, Zangger surmised, Solon assumed the Egyptian priest was referring to the ones in the west, and he misled Plato into thinking likewise. In fact, as *The Flood from Heaven* went on to argue, the description of the 'ocean' beyond the pillars of Heracles in the *Timaeus*, and in particular the statement that 'the land surrounding it may most rightly be called, in the fullest and truest sense, a continent', fitted the Black Sea better than the Atlantic Ocean. After the destruction of Atlantis, it was said that this ocean was no longer navigable, which, as Zangger went on to speculate, could have been a reference to the loss of Trojan pilots capable of guiding ships through the difficult passage into the Black Sea. Moreover, according to both Apollodorus and Homer, the Trojan kings traced their descent from Atlas, giving a positive reason why Atlantis might have been an alternative name for Troy.[73]

Many of the geographical details of Atlantis described in the *Critias* corresponded with features of the coastal plain around Troy, including the use of artificial canals (according to Zangger's interpretation of the archaeological evidence), the prevailing northerly winds (still very noticeable today) and the presence of hot and cold springs (now presumably dried up, but mentioned in Book 22 of the *Iliad*). However, if Troy was Atlantis, its influence must have been greater than hitherto realised, and it may have provoked the Trojan war by its military aggression. Also, its buildings must have extended way beyond the tiny mound of Hisarlik, so there should still be exciting archaeological discoveries to be made in that region. Zangger speculated that the lower town of Troy might have been covered by alluvial deposits during the period of exceptionally heavy rainfall which caused the flooding at Tiryns.[74]

Peter James (figure 28.3), in his 1995 book, *The Sunken Kingdom*, accepted that Zangger was right to look east rather than west for Atlantis. He pointed out that many legends and names of locations (including the Atlas Mountains and the Pillars of Heracles) were transferred westwards from their origins during the period of Greek colonisation. Nevertheless, in his view, Troy could not have been Atlantis, for there are no legends coupling the Greek conquest of Troy with a natural catastrophe. On the other hand, there *are* legends of severe earthquakes and associated floods causing devastation in the Troad, and more southerly regions of Asia Minor, some three or four generations before the Trojan war. This was in the time of the Lydian king, Tantalus, whose city, called Tantalis, was destroyed by an earthquake and then flooded by the waters of a newly created lake.[75]

James argued that Tantalus was another name for Atlas, the supposed founder of Atlantis. The identification of Tantalus with Atlas had been suggested before, for the two characters have several similar legends associated with them, in particular ones where they are made to support a huge weight (perhaps the Earth, a celestial body, or the sky itself). Both names are derived from the verb *tláo*, meaning 'to carry'. On this basis, therefore, Tantalis, the ill-fated city of Tantalus, was none other than Atlantis.[76]

Figure 28.3   Ancient historians Peter James (left) and Bob Porter during a break at the Second SIS Cambridge Conference in 1997 (see Chapter 29).

James found reasons for placing Tantalis in the district of Magnesia, on the far (northern) side of Mount Sipylus from the coastal city of Smyrna (the modern Izmir). Pausanias, a native of Magnesia, wrote in the second century A.D. that a city near Mount Sipylus had once been destroyed by an earthquake, this setting in motion events which flooded the ruins and created Lake Saloe. Pausanias did not actually name the city, but it is clear from the writings of Pliny that Tantalis was believed to lie under that particular body of water. A lake marked on a late nineteenth century map of the region by Georges Perrot and Charles Chipiez was independently identified as Saloe by Sir William Ramsay and James Frazer. It has since been drained, but other features described by Pausanias can still be recognised. These include the 'throne of Pelops', cut into the rock on the summit of Sipylus; 'weeping Niobe', a natural rock formation resembling a sorrowful woman; the 'grave of Tantalus', later called the tomb of St Charalambos, cut into the side of the mountain; the 'crack in the mountain', now known as the ravine of Yarikkaya, from which the flood water is supposed to have poured after the earthquake; and the 'Mother of the Gods', a ten-metre-high carving with Hittite inscriptions, positioned one hundred metres up the north face of Sipylus, overlooking the presumed site of Lake Saloe, and hence of Tantalis.[77]

James went on to argue that Tantalis was the capital of Zippasla, known from Hittite records as a troublesome western neighbour. With the help of the Mycenaeans, Zippasla expanded and challenged Hittite authority for a brief period in Late Bronze Age times, then disappeared from history. James suggested on philological grounds that Zippasla = Sipylus. If this identification was correct, the end of Zippasla as a powerful force could be attributed to the natural disaster which destroyed its capital city.[78]

But where did that leave the reader in the search for Atlantis? Plato claimed that his source for the Atlantis story was his ancestor, Solon, who heard it from a priest in the Egyptian Delta. However, on the same tour which took him to Egypt, Solon also visited King Croesus in Lydia. Hence, James argued, Plato might have confused the locations, with Solon actually hearing the story in the very land once ruled by Tantalus, a land (like Atlantis) famous for its gold. Presumably Solon or one of his contemporaries had translated Tantalus into Atlas, and Tantalis into Atlantis, so the association with Lydia was lost by Plato's time. Perhaps also it was the mention of Atlas, whose brother, Prometheus, was said to have taught humankind how to use fire, which persuaded Plato to move the story backwards in time. Alternatively, the Lydians may have exaggerated the antiquity of Tantalis in order to give themselves a long and impressive history, a far from uncommon tendency.[79]

James criticised other authors for trying to take Plato's words too literally, arguing that the story must be viewed as a morality tale of a great nation becoming corrupt, and meeting a deserved fate in cataclysmic fashion. In this context, it could be seen as part of the mainstream of Plato's writings, not some strange aberration. Plato had previously written *The Republic*, about an ideal state where a philosopher king ruled with absolute authority, and was later to write *The Laws*, attempting to provide an ideal legal system, which would be needed in the absence of a suitable philosopher king. In between, in the *Timaeus* and *Critias*, he was describing what might happen if an ideal state (Athens in a previous golden age, which he referred to elsewhere) was confronted by an aggressor (Atlantis). Hence, the material passed on to him from Solon must have provided an ideal starting point for his planned, but uncompleted, epic.[80]

The problem with this argument was that, if Plato's story of Atlantis was a morality tale rather than a historical account, then why should it have needed a single, or indeed any, source? Without question, the Atlantis of Plato's writings was a long way removed from the Zippasla of history, and not only in time. Atlantis was said to be 'larger than Libya and Asia [Minor] together', whereas Zippasla was just part of Asia Minor. Similarly, Atlantis 'ruled over Libya as far as Egypt, and over Europe as far as Tuscany', whilst Zippasla at its greatest was little more than a buffer state between the Mycenaeans and the Hittites. Even if James could prove his thesis, the intellectual satisfaction of seeing a puzzle solved would be countered by a profound sense of anti-climax.

Of course, although James may have had grounds for thinking that he had identified a source for Plato's Atlantis story, it was surely not the only one. In 1998, in

Atlantis Destroyed, writer Rodney Castleden, like Kukal, Schoch and others, acknowledged that Plato probably made use of several sources for his story of Atlantis. However, Castleden maintained that the catastrophic effect of the eruption of Thera on the Minoan civilisation was a very significant factor, and he suggested that the settlement found on Santorini near Akrotiri (see Chapter 13) was that known as Therassos in the Bronze Age. In Castleden's view, Plato's descriptions of Atlantis were an amalgam of features of Therassos (the Minoan capital city, as he saw it) and Crete (the religious centre). As previously pointed out by Angelos Galanopoulos, Crete is situated just beyond the two most southerly headlands of Greece, Capes Tainaron (Matapan) and Malea in the Peloponnese which, in the ancient world, were apparently another of the pairs of rock formations sometimes referred to as the 'pillars of Heracles'. However, Plato may have been making genuine use of material brought back by Solon, so was not fully aware of what he was writing about. Therefore, references to Atlantis being in the west (as Crete is from Egypt, but not from Athens), together with a translational error, suggesting that Atlantis was *larger* than Libya and Asia Minor (rather than located *between* them), may have confused Plato into placing it outside the straits of Gibraltar. To Plato, as to later generations, that was where the 'Atlantic Ocean' was to be found, but the name may have had a different meaning in the time of Solon.[81]

The various eruptions of Thera formed the subject of Walter Friedrich's 1994 book, *Feuer im Meer* (published in English, as *Fire in the Sea*, in 2000) and the 1999 Geological Society of London 'memoir', *Santorini Volcano*.[82] Friedrich devoted a full chapter of *Fire in the Sea* to the possible link between Thera and Atlantis, describing the Galanopoulos theory as 'fascinating' and 'plausible', before concluding that it left 'many unanswered questions'.[83]

Peter James gave *The Sunken Kingdom* the sub-title, 'The Atlantis Mystery Solved'. However, the Atlantis mystery has clearly *not* been solved. Indeed, it may well not have a solution, in terms of a specific geographical and historical location. Only one thing is certain: many more books and articles will be written in the years to come, devoted to the subject of Atlantis.

29 Natural catastrophes and the rise and fall of civilisations

Myths as possible recollections of catastrophes in ancient times

Regardless of possible associations with the Atlantis legend, it is an undeniable fact that Thera erupted, in catastrophic fashion, during the Bronze Age (figure 29.1). The resulting tsunami and injection of pollutants into the atmosphere must have had an extremely destabilising effect on the Minoan civilisation, albeit without bringing it to an immediate end, and the situation was exacerbated by subsequent earthquakes. The precise date of the eruption, however, is still uncertain. On the basis of archaeological evidence, linked to the chronology of Egypt, Spyridon Marinatos (see Chapter 13) placed the event at around 1500 B.C., and his arguments still receive much support, for example, by Peter Warren, Professor of Classical Archaeology at the University of Bristol. On the other hand, tree-ring and ice-core data show no indications of a major volcanic eruption or, indeed, of any environmental crisis at this time. Instead, narrow growth rings, which are evidence of frost damage, in trees in the White Mountains of eastern California and in Irish oaks, suggest that a major eruption may have occurred in 1628 B.C. Similarly, Greenland ice cores show an acidity peak at around the same time, which could have been caused by acid rain resulting from the injection of sulphur into the atmosphere by a volcanic eruption. Thus, some believe the Thera event to have occurred around 1628 B.C. However, the Thera eruption might not have thrown up enough sulphur to cause acid rain, so the acidity peak in the Greenland ice and the narrow tree rings might be indicators of another northern hemisphere volcanic eruption, possibly in Iceland. Indeed, some have argued that tephra particles found near the acidity peak in the Greenland ice do not match ones from Thera, although others disagree. Radiocarbon studies on the Minoan settlement on Thera, destroyed by the eruption, show a range of dates, the most recent ones corresponding, after calibration by reference to tree-ring dates (see Chapter 17), to around 1500 B.C. Another acidity peak in Greenland ice-cores corresponds to a date of around 1390 B.C. This is too late to correspond to Thera by conventional archaeological dating, but could be regarded as being consistent with uncalibrated radiocarbon dates for the eruption, although these are not usually regarded as having any meaning.[1]

Several writers have associated the Thera eruption with environmental upheavals in Egypt, at the time of the supposed 'Exodus'. Immanuel Velikovsky, in *Earth in Upheaval* (see Chapter 12), saw the eruption as a secondary feature of his 'Venus event',

Figure 29.1   The town of Fira above the cliffs of Santorini, which are the walls of the caldera formed by the Bronze Age eruption of Thera.

around 1500 B.C. (although, as we have just noted, there is no scientific evidence of a global catastrophe at this time). Others saw the eruption as a primary cause of the environmental crises in Egypt. So, Ian Wilson, in *The Exodus Enigma* of 1986, acknowledging detailed advice from John Bimson, an authority on the Old Testament from Trinity College, Bristol, envisaged the Exodus taking place against the background of an eruption of Thera in around 1500 B.C., whereas Graham Phillips, in his *Act of God* of 1998, had a similar scenario, but taking place later, around 1390 B.C. Those orthodox scholars who accept that the Exodus event really took place (by no means all) generally place it around the latter date, i.e. in the Late Bronze Age, but that does not fit in with subsequent events described in the Bible. In particular, the book of Joshua tells how, when the Israelites reached the city of Jericho, the walls fell down, making a conquest relatively easy. The investigations of the British archaeologist Kathleen Kenyon at Jericho during the 1950s established that the walls of Jericho had indeed collapsed, probably during an earthquake, but that was at the end of the Middle Bronze Age, long before a 1390 B.C. departure from Egypt. However, it could be consistent with a 1500 B.C. Exodus, provided some adjustments were made to the conventional stratigraphy. Another who supported the earlier date was David Rohl, in his 1995 book, *A Test of Time* (which also acknowledged contributions from Bimson). Rohl pointed out that, as well as fitting in with the archaeological evidence at Jericho, the earlier date, linked to a revised stratigraphy, would make Solomon's palace at Megiddo the impressive Late Bronze Age building at the site, rather than the far less imposing Iron Age structure that has usually been associated with one of the greatest kings of the Old Testament. His arguments, however, have received little support from mainstream scholars.[2]

If it may be considered possible, although far from certain, that these Old Testament stories were based on memories of actual events, then what can be made of the statements in *Joshua* about what took place after the fall of Jericho, as the Israelites continued their advance? The rival armies clashed at Gibeon, with the Israelites coming out on top, after which they pursued their retreating enemies into the pass of Beth-horon. It was then that 'the Lord cast great stones from heaven' upon the fleeing Amorites, and 'the Sun stood still', prolonging the day to enable the Israelites to complete the slaughter. A possible natural explanation for what the victors believed to be the intervention of God on their behalf was put forward in 1946 in the *Palestine Excavation Quarterly* by the British archaeologist John Phythian-Adams. As a youth in the summer of 1908, Phythian-Adams had experienced the unique occasion when it was still light enough after midnight for people in England to read newspapers and play cricket. Almost four decades later, when Phythian-Adams found out that this phenomenon had been linked to the Tunguska event in Siberia (see Chapter 20), he quickly realised that a similar event, perhaps involving the explosion of a bolide in the atmosphere and the fall of a shower of meteorites, could explain the apparent extension of the day after the battle of Gibeon. Regardless of whether such a battle ever took place, Phythian-Adams' hypothesis was remarkable for its time, when the cause of the Tunguska event was still far from certain. Nevertheless, as would have been entirely predictable, it was completely ignored by his colleagues. That was not simply because he produced no evidence to support his case. Catastrophist explanations, particularly for events described in the Bible or other ancient sources, were simply not acceptable, either then or for the following three decades. As well as the constraining influence of the ruling gradualistic paradigm, the efforts of creationists to establish the literal truth of the Bible by finding evidence of catastrophes, and the similar activities of those who abused logic and evidence in trying to prove the reality of Plato's Atlantis, made most scholars resistant to catastrophist arguments, no matter how rational they might have been.[3]

However, in recent years, there has been growing acceptance of the view that the history of life may have been shaped by major catastrophes to a far greater extent than previously realised. Partly that has come about because of an increased knowledge of the occurrence of immense volcanic eruptions and sustained episodes of vulcanism over wide areas. Another factor is a developing awareness of the threat from asteroids and comets in Earth-crossing orbits, together with the realisation that many of the craters at the Earth's surface, previously thought to be of volcanic origin, were in fact formed by impacts (see Chapters 14, 19 and 20). Many previously sceptical scientists started to become receptive to catastrophist arguments after 1980, when Luis Alvarez and colleagues not only claimed that a huge asteroid had struck the Earth at the time of a mass extinction episode in the Late Cretaceous Period 65 Myr ago, but produced evidence for it in the form of high iridium concentrations at sites throughout the world, as we discussed in Chapter 21. It was apparent that a much smaller impact would have been sufficient to have caused devastation to ancient civilisations, and small impact events occur much more frequently than large ones.

In contrast, Velikovsky's belief that the planets Mars and Venus passed sufficiently close to the Earth in historical times to cause global catastrophes could not be reconciled with the known laws of physics. Hence, although planetary catastrophism still receives enthusiastic support in some quarters, it has been firmly rejected by professional scientists. Some astronomers acknowledge that Velikovsky may have been correct in suggesting that a number of myths were derived from objects which had been prominent in the ancient sky, and caused catastrophes on Earth, but they nevertheless maintain that these cosmic bodies must have been asteroids or comets, not wandering planets. Consistent with that thesis, Chinese records from about 1058 B.C. noted the presence of an particularly bright comet in the sky. Perhaps even more significantly, Donald Yeomans, of the Jet Propulsion Laboratory, Pasadena, together with Tao Kiang of the Dunsink Observatory, Ireland, showed in 1981 by back-calculation that, unless some other encounter had affected its orbit in the intervening period, Comet Halley must have passed so close to the Earth in 1404 B.C. that it would have been an immensely prominent object in the sky at that time.[4]

Similarly, as we saw in Chapter 27, extrapolations of the orbits of Encke's Comet, the Taurid meteoroid stream and associated Apollo asteroids, together with other evidence, led Victor Clube and Bill Napier to conclude that all were products of a huge comet which came into an Earth-crossing orbit around 20,000 years ago and began to break up, with particular disintegration events occurring about 7500 and 2700 B.C. Fragments would have struck the Earth at intervals throughout the Bronze Age, with catastrophic consequences, which gave rise to myths of devastation coming from the heavens. These ideas were placed before the general public at a London meeting of the SIS (see Chapter 12) in 1982, and developed at another meeting of the same society in Nottingham in the following year. At around the time of the earlier meeting, Clube and Napier presented them in detail in *The Cosmic Serpent*, and they were extended in 1990 in a further book, *The Cosmic Winter*. An alternative but equally catastrophist view of the ancient world was taken by the Icelandic geologist Haraldur Sigurdsson, who pointed out in his 1999 book, *Melting the Earth*, that some myths and social upheavals could be associated with volcanic eruptions in the Bronze Age (and afterwards), and similar arguments were presented in 2000 by Stephen Harris, of California State University, Sacramento, in the *Encyclopedia of Volcanoes*. Others suggested that catastrophic inundations during the Holocene, such as the rapid expansion of the Black Sea and the tsunami which struck the North Sea at around the same time, could have been the origin of various flood myths.[5]

Again, as we saw in Chapter 12, the renowned archaeologist Claude Schaeffer argued in 1948 that on at least five occasions during the Bronze Age in the Middle East, there had been widespread destruction of cities, often associated with evidence of natural catastrophes, particularly earthquakes. Four decades later, Colin Burgess, of the University of Newcastle-upon-Tyne, writing in 1989 in *Current Archaeology*, suggested vulcanism as the common cause of the catastrophes of the late second millennium B.C. At this time there was a population collapse in northern Europe, as well as the termination of the Late Bronze Age in the Middle East and Mediterranean regions, when the Mycenean and Hittite civilisations came to an end, and there were crises in Cyprus, Syria and Egypt. This corresponded to the last of Schaeffer's five catastrophic Bronze Age events. The first of these events occurred towards the end of the Early Bronze Age, around 2300 B.C. At about the same time, settlements in the region of the Dead Sea were consumed by fire and abandoned, which may be an indication that, perhaps as part of a more widespread crisis, there was some basis of historical fact in the biblical story of the destruction of Sodom and Gomorrah (see Chapter 12). It has been suggested, for example, that the area around the Dead Sea may have suffered a major conflagration when earthquakes released inflammable gases and bitumen from the Earth's interior (a distinct possibility in that part of the world). Then, in the aftermath, the cities of Sodom and Gomorrah (if they actually existed) might have been swept away by landslides, or covered by the rising waters of the Dead Sea, so might not be amongst the sites already found. However, other than the undeniable fact that major centres of population in the region were abandoned as the Early Bronze Age drew to its close, this is largely speculation.[6]

More certainly, a detailed investigation at Tell Leilan in northern Syria, carried out by a team led by the Yale archaeologist Harvey Weiss, showed that the climate in the region of this previously thriving site had suddenly become arid at the end of the Early Bronze Age, resulting in its abandonment for a period of several centuries. Furthermore, this seemed to be typical of what had happened throughout the Middle East. A layer of tephra particles at the level of the climate change at several sites implicated a volcanic eruption as the cause. This may have initiated the collapse of the short-lived Akkad (or Agade) empire, founded by Sargon the Great, which included Tell Leilan on its northwestern fringe. As a result of his investigations in the region, Weiss became a consistent advocate of the view that climate change can cause societal collapse, without the necessary involvement of any human-centred factors. The results of his Tell Leilan study were published in *Science* in 1993.[7]

Before then, during the 1980s, archaeological, geological and climatic evidence for a worldwide catastrophic event around 2300 B.C. was presented in the pages of the *SIS Review* by the American engineer Moe Mandelkehr. At this time, for example, there were global crustal deformations, sea-level discontinuities, earthquakes, volcanic activity, a geomagnetic transient and a change in the atmospheric radiocarbon concentration. In this series of papers, Mandelkehr was not prepared to suggest any causal mechanism, but the implication was that something very unusual must have happened at the termination of the Early Bronze Age, disrupting a range of terrestrial systems.[8]

The SIS Cambridge Conferences

In 1993, at Fitzwilliam College, Cambridge, the SIS held a conference entitled, 'Evidence that the Earth has Suffered Catastrophes of Cosmic Origin in Historic Times'. At this conference, the engineer and ancient historian Bob Porter (see figure 28.3) outlined the destructions which had occurred at various sites during the Bronze Age, and concluded that only towards the end of the Early Bronze Age, the period about which Moe Mandelkehr had been writing, was there evidence of a global catastrophe. Even here, however, there were doubts about the precise dating of events at the different sites. If catastrophic events had occurred at the end of the Middle Bronze Age, as envisaged by Velikovsky, they were on a much smaller scale than he had supposed. That latter conclusion was also reached by John Bimson, with both Porter and Bimson considering that comets were a far more plausible cause of extraterrestrial catastrophes during the Bronze Age than planetary encounters.[9]

Following on from this came the Second SIS Cambridge Conference, 'Natural Catastrophes during Bronze Age Civilisations', which was held in 1997. The proceedings of this conference were published under the same title as a volume in the British *Archaeological Reports* international series in the following year. In the keynote address, science journalist Robert Matthews made two main points: (1) that observations made in the distant past may be far more accurate than we generally assume; and (2) that, because of the dangers from asteroids and comets, the Earth is not, and never has been, a safe place to live. He concluded with a quotation from George Santayana, 'Those who do not remember the past are condemned to relive it'.[10]

Then followed a series of papers by astronomers, concerned with those hazards from space. Firstly, Mark Bailey, Director of the Armagh Observatory, reviewed recent advances in knowledge, arguing that several giant comets were likely to come into the inner Solar System and break up every million years, supplementing the population of Near-Earth objects. Whatever their origin, Near-Earth objects undoubtedly pose a threat, and some have struck the Earth in the astronomically recent past. Bill Napier then assembled data from a variety of sources to present a picture of the current interactions between the Earth and its cosmic environment. In his view, the Taurid/Encke complex, referred to above, has been a regular and occasionally conspicuous hazard over the past 12,000 years or more. This has resulted in impacts such as that which devastated the Tunguska region of Siberia in 1908; in an occasional contamination of the stratosphere by cometary dust, leading to freezing episodes which may have lasted decades; and in small-body impacts into an ocean, causing catastrophic flooding of coastal areas.[11]

After these two papers came one from Duncan Steel (see Chapter 20), which was more speculative, although based on the same astronomical data and interpretations. Steel suggested that the construction around 3500 B.C. of the Great Cursus near Stonehenge, and that around 3100 B.C. of the first stage of Stonehenge itself, were intended as predictors of catastrophes, since these were the approximate times when the orbit of the giant proto-Encke comet intersected that of the Earth. Gerrit Verschuur, of the University of Memphis, then took both a scientific and a philosophical view of the Earth's place in space. In his view, impacts have been the rule rather than the exception, and will still be so in the future. The problem of humankind was that hope prevented us from seeing that the cosmic events which have destroyed civilisations in the past will continue to do so, unless we are prepared to commit a great deal of money and effort into taking preventative action.[12]

The next, and largest, group of papers in *Natural Catastrophes during Bronze Age Civilisations* were concerned with archaeology, geology and climatology. To start this section, Bruce Masse, an environmental archaeologist attached to the University of Hawaii, attempted to re-evaluate events on Earth in the light of estimates made by astronomers of the rates of impact of asteroids and comets. Believing that more than twenty impacts causing at least local catastrophes were likely to have occurred in the past 6,000 years, he examined literary traditions, together with archaeological and palaeoenvironmental data, to see if any previously unknown Bronze Age catastrophes could be identified. The most significant one appeared to be a cometary impact in the ocean around 2800 B.C., which released almost a million megatons of energy, causing widespread devastation.[13]

After Masse's contribution came three papers which were concerned, at least in part, with happenings around the time of Mandelkehr's supposed 2300 B.C. catastrophic event, close to the end of the Early Bronze Age in the Middle East. Firstly, Marie-Agnès Courty (figure 29.2), a geologist at the French Centre for Scientific Research, presented new archaeological evidence of a conflagration and a dust layer at sites in Mesopotamia and northern Syria, particularly in the Tell Leilan region. Courty had been a co-author, with Harvey Weiss, of the 1993 *Science* paper in which a volcanic hypothesis had been put forward to explain events at this time. However, she had now rejected this, because a cosmic catastrophe appeared more consistent with the evidence. In her view, the tephra particles at the sites were not from a nearby volcanic eruption, but were components of dust thrown into the atmosphere and dispersed over a wide area as a result of an impact into igneous rock.[14]

Evidence for an adverse climate change in Ireland at about the same time, and on several other occasions, was then given by palaeoecologist Mike Baillie, of Queen's University, Belfast. Narrowest-ring events in Irish oak chronologies corresponding to 2345 B.C., 1628 B.C. and 1159 B.C. lined up with similar events in other tree-ring chronologies and also large acidities in Greenland ice records. Baillie pointed out that the first of these also corresponded to the approximate age of a major eruption of the Hekla volcano in Iceland (that known as 'Hekla 4'), the second to the eruption of Thera (a controversial association, as we have discussed), and the third to another large eruption of Hekla ('Hekla 3'). Like Courty, Baillie had originally regarded these volcanic eruptions as primary causes of environmental crises. However, the narrowest-ring events were imposed on pre-existing climatic downturns, which, as with similar

Figure 29.2   Geologist Marie-Agnès Courty, speaking at the Second SIS Cambridge Conference in 1997.

events around 207 B.C. and 540 A.D., suggested a scenario of stratospheric dust-loading and bombardments from space, the latter triggering the volcanic eruptions.[15]

Benny Peiser, of Liverpool John Moores University, then summarised a survey he had made of around 500 reports of civilisation collapse and climate change in the late third millennium B.C., which showed a significant clustering around 2300 B.C. Most sites in Europe, the Middle East, India and China where civilisation collapsed at this time showed clear signs of natural disasters and/or rapid abandonment, whilst around the world there was strong evidence of water-level and vegetation changes, glacier and desert expansion, seismic activity, floods and extinctions of animal species. He concluded that only extraterrestrial bodies acting on terrestrial systems could produce the range of glaciological, geological and archaeological features reported.[16]

The next group of papers were concerned with events which were slightly more recent, occurring around the time certain Late Bronze Age cultures came to an end. Firstly, Amos Nur, a geophysicist from Stanford University, argued that large earthquakes were likely to have contributed to the physical and political collapse of Late Bronze Age civilisations around the eastern Mediterranean. It is known that, every few centuries, massive earthquakes occur in bursts that sweep across about 1000 km of the eastern Mediterranean over a time-scale of approximately 50 years. In Nur's scenario, the burst at the end of the Late Bronze Age probably began between 1225 and 1175 B.C., and made urban centres vulnerable to opportunist military attacks. Then, Lars Franzén and Thomas Larsson, of Goeteborg University, Sweden, presented evidence from sites in Tunisia and Sweden showing that a major atmospheric cooling event, accompanied by excessive precipitation, which led to flooding, occurred around 1000 B.C. Other sources indicated that the event was sudden and widespread, and the finding of small glassy spherules pointed to a possible impact origin. Frenzén and Larsson suggested that an asteroid or comet of diameter in the range 0.5–5 km may have landed in the eastern Atlantic around 1000 B.C., affecting in particular Europe, North Africa and the Middle East. After this, Bas van Geel and colleagues from the University of Amsterdam showed that a sharp rise in the carbon-14 content of the atmosphere towards the end of the Bronze Age in northwestern Europe, around 850 B.C., was accompanied by a rapid transition from a relatively warm and dry climate to one which was cooler and wetter. They suggested that a reduced sunspot activity at that time allowed more high-energy galactic cosmic rays to reach the top of the atmosphere, leading to an increased production of carbon-14, cloudiness and precipitation.[17]

The final paper in the section on archaeology, geology and climatology was by Euan MacKie (see Chapter 12), who began by warning that astronomers would have to produce clear evidence of comet swarms or the likelihood of large impacts at specific dates before most of his fellow archaeologists would be willing to re-examine their data with this in mind. He then briefly suggested some examples of instances where such a re-examination might be productive, including two around the end of the Bronze Age in northwestern Europe. One of these concerned a site ten miles west of Glasgow, where there were two phases of 'cup and ring' rock carvings, the first perhaps from the latter part of the third millennium B.C., and the other probably from the sixth or seventh century B.C. A suggestion by Clube and Napier that these could be representations of comets was not currently being taken seriously by archaeologists. The other example concerned the vitrified forts of Scotland, dating from the period after 800 B.C., whose timber-framed construction might have been intended as a protection against earthquakes stimulated by a cosmic bombardment.[18]

Natural Catastrophes during Bronze Age Civilisations was then brought to a close by five papers on the subject of history and culture. In the first paper, Gunnar Heinsohn, of the University of Bremen, considered the origins of kingship, priesthood and blood sacrifice in the Early Bronze Age. Nicolas-Antoine Boulanger, in the eighteenth century, believed they were reactions to major catastrophes taking place at the time, but that view has been disregarded almost ever since. However, in the light of increased knowledge about cosmic events, Heinsohn argued that Boulanger was correct after all. Re-enacting catastrophic events as rituals involving blood sacrifice would have had a therapeutic effect on traumatised survivors. Significantly, according to Heinsohn, there was a gradual abandonment of blood sacrifice in the Iron Age, when cosmic catastrophes were much rarer events than they had been in the Bronze Age. Similarly in the next paper, David Pankenier, an authority on ancient Chinese history from Stanford University, suggested that, contrary to what had generally been supposed, legends and rituals from Bronze Age China may reflect actual events. In particular, around the time of the transition from the Xia to the Shang dynasty in the middle of the second millennium B.C., there is a story of ten suns appearing in the sky and then, a few years later, of five planets criss-crossing, and stars falling like rain, after which there was an earthquake and then a drought. It would not be difficult to see this as an indication of the appearance of multiple comets in the sky, and impact-induced catastrophes. The same or a different cometary catastrophe could also form the basis for the legend of the battles between the Yellow Emperor and the wicked Chi You, whose descent to the Earth from the heavens brought widespread destruction and misery to the people.[19]

Finally came three papers which, accepting the hypothesis that major catastrophes were indeed a feature of the Bronze Age and the first few centuries of the first millennium B.C., considered how humankind reacted when more peaceful times came along. Firstly, William Mullen, Professor of Classical Studies at Bard College, described how the 'Milesian School' of pre-Socratic philosophers in the sixth century B.C. set out to explain terrifying phenomena such as thunder, lightning, earthquakes and eclipses in terms of the same processes which it used to explain the orderly arrangement of the Earth and the heavens, thus moving away from the old view which associated them with the unpredictable activities of the Olympian gods. World-destructions could occur, but only in cycles which stretched over long periods of time. Mullen suggested that the hidden agenda may have been a desire to re-assure the population that they were now safe from the cosmic catastrophes which had occurred in the past. In similar fashion, Irving Wolfe, of the University of Montreal, then argued that a 'cultural crisis' occurred in the sixth century B.C., with the appearance of new religions, new philosophies, new art forms, new types of games and new forms of social organisation, all of which were very different from what had existed previously. In many ways, these laid the foundations for the cultural characteristics of our modern age. According to Wolfe, the cost has been that, ever since the middle of the first millennium B.C., humankind has been suffering from a collective form of Post-traumatic Stress Disorder, denying not only past catastrophes, but also the possibility of future ones.[20]

The denial of past and future cosmic catastrophes was certainly a feature of the influential philosophy of Aristotle, and has been a characteristic feature of scientific thought over the past few centuries. However, in the concluding paper of the conference proceedings, Victor Clube argued that the situation in between was somewhat different. The relatively tranquil period in the middle of the first millennium B.C. did not last for long, and further episodes of cosmic bombardment conditioned people once again to believe that the world might come to an end in this way. Clube suggested that this provided strong support for coherent rather than stochastic catastrophism (see Chapter 20), because frequent small-scale events would keep the issues in

people's minds, which would not be the case if there were vast periods of time between impacts, even ones which were individually much larger. According to Chinese
astronomical records, there have been seven peaks of fireball activity in the past 2,000
years, at times which indicate an association with the Taurid/Encke complex, including
the periods from 150 B.C. to 50 A.D., 350 to 650 A.D., 1000 to 1200 A.D. and 1350 to
1550 A.D. However, the past two centuries have been a quiet period and this, together
with the influence of Lyell and Darwin (who established the gradualistic paradigm,
largely for philosophical reasons), and of Newton (who played down the threat from
space on religious grounds), led most people to think that the future of the Earth
was secure. We now know otherwise but, in contrast to previous generations, who
could only hope and/or pray, we may soon have the capability for defending ourselves.
However, Clube warned that the prospect of safeguarding the future of civilisation
was not being helped by those who clung to gradualistic, Earth-centred views, or by
those who adopted what he saw as erroneous forms of catastrophism.[21]

The discussion of these issues did not end at the close of the conference. The
Cambridge Conference Network was established, moderated by Benny Peiser, to enable the
contributors, and others, to share ideas and circulate new information by electronic
means. This network rapidly expanded, demonstrating the growing interest in catastrophism and related topics. Now, more than one thousand members, spread right
around the world, participate and receive mailings several times each week.[22]

Natural catastrophes in the Bronze Age

In addition to exchanges through the electronic network, many of the ideas put
forward at the Second SIS Cambridge Conference have been discussed subsequently
in books and articles, and at other conferences. So, for example, Victor Clube, Benny
Peiser and Marie-Agnès Courty participated in a meeting organised jointly by the Royal
Astronomical Society, the British Interplanetary Society and the Geological Society,
entitled 'Defining the effects of sub-critical cosmic impacts on the Earth', which was
held in London in December 1998. At this meeting, Courty produced further evidence
for a catastrophic extraterrestrial impact in the Middle East during the second half of
the third millennium B.C.[23]

Whatever the merits of the arguments for catastrophic events at various times
during the Bronze Age and afterwards, it seems that the strongest arguments can
be put forward for a catastrophe around 2300 B.C., bringing to a close the Early
Bronze Age in Mesopotamia and surrounding areas, which was the focus of Courty's
attentions.

The Old Kingdom of Egypt, the period of construction of the major pyramids, came to an end at this time. As Professor Fekri Hassan of the Institute of
Archaeology, University College London, has shown, the Old Kingdom fell apart during a period of nationwide famine, associated with failure of the Nile floods, when
the Egyptian population was reduced by about a third. The apocalyptic *Admonitions of*

Ipuwer (see Chapter 12) and the similar *Prophecies of Neferti*, although written during the Middle Kingdom, may have been making use of descriptions of events during this time of tribulation. Some scholars, including astronomers Duncan Steel and Chandra Wickramasinghe, have speculated that the building of pyramids in the Old Kingdom might have been stimulated by a sustained cosmic bombardment, and they could even have served as air-raid shelters for the living pharaoh. When stability returned to Egypt, after more than a century of chaos, no more large, stone pyramids were built: either the techniques necessary for their construction had been lost in the intervening period, or else it was no longer thought necessary or worthwhile to spend immense resources on such projects.[24]

As well as impacts, more Earth-bound mechanisms have been suggested as possible causes of the crises at this time. Some of these suggestions have already been mentioned, but there have been others. At the 22nd General Assembly of the International Union of Geodesy and Geophysics, held in Birmingham, England, in 1999, Douglas Keenan presented evidence that the environmental upheaval clearly documented at the end of the Early Bronze Age in the Middle East had in fact encompassed most of the northern hemisphere. For example, the environmental downturn in Egypt and Mesopotamia coincided with the transition of the Indus Valley civilisation to its depopulated post-Urban phase. Similarly, there were floods and droughts in different parts of China, causing the turmoil which eventually resulted in the institution of the first dynasty. All of this was consistent with what Mandelkehr, Peiser and Baillie had argued. Unlike them, however, Keenan maintained that the primary cause was a colossal volcanic eruption, perhaps in Turkey, which gave rise to the period of sulphurous yellowish mists in China associated with the reign of the Yellow Emperor. He produced a model to explain how the various climate changes around the world could have resulted from the eruption.[25]

On the other hand, Peter James and Nick Thorpe concluded in favour of an extraterrestrial mechanism in their 1999 book, *Ancient Mysteries*. Despite the impression which could be given by the title, James and Thorpe, as we noted in Chapter 28, took a sceptical view of many unorthodox theories, finding little reason to support the ideas of Hancock, Bauval, Hapgood, Velikovsky and others. The Clube–Napier theory of cometary encounters at the end of the Early Bronze Age was, however, another matter, even without any firm evidence for any significant impacts. The fact that earthquakes had occurred at many sites at this time seemed to rule out vulcanism as the primary cause, for volcanoes do not cause anything more than local tremors. James and Thorpe concluded:

> There is one cause that could explain the widespread seismic activity quite easily, as well as lending a hand in altering the climate by throwing up dust veils: the impact of large meteorites and cometary fragments. The relatively small cometary fragment that exploded above Tunguska in Siberia in 1908 caused several tremors,

and burned and devastated an enormous expanse of forest. A larger body actually impacting near a fault in the crust such as the Great Rift Valley could trigger off both earthquakes and volcanoes.[26]

Similarly, Moe Mandelkehr, who was apparently the first to suggest that there had been a worldwide catastrophe in 2300 B.C., concluded in 1999 that the most likely cause was an encounter between the Earth and the breakdown products of a giant comet, as in the Clube–Napier scenario. Again, Mike Baillie, in his 1999 book, *Exodus to Arthur*, developed ideas which he had presented at the Cambridge Conference, to argue that the event at the end of the Early Bronze Age, centred around 2345 B.C. (according to tree-ring evidence) was primarily a cometary catastrophe.[27] A possible impact structure associated with this event was located in 2001. This is a 3.4-km-diameter crater in the Al 'Amarah marshes, Iraq.[28]

In *Exodus to Arthur*, Mike Baillie also amplified the case for cometary encounters causing environmental upheavals around the years 1628 B.C., 1159 B.C., 207 B.C. and 540 A.D. Like the writers mentioned at the start of this chapter, Baillie thought it very possible that the Exodus event occurred at the same time as the Santorini eruption, although he placed this earlier in time than the others, at around 1628 B.C. More conventionally, he associated the 1159 B.C. event, linked to earthquakes and floods in eastern Mediterranean regions (and, in Baillie's view, with the massive Hekla 3 eruption in Iceland, although the precise dating of this remains uncertain), with the collapse of the Mycenaean culture of Greece and of other Bronze Age civilisations in the Middle East, as well as the depopulation of the British Isles and the end of the Shang Dynasty in China. Baillie also considered that the next major period of cometary catastrophism, around 207 B.C., could have prepared the way in China for the Han Dynasty to seize power and become established. Meanwhile, others investigated the possible links between volcanic eruptions and climate and societal changes around 1150 B.C., as the belief widened that the course of human history around that time could have been influenced by some kind of catastrophic event.[29]

Natural catastrophes in the sixth century A.D.

There has been much controversy concerning the event centred around the year 540 A.D. Baillie had demonstrated in 1994 that reduced tree-ring growth over the fifteen years following 535 A.D. indicated that the climate was more severe at this time than at any other in the entire period covered by the European tree-ring record. In *Exodus to Arthur*, he advanced the theory that stories of hardships described in the legends of King Arthur, including a 'wasteland' in the west of Britain, shrouded in mist and devastated by plague and famine, together with accounts of dragons seen in the sky, could be associated with a cometary induced catastrophe in the middle of the sixth century. Gildas, who wrote *On the Ruin and Conquest of Britain* around this time, spoke of the terrible afflictions of his fellow-countrymen, without making clear the nature of these. Procopius, the sixth century Byzantine historian, wrote that conditions in the west of Britain were so bad that it was impossible for anyone to survive there for more than a half-hour, but he was not speaking from first-hand experience. However, even in Mediterranean regions, according to the same author and others, including John the Lydian and John of Ephesus, the sun became much dimmer than usual in 535 A.D., this situation continuing throughout most of the following year.[30]

Arguments for an environmental upheaval in Europe and elsewhere during the sixth century A.D. were also put forward by the British archaeology journalist David Keys, in his 1999 book, *Catastrophe.* Following climatic disturbances during the late 530s, bubonic plague arrived in Mediterranean regions in 541 A.D. and, over the next hundred years, caused a dramatic reduction in the populations of western Europe and the Roman/Byzantine empire to the east. As a consequence of this and the various political and social convulsions which ensued, much territory was lost to invading Avars, Slavs and Persians, followed at a later stage by the armies of Islam, a religion whose formation was itself facilitated by political and social changes resulting from the plague.[31]

A similar environmental crisis also occurred in other parts of the world. So, for example, the *Nan shi* (the history of the Southern Dynasties of China) records great falls of yellow dust, beginning towards the end of 535 A.D. Extremely cold conditions prevailed for the next few years, destroying crops and causing famine. Similarly, the *Bei shi* (the annals of the northern districts of China) record a climatic deterioration and famine, starting in the mid 530s. As a consequence of these disturbances, northern China invaded the south and conquered it, bringing the whole country under the control of a single Emperor. Korea and Japan suffered much the same environmental stresses, records in both countries describing the occurrence of severe hardship, famine and disease in 536 A.D.[32]

In the Americas, as Keys went on to describe, evidence from tree-rings and lake deposits indicates that an unusually cold, dry period began between 535 and 540 A.D. and lasted for several decades. This phenomenon was widespread, stretching all the way from California in the north to Argentina and Chile in the south. In Mexico, the drought led to the collapse of the Teotihuacan civilisation in the highlands and the so-called Hiatus in the Maya Lowlands whilst, in Peru, the Nazca people began a frenzy of line-drawing in the desert, presumably to form part of rain-making rituals. More practically, they also carried out a number of engineering projects to extract water from under the ground and channel it to where it could be used. Further south, many people left the arid coastal regions of Peru to move into the mountains around the vast freshwater lake, Titicaca. This led to the hitherto small settlement of Tiahuanaco becoming a major centre of population, with major building projects being carried out to reflect this. (To modern scholars, that is the true story of Tiahuanaco, not the extremely unlikely one outlined in Chapter 28, supposed to have taken place many thousands of years earlier.)[33]

So, from these and other examples, Keys saw strong reason to think that the dramatic climate changes which began in 535 A.D. influenced the course of history around the world. In *Catastrophe*, he wrote:

> Although human invention, achievements and actions are obviously key factors in determining the course of human history, the forces of nature and other mechanisms beyond the control of individual human beings, or even states, play an even greater role – both directly, and indirectly, by conditioning the circumstances which induce, produce or permit, individual or collective human actions.[34]

Keys, in contrast to Baillie, put the blame for all this environmental turmoil on a massive eruption of the Krakatoa volcano, in the East Indies, in 535 A.D. This was based on a various pieces of evidence. An ancient caldera, one of the largest known, surrounds the site of the volcano which erupted in 1883, and the Javanese *Book of Ancient Kings* describes an event which could have been a massive volcanic eruption during the first millennium A.D. The *Nan shi* of China records that loud bangs were heard at Nanjing (Jiankang) in 535 A.D., coming from the southwest, which is the direction of Krakatoa. Finally, the presence of acidity peaks in ice-layers from the middle of the first millennium A.D. in cores from both Greenland and Antarctica suggests that, if these resulted from the same eruption, it must have involved a volcano in the tropics, like Krakatoa.[35]

However, Keys failed to convince medieval historians that a major natural catastrophe had occurred in 535 A.D., leaving its mark on cultures around the world. He was accused of the selective use of evidence, and of drawing too firm conclusions from unreliable sources. So, in a *New Scientist* review of *Catastrophe*, Nicholas Saunders, of University College, London, wrote, 'Unfortunately, the American case studies are riddled with errors of detail that undermine the general principles of his argument'. In particular, Saunders argued that some sixth century activities (e.g. water rituals) had been presented as new developments, rather than as part of an on-going tradition, and that Keys had placed too much reliance on what the Aztecs were supposed to have told the Spaniards about events which had taken place 700 years previously. For similar reasons, but from the perspective of European history, Edward James, of Reading University, warned readers of the journal *Medieval Life* to 'approach this book with *extreme* caution'. Nevertheless, James's consideration of *Catastrophe* was given prominence as a full article rather than a book review, acknowledging that 'if the thesis of this work is correct, a complete reassessment of the history of the early medieval period will be required'. Furthermore, James wrote, 'Unlike Velikovski [sic], however, he may just be right. He has not proved his case, but it would not be easy to disprove it either'.[36]

As to the proposed mechanism, Keys, in putting forward his volcanic hypothesis, rejected the involvement of asteroid and cometary impacts. However, he only considered the effects of these operating in isolation. There was no mention in the book of the Clube and Napier concept of coherent catastrophism, nor was there any consideration of the very point which convinced Baillie that the various catastrophic episodes in history were not initiated by major volcanic eruptions. This was that the volcanic eruptions, which undoubtedly occurred at these times, appeared to take place on each occasion *after* the environmental crisis had begun. To Baillie, an environmental scientist of repute, that was consistent with the Clube–Napier scenario of atmospheric dusting accompanied by a series of impacts, which in turn triggered volcanic activity. In his opinion, as expressed in *Exodus to Arthur*, the 'only remaining question is whether there has been just one close-approach comet, or several, in the last 5000 years', bearing responsibility for the series of environmental crises which changed the course of history. That view was supported by Chandra Wickramasinghe, in his 2001 book, *Cosmic Dragons*.[37]

On the other hand, despite being aware of Baillie's arguments, the American businessman and archaeologist Richardson Gill suggested in 2000 in his book, *The Great Maya Droughts*, that the main cause of the 535 event was a massive eruption of El Chichón. A relatively small eruption of this sulphur-rich Mexican volcano in 1982 killed about 2,000 people and reduced average surface temperatures by almost 1 °C over the following year, so a larger one around 535 A.D. could have had widespread and long-lasting events. El Chichón did indeed erupt during the sixth century, although the precise date and the magnitude are still uncertain. Gill focused mainly on the Maya Hiatus, which began in 536, but concluded that, regardless of whether he was right or wrong about the cause, 'The one incontrovertible fact to emerge from this discussion of the A.D. 536 mystery is that the Hiatus occurred at a time of global climatic aberrations which resulted in worldwide cold. It was not a localized Maya phenomenon'.[38]

That still remains to be established. In 1999, environmental scientists Jon Sadler, of Birmingham University, and John Grattan, from the University of Aberystwyth, argued in the journal *Global and Planetary Change* that connections which had been claimed between volcanic eruptions and disparate environmental events generally involved a number of unproven assumptions. They pointed out that the effects of a volcanic eruption on the climate are extremely complex, depending not only on the magnitude of the eruption but also on the sensitivity of the responding phenomena, which could be different at different times or in different regions, depending on circumstances. In a companion paper, Grattan, this time in collaboration with Brian Pyatt, of Nottingham Trent University, focused on just one of the environmental phenomena and suggested that the strange mists and dimming of the Sun reported in the middle of the sixth century could have been caused by a relatively low-powered eruption, under conditions which resulted in the concentration of volcanic gases in the lower (rather than upper) atmosphere. They noted that the opening of the Laki fissure in Iceland, which oozed lava and emitted sulphurous gases over a ten-month period, caused dry fogs throughout North America and Europe in the summer of 1783.[39]

Clearly, the arguments about what happened in the middle of the sixth century can only be resolved by removing the uncertainties over the time-scales of events at different locations. If (but only if) the various climate changes and atmospheric phenomena from around the world which have been linked to the 535 event really did occur within the same short period of time, it would seem likely that something much more powerful than a minor volcanic eruption must have been responsible.

In the Introduction to the multi-author work *The Years Without Summer – Tracing A.D. 536 and its Aftermath*, published in 2000, Joel Gunn, an anthropologist from the University of North Carolina at Chapel Hill, explained that, because of the worldwide scale of the changes which took place around 535 A.D., a global perspective was needed to allow the meaning and context of the situation to be properly evaluated. For this reason, *The Years Without Summer* included accounts of events in Europe, North and Central America, China and Africa. Gunn continued, 'In the chapters of this book, the authors struggle to extract bits of information from history and archaeology on an important moment in the past. Each chapter is a hallway opened with difficulty and uncertainty in a tomb of strangely forgotten information'. Nevertheless, he then made his own personal beliefs clear by adding, 'The moment was so important that the forces of history pivot on its weight. So-called modern times as distinguished from the Classical era were wrought in its forge'.[40]

The Years Without Summer was primarily concerned with evidence for climate and societal change, rather than with possible causal mechanisms. However, in the Epilogue to the work, David Anderson, of the Southeast Archaeological Center, Tallahassee, Florida, wrote:

> The changes observed in mid-sixth century cultures worldwide may eventually be linked to a specific physical mechanism, such as a volcanic eruption, meteor impact, or some other natural phenomenon. It is the belief of most of the authors of this volume that something significant, at least in terms of its effect on human cultures, did occur at this time. A fairly sudden change in global climate appears to be the most likely explanation for what has been observed. We ask our colleagues in the natural and physical sciences to seriously explore this possibility.[41]

Beyond question, only by adopting an interdisciplinary approach can scholars hope to establish exactly what occurred during the sixth century.

Natural catastrophes in the ninth century A.D.

Most surviving records of events in sixth century Europe are later copies, so may not be completely reliable. This applies, for example, to reports of widespread famine and plague in the *Annals of Ulster* and the *Welsh Annals*. Nevertheless, Swedish tree-ring data suggest that, although there were fluctuations from year to year, average temperatures in northern Europe were exceptionally low from the sixth century to the middle of the eighth. After a warm period between 750 and 780, average temperatures once again started to fall. Although it is generally accepted that a European 'Dark Age' lasted from roughly 500 to 1000 A.D., this was not completely dark, for it left behind both archaeological and written evidence. However, more trust can be given to documents from the second half of that period, when much material survived in its original form, than from the first half. Thus, it is clear from written evidence as well as from tree-rings that, whatever precisely may have happened in the sixth century, the ninth century was a time of great hardship throughout Northern Europe. In 2000, James Palmer, a postgraduate researcher in medieval history at Sheffield University (who also happens to be my son) pointed out that, although the evidence from the documents is insufficient for definite conclusions to be drawn, it could with good reason be taken to indicate that an encounter between the Earth and debris from a disintegrating giant comet (or a continuation of an ongoing, sporadic encounter) occurred at this time.[42]

The central decades of the ninth century were turbulent ones in northern Europe, with civil wars starting in the 830s between the sons of Louis the Pious, who had succeeded his father, Charlemagne, as emperor of the Franks in the year 814. These disputes eventually led to the break-up of the Carolingian empire. At the same time, Vikings were invading coastal regions and raiding inland down the rivers.

Chronicles of the period inevitably presented a political bias to the events taking place. Before the civil wars, the *Royal Frankish Annals*, which were compiled until the year 829, conveyed the prevailing attitudes of a united empire. Later, whilst various Frankish sources, written by Christians, all condemned the activities of the heathen Vikings, their individual sympathies were with different grandsons of Charlemagne, and their interpretations of history differed accordingly. Thus, the *Annals of St. Bertin* from western Francia (essentially the region of Gaul, or modern France) presented an account generally favourable to Charles the Bald, as did the *Histories* written by Nithard, who was himself a grandson of Charlemagne through his mother, Bertha. In contrast, the *Annals of Fulda*, from eastern Francia (mainly modern Germany) supported first Lothair and, later, Louis the German, whereas the *Annals of Xanten*, written close to the present-day border between Holland and Germany, remained loyal to Lothair until his death in 855.[43]

Nevertheless, despite these differences, all told a consistent story of environmental hardships, possibly associated with cosmic events. These could have played a significant role in the events taking place, people being driven by desperate circumstances into acts of conflict.

Apart from a severe frost in June 800 and earth tremors in 801 and 803, with localised outbreaks of disease occurring as a consequence of these, there were no reports in the *Royal Frankish Annals* of environmental problems during the period from 741 to 809. Indeed, several references were made to the mild weather. In contrast, the winter of 810–811 was said to be very severe, following a year in which there had been widespread loss of animals through pestilence, and which featured both solar and lunar eclipses. The opening entry of the *Annals of Xanten* in 810 confirmed these details, ending with the statement that 'there were great losses among cattle and other beasts that year, and the winter was very hard'.[44]

Reports of this nature, sporadic to start with, became increasingly common as the century progressed. The next mention of a severe winter, but without any additional hardships, came in 813. Two years later, an earthquake occurred in Gaul and floods caused damage in Germany. Then, in 817, a comet appeared in the constellation of Sagittarius whilst, during the same year, it was reported that 'rays of fire' appeared in the sky. Three years later, persistent rainfall and humidity led to the widespread loss of crops and the death of many animals, after which there was an exceptionally harsh winter.[45]

In the year 823, the emperor's palace at Aachen was shaken by an earthquake, and severe electrical storms caused much damage to people, animals and property, after which a great pestilence raged throughout Francia. In Saxony, lightning struck out of a clear sky, and twenty-three villages were burned by 'fire from heaven'. Elsewhere, crops were destroyed by hailstorms and, in some places, 'real stones of tremendous weight' were seen to fall with the hail. After another long, cold winter, famine continued to be very severe. Hailstorms broke out again in the summer, and an enormous block of ice was said to have fallen with the hail near Autun, in Gaul. According to reports, this was fifteen feet long, seven feet wide, and two feet thick. Three years later, during fighting for territory with Moors in the southwest of the kingdom, 'people were sure they saw battle lines and shifting lights in the sky and that these marvels forboded the Frankish defeat'. Two years after that, the occurrence of another earthquake and violent electrical storm at Aachen was mentioned in the final entry of the *Royal Frankish Annals*.[46]

The *Annals of St. Bertin* made their first appearance in the same year and, together with the *Annals of Xanten*, chronicled how the conditions continued to deteriorate. Extensive flooding, causing great damage, took place in Francia in 834. Two years later, strange rays of light appeared from east to west in the night sky.[47] For 837 and the transition to the following year, the *Annals of Xanten* reported:

> A mighty whirlwind kept breaking out, and a comet was seen, sending out a great tail to the east, which to human eyes looked as if it was three cubits long [one cubit being the length of a forearm] . . . The winter was wet and windy, and on 21 January thunder was heard, just as on 18 February loud thunder could be heard. And the excessive heat of the Sun scorched the Earth, and there were earthquakes in some parts of the land, and fire in the shape of a dragon was seen in the air . . . and the distress and misfortune of men grew daily in many ways.[48]

The very first entry in the *Annals of Fulda* in 838 confirmed the occurrence of an earthquake in January of that year, going on to mention that tremors were experienced in Lorsch and the region around Worms, Speyer and Ladenburg.[49]

The winter of 838–839 was similarly a very hard one. At the end of December, a great flood covered almost the whole of Frisia, with more than 2,000 deaths being reported, and further floods, together with whirlwinds, occurred in the following year. In February 839, 'an army of fiery red and other colours could often be seen in the sky, as well as shooting stars trailing fiery tails'. A comet was observed in the constellation of Aries, and strange rays of light were seen throughout this and the following year. Nithard, in his *Histories*, referred to an exceptional reddening in two parts of the sky in March and April of 840 A.D., these red patches eventually coming together to give 'the appearance of a clot of blood in the heavens directly overhead', whilst the occurrence of strange rings of light in the sky during daytime was described in the year 841.[50]

The summer of 841 was a very cold one, which delayed the harvest. The Seine flooded in March of that year, with violent tides occurring at the river mouth, and it flooded again in October, even though there had been no rain in the region for two months. The following December, a comet became visible, passing across the constellations of Aquarius and Pisces before disappearing in the vicinity of Andromeda. Around this time, much snow fell, initiating another period of extremely cold weather.[51]

An earthquake shook most of Gaul in the autumn of 842. The following winter was particularly cold and lengthy, causing severe damage to agriculture and livestock, and resulting in widespread disease amongst the human population. Nithard contrasted the terrible conditions with what seemed like the golden age of Charlemagne, and drew a moral conclusion, writing:

> In the times of Charles the Great of good memory, who died almost thirty years ago, peace and concord ruled everywhere because our people were treading the one proper way, the way of the common welfare, and thus the way of God. But now since each goes his separate way, dissension and struggle abound. Once there was abundance and happiness everywhere, now everywhere there is want and sadness. Once even the elements smiled on everything and now they threaten, as Scripture which was left to us as the gift of God, testifies: *And the world will wage war against the mad.*[52]

After snow fell on a night when there was an eclipse of the moon in March 843, Nithard brought his *Histories* to a despairing end, concluding that, with 'rapine and wrongs of all sort' rampant on all sides, the 'unseasonable weather killed the last hope of any good to come'.[53] Two years later, Nithard died fighting the invading Vikings, after spending the intervening period as lay abbot of the monastery at St. Riquier in northern France. However, the annals of St. Bertin, Fulda and Xanten continued to describe the tribulations of the people.

In 843 A.D., it was recorded that the people living in parts of Gaul were forced to eat earth mixed with a little flour in order to satisfy their hunger. Again, the winter of 844–845 was a severe one, and a famine consumed the western region of Gaul, with many thousands dying. There were two earthquakes in the Worms region in 845, after which there was an outbreak of plague.[54] Four years later, another violent earthquake occurred in Gaul. In Germany, there were winter floods and ferocious electrical storms in Germany during the winter of 849–850, followed by a scorchingly hot summer, causing a famine that persisted for three years.[55]

Twenty earth tremors were reported to have occurred in the Mainz region during 855. The weather was unusually changeable, with whirlwinds and hailstorms appearing without warning. Many buildings were struck by lightning, including the church of St. Kilian the Martyr in Würzburg in June. A month later, those walls which had escaped being burnt by lightning collapsed during a violent, sudden storm. Shooting stars were observed in August of that year, a large one and a small one appearing alternately, and this was followed in October by an intense shower of small fireballs.[56]

The winter of 855–856 was excessively cold and dry, and a pestilence carried off a sizeable proportion of the population. In the following winter, the *Annals of Xanten* reported a plague characterised by swollen abscesses, rotting flesh and loss of limbs. Destructive electrical storms also occurred at the time.[57]

There were violent earthquakes in Mainz and neighbouring regions in 858 and 859, causing a great pestilence, and great floods were reported at Liège in May 858. After that, the *Annals of St. Bertin* described how, for three months in the autumn of 859 A.D., 'armies were seen in the sky at night: a brightness like that of daylight shone out unbroken from the east right to the north and bloody columns came streaming out from it'.[58]

The three annals continued to present a consistent account, reporting that the winter of 859–860 was very severe, and that it lasted longer than usual. According to the *Annals of Fulda*, blood-red snow fell in many places, and the Adriatic region was so cold that merchants were able to cross the sea and visit Venice by horse and cart. When winter was finally coming to a close, the *Annals of St. Bertin* described how part of the Moon was obscured one night in a strange fashion. Then, just a few days later, something similar happened to the Sun.[59]

Shortly after this, the authorship of the *Annals of St. Bertin* passed from Prudentius (on his death) to Hincmar, who concentrated on political and ecclesiastical matters, showing little interest in the physical environment. However, in the other two sources, reports of floods, plagues and other catastrophic events continued as before. For example, the entry in the *Annals of Fulda* for 868 A.D. described the sighting of comets in the sky, after which there was exceptionally heavy rainfall, which caused serious flooding. Later in the same year, famine became widespread throughout Germany and Gaul. Confirming this, the *Annals of Xanten* recorded:

> In the month of February peals of thunder were heard from the dark waters in the clouds in the air, and on 15 September, that is the holy night of Septuagesima, a comet was seen in the north-west, followed immediately by very strong winds and

an enormous deluge of water, in which very many were caught unawares and perished. And then in the summer a very severe famine ensued in many provinces, but above all in Burgundy and Gaul, in which a large number of people suffered an untimely death, so that some people are said to have eaten human corpses, while others are supposed to have lived off dogmeat.

During the previous September, according to the same source, 'fire was seen flying through the air with the speed of an arrow, as thick as a pitchfork and shooting off sparks like an iron block in a furnace', before exploding in a cloud of oily smoke over Saxony.[60]

In 870 A.D., the *Annals of Fulda* recorded, 'At Mainz, the sky shone red like blood for many nights, and other portents were seen in the heavens . . . The lands around the same city were struck by two earthquakes . . . Several men gathering in the harvest in the district of Worms were found dead because of the heat of the Sun, which was fiercer than usual. Many were also drowned in the Rhine . . . There was also a serious cattle pestilence in many parts of Francia, which caused irretrievable loss to many'.

Two years later, the Fulda and Xanten annals both reported a summer ruined by persistent thunder, lightning, rain and hailstorms, which caused considerable damage to animals, crops and property. The cathedral of St Peter in Worms was amongst the buildings set alight during electrical storms. In the following December, the city of Mainz was shaken by an earthquake.[61]

The *Annals of Xanten* were brought to an end in 873 A.D. with the words, 'And from 1 November right up until Sexagesima [the Sunday after Septuagesima, falling two weeks before Lent] snow covered the whole surface of the earth, and the Lord constantly distressed his people with various plagues, visiting their transgressions upon them with the rod, and their sins upon them with the whip'. The entry for the same year in the *Annals of Fulda* also spoke of famine and plague throughout Germany and Italy, as well as a period when 'blood rained from the sky for three days and nights' in the county of Brescia. Then, after a very long, hard winter, during which the Rhine and the Main froze over, hunger and pestilence raged through the whole of Gaul and Germany, killing nearly a third of the population. Conditions had clearly not become any easier since the distraught Nithard found himself unable to carry on writing about the problems facing the people thirty years earlier.[62]

Nor was that the end of the suffering, which was often linked, in the minds of the people, to portents in the sky. In June of 875 A.D., according to the *Annals of Fulda*, a particularly bright comet with a lengthy tail was observed in the north and, shortly afterwards, a flash flood destroyed buildings at Eschborn, far from any river or stream, killing eighty-eight people. The annalist commented that the comet foretold 'by its appearance the remarkable and indeed tragic event which quickly followed, although for our sins it may be feared that it signified still more serious matters'. Again, in 882, a conspicuous comet 'prefigured by its appearance the disaster which quickly followed', the death of King Louis the Younger (son of Louis the

German) and another outbreak of civil war. Later in the year there was a 'great and terrible plague' in Bavaria and, further north, the worst storm within living memory, during which hailstones of unequal size, with jagged edges, fell in the area of the Rhine.[63]

During the closing years of the ninth century, the *Annals of Fulda* included several more references to exceptionally harsh winters, floods and episodes of plague and famine, although there was no further mention of celestial phenomena. The final entry was completed in 901 A.D.

Of course, it goes without saying that that the details given by these various sources are too imprecise, and probably too unreliable in some aspects, for any definite conclusions to be drawn about an underlying mechanism for the natural catastrophes described. After all, floods, earthquakes, temperature fluctuations, disease and famine can have a variety of causes, and the appearance of comets in the sky at the present time does not usually lead to any adverse effects on Earth. Nevertheless, the repeated occurrence of these and other features over a number of decades suggests the possibility of some sort of association between them. The evidence, taken as a whole, is consistent, if no more, with an episode of 'coherent catastrophism', an encounter between the Earth and the disintegrating remains of a giant comet, as in the Clube–Napier scenario. The dusting of the upper atmosphere could lead to reduced temperatures at the Earth's surface, and impacts of debris could initiate earthquakes or floods, with these various crises leading in turn to the occurrence of famine and disease.[64] Such a mechanism provides a plausible mechanism for the phenomena observed and the hardships experienced during these extraordinary years.

Yet, the case should not be overstated. Even if there had been an encounter with cometary debris over this period, it would be unrealistic to suppose that all the observed phenomena and disasters were directly related to this. The appearance of a comet in the sky could hardly have led directly to the death of a king, and cannot necessarily be assumed to have been linked to subsequent environmental crises. The very conspicuous comet of 837 A.D. was almost certainly that which we now know as Comet Halley, which travels harmlessly past the Earth every 76 years, although calculations of its orbit suggest that it has never again been so close to us as it was on that occasion.[65] Some of the other lights in the sky described in the annals might simply have been aurorae, and some of the catastrophes could have had a purely earthbound explanation.[66]

For all that, the possibility of a linkage between many of the phenomena observed and experienced during the ninth century has to be taken seriously, particularly since evidence of unusual hardship at this time is not confined to northern Europe. Even the Nile was reported to have frozen over in 829 A.D. At the other side of the world, Central America was experiencing its most arid period for 7,000 years, a significant factor in the collapse of the Classic Maya civilisation, when the population fell by at least 67% and possibly more than 90%.[67]

David Webster, an anthropologist from Pennsylvania State University, wrote in his 2002 book, *The Fall of the Ancient Maya*, 'Right now megadrought is the "hot" explanation for the Classic Collapse, and the usual bandwagon effect is in full career among many of my colleagues, although others remain properly suspicious of drought as the triggering mechanism. Generally speaking, the drought explanation seems to be most enthusiastically accepted by natural scientists, and I suspect that they derive a considerable cachet from linking their observations of chemical or astronomical phenomena to the great mystery of the Maya collapse'. Nevertheless, regardless of his suspicions about the motivation and judgement of other investigators, Webster acknowledged that a drought had occurred in Central America at the time, whilst stressing that, in his view, it was only one of several factors responsible for bringing to an end the Classic Maya civilisation. As to the cause of the drought, David Hodell and other environmental scientists from the University of Florida have argued that it could have been linked to changes in the energy output of the Sun. Alternatively, Richardson Gill suggested, in *The Great Maya Droughts*, that the environmental downturn which precipitated the Classic Maya Collapse might have been caused by the accumulation of dust and gases in the upper atmosphere as a result of major volcanic eruptions in and around Mexico. Although precise details are uncertain, it seems that Popocatépetl exploded (as a VEI 5 or 6 event) around the start of the ninth century, as did El Chichón (following its earlier eruption), with Mt Pelée and La Soufrière–St Vincent erupting in the Caribbean region shortly before or after the century's end.[68] It is also known that a major eruption from the Vatnaöldur fissure in southern Iceland, linked to the Torfajökull volcano, occurred around the year 871, depositing ash over the surrounding countryside, and as far away as Ireland.[69]

In fact, according to Greenland ice-core data, the ninth century was a particularly active period for volcanoes, for acidity peaks indicate that major eruptions took place in 822, 823, 853, 875 and 900, with another in 902. Furthermore, because of location and/or low acidity of emissions, some large eruptions might not have left any trace in the Greenland ice. So, for example, Swedish tree-ring data suggest that, even against the backdrop of a cold century (confirming the accounts in the annals), the year 860 was a particularly harsh one, indicating the possibility of a major eruption at that time.[70]

Without question, massive and sustained volcanic activity could have caused the environmental hardships experienced in Central America and Northern Europe throughout the ninth century (figure 29.3). However, once again the question arises as to whether the volcanic eruptions (even if they occurred on the scale required, which is yet to be established) were the primary cause of the environmental crises, or whether they were secondary to a cometary bombardment. On the basis of Chinese and European records (mainly the former), Barry Hetherington, in *A Chronicle of Pre-Telescopic Astronomy*, noted the observation of thirty comets during the middle half of the ninth century (826–875), some of them in association with meteor showers.

Figure 29.3   The Kerid volcanic crater (now filled with water) in southern Iceland, a region where major eruptions, characterised by extensive lava flows and the release of sulphurous gases, are far from uncommon on a time-scale of centuries, and sometimes occur in clusters.

Whether significant or not, that constituted a peak in recorded sightings of comets, for there were only eight in the previous half-century (all between 813 and 823), whilst the number of comets reported in each of the next two fifty-year periods showed a fall of over 25% from the maximum.[71]

Needless to say, the evidence is far from conclusive as to the precise causal mechanism. Nevertheless, it is reasonable to suppose that a long-enduring environmental downturn on two continents, starting around the same time, could have been the consequence of a catastrophic event, whether of terrestrial or extraterrestrial origin.

Natural catastrophes in more recent times

In Europe, there was another period of particularly severe winters, famine and pestilence around 940 A.D., possibly associated with the great Eldgjá ('Fire Chasm') eruption in southern Iceland, as suggested by Richard Stothers of NASA. After that, average temperatures generally moved upwards for several hundred years, but then the trend went into reverse. Mike Baillie, in *From Exodus to Arthur*, argued that, according to the tree-ring evidence, a significant and widespread environmental downturn occurred in the decade prior to 1347 A.D., when the plague known as the Black Death began to ravage the populations of Europe. Going along with a suggestion made by Clube and Napier in *The Cosmic Winter*, Baillie thought that the primary cause, once again, was an encounter between the Earth and a cluster of cosmic bodies, the breakdown products of a giant comet. Numerous 'portents' were seen in the sky throughout this period, in particular a great comet, which appeared in 1337. Ten years after that, in the year the plague arrived, a severe earthquake struck the eastern Mediterranean region, accompanied by a tidal wave and clouds of poisonous gases. At around the same time, 'fire from heaven' fell on Turkey over a sixteen-day period.[72]

From the fourteenth century until the seventeenth, the climate continued to deteriorate, leading to what has been called the 'Little Ice Age', when rivers in northern Europe and the U.S.A. regularly froze over. Not until the middle of the nineteenth century was there a significant improvement in conditions. Amongst the victims of the downturn in temperature were the Vikings of western Greenland, who had settled there in the tenth century, and thrived for 500 years, but disappeared rapidly between 1340 and 1360. Some attribute the Little Ice Age to an accumulation of dust in the upper atmosphere, due either to volcanic activity or to the influence of a comet. Certainly, particularly large eruptions of Hekla, in Iceland, occurred in 1104 ('Hekla 2') and 1158 ('Hekla 1'), with slightly lesser ones in 1510, 1693 and 1766, and others in 1206, 1222, 1300, 1341, 1389, 1597 and 1636. In addition, according to Stothers, writing in the journal *Climatic Change* in 2000, the largest volcanic eruption of the past millennium occurred in the tropics early in 1258, causing dry fogs, reduced temperatures, crop failure and disease in Europe and elsewhere. Also, as Richardson Gill pointed out in *The Great Maya Droughts*, there were large tropical eruptions, each with a VEI of 4 or more, at Kuwae (Indonesia) in 1452, Cotopaxi (Ecuador) in 1534, San Salvador (El Salvador) around 1572, Makian (Indonesia) in 1646, Santorini (the Aegean) around 1650, Jorullo (Mexico) in 1764 and San Martin (Mexico) in 1793, all of which were followed by a few years of drought and famine in the Maya regions of Mexico. There was also a significant eruption of Huaynaputina (Peru) in 1600.[73] Furthermore, the period between 1350 to 1550 was noted by Chinese astronomers as one in which fireball activity was particularly high, and there were many reports of strange lights, noises and 'battles in the sky' from northern Europe throughout the sixteenth and seventeenth centuries.[74]

As we discussed in Chapter 20, there have been many devastating natural catastrophes on Earth since then. These include the Lisbon earthquake of 1755, which measured 8.75 on the Richter scale and, through its direct and indirect effects, killed tens of thousands of people, and the Messina earthquake of 1908, when the death toll was even higher. In the same year, the explosion of an asteroid or cometary fragment in the atmosphere devastated the Tunguska region of Siberia. A few decades earlier, in 1883, the VEI 6 Krakatoa eruption produced a tsunami which killed 40,000 people, whilst the VEI 8 Tambora eruption of 1815 lowered temperatures around the world during the following year. However, none of these phenomena can be said to have changed the course of history to any significant extent. Nevertheless, that outcome seems to have been far from inevitable. For instance, had the Tunguska object arrived just four hours later, destroying St Petersburg, the history of the twentieth century could have been very different from what it was. At the time, St Petersburg was the main focus for

Russian politics, with Tsar Nicholas II becoming increasingly unpopular, and events taking place which eventually led to the October Revolution in 1917.

For all that, many archaeologists, ancient historians and Earth scientists continue to look for traditional, gradualistic mechanisms for all climate and cultural changes, whether during the Little Ice Age, the ninth or sixth centuries, or at other times. In that view, major changes of climate could simply be due to natural fluctuations.[75] The gradualist position is often justified by reference to evidence which seems to suggest that similar events at different locations may have taken place hundreds of years apart. On the other hand, few chronologies are absolutely secure, for fine-tuning is taking place all the time, and it is becoming to appear increasingly possible that such events may have a linked cause. A number of investigators implicate the Sun itself, drawing attention to the existence of natural cycles involving sunspot activity and total energy output. Some of these cycles have a periodicity of a few years, others of several hundred years. However, the fluctuations in solar intensity are only in the order of a fraction of one per cent so, although they must have had some effect on the Earth's climate, it is unlikely that, in themselves, they could have caused environmental changes on the scale observed over the past 1,500 years.[76] Hence, many continue to suspect the involvement of a catastrophic event, such as a massive volcanic eruption or a cometary encounter.

The issues are far from resolved, and very different points of view are being maintained. All that is certain is that the arguments will continue for many years.

30 Conclusions

Changing perceptions of Earth history

Our account of the development of life on Earth has reached the present day, but is that the end of the story? The human race has a today, but will it have a tomorrow? We fervently hope it will, but before addressing the question, let us recall some important aspects of our journey through time.

Even in the days of the philosophers of ancient Greece there were major disagreements about the nature of change at the Earth's surface. Plato accepted that some myths could have been derived from catastrophic events, including ones of extraterrestrial origin, whereas his pupil, Aristotle, denied that possibility and maintained that changes at the surface of the Earth all took place in gradual fashion.

In the medieval period, the Christian Church adopted the philosophy of Aristotle, insisting that natural catastrophes could never occur. Of course, on very rare occasions, such as the Flood in the time of Noah, God had devastated the Earth to punish the wicked, but that was a very different matter. Up until the middle of the nineteenth century, all scientists in the western world were constrained by Christian philosophy, although as time went by they showed greater and greater flexibility of interpretation, in attempts to reconcile stories in the Bible with field observations. God could not be excluded but, in the eighteenth and early nineteenth centuries, there were two main views about the role of the deity in the development of the Earth. The older Anglican/Roman Catholic tradition maintained that God could intervene as required (and what was the point of prayer if that never happened?) whereas, according to the more recent Deist/Unitarian school of thought, God had set up laws of nature at the time of Creation and subsequently let them operate without interference. Because of their different underlying philosophies, scientists in the first group (which was still very much the dominant one amongst society in general) accepted the possibility of sudden and profound change, whereas those in the second group generally preferred a belief in a more gradual, continuous process. Contrary to what is often stated, scientists advocating gradualism did not, in general, display a more scientific approach in their work than their catastrophist contemporaries.

In Britain, by the early nineteenth century, scientists in both groups had, by-and-large, rejected supernatural explanations for geological change. Equally, both were opposed to the possibility of the evolution of biological forms by natural processes.

The balance of opinion shifted towards gradualism when Charles Lyell, a geologist in the Deist/Unitarian tradition, introduced confusion by using the same term, 'uniformity', to describe both a methodological principle and a world-view. Because of this linkage, young geologists entering the profession were easily persuaded that, because they accepted his methodological principle that the same laws and processes operated throughout time (an essential requirement of science, as acknowledged by gradualists and catastrophists alike), they must also accept his world-view that processes had always taken place at the same rate as they did at the present time, i.e. in gradualistic fashion. Lyell believed in the fixity of species, and was opposed to any form of biological progression, but his friend Charles Darwin produced a theory of evolution which extended Lyell's geological gradualism to the realm of living organisms. Darwin's evolutionary gradualism was derived from a mixture of philosophy and science, but not on any evidence from the fossil record. Indeed, this indicated that there had been abrupt transformations of species, particularly at the end of geological periods. Only by stressing the imperfections of the fossil record was Darwin able to justify his gradualistic views.

Nevertheless, Darwin's theory of evolution was seized upon and used by a new generation of scientists as a means of separating science from religion. In that context, as one of their number, Thomas Huxley, fully appreciated, gradualism could not be challenged without providing ammunition for the forces of reaction. In any case, the only plausible catastrophist hypothesis to be put forward during the nineteenth century, the cooling Earth model of Léonce Élie de Beaumont, was found to be inconsistent with accumulating evidence, and so had to be rejected. Hence gradualism, in both geology and evolution, was the ruling paradigm throughout the closing decades of the century. Catastrophism was out in the cold.

That situation continued for most of the twentieth century, because of a general lack of appreciation of hazards from space and of the fact that episodes of vulcanism could occasionally be much more violent and extensive than those normally observed. Also, it came to be believed, quite wrongly, that the catastrophists of previous centuries had been driven by religious dogma, whereas the early gradualists were objective scientists, formulating theories solely on the basis of evidence. For decades, the dominance of gradualism over catastrophism in mainstream scientific circles could hardly have been more complete.

That did not deter a small number of reputable scientists (such as Otto Schindewolf and Digby McLaren), and rather more non-scientists, from putting forward catastrophist theories. In retrospect, it can be seen that some of these showed considerable perception in identifying problems with conventional views about what happened at certain times, suggesting both plausible and implausible alternatives. However, none of these challenges to orthodoxy made much of an impression on mainstream thought. At best, there was little real evidence to support the alternative theories, other than indications of abrupt evolutionary change or civilisation collapse at certain times. At worst, catastrophist arguments sometimes depended largely on a subjective interpretation of ancient myths, an occult revelation or the literal acceptance of either Plato's account of the destruction of Atlantis or a story from the Old Testament. Without question, despite the rational arguments which were also being put forward, dubious and irrational catastrophist claims in the 'fringe' literature reinforced the belief in mainstream circles that gradualistic theories were the only ones with any scientific merit.

However, during the third quarter of the twentieth century, the background had begun to change. The existence of asteroids and comets in Earth-crossing orbits became known, and it was realised that many craters at the surface of the Earth had been formed by the impact of such objects. Furthermore, it became apparent that a nearby supernova explosion could devastate life on Earth, and the full destructive potential of volcanoes and earthquakes also came to be appreciated. Palaeontologists such as Niles Eldredge and Stephen Jay Gould began to argue that evolutionary change was generally punctuational, not evenly paced, whilst, following the example of Norman Newell, there was increasing acceptance that mass extinctions of species, particularly those in the Late Ordovician, Devonian, Permian, Triassic and Cretaceous Periods, were real events, not artifacts of the fossil record. Even so, there was strong resistance to any suggestion that catastrophist mechanisms had been involved, explanations based on gradual environmental change related to continental drift and sea-level fluctuation being widely supported. Despite the changing background, the vast majority of palaeontologists (and archaeologists) continued to believe that catastrophic events had made no significant contribution to the course of life on Earth.

The return of catastrophism

The catalyst for a fundamental re-appraisal of the situation was provided in 1980, when Luis and Walter Alvarez not only proposed that the mass extinction episode at the end of the Cretaceous Period had been caused by the impact of a 10-km-diameter asteroid, but also produced evidence in the form of high iridium concentrations at sites from around the world. This excess iridium could only have come from the explosion of an extraterrestrial object or from the Earth's core, released by volcanic activity. That stimulated an intense and long-lasting debate about whether an impact or volcanic activity was responsible for the iridium, and for the extinctions of species. Due to the work of Vincent Courtillot and others, it was clear that vulcanism had occurred in India in the Late Cretaceous on a scale which could have had worldwide effects. However, during the 1990s, it also became established that a 10-km projectile had struck Chicxulub, in Mexico, at the end of the Period. Yet volcanic eruptions and impacts were only part of the story, for there was evidence that the shallow seas which had been covering large areas of the continents drained away during the Late Cretaceous, changing environments, whilst average temperatures fell significantly over the same time-interval.

It is now generally agreed that the Late Cretaceous extinction event was a complex affair, possibly lasting several million years, but culminating in a relatively brief, intense episode which may have been linked to the impact. Both gradualist and catastrophist mechanisms appear to have played significant roles. Michael Rampino, in his Shiva hypothesis, has suggested that, on this and other occasions, there may have been a causal link between an extraterrestrial impact and tectonic events at the Earth's surface. However, that remains a minority view, in the absence of a precise mechanism, and the indications that the Late Cretaceous vulcanism in India, as well as other tectonic events, began before the impact at Chicxulub. On the other hand, the Chicxulub event may not have been an isolated one and, indeed, we have evidence of a cluster of moderately sized impacts ten to fifteen million years earlier. There is reason to think that impacts may often occur in clusters, either because of an encounter with a cometary shower or the disintegrating nucleus of a giant comet.

The other major mass extinction events were similarly complex, each apparently having a variety of causes. Continental drift played a significant role in the Late Ordovician extinctions, which occurred at a time of climate changes (including episodes of glaciation), sea-level fluctuations and periods of anoxia. No evidence of the involvement of large-scale vulcanism or major impacts has yet been produced. In contrast, the anoxia and the fluctuations in climate and sea-level at the time of the Late Devonian (Frasnian–Famennian) extinctions were accompanied by extensive volcanic activity in Eastern Europe, and there was also a cluster of impacts at around this time. In particular, an extraterrestrial object more than 5 km in diameter struck Woodleigh, Western Australia, in the Late Devonian, although this event has yet to be linked precisely to the Frasnian–Famennian extinctions.

At the end of the Permian Period, when the continental plates locked together with the formation of the supercontinent, Pangaea, and the greatest mass extinction event of all took place, there were again episodes of anoxia and significant changes of climate and sea-level, together with major outpourings of lava, which resulted in the laying down of the Siberian traps. As with the Late Cretaceous extinctions, a significant increase in the intensity of extinctions at the very end of the Period is consistent with the occurrence of a major impact, and there are some indications of this. However, direct evidence, in the form of an appropriate crater (the 'smoking gun'), is still lacking. Similarly, the extinctions at the end of the Triassic Period took place during a time of anoxic episodes and fluctuations in climate and sea-level, when extensive volcanic activity produced the basalts of the Central Atlantic Magmatic Province. Shocked quartz grains and other findings in boundary rocks suggest the involvement of an extraterrestrial impact but, again, no appropriate crater is known. A cluster of missiles, including the 5-km one which produced the Manicouagan crater, struck the Earth during the Late Triassic, but these impacts were too early to be linked directly to the extinctions at the end of the Period.

Regardless of the causes of these various mass extinction events, it is now clear that they played an important role in evolution. Darwin believed that all evolutionary change was the result of competition, with the winners thriving and the losers being forced into extinction. We now know that, on at least some occasions, mass extinctions took place in ways that cut across ongoing trends, presenting the lucky survivors with an unexpected opportunity to proliferate without challenge and to provide the basis for subsequent evolution. Thus, without the destruction of the dinosaurs in the Late Cretaceous extinctions, it is extremely unlikely that the mammals, and hence ourselves, would have come to dominate the Earth.

Even with the dinosaurs out of the way, the triumph of the mammals was far from assured, as environmental conditions fluctuated. Also, even after the triumph of the mammals, *Homo sapiens* may never have emerged had circumstances been slightly different.

The fall in average temperatures in the Late Cretaceous was reversed in the Early Tertiary but then, starting in the Eocene Epoch, came a more sustained period of cooling which culminated in the Pleistocene Ice Ages. A significant deterioration of the climate could be caused by a catastrophic event, or by something much more mundane. Climates can be extremely unstable, so a change in a single parameter (say, the atmospheric carbon dioxide concentration) could sometimes upset a delicate balance of factors and result in an abrupt rise or fall in average temperatures. Climate changes could also occur in a more gradual fashion, for example, because of the effects of continental drift. It seems that continental drift played a significant role in the climatic downturn after the Early Eocene, but other factors are also likely to have been involved. At the end of the Eocene Epoch, two 5-km objects from space struck in quick succession at Popigai, Russia, and Chesapeake Bay, U.S.A., whilst in the Early Oligocene, the Ethiopian traps were laid down during an extensive episode of vulcanism. Around the time of the Eocene–Oligocene transition, there was a mass extinction episode, albeit one which was protracted and on a smaller scale than the 'big five'. Much later, just 75,000 years ago, the emerging species of *Homo sapiens* was almost wiped out by the onset of the Würm glaciation, which coincided with the super-eruption of Toba, in Indonesia.

The last of the Pleistocene Ice Ages came to an abrupt end around 11,500 years ago, causing catastrophic flooding around the world. Human beings thrived under the new climatic conditions and expanded into new areas, whilst many species of large animals became extinct. The development of farming led to the development of towns to serve as regional centres and later, to the introduction of more complex cities. However, as well as the advances, there were also some set-backs, as cultures which appeared well established came to a sudden end. That could have been due to human factors, such as wars, but on some occasions, although human factors may still have played a part, societal collapse seems to have been associated with an environmental crisis in the region. Again, that could have resulted from a 'normal' climatic fluctuation, or it might have been the consequence of a natural catastrophe, such as an earthquake, volcanic eruption or encounter with cometary debris (as proposed by Victor Clube and Bill Napier). A catastrophic event capable of disrupting an early civilisation would not need to be anything like as powerful as one which could cause a mass extinction episode, so would occur much more frequently.

Such a time of crisis occurred around 2300 B.C., at the end of the Early Bronze Age in the Middle East, when the Old Kingdom of Egypt collapsed. There was another around 1150 B.C., when Late Bronze Age civilisations such as the Mycenaean disappeared. In between, Thera erupted in catastrophic fashion in the Aegean, destabilising (although not destroying) the Minoan civilisation. Another environmental downturn began about 535 A.D., affecting many locations, including South America, Europe and China. Tree-ring evidence shows that conditions were unusually cold at this time. Very low temperatures were also a characteristic feature of the ninth century, when there was turmoil in northern Europe, and the Classic Maya civilisation of Central America came to an end during the most arid period experienced in that region for thousands of years. After a reversion to warmer climates, a further environmental downturn began during the fourteenth century, around the time of the Black Death, and culminated in the 'Little Ice Age' of the seventeenth and eighteenth centuries, when rivers often froze over in winter in northern Europe and the U.S.A.

Past, present and future

Regardless of the details of what happened at any particular time, it is clear that a major change of thinking has taken place. In contrast to the situation only twenty years ago, mainstream scholars are now prepared to take seriously the possibility that catastrophic events, including ones of extraterrestrial origin, may have shaped the course of life on Earth. Such possibilities have been widely discussed for events which took place millions of years in the past, particularly the mass extinction episodes in the stratigraphic record, but they could also be of relevance to events in more recent times, affecting human history or pre-history. The very mention of Atlantis, Noah's Flood or Sodom and Gomorrah still tends to suggest to some scholars an inevitable association with unscientific nonsense or religious dogma. However, that does not necessarily follow, despite the fact that some writers on these topics continue to pursue their own agenda, whether through self-delusion or the desire to make money out of a gullible public. A significant number of serious scholars now accept that these and other ancient legends could be based on some genuine memory of a catastrophic event, and it would be just as unscientific to deny that possibility as to claim that the legends must be correct in every particular. With the developments of the past few years, there is no longer any intellectual justification for dismissing catastrophist arguments out-of-hand, so these, whether right or wrong, should now be considered on exactly the same basis as gradualist ones.

Several important contributions to our current understanding have come from scientists who crossed disciplinary lines. Thus, for example, physicists and chemists played a major role in stimulating a re-assessment by palaeontologists of events at the K–T boundary, observations by biochemists caused palaeoanthropologists to change their views about the date of the ape–human split, whilst Earth scientists and astronomers provided ancient historians and social anthropologists with additional factors which may have been of significance in the collapse of the Classic Maya and other civilisations. Often, such incursions have been resented by those who believed that the theories they had established after years of hard work and careful thought were being questioned by outsiders with only a superficial knowledge of the subject area. Indeed, specialists must almost always have a much better understanding of their particular disciplines than anyone else. Nevertheless, very occasionally, an outsider can introduce an important new piece of evidence, or a way at looking at a situation that would never occur to a specialist schooled in a particular way of thinking. Even then, intruders should be wary of thinking they have found a simple solution to a complex, long-standing problem, just as insiders should avoid the trap of believing that no-one without their specialist knowledge can possibly have anything useful to contribute. Apart from the peculiarities of human nature, there can be no reason for antagonism between groups with different specialities who find themselves working in the same area. After all, it would be expected that an interdisciplinary approach, involving people with different skills, knowledge and perspectives, working together in a spirit of genuine co-operation, would offer the best chance of being able to come to an understanding of complex events such as episodes of mass extinction or civilisation collapse. Such an approach has worked well on occasions in the past, and is likely to be even more important in the future, as scholars try to find more complete explanations, distinguishing between cause and effect.

However, the search for the truth about events at times of drastic change is more than just a quest to find a satisfactory solution to an academic puzzle. Rather, it could be a matter of life and death for populations, cultures or even the human race. Although we may try to put it out of our minds, we know with near certainty of forthcoming events which could kill thousands, if not millions, of people. The only question is, 'When?' We know, for example, that it is only a matter of time before a major earthquake hits California. Similarly, we have strong reason to think that, within a few years or, at most, a few decades, Vesuvius will erupt again, putting the population of Naples and surrounding regions at great risk. Perhaps on a longer time-scale (or perhaps not, for we can only guess), there will be another Tunguska-style explosion, caused by the arrival of a moderately sized object from space. Depending on sheer chance, this could occur in a sparsely populated region, as before, or it could wipe out a major centre of population or even an entire country. If the explosion occurred over or within an ocean, a tsunami would be produced which could devastate coastal regions.

Such events would be bad enough, but we know that, on very rare occasions, eruptions and impacts can occur on a much larger scale, causing mayhem over a wide area and, possibly, changing environments right around the world. It seems likely that events of this nature, perhaps coupled with stresses produced by more gradual processes, were responsible for at least some of the major mass extinction episodes in the history of the Earth, when over half of existing species became extinct on each occasion. Whatever the precise causes, and regardless of arguments about an exact periodicity, it is now clear that major mass extinction episodes occur, on average, about once every 30 Myr. And, although there have been waves of extinctions since, particularly those concentrated around the end of the Eocene and the beginning of the Oligocene, the last of the 'big five' mass extinction events took place at the end of the Cretaceous, 65 Myr ago.

Of course, that does not necessarily mean that a Chicxulub-style impact is overdue, for it could be millions of years (although we cannot guarantee it) before another impact on anything like the same scale takes place. In any case, it is far from certain that impacts were involved in many of the previous mass extinction events. Even in the case of the Late Cretaceous extinctions, the precise significance of the Chicxulub impact remains uncertain. Did it merely provide the *coup de grâce* to species which were already struggling to survive, simply accelerating progress towards an outcome which would have happened anyway? Alternatively, did it cause such devastation, directly and indirectly, that it would have wiped out many species even if it had occurred under very different circumstances?

Opinions differ about that but, given what we know about the theoretical consequences of the impact of a 10-km-diameter asteroid or comet, and the actual destruction caused by the arrival of a much smaller cosmic body over the Tunguska region in 1908, it seems difficult to play down the likely importance of the Chicxulub event. And what of the future? We human beings are remarkably resourceful and adaptable. We can live in a wide range of environments, finding a variety of ways of protecting ourselves against extreme conditions. However, could our species survive the impact of another 10-km body? It is by no means certain. At best, we might hope that a small population could be maintained throughout the period of crisis by being cocooned in a well-equipped underground bunker (and then what?), or that scattered communities could cling onto existence in the most basic of fashions. Even a much smaller, and hence more likely, impact could destroy nations, cultures and the systems on which our civilisation relies.

Now that we are aware of the threat to Earth from space, it is clearly important for us to identify catastrophes which have taken place in far-distant times, or more recently, where the primary causal mechanism was an extraterrestrial one. The same applies to catastrophes resulting from explosive events of great magnitude at the Earth's surface. If we are to protect ourselves from the possibility of such catastrophes occurring again, we also need to know exactly what caused them. If a catastrophe resulted from a massive volcanic eruption, then did something trigger this, and were there any warning signs? If a catastrophe resulted from an extraterrestrial impact, then did this occur in isolation, or was it part of an encounter with cometary fragments? What changes in terrestrial processes could be stimulated by the shock of an extraterrestrial impact? And what about those episodes of mass extinction or sudden climate change which were apparently caused by an interplay of terrestrial factors, without the direct involvement of catastrophist mechanism? Can any pattern be detected which could help with prediction of the future?

As with many situations, knowledge is of paramount importance. Only if we know with reasonable certainty what happened in the past, and have developed a good appreciation of the nature of all actual and potential agents of destruction, can we start to address the issues. To produce the best answer to threats against the future of our civilisation, we must fully understand the nature of the problem. The Lyellian dictum of the present being the key to the past has served us well for over a century but, perhaps, there is now a need for it to be re-cast. In a very real sense, the past could be the key to the future.

Notes

Because of the nature of this book, the references include a mixture of primary and secondary sources, to help both the specialist and the interested layperson. As will be apparent, the range of topics considered will mean that specialists in some of them are likely to be non-specialists in others. Regardless of that, some of the older primary sources are not readily accessible, even to professional academics. Also, secondary sources are sometimes used in the book in primary fashion, to provide evidence of changing attitudes. Hence, whilst it might be good practice in a work dealing with a much narrower topic, and aimed exclusively at specialists, to restrict the citation of references to primary sources, that would not be appropriate in this case. Hopefully, the approach adopted will be clear to everyone, and provide specialists and non-specialists alike with a useful range of sources.

Chapter 1

1. *Genesis* Chapters 5–8. Quotations here and elsewhere from the King James (1611) version of the Bible.
2. D.R. Dean, The age of the Earth controversy, *Annals of Science*, 38, 1981, 435–456; S.J. Gould, *Eight Little Piggies*, Penguin Books, Harmondsworth, 1994, 181–193; S.J. Gould, *Questioning the Millennium*, Jonathan Cape, London, 1997, 89–98; D.E. Duncan, *The Calendar*, Fourth Estate, London, 1998, 120–123, 285–286; E.G. Richards, *Mapping Time*, Oxford University Press, 1998, 11, 220–226; K. Lippincott, *The Story of Time*, Merrell Holberton, London, 1999, 48, 258.
3. *New Larousse Encyclopedia of Mythology*, Hamlyn, London, 1968, 54–55, 62–63, 93–94, 345, 438, 445, 466, 481; R. Huggett, *Cataclysms and Earth History*, Oxford University Press, 1989, 12–17; H.A. Guerber, *Greece and Rome*, Senate, London, 1994, 22–26; D. Ferguson, *History of Myths Retold*, Chancellor Press, London, 2000, 35; A. Cotterell (ed.), *Encyclopedia of World Mythology*, Parragon, Bath, 1999, 20, 33, 57; R. Storm, *The Encyclopedia of Eastern Mythology*, Lorenz, London, 1999, 30–31, 56–57, 81; V. Irons, *History of Mythology*, Chancellor Press, London, 2000, 170–171.
4. *New Larousse Encyclopedia of Mythology*, 92–93; A. Cotterell (ed.), *Encyclopedia of World Mythology*, 57, 66, 285; V. Irons, *History of Mythology*, 118, 180–183; R. Graves, *The Greek Myths*, Penguin Books, Harmondsworth, revised edn, 1960, Chapters 7, 35, 36; D.A. Mackenzie, *Pre-Columbian America*, Senate, London, 1996, 67–69; D.M. Jones and B.L. Molyneux, *The Mythology of the Americas*, Anness, London, 2002, 110–111; H.A. Guerber, *Greece and Rome*, 1–14, 122–125, 364; D.A. Hardy and J. Murray, *The Fires Within*, Dragons World, Limpsfield, Surrey, 1991, 31, 106–107; H. Sigurdsson,

Melting the Earth, Oxford University Press, 1999, 14–17, 21–22, 34–50; H. Sigurdsson (ed.), *Encyclopedia of Volcanoes*, Academic Press, San Diego, Calif., 2000, 1312; A. Scarth, *Savage Earth*, HarperCollins, 2001, 34, 122–123.

5. *New Larousse Encyclopedia of Mythology*, 49–54, 142–143, 320–321; D. Ferguson, *History of Myths Retold*, 18–23, 44–45, 52–53; A. Cotterell (ed.), *Encyclopedia of World Mythology*, 18, 40, 288–289; V. Irons, *History of Mythology*, 19, 174–175; R. Storm, *The Encyclopedia of Eastern Mythology*, 34, 40–41, 80; D.M. Jones and B.L. Molyneux, *The Mythology of the Americas*, 109; R. Graves, *The Greek Myths*, Chapter 42; K.E. Sullivan, *Viking Myths and Legends*, Brockhampton Press, London, 1998, 119–123; H. Sigurdsson, *Melting the Earth*, 17; H.A. Guerber, *Greece and Rome*, 64–68.

6. Plato, *Timaeus*, chapter 22c,d. Available, for example, in R.G. Bury (transl.), *Timaeus, Critias, Cleitophon, Menexenus, Epistles*, Loeb Classical Library, Harvard University Press, Cambridge, Mass., 1929.

7. E. Zangger, *The Flood from Heaven*, Sidgwick and Jackson/Book Club Associates, London, 1992, 37–62; P. James, *The Sunken Kingdom*, Jonathan Cape, London, 1995, 1–20; P. Jordan, *The Atlantis Syndrome*, Sutton Publishing, Stroud, Gloucs., 2001, 8–30.

8. *New Larousse Encyclopedia of Mythology*, Introduction; R. Graves, *The Greek Myths*, Introduction.

9. *Larousse Encyclopedia of Archaeology*, Hamlyn, London, 1972, 245–265; C.W. Ceram, *Gods, Graves and Scholars*, Penguin Books, Harmondsworth, 1974; J.V. Luce, *Homer and the Heroic Age*, Thames and Hudson, London, 1975; R. Etienne and F. Etienne, *The Search for Ancient Greece*, Thames and Hudson, London, 1992, 101–115; H. Duchêne, *The Golden Treasures of Troy*, Thames and Hudson, London, 1996, 38–115; J. Fleischman, Homer's bones, *Discover*, 23(7), 2002, 58–65.

10. *Genesis* 19:24.

11. *Psalm* 11:6.

12. *Malachi* 4:1.

13. *The Illustrated Guide to Egyptian Mythology*, Studio Editions, London, 1996, 87–88; V. Irons, *Egyptian Mythology*, Chancellor Press, London, 1997, 37; A. Cotterell (ed.), *Encyclopedia of World Mythology*, 46; R. Storm, *The Encyclopedia of Eastern Mythology*, 38.

14. A.S. Murray, *Who's Who in Mythology*, Studio Editions, London, 1994, 45–46, 50–54; G. Howe and G.A. Harrer, *A Handbook of Classical Mythology*, Oracle, Royston, Hertfordshire, 1996, 230–231, 297.

15. Aristotle, *Meteorologica*, Book 1, Chapter 14. Available, for example, in D. Ross (ed.), *The Works of Aristotle*, Volume 3, Clarendon Press, Oxford, 1931.

16. Aristotle, *De Caelo* ("On the Heavens"), Books 1 and 2. Available, for example, in D. Ross (ed.), *The Works of Aristotle*, Volume 2, Clarendon Press, Oxford, 1930. See also A. Chapman, *Gods in the Sky*, Channel 4 Books, London, 2002, 99–127.

17. Aristotle, *Meteorologica*, Book 1, Chapters 4–7.

18. A. Koestler, *The Sleepwalkers*, Penguin Books, Harmondsworth, 1964, 107–116; P. James, *The Sunken Kingdom*, Chapter 5; A. Chapman, *Gods in the Sky*, 183–248.

19. *2 Peter* 3:10.

20. V. Clube and B. Napier, *The Cosmic Serpent*, Faber and Faber, London, 1982, 157–189; C. Sagan and A. Druyan, *Comet*, Headline Books, London, 1997, 13–38; A. Cook, *Edmond Halley*, Clarendon Press, Oxford, 1998, 205; J. Man, *Comets, Meteors and Asteroids*, BBC Books, London, 2001, 52–55.

21. Bede, *The Ecclesiastical History of the English People*, Book 5, Chapter 23.

22. W. Shakespeare, *Julius Caesar*, Act 2, Scene 2.

23. D. Crystal (ed.), *Cambridge Encyclopedia*, Cambridge University Press, 4th edn, 2000; *The McGraw-Hill Dictionary of Scientific and Technical Terms*, McGraw-Hill, New York, 1989; *The Oxford English Reference Dictionary*, Oxford University Press, 2nd edn, 1996; *Chambers Dictionary of Science and Technology*, Chambers Harrap, Edinburgh and New York, 1999.

24. M. Ruse, *Darwinism Defended*, Addison-Wesley, Reading, Mass., 1982, 13.

25. R. Dawkins, *The Blind Watchmaker*, Penguin Books, Harmondsworth, 1988, 241.

26. S.M. Stanley, *Earth and Life Through Time*, Freeman, New York, 1986, 2.

27. A. Koestler, *The Sleepwalkers*, 431–503; A. Chapman, *Gods in the Sky*, 249–280; J. Bronowski, *The Ascent of Man*, BBC Publications, London, 1973, 194–218; H. Hellman, *Great Feuds in Science*, Wiley, New York, 1998, 1–19.

28. A. Koestler, *The Sleepwalkers*, 509–512; J. Bronowski, *The Ascent of Man*, 233–234; M. White, *Isaac Newton – The Last Sorcerer*, Fourth Estate, London, 1997.

29. *The Correspondence of Isaac Newton*, Cambridge University Press, 1961, Volume 3, 246–256.

30. E. Mayr, *The Growth of Biological Thought*, Harvard University Press, Cambridge, Mass., 1982, 91–94; R. Huggett, *Cataclysms and Earth History*, 22–25.

31. *Chambers Enclyclopedia*, Pergamon Press, Oxford, 1967, Volume 13, 779–780; A.D. Wright, *The Counter-Reformation*, Weidenfeld and Nicolson, London, 1982, 112.

32. M.J.S. Rudwick, *The Meaning of Fossils*, Macdonald, London, 1972, 39–41; R. Huggett, *Cataclysms and Earth History*, 26–31; S.J. Gould, *Leonardo's Mountain of Clams and the Diet of Worms*, Vintage, London, 1999, 17–44.

33. S.J. Gould, *Time's Arrow, Time's Cycle*, Harvard University Press, Cambridge, Mass., 1987, Penguin Books, Harmondsworth, 1988, 51–59; R. Huggett, *Catastrophism*, Verso, London and New York, 1997, 60–64.

34. *The Correspondence of Isaac Newton*, Cambridge University Press, 1960, Volume 2, 321–327; S.J. Gould, *Time's Arrow, Time's Cycle*, 21–59; C.C. Albritton, *Catastrophic Episodes in Earth History*, Chapman and Hall, London, 1989, 3–6.

35. C.C. Albritton, *Catastrophic Episodes in Earth History*, 14–17; R. Huggett, *Catastrophism*, 60–62.

36. M.J.S. Rudwick, *The Meaning of Fossils*, 82–90; R. Huggett, *Catastrophism*, 48–50; G.L. Davies, *The Earth in Decay – A History of British Geomorphology 1578–1878*, Macdonald, London, 1969, 74–83.

37. R. Huggett, *Catastrophism*, 50–52; G.L. Davies, *The Earth in Decay*, 83–86; S.J. Gould, *Bully for Brontosaurus*, Hutchinson Radius, London, 1991, 367–381.

38. A. Cook, *Edmond Halley*, 203–217; V. Clube and B. Napier, *The Cosmic Winter*, Basil Blackwell, Oxford, 1990, 116–119; G.L. Vershuur, *Impact!*, Oxford University Press,

1996, 84–85; G. Vanin, *Cosmic Phenomena*, Firefly Books, Buffalo, New York, 1999, 23–24.

39. M. White, *Isaac Newton – The Last Sorcerer*, 301–302; A. Cook, *Edmond Halley*, 399; G.L. Verschuur, *Impact!*, 85; D. Steel, *Rogue Asteroids and Doomsday Comets*, Wiley, New York, 1995, 16.

40. V. Clube and B. Napier, *The Cosmic Serpent*, 247–248; C. Sagan and A. Druyan, *Comet*, 54–59, 108–109, 306–307; I. Newton (translated by I. Cohen and A. Whitman), *The Principia*, University of California Press, Berkeley, Los Angeles and London, 1999, 888–938 (quotation from page 926).

41. P.J. Bowler, *Evolution – The History of an Idea*, University of California Press, Berkeley, Los Angeles and London, revised edition, 1989, 28–30; *The Hutchinson Dictionary of Scientific Biography*, Helicon, London, 1994, 177–179; D. Goodman and C.A. Russell (eds.), *The Rise of Scientific Europe 1500–1800*, Open University, Milton Keynes, 1991, 179–183.

42. M.J.S. Rudwick, *The Meaning of Fossils*, 91; R. Huggett, *Catastrophism*, 52–53.

43. C.C. Albritton, *Catastrophic Episodes in Earth History*, 20–23; P.J. Bowler, *Evolution – The History of an Idea*, 36–39; R. Huggett, *Catastrophism*, 82–84; S.J. Gould, *The Lying Stones of Marrakech*, Jonathan Cape, London, 2000, 75–90.

44. C.C. Albritton, *Catastrophic Episodes in Earth History*, 18–20; P.J. Bowler, *Evolution – The History of an Idea*, 32–33, 69–70.

45. C.C. Albritton, *Catastrophic Episodes in Earth History*, 23–25; A. Hallam, *Great Geological Controversies*, Oxford University Press, 1989, 2–9; D. Goodman and C.A. Russell (eds.), *The Rise of Scientific Europe 1500–1800*, 379–380.

46. P.J. Bowler, *Evolution – The History of an Idea*, 40–44, 48–49; R. Huggett, *Catastrophism*, 53–58, 68–71; C.C. Gillespie, *Genesis and Geology*, Harper and Row, New York, 1959, 41–72.

47. R. Huggett, *Catastrophism*, 64–68; G.L. Davies, *The Earth in Decay*, 108–110, 129–153.

Chapter 2

1. G.L. Davies, *The Earth in Decay – A History of British Geomorphology 1578–1878*, Macdonald, London, 1969, 154–199; S.J. Gould, *Time's Arrow, Time's Cycle*, Harvard University Press, Cambridge, Mass., 1987, Penguin Books, Harmondsworth, 1988, 61–97; A. Hallam, *Great Geological Controversies*, Oxford University Press, 2nd edn, 1989, 30–37.

2. A. Geikie, *The Founders of Geology*, Macmillan, London and New York, 2nd edn, 1905, 66.

3. A. Geikie, *The Founders of Geology*, 288.

4. A. Geikie, *The Founders of Geology*, 314–315.

5. W.L. Stokes and S. Judson, *Introduction to Geology*, Prentice-Hall, New Jersey, 1968, 34; S.J. Gould, *Time's Arrow, Time's Cycle*, 68; L.D. Leet and S. Judson, *Physical Geology*, Prentice-Hall, New Jersey, 4th edn, 1971, 2; *Geology Today*, CRM Books, Del Mar, Calif., 1973.

6. L.B. Halstead, *Hunting the Past*, Hamish Hamilton, London, 1982, 10.

7. *The Hutchinson Dictionary of Scientific Biography*, Helicon, Oxford, 1994, 355–356; D. Goodman and C.A. Russell (eds.), *The Rise of Scientific Europe 1500–1800*, Open University, Milton Keynes, 1991, 290–292.

8. S.J. Gould, *Time's Arrow, Time's Cycle*, 61–73.

9. J. Hutton, Theory of the Earth; or an investigation of the laws observable in the composition, dissolution, and restitution of land upon the globe, *Transactions of the Royal Society of Edinburgh*, 1, 1788, 209–305.

10. J. Hutton, *Theory of the Earth with Proofs and Illustrations*, William Creech, Edinburgh, 1795.

11. G.L. Davies, *The Earth in Decay*, 163; A. Hallam, *Great Geological Controversies*, 10–11; D. Goodman and C.A. Russell (eds.), *The Rise of Scientific Europe 1500–1800*, 292–295.

12. A. Hallam, *Great Geological Catastrophes*, 5–10; R. Huggett, *Catastrophism*, Verso, London and New York, 1997, 65–66.

13. A. Hallam, *Great Geological Controversies*, 18–19; *The Hutchinson Dictionary of Scientific Biography*, 301.

14. A. Hallam, *Great Geological Controversies*, 10–13; S.J. Gould, *Time's Arrow, Time's Cycle*, 63–66; P.J. Bowler, *Evolution – The History of an Idea*, University of California Press, Berkeley, Los Angeles and London, 1989, 45–49.

15. J. Hutton, *Transactions of the Royal Society of Edinburgh*, 1, 215.

16. J. Hutton, *Transactions of the Royal Society of Edinburgh*, 1, 304.

17. R. Hooykaas, *The Principle of Uniformity in Geology, Biology and Theology*, Brill, Leiden, 1963, v, 1–32; J. Hutton, *Theory of the Earth with Proofs and Illustrations*, II, 547.

18. A. Hallam, *Great Geological Controversies*, 30–31; R. Huggett. *Catastrophism*, 59–71.

19. J. Hutton, *Transactions of the Royal Society of Edinburgh*, 1, 217.

20. G.L. Davies, *The Earth in Decay*, 176–190; S.J. Gould, *Time's Arrow, Time's Cycle*, 70–73.

21. M.J.S. Rudwick, *The Meaning of Fossils*, Macdonalds, London, 1972, 119.

22. J. Hutton, *Transactions of the Royal Society of Edinburgh*, 1, 211.

23. J. Hutton, *Transactions of the Royal Society of Edinburgh*, 1, 275.

24. G.L. Davies, *The Earth in Decay*, 169–173; P.J. Bowler, *Evolution – The History of an Idea*, 48; C.C. Gillespie, *Genesis and Geology*, Harper and Row, New York, 1959, 73–79; D.R. Dean, James Hutton on religion and geology, *Annals of Science*, 32, 1975, 187–193; R. Porter, Philosophy and politics of a geologist: G.H.Toulmin (1754–1817), *Journal of the History of Ideas*, 39, 1978, 435–450; C.C. Albritton, *Catastrophic Episodes in Earth History*, Chapman and Hall, London, 1989, 31–32; D. Goodman and C.A. Russell (eds), *The Rise of Scientific Europe 1500–1800*, 291–292.

25. G.L. Davies, *The Earth in Decay*, 159–165; S.J. Gould, *Time's Arrow, Time's Cycle*, 70–73.

26. A. Hallam, *Great Geological Controversies*, 14–15; R. Huggett, *Catastrophism*, 67–68, 78.

27. J. Hutton, *Theory of the Earth with Proofs and Illustrations*, II, 547.

28. R. Hooykaas, *The Principle of Uniformity in Geology, Biology and Theology*, 20–23; G.L. Davies, *The Earth in Decay*, 226–227, 249–251; R. Huggett, *Catastrophism*, 59–69.

29. S.J. Gould, *Time's Arrow, Time's Cycle*, 73–80; A. Hallam, *Great Geological Controversies*, 11; C.C. Albritton, *Catastrophic Episodes in Earth History*, 25–29.

30. J. Hutton, *Transactions of the Royal Society of Edinburgh*, 1, 216.

31. S. Lamb and D. Sington, *Earth Story*, BBC Books, London, 1998, 13–14.

32. G.L. Davies, *The Earth in Decay*, 176–178; S.J. Gould, *Time's Arrow, Time's Cycle*, 70–91.
33. A. Hallam, *Great Geological Controversies*, 60.
34. J. Playfair, *Illustrations of the Huttonian Theory of the Earth*, William Creech, Edinburgh, 1802.
35. C.C. Gillespie, *Genesis and Geology*, 74.
36. G.L. Davies, *The Earth in Decay*, 190–196; S.J. Gould, *Time's Arrow, Time's Cycle*, 93–96.

Chapter 3

1. P. Corsi, *The Age of Lamarck*, University of California Press, Berkeley, Los Angeles and London, 1988, 1–7; M.J.S. Rudwick, *Georges Cuvier, Fossil Bones, and Geological Catastrophes*, University of Chicago Press, Chicago and London, 1997, 1–6, 13–18.
2. R.W. Burkhardt, *The Spirit of System*, Harvard University Press, Cambridge, Mass., 1977, 38–39; P. Corsi, *The Age of Lamarck*, 4.
3. R.W. Burkhardt, *The Spirit of System*, 29–38; P. Corsi, *The Age of Lamarck*, 7–17; M.J.S. Rudwick, *The Meaning of Fossils*, Macdonald, London, 1972, 101–104; S.J. Gould, *Bully for Brontosaurus*, Hutchinson Radius, London, 1991, 354–366; S.J.Gould, *The Lying Stones of Marrakech*, Jonathan Cape, London, 2000, 91–143.
4. P. Corsi, *The Age of Lamarck*, 15–17; M.J.S. Rudwick, *Georges Cuvier, Fossil Bones, and Geological Catastrophes*, 15–17.
5. R. Huggett, *Cataclysms and Earth History*, Oxford University Press, 1989, 79; R.W. Burkhardt, *The Spirit of System*, 38–39, 198–199.
6. R. Lewin, *Complexity*, Dent, London, 1993, 75.
7. A. Atkinson, *Impact Earth*, Virgin, London, 1999, xxvi.
8. R. Fortey, *Life – An Unauthorised Biography*, HarperCollins, London, 1997, 275.
9. M.J.S. Rudwick, *Georges Cuvier, Fossil Bones, and Geological Catastrophes*, 127–128; M.J.S. Rudwick, *The Meaning of Fossils*, 127–133.
10. G. Cuvier, *Recherches sur les ossemens fossiles de quadrupèdes*, Deterville, Paris, 1812.
11. G. Cuvier, *Discours sur les révolutions de la surface du globe, et sur les changemens qu'elles ont produit dans le règne animal*, Dufour et d'Ocagne, Paris, 1826. For English translation see M.J.S. Rudwick, *Georges Cuvier, Fossil Bones, and Geological Catastrophes*, 183–252.
12. C.C. Albritton, *Catastrophic Episodes in Earth History*, Chapman and Hall, London, 1989, 36–40; M.J.S. Rudwick, *Georges Cuvier, Fossil Bones, and Geological Catastrophes*, 127–156.
13. R.W. Burkhardt, *The Spirit of System*, 109; P. Corsi, *The Age of Lamarck*, 102.
14. J.-B. Lamarck, *Hydrogéologie, ou recherches sur l'influence qu'ont les eaux sur la surface du globe terrestre*, Chez l'Auteur et Agasse, Maillard, Paris, 1802.
15. R.W. Burkhardt, *The Spirit of System*, 105–106; P. Corsi, *The Age of Lamarck*, 103–106.
16. J.-B. Lamarck, *Hydrogéologie*, 54. English translation by A.V. Carozzi, University of Illinois Press, Urbana, 1964.
17. P. Corsi, *The Age of Lamarck*, 107–110.
18. R. Huggett, *Cataclysms and Earth History*, 81.
19. S.M. Stanley, *Earth and Life Through Time*, Freeman, New York, 1986, 209–215.
20. R.W. Burkhardt, *The Spirit of System*, 193; P.J. Bowler, *Evolution – The History of an Idea*, University of California Press, Berkeley, Los Angeles and London, 1989, 112–113.

21. J.N. Wilford, *The Riddle of the Dinosaur*, Vintage Books, New York, 1987, 25.

22. E. Mayr, *The Growth of Biological Thought*, Harvard University Press, Cambridge, Mass., 1982, 128–129, 201–202, 319, 326.

23. R.W. Burkhardt, *The Spirit of System*, 112, 128–138, 186–187, 191–195; P. Corsi, *The Age of Lamarck*, 78, 113–114, 160–169, 201–202.

24. E. Mayr, *The Growth of Biological Thought*, 309–342; P.J. Bowler, *Evolution – The History of an Idea*, 50–81.

25. V. Blackmore and A. Page, *Evolution – The Great Debate*, Lion Books, Oxford, 1989, 10–14.

26. E. Mayr, *The Growth of Biological Thought*, 264, 327; P.J. Bowler, *Evolution – The History of an Idea*, 63.

27. E. Mayr, *The Growth of Biological Thought*, 326–327; P.J. Bowler, *Evolution – The History of an Idea*, 60–63.

28. E. Mayr, *The Growth of Biological Thought*, 256–258; V. Blackmore and A. Page, *Evolution – The Great Debate*, 14–20; P.J. Bowler, *Evolution – The History of an Idea*, 53–54.

29. C. Linnaeus, *Species Plantarum*, Salvii, Holmiae, 1753 (reprinted by the Ray Society, London, 1955); C. Linnaeus, *Systema Naturae, Volume 1*, 10th edition, 1758 (reprinted by the Ray Society, London, 1956); E. Mayr, *The Growth of Biological Thought*, 258–260; P.J. Bowler, *Evolution – The History of an Idea*, 64–68; D. Goodman and C.A. Russell (eds.), *The Rise of Scientific Europe 1500–1800*, Open University, Milton Keynes, 1991, 319–323.

30. E. Mayr, *The Growth of Biological Thought*, 182–184; P.J. Bowler, *Evolution – The History of an Idea*, 112–118.

31. R.W. Burkhardt, *The Spirit of System*, 79–82; E. Mayr, *The Growth of Biological Thought*, 180–182; P.J. Bowler, *Evolution – The History of an Idea*, 72–77; D. Goodman and C.A. Russell (eds.), *The Rise of Scientific Europe 1500–1800*, 322–323; S.J. Gould, *The Lying Stones of Marrakech*, 75–90.

32. E. Mayr, *The Growth of Biological Thought*, 343–360; P.J. Bowler, *Evolution – The History of an Idea*, 81–89.

33. R.W. Burkhardt, *The Spirit of System*, 143–185; P. Corsi, *The Age of Lamarck*, 85–206; S.J. Gould, *The Lying Stones of Marrakech*, 115–143.

34. R.W. Burkhardt, *The Spirit of System*, 131–135; P. Corsi, *The Age of Lamarck*, 201–203.

35. P. Corsi, *The Age of Lamarck*, 211–213.

36. M.J.S. Rudwick, *Georges Cuvier, Fossil Bones, and Geological Catastrophes*, 190.

37. A. Hallam, *Great Geological Controversies*, Oxford University Press, 1989, 37–39; P.J. Bowler, *Evolution – The History of an Idea*, 116–117.

38. M.J.S. Rudwick, *Georges Cuvier, Fossil Bones, and Geological Catastrophes*, 248.

39. S.M. Stanley, *Extinction*, Scientific American Books, New York, 1987, 3.

40. M.J.S. Rudwick, *Georges Cuvier, Fossil Bones, and Geological Catastrophes*, 261.

41. D. Ager, *The New Catastrophism*, Cambridge University Press, 1993, 1–5.

42. M.J.S. Rudwick, *Georges Cuvier, Fossil Bones, and Geological Catastrophes*, 254–260; M.J.S. Rudwick, *The Meaning of Fossils*, 133–135.

43. M.J.S. Rudwick, *Georges Cuvier, Fossil Bones, and Geological Catastrophes*, 77.

44. G.L. Verschuur, *Impact*, Oxford University Press, 1996, 82–83.

45. W. Glen (ed.), *Mass-Extinction Debates*, Stanford University Press, Stanford, Calif., 1994, 41.

46. C. Sagan and A. Druyan, *Comet*, Headline Books, London, 1997, 97–104, 276–281.

47. G.L. Verschuur, *Impact*, 81; C. Sagan and A. Druyan, *Comet*, 279.

48. R. Fortey, *Life – An Unauthorised Biography*, 275.

49. G.L. Verschuur, *Impact*, 76.

50. *The Hutchinson Dictionary of Scientific Biography*, Helicon, Oxford, 1994, 410–411; M.J.S. Rudwick, *Georges Cuvier, Fossil Bones, and Geological Catastrophes*, 166–168.

51. R.W. Burkhardt, *The Spirit of System*, 129.

52. *Dictionary of Scientific Biography*, Charles Scribner's Sons, New York, 1972, Volume IV, 347–350; A. Hallam, *Great Geological Controversies*, 40–41.

53. R.W. Burkhardt, *The Spirit of System*, 38–45, 191–218; P. Corsi, *The Age of Lamarck*, 32–35, 64–65, 136–139, 157–164.

54. *The Hutchinson Dictionary of Scientific Biography*, 157–158.

55. H.G. Cannon, *Lamarck and Modern Genetics*, Greenwood Press, Westport, Conn., 1959, 134. See also S.J. Gould, *The Lying Stones of Marrakech*, 116–118.

56. R.W. Burkhardt, *The Spirit of System*, 171–172; P. Corsi, *The Age of Lamarck*, 93–94.

57. P. Corsi, *The Age of Lamarck*, 260–262; P.J. Bowler, *Evolution – The History of an Idea*, 88–89, 117–118.

58. R.W. Burkhardt, *The Spirit of System*, 199; P. Corsi, *The Age of Lamarck*, 94.

59. P. Corsi, *The Age of Lamarck*, 180–185; M.J.S. Rudwick, *Georges Cuvier, Fossil Bones, and Geological Catastrophes*, 74–88.

60. M.J.S. Rudwick, *Georges Cuvier, Fossil Bones, and Geological Catastrophes*, 87.

61. M.J.S. Rudwick, *Georges Cuvier, Fossil Bones, and Geological Catastrophes*, 248.

62. M.J.S. Rudwick, *Georges Cuvier. Fossil Bones, and Geological Catastrophes*, 70.

63. E. Mayr, *The Growth of Biological Thought*, 363–371; M.J.S. Rudwick, *Georges Cuvier, Fossil Bones, and Geological Catastrophes*, 257–264.

64. R.W. Burkhardt, *The Spirit of System*, 191–195; P. Corsi, *The Age of Lamarck*, 164–168.

65. J.N. Wilford, *The Riddle of the Dinosaur*, 19–26.

Chapter 4

1. E. Mayr, *The Growth of Biological Thought*, Harvard University Press, Cambridge, Mass., 1982, 103–105, 371–375; P.J. Bowler, *Evolution – The History of an Idea*, University of California Press, Berkeley, Los Angeles and London, 1989, 52–59, 123–126; D. Goodman and C.A. Russell (eds.), *The Rise of Scientific Europe 1500–1800*, Open University, 1991, 257–258, 326–329.

2. R. Dawkins, *The Blind Watchmaker*, Penguin Books, Harmondsworth, 1988, 4–5; V. Blackmore and A. Page, *Evolution – The Great Debate*, Lion Books, Oxford, 1989, 24–26.

3. E. Mayr, *The Growth of Biological Thought*, 373; C.C. Gillespie, *Genesis and Geology*, Harper and Row, New York, 1959, 184–216; A. Desmond and J. Moore, *Darwin*, Penguin Books, Harmondsworth, 1992, 213.

4. V. Blackmore and A. Page, *Evolution – The Great Debate*, 37; M. White, *Isaac Newton – The Last Sorcerer*, Fourth Estate, London, 1997, 150–151. (Note that Newton agreed

to take holy orders as a condition of being appointed the Lucasian Professor at Cambridge, but later obtained special dispensation from King Charles II to avoid doing so.)

5. A. Desmond and J. Moore, *Darwin*, 47–97.
6. *Dictionary of Scientific Biography*, Charles Scribner's Sons, New York, 1970–2, Volume XII, 275–279.
7. *Dictionary of Scientific Biography*, Volume XIV, 292–295.
8. *Dictionary of Scientific Biography*, Volume II, 566–573.
9. V. Blackmore and A. Page, *Evolution – The Great Debate*, 35–36; A. Hallam, *Great Geological Controversies*, Oxford University Press, 1989, 54.
10. P.J. Bowler, *Evolution – The History of an Idea*, 119–121.
11. C.C. Gillespie, *Genesis and Geology*, 98–120; R. Huggett, *Catastrophism*, Verso, London and New York, 1997, 55–58; S.J. Gould, *The Flamingo's Smile*, Penguin Books, Harmondsworth, 1986, 114–121.
12. C.C. Gillespie, *Genesis and Geology*, 111; R. Huggett, *Catastrophism*, 57.
13. S.J. Gould, *The Flamingo's Smile*, 114–120; G.L. Davies, *The Earth in Decay – A History of British Geomorphology 1578–1878*, Macdonald, London, 251.
14. S.J. Gould, *The Flamingo's Smile*, 119.
15. A. Hallam, *Great Geological Controversies*, 41–43; S.J. Gould, *The Flamingo's Smile*, 121–122.
16. A. Sedgwick, *Annals of Philosophy*, 10, 1825, 34.
17. S.J. Gould, *The Flamingo's Smile*, 122–125; G.L. Davies, *The Earth in Decay*, 251–252.
18. A. Sedgwick, Address to the Geological Society, delivered on the Evening of the 18th of February 1831, by the Rev. Professor Sedgwick, MA, FRS &c, on retiring from the President's Chair, *Proceedings of the Geological Society of London*, 1, 1831, 281–316 (quotation from page 313).
19. G.L. Davies, *The Earth in Decay*, 252–254; R. Huggett, *Cataclysms and Earth History*, Oxford University Press, 1989, 92–99; R. Huggett, *Catastrophism*, 56–58; *Dictionary of Scientific Biography*, Volume III, 395–396; W.D. Conybeare, On the hydrographical basin of the Thames, *Proceedings of the Geological Society of London*, 1(12), 1829, 145–149.
20. A. Hallam, *Great Geological Controversies*, 40–41; C. Lyell, *Principles of Geology*, Volume 3, John Murray, London, 1833 (Reprinted by University of Chicago Press, Chicago and London, 1991), 337–351.
21. A. Sedgwick, *Proceedings of the Geological Society of London*, 1, 1831, 308.
22. *Dictionary of Scientific Biography*, Volume VIII, 563–576; *The Hutchinson Dictionary of Scientific Biography*, Helicon, Oxford, 1994, 447.
23. G.L. Davies, *The Earth in Decay*, 254–259, 317–333; R. Huggett, *Cataclysms and Earth History*, 108–116.
24. A. Hallam, *Great Geological Controversies*, 87–104; G.L. Davies, *The Earth in Decay*, 263–316; *Dictionary of Scientific Biography*, Volume I, 72–74; E. Lurie, *Louis Agassiz*, The Johns Hopkins University Press, Baltimore and London, 1988, 94–109.
25. R. Huggett, *Catastrophism*, 58; A. Hallam, *Great Geological Controversies*, 94–95.

Chapter 5

1. A. Hallam, *Great Geological Controversies*, Oxford University Press, 1989, 65–68; P.J. Bowler, *Evolution – The History of an Idea*, University of California Press, Berkeley, Los Angeles and London, 1989, 121.

2. M.J.S. Rudwick, *The Meaning of Fossils*, Macdonald, London, 1972, 139–140; V. Blackmore and A. Page, *Evolution – The Great Debate*, Lion Books, Oxford, 1989, 34; *The Hutchinson Dictionary of Scientific Biography*, Helicon, Oxford, 1994, 630; S. Winchester, *The Map that Changed the World*, Viking, London, 2001, 7–223.

3. A. Hallam, *Great Geological Controversies*, 65–66; M.J.S. Rudwick, *The Meaning of Fossils*, 126–131; *Dictionary of Scientific Biography*, Volume I, Charles Scribner's Sons, New York, 1970, 233–236.

4. A. Hallam, *Great Geological Controversies*, 66–83; P.J. Bowler, *Evolution – The History of an Idea*, 121–122; M.J.S. Rudwick, *The Meaning of Fossils*, 191–194.

5. A. Hallam, *Great Geological Controversies*, 84; P.J. Bowler, *Evolution – The History of an Idea*, 124.

6. A. Hallam, *Great Geological Controversies*, 51; S.J. Gould, *Time's Arrow, Time's Cycle*, Harvard University Press, 1987, Penguin Books, Harmondsworth, 1998, 155–167, C. Lyell, *Principles of Geology*, Volume 3, John Murray, London, 1833 (Reprinted by the University of Chicago Press, Chicago and London, 1991), 52–61.

7. A.N. Strahler, *Principles of Earth Science*, Harper and Row, New York, 1976, 217.

8. A. McLeish, *Geological Science*, Blackie, London, 1986, Nelson, Walton-on-Thames, Surrey, 1992, 231.

9. R.M. Young, *Darwin's Metaphor*, Cambridge University Press, 1985, 136.

10. C. Darwin, *On the Origin of Species by Means of Natural Selection*, John Murray, London, 1859, Penguin Books, Harmondsworth, 1968.

11. C.C. Gillespie, *Genesis and Geology*, Harper and Row, New York, 1959, 120.

12. M.J.S. Rudwick, *The Meaning of Fossils*, 179–191; S.J. Gould, *Time's Arrow, Time's Cycle*, 117–126; A. Hallam, *Great Geological Controversies*, 30–31; R. Hooykaas, *The Principle of Uniformity in Geology, Biology and Theology*, Brill, Leiden, 1963, v–ix, 1–47; M.J.S. Rudwick, The strategy of Lyell's *Principles of Geology*, *Isis*, 61, 1970, 5–33; R. Porter, Charles Lyell and the principles of the history of geology, *British Journal for the History of Science*, 9, 1976, 91–103.

13. S.J. Gould, *Time's Arrow, Time's Cycle*, 118–119; S.J. Gould, *The Panda's Thumb*, Penguin Books, Harmondsworth, 1983, 168; C.C. Albritton, *Catastrophic Episodes in Earth History*, Chapman and Hall, 1989, 45–50; R. Huggett, *Catastrophism*, Verso, London and New York, 1997, 85–87; T. Palmer, *Controversy – Catastrophism and Evolution: The Ongoing Debate*, Kluwer/Plenum, New York, 1999, 31–37; S.J. Gould, *The Lying Stones of Marrakech*, Jonathan Cape, London, 2000, 147–168; S.J. Gould, *The Structure of Evolutionary Theory*, Harvard University Press, Cambridge, Mass., 2002, 467–502.

14. A. Geikie, *The Founders of Geology*, Macmillan, London and New York, 1905, 403.

15. C. Lyell, *Principles of Geology*, Volume 1, John Murray, London, 1830 (Reprinted by the University of Chicago Press, Chicago and London, 1990), 4.

16. C. Lyell, *Principles of Geology*, Volume 1, 37.

17. C. Lyell, *Principles of Geology*, Volume 1, 39.

18. C. Lyell, *Principles of Geology*, Volume 1, 39.

19. C. Lyell, *Principles of Geology*, Volume 3, 6.

20. C. Lyell, *Principles of Geology*, Volume 3, 339.

21. C. Lyell, *Principles of Geology*, Volume 3, 347.

22. S.J. Gould, *Time's Arrow, Time's Cycle*, 120–142; A. Hallam, *Great Geological Controversies*, 45–51; P.J. Bowler, *Evolution – The History of an Idea*, 134–136.

23. R. Porter, *British Journal for the History of Science*, 9, 98.

24. C. Lyell, *Principles of Geology*, Volume 1, 61.

25. S.J. Gould, *Time's Arrow, Time's Cycle*, 66–67, 128–129; C.C. Albritton, *Catastrophic Episodes in Earth History*, 46–48.

26. C. Lyell, *Principles of Geology*, Volume 1, 79–80.

27. A. Hallam, *Great Geological Controversies*, 44–48; P.J. Bowler, *Evolution – The History of an Idea*, 134–135; M.J.S. Rudwick, *The Meaning of Fossils*, 164–179.

28. C. Lyell, *Principles of Geology*, Volume 1, 361–371.

29. A. Hallam, *Great Geological Controversies*, 55.

30. S.J. Gould, *Time's Arrow, Time's Cycle*, 115–117, 126–128.

31. R. Huggett, *Catastrophism*, 25; A. Hallam, *Great Geological Controversies*, 53; A. Sedgwick, Address to the Geological Society, delivered on the Evening of the 18th of February 1831, by the Rev. Professor Sedgwick, M.A., F.R.S., &c, on retiring from the President's Chair, *Proceedings of the Geological Society of London*, 1, 1831, 281–316 (in particular, pages 298–307).

32. W. Whewell (published anonymously), Art. IV – *Principles of Geology, being an Attempt to Explain the Former Changes of the Earth's Surface, by Reference to Causes now in Operation*. By Charles Lyell, Esq., F.R.S., Professor of Geology in King's College, London. Vol. II. London. 1832. *Quarterly Review*, 47, 1832, 103–132. (Quotation from page 126).

33. S.J. Gould, *Time's Arrow, Time's Cycle*, 104–112; R. Huggett, *Catastrophism*, 85–87.

34. W. Whewell, *History of the Inductive Sciences*, Volume 2, Parker, London, 1837, 120 (quoted in A. Hallam, *Great Geological Controversies*, 54).

35. C.C. Albritton, *Catastrophic Episodes in Earth History*, 49–50.

36. A. Hallam, *Great Geological Controversies*, 49–55; S.J. Gould, *Time's Arrow, Time's Cycle*, 123–132.

37. C. Lyell, *Principles of Geology*, Volume 1, 123.

38. C. Lyell, *Principles of Geology*, Volume 1, 64.

39. C. Lyell, *Principles of Geology*, Volume 1, 165.

40. P.J. Bowler, *Evolution – The History of an Idea*, 136–141; E. Mayr, *The Growth of Biological Thought*, Harvard University Press, Cambridge, Mass., 1982, 371–373, 404–408; R. Hooykaas, *The Principle of Uniformity in Geology, Biology and Theology*, 94–100, 193–196.

41. W. Whewell, *History of the Inductive Sciences*, Volume 2, 120 (Quoted in A. Hallam, *Great Geological Controversies*, 54).

42. P.J. Bowler, *Evolution – The History of an Idea*, 139–141; E. Mayr, *The Growth of Biological Thought*, 404–407, 875, 881–882.

43. C. Lyell, *Principles of Geology*, Volume 2, John Murray, London, 1832 (Reprinted by the University of Chicago Press, Chicago and London, 1991), 23.

44. C. Lyell, *Principles of Geology*, Volume 2, 23–24.

45. C. Lyell, *Principles of Geology*, Volume 2, 180.

46. P.J. Bowler, *Evolution – The History of an Idea*, 104–108, 126–129; E. Mayr, *The Growth of Biological Thought*, 338–339, 387–390; T. Honderich (ed.), *The Oxford Companion to Philosophy*, Oxford University Press, 1995, 386–388.

47. *Dictionary of Scientific Biography*, Charles Scribner's Sons, New York, 1979, Volume I, 72–74; E. Lurie, *Louis Agassiz*, The Johns Hopkins University Press, Baltimore and London, 1988, 27–28, 50–52, 62–63, 82–87, 283–288.

48. A. Agassiz, On the succession and development of organized beings at the surface of the terrestrial globe, *Edinburgh New Philosophical Journal*, 33, 1842, 388–399 (quotation from final page).

49. E. Lurie, *Louis Agassiz*, 55–57, 83–88, 97–100, 284–286; C.C. Albritton, *Catastrophic Episodes in Earth History*, 53–56; P.J. Bowler, *Evolution – The History of an Idea*, 128–129; *The Hutchinson Dictionary of Scientific Biography*, 5–6.

50. C.C. Gillespie, *Genesis and Geology*, 149–183; P.J. Bowler, *Evolution – The History of an Idea*, 129, 141–147; E. Mayr, *The Growth of Biological Thought*, 381–385.

51. P.J. Bowler, *Evolution – The History of an Idea*, 129–134; E. Mayr, *The Growth of Biological Thought*, 458, 464, 472–473.

Chapter 6

1. A. Hallam, *Great Geological Controversies*, Oxford University Press, 2nd edn, 1989, 56–58.

2. B. Kobres, electronic communication, http://abob.libs.uga.edu/bobk/eoc/tovueoc.html

3. E.J. Lovell (ed.), *Medwin's Conversations of Lord Byron*, Princeton University Press, Princeton, New Jersey, 1966, 188. See also D. Steel, *Target Earth*, Quarto, London, 2000, 91.

4. V. Clube and B. Napier, *The Cosmic Serpent*, Faber and Faber, London, 1982, 206.

5. *The Cambridge Biographical Encyclopedia*, Cambridge University Press, 1998, 382.

6. V. Clube and B. Napier, *The Cosmic Winter*, Basil Blackwell, Oxford, 1990, 123–125; R.M. Schoch, *Voices of the Rocks*, Thorsons, London, 1999, 25.

7. *The New Larousse Encyclopedia of Mythology*, Hamlyn, London, 1968, 264, 275; A. Cotterell, *Norse Mythology*, Ultimate Editions, London, 1997, 52–53, 78–79; K.E. Sullivan, *Viking Myths and Legends*, Brockhampton Press, London, 1998, 108–123; D. Ferguson, *History of Myths Retold*, Chancellor Press, London, 2000, 44–45.

8. I. Donnelly, *Ragnarok: The Age of Fire and Gravel*, Sampson Low, Marston, Searle and Rivington, London, 1883 (quotation from page 438).

9. G.L. Verschuur, *Impact*, Oxford University Press, 1996, 75–86 (particularly pages 76–77 and 85); C. Sagan and A. Druyan, *Comet*, Headline, London, 1997, 291, 337; S. Newcomb, *Popular Astronomy*, Harper Brothers, New York, 1879; T. Dick, *The Siderial Heavens*, Worthington, New York, 1840.

10. I. Donnelly, *Ragnarok: The Age of Fire and Gravel*, 439–441.

11. G.L. Verschuur, *Impact*, 80–82; T. Milner, *The Gallery of Nature*, W. and R. Chambers, London, 1860.

12. V. Clube and B. Napier, *The Cosmic Winter*, 138–149; G.L. Verschuur, *Impact*, 133; C. Sagan and A. Druyan, *Comet*, 95–97; G. Vanin, *Cosmic Phenomena*, Firefly Books, Buffalo, New York, 1999, 78–79.

13. P. Lancaster Brown, *Comets, Meteorites and Men*, Hale, London, 1973, 154; F. Heide and F. Wlotzka, *Meteorites*, Springer-Verlag, Berlin, Heidelberg and New York, 1995, 87.

14. F. Heide and F. Wlotzka, *Meteorites*, 80, 87; J.S. Lewis, *Rain of Iron and Ice*, Addison-Wesley, Reading, Mass., 1995, 24, 177.

15. P. Lancaster Brown, *Comets, Meteorites and Men*, 154–156; F. Heide and F. Wlotzka, *Meteorites*, 87–88.

16. P. Lancaster Brown, *Comets, Meteorites and Men*, 156–158; J.S. Lewis, *Rain of Iron and Ice*, 24–25; Wold Cottage meteorite home for the day, *Astronomy Now*, 14(5), 2000, 72.

17. F. Heide and F. Wlotzka, *Meteorites*, 1, 83–86; J.S. Lewis, *Rain of Iron and Ice*, 10, 18–19, 24–25; G. Vanin, *Cosmic Phenomena*, 72–73.

18. P. Lancaster Brown, *Comets, Meteorites and Men*, 158–159; J.S. Lewis, *Rain of Iron and Ice*, 25.

19. J.S. Lewis, *Rain of Iron and Ice*, 26; C. Officer and J. Page, *Tales of the Earth*, Oxford University Press, 1993, 89.

20. P. Lancaster Brown, *Comets, Meteorites and Men*, 165; J.S. Lewis, *Rain of Iron and Ice*, 162–173, 176–178.

21. F. Heide and F. Wlotzka, *Meteorites*, 79–81; J.S. Lewis, *Rain of Iron and Ice*, 168–174, 176–178; P. Lancaster Brown, *Comets, Meteorites and Men*, 169.

22. R. Huggett, *Cataclysms and Earth History*, Oxford University Press, 1989, 141–145.

23. A. Hallam, *Great Geological Controversies*, 101–102, 112, 116; F. Hoyle, *Ice*, New English Library, Sevenoaks, Kent, 1982, 61–62; *Oxford Dictionary of Scientists*, Oxford University Press, 1999, 115–116.

24. A. Hallam, *Great Geological Controversies*, 30–31, 49–60, 129–132, 226; S.J. Gould, *Ever Since Darwin*, Burnett Books, London, 1978, Penguin Books, Harmondsworth, 1980, Chapter 18, 147–152; S.J. Gould, *Time's Arrow, Time's Cycle*, Harvard University Press, Cambridge, Mass., 1987, Penguin Books, Harmondsworth, 1988, 117–126, 167–179; M.J.S. Rudwick, Introduction, in C. Lyell, *Principles of Geology*, Volume 1, University of Chicago Press, Chicago and London, 1990, liii–lv.

Chapter 7

1. A. Desmond and J. Moore, *Darwin*, Michael Joseph, London, 1991, Penguin Books, Harmondsworth, 1992, 90, 101–114, 220; M. White and J. Gribbin, *Darwin – A Life in Science*, Simon and Schuster, London, 1996, 42–46, 49–53, 104, 122–123.

2. D. King-Hele, *Erasmus Darwin*, Giles de la Mare, London, 1999, 54–89, 297–301, 346–351; N. Priestland, *Erasmus Darwin*, Ashbracken, Nottingham, 1990, 39–56; P.J. Bowler, *Evolution – The History of an Idea*, University of California Press, Berkeley, Los Angeles and London, 1989, 81–82; D. King-Hele, Erasmus Darwin, master of interdisciplinary science, *Interdisciplinary Science Review*, 10, 1985, 170–191.

3. D. King-Hele, *Erasmus Darwin*, 297–301; D. King-Hele, *Interdisciplinary Science Review*, 170–191.

4. D. King-Hele, *Erasmus Darwin*, 347; N. Priestland, *Erasmus Darwin*, 54–55.

5. D. King-Hele, *Erasmus Darwin*, 301, 314–320; D. King-Hele, *Interdisciplinary Science Review*, 170–191.

6. A. Desmond and J. Moore, *Darwin*, 31–41; M. White and J. Gribbin, *Darwin – A Life in Science*, 13–20.

7. A. Desmond and J. Moore, *Darwin*, 49–82; M. White and J. Gribbin, *Darwin – A Life in Science*, 23–24, 121–125.

8. P.J. Bowler, *Evolution – The History of an Idea*, 164–167; E. Mayr, *The Growth of Biological Thought*, Harvard University Press, Cambridge, Mass., 1982, 116–117.

9. E. Mayr, *The Growth of Biological Thought*, 505–510.

10. P.J. Bowler, *Evolution – The History of an Idea*, 156–158, A. Desmond and J. Moore, *Darwin*, 131; V. Blackmore and A. Page, *Evolution – The Great Debate*, Lion Books, Oxford, 1989, 66.

11. A. Desmond and J. Moore, *Darwin*, 169–172, 202–228; S.J. Gould, *The Flamingo's Smile*, Norton, New York and London, 1985, Penguin Books, Harmondsworth, 1986, 347–359; V. Blackmore and A. Page, *Evolution – The Great Debate*, 60–72; S.J. Gould, *The Lying Stones of Marrakech*, Jonathan Cape, London, 2000, 169–181.

12. A. Desmond and J. Moore, *Darwin*, 116–118; A. Hallam, *Great Geological Controversies*, Oxford University Press, 2nd edn, 1989, 55.

13. A. Desmond and J. Moore, *Darwin*, 262–276; V. Blackmore and A. Page, *Evolution – The Great Debate*, 72–73; P.J. Bowler, *Evolution – The History of an Idea*, 101–102.

14. C. Darwin, *The Origin of Species*, John Murray, London, 6th edn., 1872 (Reprinted by Studio Editions, London, 1994), Historical Sketch, xiii–xxi; F. Darwin (ed.), *The Life of Charles Darwin*, John Murray, London, 1902 (Reprinted by Studio Editions, London, 1995), 232–233.

15. D. King-Hele, *Erasmus Darwin*, 299, 350; D. King-Hele, *Interdisciplinary Science Review*, 170–191.

16. E. Mayr, *The Growth of Biological Thought*, 488–491; V. Blackmore and A. Page, *Evolution – The Great Debate*, 70–71; W.J. Dempster, *Evolutionary Concepts in the Nineteenth Century*, Pentland Press, Edinburgh, Cambridge and Durham, 1996, 34–35, 141, 184–195.

17. W.J. Dempster, *Evolutionary Concepts in the Nineteenth Century*, 9–25, 244–260; C.L. Harris, *Evolution – Genesis and Revelations*, State University of New York Press, Albany, 1981, 129–132, 140–144.

18. W.J. Dempster, *Evolutionary Concepts in the Nineteenth Century*, 255; C.L. Harris, *Evolution – Genesis and Revelations*, 140–141.

19. W.J. Dempster, *Evolutionary Concepts in the Nineteenth Century*, 1–8; C.L. Harris, *Evolution – Genesis and Revelations*, 130–131.

20. W.J. Dempster, *Evolutionary Concepts in the Nineteenth Century*, 255; C.L. Harris, *Evolution – Genesis and Revelations*, 141. (Note that the word "nearly" is included by Harris but not by Dempster. Regardless of whether "nearly" was actually present in the original, Matthew clearly stated that "new, diverging ramifications of life" were able to take advantage of the situation, so he could hardly have been suggesting a total extinction of living things.)

21. W.J. Dempster, *Evolutionary Concepts in the Nineteenth Century*, 256–257; C.L. Harris, *Evolution – Genesis and Revelations*, 142.

22. W.J. Dempster, *Evolutionary Concepts in the Nineteenth Century*, 257; C.L. Harris, *Evolution – Genesis and Revelations*, 143.

23. A. Desmond and J.Moore, *Darwin*, 40, 266, 286; W.J. Dempster, *Evolutionary Concepts in the Nineteenth Century*, 1–8; C.L. Harris, *Evolution – Genesis and Revelations*, 129–131.

24. W.J. Dempster, *Evolutionary Concepts in the Nineteenth Century*, 16–18, 351–354.

25. F. Darwin (ed.), *The Life of Charles Darwin*, 232–233.

26. A. Desmond and J. Moore, *Darwin*, 195–387; M. White and J. Gribbin, *Darwin – A Life in Science*, 99–174; S.J. Gould, *The Lying Stones of Marrakech*, 169–181.

27. A. Desmond and J. Moore, *Darwin*, 84–87; L. Woodward, *The Age of Reform, 1815–1870*, Oxford University Press, second edition, 1962, 52–78.

28. A. Desmond and J. Moore, *Darwin*, 195–197; L. Woodward, *The Age of Reform, 1815–1870*, 78–103, 126–153.

29. A. Desmond and J. Moore, *Darwin*, 197–220; M. White and J. Gribbin, *Darwin – A Life in Science*, 127–132.

30. E. Mayr, *The Growth of Biological Thought*, 401–410; A. Desmond and J. Moore, *Darwin*, 220–239.

31. P.J. Bowler, *Evolution – The History of an Idea*, 140–141.

32. E. Mayr, *The Growth of Biological Thought*, 881–882.

33. A. Desmond and J. Moore, *Darwin*, 202–228; P.J. Bowler, *Evolution – The History of an Idea*, 162–175.

34. A. Desmond and J. Moore, *Darwin*, 227–253; P.J. Bowler, *Evolution – The History of an Idea*, 90–94, 99–102, 168–169; E. Mayr, *The Growth of Biological Thought*, 620.

35. Mrs C.F. Alexander, *Hymns for Little Children*, 1848. (The "missing" verse from *All things bright and beautiful* is quoted in P. Dearmer, *Songs of Praise Discussed*, Oxford University Press, 1933, 239.)

36. P.H. Barrett (ed.), *The Collected Papers of Charles Darwin*, Volume 2, University of Chicago Press, Chicago and London, 1977, 3–8.

37. A. Desmond and J. Moore, *Darwin*, 313–332, 343–345; M. White and J. Gribbin, *Darwin – A Life in Science*, 141–144; F. Burkhardt (ed.), *Charles Darwin's Letters*, Cambridge University Press, 1996, 80–81.

38. A. Desmond and J. Moore, *Darwin*, 339–409; M. White and J. Gribbin, *Darwin – A Life in Science*, 157–174.

39. A. Desmond and J. Moore, *Darwin*, 403–411; *The Hutchinson Dictionary of Scientific Biography*, Helicon, Oxford, 1994, 357–358; A. Desmond, *Huxley*, Addison-Wesley, Reading, Mass., 1997, Penguin Books, Harmondsworth, 1998, 3–215.

40. A. Desmond and J. Moore, *Darwin*, 412–416, 442; R. Hooykaas, *The Principle of Uniformity in Geology, Biology and Theology*, Brill, Leiden, 1963, 109–115; R.M. Young, *Darwin's Metaphor*, Cambridge University Press, 1985, 101–102.

41. C. Lyell, Anniversary address of the President, *Quarterly Journal of the Geological Society of London*, 7, 1851, xxxii–lxxvi.

42. A. Desmond and J. Moore, *Darwin*, 438–466; M. White and J. Gribbin, *Darwin – A Life in Science*, 175–188.

43. A. Williams-Ellis, *Darwin's Moon*, Blackie, London and Glasgow, 1966, 1–138; H. Clements, *Alfred Russel Wallace*, Hutchinson, London, 1983, 1–40; P. Raby, *Alfred Russel Wallace – A Life*, Princeton University Press, Princeton, New Jersey, 2001.

44. A. Desmond and J. Moore, *Darwin*, 467–481; M. White and J. Gribbin, *Darwin – A Life in Science*, 188–218; F. Burkhardt (ed.), *Charles Darwin's Letters*, 177–180; P.H. Barrett (ed.), *The Collected Papers of Charles Darwin, Volume 2*, 3–19.

45. A. Desmond and J. Moore, *Darwin*, 456–466; D. Ospovat, *The Development of Darwin's Theory*, Cambridge University Press, 1981, 2, 87–89.

46. D. Ospovat, *The Development of Darwin's Theory*, 39–114; P.J. Bowler, *Evolution – The History of an Idea*, 164–184; M. Ghiselin, The path to natural selection, in J. Cherfas (ed.), *Darwin up to Date*, New Science Publications, London, 1982, 64–66; C. Darwin (ed. R.C. Stauffer), *Natural Selection*, Cambridge University Press, 1975.

47. D. Ospovat, *The Development of Darwin's Theory*, 33–83; R.M. Young, *Darwin's Metaphor*, 189–181; E. Mayr, *One Long Argument*, Penguin Books, Harmondsworth, 1991, 55–57.

48. D. Ospovat, *The Development of Darwin's Theory*, 83–86; E. Mayr, *One Long Argument*, 40–47; P.J. Bowler, *Evolution – The History of an Idea*, 175–184.

49. S.M. Stanley, *The New Evolutionary Timetable*, Basic Books, New York, 1981, 35–53; E. Mayr, *One Long Argument*, 44–46; S.J. Gould, *The Panda's Thumb*, Norrton, New York, 1980, Penguin Books, Harmondsworth, 1983, 149–154, 187–188.

50. C. Darwin, *On the Origin of Species by Means of Natural Selection*, John Murray, London, 1859, Penguin Books, Harmondsworth, 1968, 439–440. (Note that page references, here and elsewhere, are to the Penguin edition.)

51. C. Darwin, *On the Origin of Species*, 440.

52. S.M. Stanley, *The New Evolutionary Timetable*, 99–100; A. Lister and P. Bahn, *Mammoths*, Macmillan, New York, 1994, 22–23.

53. F. Darwin, *The Life of Charles Darwin*, 213–214.

54. A. Desmond, *Huxley*, 254–263; S.J. Gould, *The Panda's Thumb*, chapter 17, 149–154.

Chapter 8

1. D.R. Oldroyd, *Darwinian Impacts*, Open University Press, Milton Keynes, 1980, 193–203; P.J. Bowler, *Evolution – The History of an Idea*, University of California Press, Berkeley, Los Angeles and London, 1989, 187–199; A. Desmond and J. Moore, *Darwin*, Michael Joseph, London, 1991, Penguin Books, Harmondsworth, 1992, 477–499; C. Overy, *Charles Darwin*, English Heritage, London, 1997, 55–57.

2. F. Darwin, *The Life of Charles Darwin*, John Murray, London, 1902 (reprinted by Studio Editions, London, 1995), 227–228.

3. A. Desmond and J. Moore, *Darwin*, 487.

4. R.M. Young, *Darwin's Metaphor*, Cambridge University Press, 1985, 144; P.J. Bowler, *Evolution – The History of an Idea*, 214–217.

5. F. Darwin, *The Life of Charles Darwin*, 216–218.

6. E. Lurie, *Louis Agassiz*, Johns Hopkins University Press, Baltimore and London, 1988, 298.

7. A. Desmond and J. Moore, *Darwin*, 478–492, 534; P.J. Bowler, *Evolution – The History of an Idea*, 222–226; R.M. Young, *Darwin's Metaphor*, 99–112.

8. F. Darwin, *The Life of Charles Darwin*, 235–236.

9. A. Desmond and J. Moore, *Darwin*, 470; V. Blackmore and A. Page, *Evolution – The Great Debate*, Lion Books, Oxford, 1989, 77.

10. P.J. Bowler, *Evolution – The History of an Idea*, 195–196, 214–217; V. Blackmore and A. Page, *Evolution – The History of an Idea*, 163–165; A. Desmond, *Huxley*, Addison-Wesley, Reading, Mass., 1997, Penguin Books, Harmondsworth, 1998, 252–263, 385–392.

11. C. Overy, *Charles Darwin*, 65.

12. S. Wilberforce (published anonymously), Art. VII – *On the Origin of Species, by Means of Natural Selection; or the Preservation of Favoured Races in the Struggle for Life. By Charles Darwin, M.A., F.R.S., London, 1860. The Quarterly Review*, 108, 1860, 225–264 (quotation from page 237).

13. R.M. Young, *Darwin's Metaphor*, 198–199; D.R. Oldroyd, *Darwinian Impacts*, 244–247.

14. C. Darwin, *On the Origin of Species by Means of Natural Selection*, John Murray, London, 1859, Penguin Books, Harmondsworth, 1968, 458. (Note that page references, here and elsewhere, are to the Penguin edition.)

15. C. Overy, *Charles Darwin*, 66–67; V. Blackmore and A. Page, *Evolution – The Great Debate*, 105–116; R. Leakey, *The Making of Mankind*, Michael Joseph, London, 1981, 10.

16. A. Desmond and J. Moore, *Darwin*, 521–527; P.J. Bowler, *Evolution – The History of an Idea*, 228–230; H. Clements, *Alfred Russel Wallace*, Hutchinson, London, 1983, 163–176.

17. S. Wilberforce (published anonymously), *The Quarterly Review* 108, 1860, 225–264 (quotation from page 258).

18. A. Desmond and J. Moore, *Darwin*, 488; M. White and J. Gribbin, *Darwin – A Life in Science*, Simon and Schuster, London, 1995, 230–231.

19. A. Desmond and J. Moore, *Darwin*, 492–498; F. Darwin, *The Life of Charles Darwin*, 236–242; A. Desmond, *Huxley*, 275–281; H. Hellman, *Great Feuds in Science*, Wiley, New York, 1998, 81–92.

20. R.M. Young, *Darwin's Metaphor*, 126–163; R. Hooykaas, *The Principle of Uniformity in Geology, Biology and Theology*, Brill, Leiden, 1963, 169–229; M.J.S. Rudwick, *The Meaning of Fossils*, Macdonald, London, 1972, 200–214; C. Darwin, *On the Origin of Species*, 458.

21. A. Desmond and J. Moore, *Darwin*, 459, 472, 488–493; E. Mayr, *One Long Argument*, Penguin Books, Harmondsworth, 1991, 98–107.

22. A. Desmond and J. Moore, *Darwin*, 500–515; B. Gardiner (published anonymously), Editorial, *The Linnean*, 13(3), 1997, 1–4.

23. C. Lyell, *The Geological Evidences for the Antiquity of Man*, John Murray, London, 1863. (Note that the fourth edition, 1873, was reprinted by AMS Press, New York, 1973.)

24. A. Desmond and J. Moore, *Darwin*, 515; D.R. Oldroyd, *Darwinian Impacts*, 144; M. White and J. Gribbin, *Darwin – A Life in Science*, 227–228.

25. F. Darwin, *The Life of Charles Darwin*, 253–255.

26. R.M. Young, *Darwin's Metaphor*, 103.

27. A. Desmond and J. Moore, *Darwin*, 547–548.

28. B. Gardiner (published anonymously), *The Linnean*, 13(3), 1997, 1–4.

29. A. Desmond and J. Moore, *Darwin*, 613–614, 664–677; M. White and J. Gribbin, *Darwin – A Life in Science*, 269–279.

30. P.J. Bowler, *Evolution – The History of an Idea*, 109.

31. R. Lewin, *Complexity*, Dent, London, 1993, 75.

32. M.J.S. Rudwick, *The Meaning of Fossils*, 187–201; P.J. Bowler, *Evolution – The History of an Idea*, 109–110; E. Mayr, *The Growth of Biological Thought*, Harvard University Press, Cambridge, Mass., 1982, 379–381.

33. T.H. Huxley, Anniversary Address of the President, *Quarterly Journal of the Geological Society of London*, 25, 1869, xxxviii–liii (quotation from pages xlvi–xlvii).

34. E. Mayr, *The Growth of Biological Thought*, 379.

35. C. Darwin, *On the Origin of Species*, 152–153.

36. C. Darwin, *On the Origin of Species*, 223–224.

37. C. Darwin, *On the Origin of Species*, 119.

38. C. Darwin, *On the Origin of Species*, 171–172, 321.

39. C. Darwin, *On the Origin of Species*, 154.

40. D.R. Oldroyd, *Darwinian Impacts*, 125, 130–131, 207–224; P.J. Bowler, *Evolution – The History of an Idea*, 237–245; M. White and J. Gribbin, *Darwin – A Life in Science*, 229–230, 240.

41. C. Darwin, *On the Origin of Species*, 343.

42. C. Darwin, *On the Origin of Species*, 444–445, 448, 459.

43. J. Reader, *Missing Links*, Penguin Books, Harmondsworth, 2nd edn, 1988, 17–19; E. Trinkaus and P. Shipman, *The Neandertals*, Jonathan Cape, London, 1993, 33–45, 110; I. Tattersall, *The Fossil Trail*, Oxford University Press, 1995, 7–11, 24–26.

44. J. Reader, *Missing Links*, 6–13; E. Trinkaus and P. Shipman, *The Neandertals*, 3–7, 45–90; I.Tattersall, *The Fossil Trail*, 13–16, 20–22.

45. A. Desmond, *Huxley*, 304–335; J. Reader, *Missing Links*, 10; I. Tattersall, *The Fossil Trail*, 20.

46. J. Reader, *Missing Links*, 33–36; E. Trinkaus and P. Shipman, *The Neandertals*, 109–118; I. Tattersall, *The Fossil Trail*, 24, 28–29.

47. E. Haeckel, *Generelle Morphologie der Organismen*, Georg Reimer, Berlin, 1866; E. Haeckel, *Natürliche Schöpfungsgeschichte*, Georg Reimer, Berlin, 1868 (published in English as *The History of Creation*, London and New York, 1876); H. Wendt, *From Ape to Adam*, Thames and Hudson, London, 1971, 1971, 77–82.

48. C. Darwin, *The Descent of Man*, John Murray, London, 1871 (reprinted by Princeton University Press, Princeton, New Jersey, 1981).

49. A. Desmond and J. Moore, *Darwin*, 577–581; P.J. Bowler, *Evolution – The History of an Idea*, 228–237; I. Tattersall, *The Fossil Trail*, 29–30.

50. C. Darwin, *The Descent of Man*, 135–136.

51. C. Darwin, *The Descent of Man*, 140–141.

52. C. Darwin, *The Descent of Man*, Chapter IV, 107–157. Note that this, with some slight changes of wording, became Chapter II, 38–97, in the second edition of *The Descent of Man*, published by John Murray in 1874. The quotation is from page 97 of the second edition (1913 reprint) and corresponds to a passage on page 157 of the first edition.

53. A. Desmond and J. Moore, *Darwin*, 547; E. Mayr, *The Growth of Biological Thought*, 512–514; M.A. Edey and D.C. Johanson, *Blueprints*, Oxford University Press, 1990, 98–99; S.J. Gould, *Bully for Brontosaurus*, Hutchinson Radius, London, 1991, 340–353.

54. P.J. Bowler, *Evolution – The History of an Idea*, 270–274; E. Mayr, *The Growth of Biological Thought*, 710–726; V. Blackmore and A. Page, *Evolution – The Great Debate*, 124–127; M.A. Edey and D.J. Johanson, *Blueprints*, 105–122; M. White and J. Gribbin, *Darwin – A Life in Science*, 284–291; V. Orel, *Gregor Mendel*, Oxford University Press, 1996, 92–209, 256–259, 271–282; C. Tudge, *In Mendel's Footnotes*, Jonathan Cape, London, 2000, 10–97.

55. D.R. Oldroyd, *Darwinian Impacts*, 193–203; E. Mayr, *The Growth of Biological Thought*, 501–514; P.J. Bowler, *Evolution – The History of an Idea*, 187–217.

56. C. Darwin, *On the Origin of Species*, 293–297.

57. P.J. Bowler, *Evolution – The History of an Idea*, 205–208; M.A. Edey and D.C. Johanson, *Blueprints*, 99–101; H. Hellman, *Great Feuds in Science*, 105–119; A. Hallam, *Great Geological Controversies*, Oxford University Press, 2nd edn, 1989, 105–134; L. Badash, The age-of-the-Earth debate, *Scientific American*, 261(2), 1989, 78–83.

58. A. Hallam, *Great Geological Controversies*, 125–129; *The Hutchinson Dictionary of Scientific Biography*, Helicon, Oxford, 1994, 343; S.M. Stanley, *Earth and Life Through Time*, Freeman, New York, 1986, 119–132; G.B. Dalrymple, *The Age of the Earth*, Stanford University Press, Palo Alto, Calif., 1991; C. Zimmer, How old is it?, *National Geographic*, 200(3), 2001, 78–101.

Chapter 9

1. D.T. Parkin, *An Introduction to Evolutionary Genetics*, Edward Arnold, London, 1979, 13–22; E. Mayr, *The Growth of Biological Thought*, Harvard University Press, Cambridge, Mass., 1982, 553–554, 710–731; P.J. Bowler, *Evolution – The History of an Idea*, University of California Press, Berkeley, Los Angeles and London, 1989, 270–275; M.A. Edey and D.C. Johanson, *Blueprints*, Oxford University Press, 1990, 105–132, 180–181; C. Tudge, *In Mendel's Footnotes*, Jonathan Cape, London, 2000, 98–119.

2. E. Mayr, *The Growth of Biological Thought*, 731–760; P.J. Bowler, *Evolution – The History of an Idea*, 275–280; M.A. Edey and D.C. Johanson, *Blueprints*, 126–127, 135–182.

3. P.J. Bowler, *Evolution – The History of an Idea*, 279–281; G.L. Stebbins, *Darwin to DNA, Molecules to Humanity*, Freeman, San Francisco, 1982, 37–46; V. Blackmore and A. Page, *Evolution – The Great Debate*, Lion Books, Oxford, 1989, 122–127.

4. E. Mayr, *The Growth of Biological Thought*, 526–531; P.J. Bowler, *Evolution – The History of an Idea*, 226–228, 257–268, 296–299, 341–342; S.J. Gould, *The Panda's Thumb*, Norton, New York, 1980; Penguin Books, Harmondsworth, 1983, Chapter 7, 65–71.

5. P.J. Bowler, *Evolution – The History of an Idea*, 262–263; J. Maynard Smith, *The Theory of Evolution*, Penguin Books, Harmondsworth, 1958, 284–286; R. Sheldrake, *The Presence of the Past*, Collins, London, 1988, 275–279; E.J. Steele, R.A. Lindley and R.V. Blanden, *Lamarck's Signature*, Perseus Books, Reading, Mass., 1998, 190–192.

6. E. Mayr, *The Growth of Biological Thought*, 698–701; P.J. Bowler, *Evolution – The History of an Idea*, 250–253.

7. P.J. Bowler, *Evolution – The History of an Idea*, 251–252, 264; V. Blackmore and A. Page, *Evolution – The Great Debate*, 127; H.G. Cannon, *Lamarck and Modern Genetics*, Greenwood Press, Westport, Conn., 1959, 38–40; E.J. Steele, R.A. Lindley and R.V. Blanden, *Lamarck's Signature*, 12–13.

8. P.J. Bowler, *Evolution – The History of an Idea*, 265–267; E.J. Steele, R.A. Lindley and R.V. Blanden, *Lamarck's Signature*, 13–14; A. Koestler, *The Case of the Midwife Toad*, Hutchinson, London, 1971; *Oxford Dictionary of Scientists*, Oxford University Press, 1999, 290–291.

9. D.T. Parkin, *An Introduction to Evolutionary Genetics*, 10–13; S.J. Gould, *The Panda's Thumb*, 68–70.

10. P.J. Bowler, *Evolution – The History of an Idea*, 267–268; E.J. Steele, R.A. Lindley and R.V. Blanden, *Lamarck's Signature*, 14–15; V. Blackmore and A. Page, *Evolution – The Great Debate*, 138; R. Dawkins, *The Blind Watchmaker*, Longman, London, 1986, Penguin Books, Harmondsworth, 1988, 273–274.

11. E. Mayr, *The Growth of Biological Thought*, 777–779; R. Dawkins, *The Blind Watchmaker*, 304–306; S.M. Stanley, *The New Evolutionary Timetable*, Basic Books, New York, 1981, 65–67; J.H. Schwartz, *Sudden Origins*, Wiley, New York, 1999, 187–242; S.J. Gould, *The Structure of Evolutionary Theory*, Harvard University Press, Cambridge, Mass., 2002, 342–466.

12. E. Mayr, *The Growth of Biological Thought*, 550–559; P.J. Bowler, *Evolution – The History of an Idea*, 308–314; G.L. Stebbins, *Darwin to DNA, Molecules to Humanity*, 46–64; J.H.Schwartz, *Sudden Origins*, 243–275.

13. R.A. Fisher, *The Genetical Theory of Natural Selection*, Clarendon Press, Oxford, 1930.

14. R.C. Punnett, Genetics, mathematics and natural selection, *Nature*, 126, 1930, 595–597 (quotation from page 596).

15. T.H. Morgan, *The Scientific Basis of Evolution*, Norton, New York, 1932.

16. E. Mayr, *The Growth of Biological Thought*, 555–556; S.M. Stanley, *Macroevolution*, Freeman, San Francisco, 1979, 23–25; M. Kimura, *The Neutral Theory of Molecular Evolution*, Cambridge University Press, 1983, 12–14; N. Eldredge, *Time Frames*, Simon and Schuster, New York, 1985, 131–132.

17. S. Wright, Evolution in Mendelian populations, *Genetics*, 16, 1931, 97–159; S. Wright, The rules of mutation, inbreeding, crossbreeding and selection in evolution, *Proceedings of the Sixth International Congress of Genetics*, 1, 1932, 356–366.

18. M. Kimura, *The Neutral Theory of Molecular Evolution*, 21.

19. E. Mayr, *The Growth of Biological Thought*, 535–570; P.J. Bowler, *Evolution – The History of an Idea*, 307–318; G.L. Stebbins, *Darwin to DNA, Molecules to Humanity*, 64–66; J.H. Schwartz, *Sudden Origins*, 276–309; S.J. Gould, *The Structure of Evolutionary Theory*, 503–584.

20. T. Dobzhansky, *Genetics and the Origin of Species*, Columbia University Press, New York, 1937.

21. T. Dobzhansky, *Genetics and the Origin of Species*, 176–191 (quotations from pages 186 and 190).

22. T. Dobzhansky, *Genetics and the Origin of Species*, 187–190; S.M. Stanley, *The New Evolutionary Timetable*, 68–69; N. Eldredge, *Time Frames*, 132.

23. C. Darwin, *On the Origin of Species by Means of Natural Selection*, John Murray, London, 1859, Penguin Books, Harmondsworth, 1968, 132.

24. J.S. Huxley, *Evolution – The Modern Synthesis*, Allen and Unwin, London, 1942 (quotation from page 28).

25. E. Mayr, *Systematics and the Origin of Species*, Columbia University Press, New York, 1942.

26. B. Rensch, *Neuere Probleme der Abstammungslehre*, Enke, Stuttgart, 1947; G.L. Stebbins, *Variation and Evolution in Plants*, Columbia University Press, New York, 1950.

27. S.M. Stanley, *The New Evolutionary Timetable*, 66–71; M.A. Edey and D.C. Johanson, *Blueprints*, 126–127, 179–182; E. Mayr, *One Long Argument*, Penguin Books, Harmondsworth, 1991, 132–140.

28. E. Mayr, *The Growth of Biological Thought*, 528–534; G.L. Stebbins, *Darwin to DNA, Molecules to Humanity*, 140–141.

29. E. Mayr, *The Growth of Biological Thought*, 532.

30. C. Darwin, *On the Origin of Species*, 173.

31. F. Hitching, *The Neck of the Giraffe – or Where Darwin Went Wrong*, Pan Books, London, 1982, 8–84, 103; J. Rifkin, *Algeny*, Viking Press, New York, 1983, 151–156; G.R. Taylor, *The Great Evolution Mystery*, Secker and Warburg, London, 1983, 2–5, 17–18; R. Milton, *The Facts of Life*, Fourth Estate, London, 1992, 116–152.

32. S.M. Stanley, *The New Evolutionary Timetable*, 9–12; E. Mayr, *One Long Argument*, 48–67; R. Dawkins, The necessity of Darwinism, in J. Cherfas (ed.), *Darwin up to Date*, New Science Publications, London, 1982, 61–63.

33. E. Mayr, *The Growth of Biological Thought*, 756–784; G.L. Stebbins, *Darwin to DNA, Molecules to Humanity*, 37–59; C. Tudge, *In Mendel's Footnotes*, 112–116; J. Maynard Smith, *Evolutionary Genetics*, Oxford University Press, 1989, 231 254.

34. M.A. Edey and D.C. Johanson, *Blueprints*, 205–214; O.T. Avery, C.M. MacLeod and M. McCarty, Studies on the chemical nature of the substance inducing transformation of pneumococcal types, *Journal of Experimental Medicine*, 79, 1944, 137–158.

35. F.H.C. Crick and J.D. Watson, Molecular structure of nucleic acids – a structure for DNA, *Nature*, 171, 1953, 737–738; J.D. Watson, *The Double Helix*, Atheneum, New York, 1968; R. Olby, *The Path to the Double Helix*, University of Washington Press, Seattle, 1974.

36. M.A. Edey and D.C. Johanson, *Blueprints*, 215–278; C. Tudge, *In Mendel's Footnotes*, 116–132; D.J. Futuyma, *Evolutionary Biology*, Sinauer Associates, Sunderland, Mass., 2nd edn, 1986, 42–56; D.L. Nelson and M.M. Cox, *Lehninger Principles of Biochemistry*, Worth, New York, 3rd edn, 2000, 931–1069; T. Palmer, *Enzymes – Biochemistry, Biotechnology, Clinical Chemistry*, Horwood, Chichester, 2001, 44–54.

37. F.H.C. Crick, On protein synthesis, *Symposium of the Society for Experimental Biology*, 12, 1958, 138–163; F.H.C. Crick, Central dogma of molecular biology, *Nature*, 227, 1970, 561–563; F. Crick, *What Mad Pursuit*, Penguin Books, Harmondsworth, 1990, 109, 168.

38. E. Mayr, *The Growth of Biological Thought*, 808–828; V. Blackmore and A. Page, *Evolution – The Great Debate*, 139–144.

39. H.M. Temin, The DNA provirus hypothesis, *Science*, 192, 1976, 1075–1080.

40. J. Cherfas, Proving the pattern of life, in J. Cherfas (ed.), *Darwin up to Date*, 39–42; D.T. Parkin, *An Introduction to Evolutionary Genetics*, 50–64; T. Palmer, *Enzymes: Biochemistry, Biotechnology, Clinical Chemistry*, 345–347, 369–378; J. Phillips and P. Murray (eds.), *The Biology of Disease*, Blackwell, Oxford, 1995, 207–218.

Chapter 10

1. E. Mayr, *Systematics and the Origin of Species*, Columbia University Press, New York, 1942, 147; N.Eldredge, *Time Frames*, Simon and Schuster, New York, 1985, 32–34, 98, 109.

2. C. Darwin, *On the Origin of Species by Means of Natural Selection*, John Murray, London, 1859, Penguin Books, Harmondsworth, 1968, 107.

3. N. Eldredge, *Time Frames*, 109–110; E. Mayr, *One Long Argument*, Penguin Books, Harmondsworth, 1991, 40–42.

4. W. Whewell, *Philosophy of the Inductive Sciences, Volume 3*, Parker, London, 1840, 626.

5. C. Darwin, *On the Origin of Species*, 159–165 (quotations from pages 163 and 164).

6. C. Darwin, *On the Origin of Species*, 341.

7. L.B. Halstead, *Hunting the Past*, Hamish Hamilton, London, 1982, 38–45; S.M. Stanley, *Earth and Life Through Time*, Freeman, New York, 1986, 9–13; D. Dathe, *Fundamentals of Historical Geology*, Brown, Dubuque, Iowa, 1993, 33–39.

8. T. Dobzhansky, *Genetics and the Origin of Species*, Columbia University Press, New York, 1937, 312.

9. E. Mayr, *Systematics and the Origin of Species*, 120.

10. E. Mayr, *Systematics and the Origin of Species*, 147–162; N. Eldredge, *Time Frames*, 69, 98–115; S.M. Stanley, *The New Evolutionary Timetable*, Basic Books, New York, 1981, 68–70.

11. T. Dobzhansky, *Genetics and the Origin of Species*, 312.

12. H.J. Müller, Bearings of the 'Drosophila' work on systematics, in J. Huxley (ed.), *The New Systematics*, Oxford University Press, 1940, 185–268 (quotation from page 258).

13. E. Mayr, *Systematics and the Origin of Species*, 153.

14. J. Maynard Smith, *The Theory of Evolution*, Penguin Books, Harmondsworth, 3rd edn, 1975, 201–202.

15. S.M. Stanley, *The New Evolutionary Timetable*, 101–106; E. Mayr, *The Growth of Biological Thought*, Harvard University Press, Cambridge, Mass., 1982, 607; M.J.S. Rudwick, *The Meaning of Fossils*, Macdonald, London, 1972, 218–266; S.J. Gould, *The Structure of Evolutionary Theory*, Harvard University Press, Cambridge, Mass., 2002, 342–466.

16. S.M. Stanley, *The New Evolutionary Timetable*, 107.

17. G.G. Simpson, *Tempo and Mode in Evolution*, Columbia University Press, New York, 1944.

18. E. Mayr, *The Growth of Biological Thought*, 608–610; P.J. Bowler, *Evolution – The History of an Idea*, University of California Press, Berkeley, Los Angeles and London, 1989, 317; S.M. Stanley, *The New Evolutionary Timetable*, 107; L.K. Laporte, *George Gaylord Simpson*, Columbia University Press, New York, 2000.

19. P.J. Bowler, *Evolution – The History of an Idea*, 317–318; S.M. Stanley, *The New Evolutionary Timetable*, 107–108; N. Eldredge, *Time Frames*, 146.

20. H.B.D. Kettlewell, The phenomenon of industrial melanism in Lepidoptera, *Annual Review of Entomology*, 6, 1961, 245–262; H.B.D. Kettlewell, *The Evolution of Melanism*, Clarendon Press, Oxford, 1973; H.B.D. Kettlewell, Evolution and the environment, in J. Cherfas (ed.), *Darwin up to Date*, New Science Publications, London, 1982 21–22; R. Osborne and M. Benton, *The Viking Atlas of Evolution*, Viking, London, 1996, 38–39.

21. D.T. Parkin, *An Introduction to Evolutionary Genetics*, Edward Arnold, London, 1979, 38–41, 162–164, 176–177; J. Maynard Smith, *Evolutionary Genetics*, Oxford University Press, 1989, 42–44; V. Blackmore and A. Page, *Evolution – The Great Debate*, Lion Books, Oxford, 1989, 94–95.

22. J. Maynard Smith, *Evolutionary Genetics*, 273–286; N. Eldredge, *Time Frames*, 196–201; E. Mayr, *Populations, Species and Evolution*, Harvard University Press, Cambridge, Mass., 1970, 351–374.

23. E. Mayr, *Populations, Species and Evolution*, 291–293; D.T. Parkin, *An Introduction to Evolutionary Genetics*, 191–192; R. Osborne and M. Benton, *The Viking Atlas of Evolution*, 56–57; G. de Beer, *Evolution*, British Museum (Natural History), London, 3rd edn, 1964, 28–29, 73.

24. R. Bakker, *The Dinosaur Heresies*, Penguin Books, Harmondsworth, 1988, 298–322; D. Norman, *Dinosaur!*, Boxtree, London, 1991, 81–82, 130–143; D. Norman, *The Illustrated Encyclopedia of Dinosaurs*, Greenwich Editions, London, 1998, 191–193; P. Shipman, *Taking Wing*. Weidenfeld and Nicolson, London, 1998, 13–46.

25. N. Eldredge, *Time Frames*, 72–75; S.M. Stanley, *The New Evolutionary Timetable*, 71, 108, 135.

26. R. Goldschmidt, *The Material Basis of Evolution*, Yale University Press, New Haven, 1940.

27. E. Mayr, *Populations, Species and Evolution*, 252–254; S.M. Stanley, *Macroevolution*, Freeman, San Francisco, 1979, 20–22, 159; S.J. Gould, *The Panda's Thumb*, Norton, New York, 1980, Penguin Books, Harmondsworth, 1983, Chapter 18, 155–161.

28. O.H. Schindewolf, *Grundfragen der Paläontologie*, Schweizerbart, Stuttgart, 1950; S.M. Stanley, *Macroevolution*, 35.

29. J. Huxley, *Evolution – The Modern Synthesis*, Allen and Unwin, London, 1942, 389.

30. T. Dobzhansky, *Mankind Evolving – The Evolution of the Human Species*, Yale University Press, New Haven, 1962, 181.

31. G. de Beer, *Evolution*, 31–32.

32. G. de Beer, *Evolution*, 29–31 (quotation from page 31).

33. S.A. Kauffman, *The Origins of Order*, Oxford University Press, 1993, 3–26; B. Goodwin, *How the Leopard Changed its Spots*, Weidenfeld and Nicolson, London, 1994, 1–39; C.D. Rollo, *Phenotypes – Their Epigenetics, Ecology and Evolution*, Chapman and Hall, London, 1995, 1–31.

34. M.A. Edey and D.C. Johanson, *Blueprints*, Oxford University Press, 1990, 98; F. Darwin (ed.), *The Life of Charles Darwin*, John Murray, London, 1902 (reprinted by Studio Editions, London, 1995), 208.

35. C. Darwin, *On the Origin of Species*, 217.

36. P.J. Bowler, *Evolution – The History of an Idea*, 112–118; S.A. Kauffman, *The Origins of Order*, 3–5; B. Goodwin, *How the Leopard Changed its Spots*, 131–132.

37. E. Mayr, *The Growth of Biological Thought*, 455–469; S.A. Kauffman, *The Origins of Order*, 537–538; D'A.W. Thompson, *On Growth and Form*, Cambridge University Press, 1917.

38. F. Hitching, *The Neck of the Giraffe, or Where Darwin Went Wrong*, Pan Books, London, 1982, 173–196; E. Mayr, *The Growth of Biological Thought*, 51–67; R. Sheldrake, *The Presence of the Past*, Collins, London, 1988, 87–96; N. Eldredge, *Reinventing Darwin*, Weidenfeld and Nicolson, London, 1995, 167–179.

39. E. Mayr, *The Growth of Biological Thought*, 51–53, 105–107, 118; R. Shedrake, *The Presence of the Past*, 72–87; H. Driesch, *The History and Theory of Vitalism*, Macmillan, London, 1914; H. Hellman, *Great Feuds in Science*, Wiley, New York, 1998, 77–79.

40. R. Sheldrake, *The Presence of the Past*, 97–101; P. Weiss, *Principles of Development*, Holt, New York, 1939.

41. R. Sheldrake, *The Presence of the Past*, 100–106; R. Thom, *Mathematical Models of Morphogenesis*, Ellis Horwood, Chichester, 1983.

42. S.A. Kauffman, *The Origins of Order*, 549–556; L. Wolpert, The shape of things to come, *New Scientist*, 27 June 1992, 38–42; S.F. Gilbert, *Developmental Biology*, Sinauer Associates, Sunderland, Mass., 4th edn, 1994, 535–544, 562–568, 614–616; S.H. Lee, K.K. Fu, J.N. Hui and J.M. Richman, Noggin and retinoic acid transforms the identity of avian facial prominences, *Nature*, 414, 2001, 909–912.

43. C.D. Rollo, *Phenotypes*, 130–132, 218–225; D.J. Pritchard, The missing chapter in evolution theory, *The Biologist*, 37, 1990, 149–152; S.F. Gilbert, *Developmental Biology*, 638–640.

44. C.H. Waddington, *The Strategy of the Genes*, Allen and Unwin, London, 1957.

45. C.D. Rollo, *Phenotypes*, 223–229; D.J. Pritchard, *The Biologist*, 37, 1990, 149–152.

46. G.C. Williams, *Adaptation and Natural Selection*, Princeton University Press, New Jersey, 1966 (reprinted by Princeton Science Library, New Jersey, 1996), 80.

47. G.C. Williams, *Adaptation and Natural Selection*, 3–7.

48. J. Monod, *Chance and Necessity*, Collins, London, 1972, Fontana Books, London, 1974, 96.

49. E. Mayr, *Populations, Species and Evolution*, 351.

50. T. Dobzhansky, F.J. Ayala, G.L. Stebbins and J.W. Valentine, *Evolution*, Freeman, San Francisco, 1977, 233.

51. E.O. Wilson, T. Eisner, W.R. Briggs, *et al.*, *Life on Earth*, Sinauer Associates, Sunderland, Mass., 2nd edn, 1978, 653.

Chapter 11

1. J. Reader, *Missing Links*, Penguin Books, Harmondsworth, 2nd edn, 1988, 33–53; E. Trinkaus and P. Shipman, *The Neandertals*, Jonathan Cape, London, 1993, 133–158; I. Tattersall, *The Fossil Trail*, Oxford University Press, 1995, 31–40; G. Curtis, C. Swisher and R. Lewin, *Java Man*, Little, Brown & Co., London, 2000, 49–89.

2. E. Trinkaus and P. Shipman, *The Neandertals*, 126–133, 159–190; I. Tattersall, *The Fossil Trail*, 31–32, 41–47; C. Stringer and C. Gamble, *In Search of the Neanderthals*, Thames and Hudson, London, 1993, 12–26, 127.

3. J. Reader, *Missing Links*, 20–26; A. Hrdlička, The Neanderthal phase of man, *Journal of the Royal Anthropological Institute*, 57, 1927, 249–274; J. Shreeve, *The Neandertal Enigma*, William Morrow, New York, 1995, 41–42.

4. C. Stringer and C. Gamble, *In Search of the Neanderthals*, 16–30; E. Trinkaus and P. Shipman, *The Neandertals*, 190–196; J. Shreeve, *The Neandertal Enigma*, 36–39.

5. J. Reader, *Missing Links*, 54–78; I. Tattersall, *The Fossil Trail*, 47–51; J.E. Walsh, *Unravelling Piltdown*, Random House, New York, 1996.

6. J. Reader, *Missing Links*, 91–111; I. Tattersall, *The Fossil Trail*, 59–63; A. Walker and P. Shipman, *The Wisdom of Bones*, Weidenfeld and Nicolson, London, 1996, 45–52.

7. E. Trinkaus and P. Shipman, *The Neandertals*, 237–245, 250–251; I. Tattersall, *The Fossil Trail*, 63–67; A. Walker and P. Shipman, *The Wisdom of Bones*, 53–62; G. Curtis, C. Swisher and R. Lewin, *Java Man*, 75–80, 201–214.

8. E. Trinkaus and P. Shipman, *The Neandertals*, 271–289; I. Tattersall, *The Fossil Trail*, 65–66, 213–214; F. Weidenreich, Facts and speculations concerning the origin of *Homo sapiens*, *American Journal of Physical Anthropology*, 49, 1947, 187–203.

9. J. Reader, *Missing Links*, 79–90, 112–118; I. Tattersall, *The Fossil Trail*, 53–58, 69–70; R.A. Dart, *Australopithecus africanus*: the man-ape of South Africa, *Nature*, 115, 1925, 195–199.

10. J. Reader, *Missing Links*, 118–121; I. Tattersall, *The Fossil Trail*, 70–76; R. Broom, *Finding the Missing Link*, Watts, London, 1950.

11. J. Reader, *Missing Links*, 122–131; I. Tattersall, *The Fossil Trail*, 71–73; R. Lewin, *Bones of Contention*, Simon and Schuster, New York, 1987, Penguin Books, Harmondsworth, 1989, 63–84.

12. W.E. Le Gros Clark, Letter, *Nature*, 157, 1946, 863–865.

13. A. Keith, *A New Theory of Human Evolution*, Watts, London, 1948.

14. J. Reader, *Missing Links*, 137–156; I. Tattersall, *The Fossil Trail*, 105–107; M. Leakey, *Olduvai Gorge*, Collins, London, 1979, 74–76.

15. R. Lewin, *Bones of Contention*, 138–141; D.C. Johanson and M.A. Edey, *Lucy – The Beginnings of Humankind*, Granada, London, 1981, 92–100.

16. D.C. Johanson and M.A. Edey, *Lucy – The Beginnings of Humankind*, 111–256; J. Reader, *Missing Links*, 205–217; I. Tattersall, *The Fossil Trail*, 141–158.

17. T. Dobzhansky, On species of races of living and fossil man, *American Journal of Physical Anthropology*, 2, 1944, 251–265; E. Mayr, Taxonomic categories in fossil hominids, *Cold Spring Harbor Symposia on Quantitative Biology*, 15, 1950, 109–118; T. Dobzhansky, *Mankind Evolving*, Yale University Press, New Haven, 1962.

18. I. Tattersall, *The Fossil Trail*, 89–96, 127–128; A. Walker and P. Shipman, *The Wisdom of Bones*, 114–115; C.L. Brace and A. Montague, *Human Evolution*, Macmillan, New York, 2nd edn, 1977, 276–357; M.H. Wolpoff, Competitive exclusion among Lower Pleistocene hominids – the single species hypothesis, *Man*, 6, 1971, 601–614.

19. C.L. Brace, The fate of the 'classic' Neanderthals – a consideration of hominid catastrophism, *Current Anthropology*, 5, 1964, 3–43.

20. C.L. Brace and A. Montague, *Human Evolution*, 278.

21. W.E. Le Gros Clark, *History of the Primates*, British Museum (Natural History), London, 1970, 10–11.

22. J.S. Weiner, *Man's Natural History*, Weidenfeld and Nicolson, London, 1971, 106.

23. B. Campbell, *Human Evolution*, Aldine Publishing, Chicago, 2nd edn, 1974, 95.

24. F.C. Howell, *Early Man*, Time-Life International, Amsterdam, 1970, 41–45.

25. R. Moore, *Man, Time and Fossils*, Jonathan Cape, London, 1962, 374–375; K. Petersen, *Prehistoric Life on Earth*, Eyre Methuen, London, 1963, 156.

26. J. Reader, *Missing Links*, 132–141; R. Lewin, *Bones of Contention*, 129–137; V. Morell, *Ancestral Passions*, Simon and Schuster, New York, 1995, 48–93.

27. J. Reader, *Missing Links*, 179–187; V. Morell, *Ancestral Passions*, 204–207, 219–236; R. Leakey, *The Origin of Humankind*, Weidenfeld and Nicolson. London, 1994, 26–28.

28. I. Tattersall, *The Fossil Trail*, 108–110, 112–115; M. Leakey, *Olduvai Gorge*, Collins, London, 1979, 54–58, 76–81; D. Johanson and J. Shreeve, *Lucy's Child*, Penguin Books, Harmondsworth, 1990, 69–71, 152–154.

29. L.S.B. Leakey, P.V. Tobias and J.R. Napier, A new species of the genus *Homo* from Olduvai Gorge, *Nature*, 202, 1964, 7–9.

30. D.C. Johanson and M.A. Edey, *Lucy – The Beginnings of Humankind*, 103–105; I. Tattersall, *The Fossil Trail*, 116–117.

31. R.E.F. Leakey, Skull 1470, *National Geographic*, 143, 1973, 819–829; R.E. Leakey, *The Making of Mankind*, Michael Joseph, London, 1981, 16–19; R. Lewin, *Bones of Contention*, 152–252.

32. J. Reader, *Missing Links*, 190–204; V. Morell, *Ancestral Passions*, 398–400; I. Tattersall, *The Fossil Trail*, 132–136.

33. R. Lewin, *Bones of Contention*, 128–167; D.J. Johanson and J. Shreeve, *Lucy's Child*, 70–71; A. Walker and P. Shipman, *The Wisdom of Bones*, 85–101.

34. F.C. Howell, *Early Man*, 36–38; R. Lewin, *Bones of Contention*, 85–98; I. Tattersall, *The Fossil Trail*, 119–121.

35. E.L. Simons, On the mandible of *Ramapithecus*, *Proceedings of the National Academy of Sciences*, 51, 1964, 528–535; D.R. Pilbeam and E.L. Simons, Some problems of hominid classification, *American Scientist*, 53, 1965, 237–259; D. Pilbeam, Notes on *Ramapithecus*, the earliest known hominid, and *Dryopithecus*, *American Journal of Physical Anthropology*, 25, 1966, 1–5; D. Pilbeam, *The Evolution of Man*, Thames and Hudson, London, 1970, 99–113.

36. E.L. Simons, The earliest apes, *Scientific American*, 217(6), 1967, 28–35.

37. D. Pilbeam, *The Evolution of Man*, 113.

38. W.E. Le Gros Clark, *The Antecedents of Man*, Edinburgh University Press, 3rd edn, 1971, 346.

39. D. Pilbeam, *The Evolution of Man*, 110.

40. F.C. Howell, *Early Man*, 38.

41. W.E. Le Gros Clark, *History of the Primates*, 63.

42. H. Wendt, *From Ape to Adam*, Thames and Hudson, London, 1972, 266.

43. B. Campbell, *Human Evolution*, 90.

44. R. Leakey, *The Making of Mankind*, 42–48; R. Lewin, *Bones of Contention*, 91–93; M. Pickford, Geochronology of the Hominoidea – a summary, in J.G. Else and P.C. Lee (eds.), *Primate Evolution*, Cambridge University Press, 1986, 123–128.

45. I. Tattersall, *The Fossil Trail*, 77–80; R. Dart, The osteodontokeratic culture of *Australopithecus prometheus*, *Transvaal Museum Memoir*, 10, 1957, 1–105; R. Ardrey, *African Genesis*, Collins, London, 1961.

46. R. Lewin, *Bones of Contention*, 301–319.

Chapter 12

1. A. de Grazia, *The Velikovsky Affair*, Sphere Books, London, 1978, 13–152; R. Huggett, *Cataclysms and Earth History*, Oxford University Press, 1989, 160–174; R. Huggett, *Catastrophism*, Verso, London and New York, 1997, 119–124.

2. H.A. Brown, *Cataclysms of the Earth*, Twayne, New York, 1967.

3. H. Jeffreys, *The Earth – Its Origin. History and Physical Constitution*, Cambridge University Press, 2nd edn, 1929, 303–307; K.A. Pauly, The cause of the Great Ice Ages, *The Scientific Monthly*, August 1952, 89–98; D.S. Allan and J.B. Delair, *When the Earth Nearly Died*, Gateway Books, Bath, 1995, 193–195.

4. C.H. Hapgood, *Earth's Shifting Crust*, Pantheon Books, New York, 1958; C.H. Hapgood, *The Path of the Pole*, Chilton Books, New York, 1970 (re-published by Souvenir Press, London, 1999). The context of the theory is described in *The Path of the Pole*, 1–44 (and in Hapgood's acknowledgments).

5. C.H. Hapgood, *The Path of the Pole*, 328 (not included in Souvenir Press edition).

6. C.H. Hapgood, *The Path of the Pole*, xiv–xv (pages xii–xiii in Souvenir Press edition).

7. C.H. Hapgood, *The Path of the Pole*, xi (ix in Souvenir Press edition), 40–43.

8. F.X. Kugler, *Sybillinischer Sternkampf und Phaëthon in naturgeschichtlicher Beleuchtung*, Aschendorffsche, Münster, 1927.

9. A. de Grazia, *The Velikovsky Affair*, 120–152; V. Clube and B. Napier, *The Cosmic Serpent*, Faber and Faber, London, 1982, 206–209.

10. C. Beaumont, *Mysterious Comet*, Rider, London, 1932, 9–12, 75–228.

11. I. Velikovsky, *Worlds in Collision*, Gollancz, London, 1950, Sphere Books, London, 1972. (Note that, as pointed out by the British industrial physicist and freelance writer Peter Warlow, in his book *The Reversing Earth*, published in 1982 by Dent, London, if the Earth flipped over like a tippe-top, without undergoing any change in its rotational motion, both North–South and East–West reversals would simultaneously occur.)

12. D. Vorhees, Velikovsky in America, *Aeon*, 3(4), 1993, 32–58 (quotation from page 44).

13. I. Velikovsky, *Ages in Chaos*, Sidgwick and Jackson, London, 1953, Sphere Books, London, 1973.

14. C. Torr, *Memphis and Mycenae*, Cambridge University Press, 1896 (reprinted by the Institute for Interdisciplinary Science (ISIS), 1988, with an introduction by M. Durkin and various other writings by Torr, W.M. Flinders Petrie and J.L. Myres).

15. I. Velikovsky, *Earth in Upheaval*, Gollancz, London, 1956, Sphere Books, London, 1973.

16. A. de Grazia, *The Velikovsky Affair*; I. Velikovsky, *Stargazers and Gravediggers*, William Morrow, New York, 1983; C. Ginenthal, *Carl Sagan and Immanuel Velikovsky*, New Falcon Publications, Tempe, Arizona, 1995.

17. I. Velikovsky, *Earth in Upheaval*, 262–263.

18. Editors of Pensée, *Velikovsky Reconsidered*, Sidgwick and Jackson, London, 1976.

19. E.W. MacKie, A challenge to the integrity of science, *New Scientist*, 11 January 1973, 76–77.

20. E.W. MacKie, A quantitative test for catastrophic theories, *Pensée IVR*, 3, Winter 1973, 6–9.

21. H. Tresman, The SIS, its history and achievements – a personal perspective, *Proceedings of the 1993 Cambridge Conference*, Society for Interdisciplinary Studies, 1994, 2–6; B.J. Peiser, T. Palmer and M.E. Bailey (eds.), *Natural Catastrophes During Bronze Age Civilisations*, Archaeopress, Oxford, 1998, 1–3. See also http://www.knowledge.co.uk/sis/

22. M. Magnusson, *BC – The Archaeology of the Bible Lands*, BBC/Bodley Head, London, 1977, 7–24; D. Rohl, *A Test of Time*, Century, London, 1995, i–ii; P. James and N. Thorpe, *Ancient Mysteries*, Ballantine, New York, 1999, 2–13.

23. M. Magnusson, *BC – The Archaeology of the Bible Lands*, 31–35; Z. Kukal, *Atlantis in the Light of Modern Research*, Elsevier, Amsterdam, 1984, 175–182; D. Rohl, *Legend*, Century, London, 1998, 167–171; G. Leick, *Mesopotamia*, Allen Lane, London, 2001, 82–84, 110–112.

24. P. James and N. Thorpe, *Ancient Mysteries*, 50–58.

25. C.F.A. Schaeffer, *Stratigraphie comparée et chronologie de l'Asie occidentale (IIIe et IIe millénaires)*, Oxford University Press, 1948.

26. I. Velikovsky, *Stargazers and Gravediggers*, 256–257, 318–322; G.J. Gammon, Dr Claude Schaeffer-Forrer, *SIS Review*, 5, 1980/81, 70–71; W. Bray and D. Tramp, *The Penguin Dictionary of Archaeology*, Penguin Books, Harmondsworth, 1970, 242–243.

27. B.J. Peiser, T. Palmer and M.E. Bailey (eds.), *Natural Catastrophes During Bronze Age Civilisations*, 1, 117–118; P. James and N. Thorpe, *Ancient Mysteries*, 10–11; J.J. Gammon, Bronze Age destructions in the Near East, *SIS Review*, 4, 1980, 104–108.

28. R. Gallant, *Bombarded Earth*, John Baker, London, 1964, 214–216.

29. R. Huggett, *Catastrophism*, 124; René Gallant, a pioneer of modern catastrophism, *SIS Review*, 7, 1982/83, 3.

30. E.W. MacKie, Radiocarbon dates and cultural change, *SIS Review*, 3, 1979, 98–100.

31. *Ages in Chaos? Proceedings of the Glasgow Conference, 1978*, Society for Interdisciplinary Studies, 1982.

32. G. Heinsohn, *Ghost Empires of the Past*, Society for Interdisciplinary Studies, 1988; P. James, *Centuries of Darkness*, Jonathan Cape, London, 1991; D. Rohl, *A Test of Time*, Century, London, 1995; D. Rohl, *Legend*, Century, London, 1998; E.J. Sweeney, *The Genesis of Israel and Egypt*, Janus, London, 1997; E.J. Sweeney, *The Pyramid Age*, Domra Publications, Corby, Northants., 1999.

33. V. Clube, Cometary catastrophes and the ideas of Immanuel Velikovsky, *SIS Review*, 5, 1980/81, 106–111; J.B. Moore, T. Palmer and P.J. James, Comets, meteorites and Earth history, *SIS Review*, 7, 1982/3, 2–5.

34. R.E. Juergens, Plasma in interplanetary space – reconciling celestial mechanics and Velikovskian catastrophism, in *Velikovsky Reconsidered*, 148–165.

35. E.W. Crew, Orbits of material ejected from gaseous planets, *Kronos*, 10(2), 1985, 13–31; E.W. Crew, Erratic events in the Solar System, *Chronology and Catastrophism Review*, 10, 1988, 43–48.

36. E.W. Crew, Note added in proof, *Chronology and Catastrophism Review*, 10, 1988, 48; E.W. Crew, Light on Venus, *Chronology and Catastrophism Review*, 1997(1), 58.

37. I. Velikovsky, *Saturn and the Flood* (unpublished, but see http://velikovsky.collision. org); H. Tresman and B. O'Gheoghan, The primordial light?, *SIS Review*, 2, 1977, 35–40; I. Velikovsky, On Saturn and the Flood, *Kronos* 5(1), 1979, 3–11.

38. D. Cardona, Intimations of an alien sky, *Aeon*, 2(5), 1991, 5–34; D. Talbott, From myth to a physical model, *Aeon*, 3(3), 1993, 5–38; E. Cochrane, Venus, Mars . . . and Saturn, *Chronology and Catastrophism Review*, 1998:2, 16–18.

39. A. de Grazia, *The Velikovsky Affair*, 215–230; C. Ginenthal, *Carl Sagan and Immanuel Velikovsky*, 85–100.

40. A. de Grazia, *The Velikovsky Affair*, 44–45; I. Velikovsky, *Stargazers and Gravediggers*, 287–295.

41. C. Sagan, *Broca's Brain*, Hodder and Stoughton, London, 1979, Coronet Books, London, 1980, Chapter 7, 100–159 (quotations from pages 156 and 158). (Note that, although Sagan's overall conclusions were supported by the vast majority of scientists, some considered that his attacks on Velikovsky were far from objective and included some inaccuracies on matters of detail – see for example C. Ginenthal, *Carl Sagan and Immanuel Velikovsky*).

42. M. Jones, Some detailed evidence from Egypt against Velikovsky's revised chronology, in *Ages in Chaos? Proceedings of the Glasgow Conference*, 1978, 27–33 (quotation from page 32).

43. S.J. Gould, *Ever Since Darwin*, Burnett Books, New York, 1978, Penguin Books, Harmondsworth, 1980, Chapter 19, 153–159 (quotations from pages 154 and 159).

44. A. de Grazia, *The Velikovsky Affair*, 7–8; C. Sagan, *Broca's Brain*, 103–104; D. Goldsmith, *Scientists Confront Velikovsky*, Cornell University, Ithaca, New York, 1977.

Chapter 13

1. Plato, *Timaeus*, sections 21e–25d (available, for example, in R.G. Bury (transl.), *Timaeus, Critias, Cleitophon, Menexenus, Epistles*, Loeb Classical Library, Harvard University Press, Cambridge, Mass., 1929); J.V. Luce, *The End of Atlantis*, Thames and Hudson, London, 1969, Paladin, London, 1970, 16–21, 160–163; E. Zangger, *The Flood From Heaven*, Book Club Associates/ Sidgwick and Jackson, London, 1992, 17–21; R. Castleden, *Atlantis Destroyed*, Routledge, London, 1998, 1–4; P. Jordan, *The Atlantis Syndrome*, Sutton Publishing, Stroud, Gloucs., 2001, 14–18.

2. E. Zangger, *The Flood From Heaven*, 42–49; P. Jordan, *The Atlantis Syndrome*, 43–45; P. James, *The Sunken Kingdom*, Jonathan Cape, London, 1995, 10–17; S. Hodge, *Atlantis*, Piatkus, London, 2000, 38–41.

3. P. James, *The Sunken Kingdom*, 3, 22; S. Hodge, *Atlantis*, 43–48; P. Jordan, *The Atlantis Syndrome*, 53–54, 64–65; L.S. de Camp and C.C. de Camp, *Citadels of Mystery*,

Fontana Books, London, 1972, 6–7; R. Flem-Ath and R. Flem-Ath, *When the Sky Fell*, Weidenfeld and Nicolson, London, 1995, 130–133; R. Ellis, *Imagining Atlantis*, Alfred A. Knopf, New York, 1998, 30–33.

4. I. Donnelly, *Atlantis – The Antediluvian World*, Harper, New York, 1882; I. Donnelly, *Ragnarok: The Age of Fire and Gravel*, Sampson Low, Marston, Searle and Rivington, London, 1883, 366–407.

5. L.S. de Camp and C.C. de Camp, *Citadels of Mystery*, 9–13; P. James, *The Sunken Kingdom*, 21–31, 33; R. Castleden, *Atlantis Destroyed*, 182–188; R. Ellis, *Imagining Atlantis*, 38–43, 68; S. Hodge, *Atlantis*, 48–65; P. Jordan, *The Atlantis Syndrome*, 56–79; G. Ashe, *Atlantis*, Thames and Hudson, London, 1992, 76–77; K.L. Feder, *Frauds, Myths and Mysteries*, Mountain View, California, 3rd edn, 1999, 169–178; P. James and N. Thorpe, *Ancient Mysteries*, Ballantine, New York, 1999, 5, 21–26.

6. L.S. de Camp and C.C. de Camp, *Citadels of Mystery*, 228–232; P. James and N. Thorpe, *Ancient Mysteries*, 5–6; R. Ellis, *Imagining Atlantis*, 65–66; S. Hodge, *Atlantis*, 70–77; P. Jordan, *The Atlantis Syndrome*, 59–60; C. Wilson, *The Occult*, Granada, London, 1979, 209, 430–440.

7. L.S. de Camp and C.C. de Camp, *Citadels of Mystery*, 13–15; R. Ellis, *Imagining Atlantis*, 4, 221; P. Jordan, *The Atlantis Syndrome*, 270–271; J. Wellard, *The Search for Lost Worlds*, Pan, London, 1975, 13–61.

8. L.S. de Camp and C.C. de Camp, *Citadels of Mystery*, 216–218; J. Wellard, *The Search for Lost Worlds*, 50–52; P. Jordan, *The Atlantis Syndrome*, 183; N. Zhirov, *Atlantis – Basic Problems*, Progress Publishers, Moscow, 1970, 371–372; C. Wilson, *From Atlantis to the Sphinx*, Virgin, London, 1997, 142–143.

9. C. Wilson, *The Occult*, 202–209, 218; P. James, *The Sunken Kingdom*, 132–133; R. Castleden, *Atlantis Destroyed*, 1, 189; R. Ellis, *Imagining Atlantis*, 70; S. Hodge, *Atlantis*, 77–78; P. Jordan, *The Atlantis Syndrome*, 183; L. Pauwels and J. Bergier, *The Morning of the Magicians*, Granada, London, 1971, 146–170; D. Wilson, *Lost Worlds*, Parragon, Bristol, 1998, 409–411.

10. L.S. de Camp and C.C. de Camp, *Citadels of Mystery*, 231–232; J. Wellard, *The Search for Lost Worlds*, 61–70; D. Wilson, *Lost Worlds*, 400–409; G. Ashe, *Atlantis*, 84–85; S. Hodge, *Atlantis*, 72–77; P. Jordan, *The Atlantis Syndrome*, 60–61.

11. C. Wilson, *The Occult*, 202–209; C. Wilson and D. Wilson, *The Encyclopedia of Unsolved Mysteries*, Zachary Kwinter, London, 1987, 22–29; A. Collins, *Gateway to Atlantis*, Headline, London, 2000, 274–275.

12. D. Wilson, *Lost Worlds*, 383–399; C. Wilson, *From Atlantis to the Sphinx*, 121–135; R. Ellis, *Imagining Atlantis*, 68–69; S. Hodge, *Atlantis*, 67–70; P. Jordan, *The Atlantis Syndrome*, 60–62; J. Churchward, *The Lost Continent of Mu*, Spearman, London, 1959, Futura, London, 1974.

13. C. Wilson, *The Occult*, 213–216; P. James and N. Thorpe, *Ancient Mysteries*, 598–604; C. Wilson, *From Atlantis to the Sphinx*, 87–90; R. Ellis, *Imagining Atlantis*, 71–75; S. Hodge, *Atlantis*, 109–111; P. Jordan, *The Atlantis Syndrome*, 96–100.

14. L. Spence, *The Problem of Atlantis*, Rider, London, 1924; L. Spence, *History of Atlantis*, Rider, London, 1926 (reprinted by Studio Editions, London, 1995).

15. C. Wilson and D. Wilson, *The Encyclopedia of Unsolved Mysteries*, 25–28; P. James, *The Sunken Kingdom*, 42; K.L. Feder, *Frauds, Myths and Mysteries*, 178; R. Ellis, *Imagining Atlantis*, 43–49; A. Collins, *Gateway to Atlantis*, 108–110; P. Jordan, *The Atlantis Syndrome*, 82–103.

16. L.S. de Camp and C.C. de Camp, *Citadels of Mystery*, 13–14; J. Wellard, *The Search for Lost Worlds*, 35–38; E. Zangger, *The Flood From Heaven*, 218–219; R. Ellis, *Imagining Atlantis*, 49–51; S. Hodge, *Atlantis*, 66–67.

17. N. Zhirov, *Atlantis – Basic Problems*, 357–359; R. Ellis, *Imagining Atlantis*, 34–36, 60–62; A. Collins, *Gateway to Atlantis*, 36–49; R. Ellis, *Imagining Atlantis*, 52, 60–62; P. Jordan, *The Atlantis Syndrome*, 80–82; O. Muck, *The Secret of Atlantis*, Fontana/Collins, London, 1978 45–51; C. Berlitz, *Atlantis*, Fontana/Collins, London, 1985, 51–75.

18. O. Muck, *Atlantis, die Welt von der Sinflut*, Olter, Freiburg, 2nd edn, 1956.

19. O. Muck, *The Secret of Atlantis*, 96–114, 159–200.

20. L. Zajdler, *Atlantyda*, Wiedza Powszechna, Warsaw, 1972 (in Polish, referred to in Z. Kukal, *Atlantis in the Light of Modern Research*, Elsevier, Amsterdam, 1984, 171, 175, 182).

21. O. Muck, *The Secret of Atlantis*, 178–222; R. Ellis, *Imagining Atlantis*, 58–60; A. Collins, *Gateway to Atlantis*, 275–284.

22. O. Muck, *The Secret of Atlantis*, 225–297.

23. P. James, *The Sunken Kingdom*, 28–37; K.L. Feder, *Frauds, Myths and Mysteries*, 180–182; A. Collins, *Gateway to Atlantis*, 47–50, 284–289.

24. O. Muck, *The Secret of Atlantis*, 139–145.

25. N. Zhirov, *Atlantis – Basic Problems*, 357–389.

26. N. Zhirov, *Atlantis – Basic Problems*, 43–53.

27. L.S. de Camp and C.C. de Camp, *Citadels of Mystery*, 20–25; J. Wellard, *The Search for Lost Worlds*, 53–54; C. Berlitz, *Atlantis*, 66–69; R. Ellis, *Imagining Atlantis*, 34–36, 61–62.

28. R. Carpenter, *Beyond the Pillars of Hercules*, Tandem, London, 1973, 50–67; A. Collins, *Gateway to Atlantis*, 139, 155–157, 349–351; S. Hodge, *Atlantis*, 85–88; P. Jordan, *The Atlantis Syndrome*, 18; *The Cassell Atlas of World History, Volume 1 – The Ancient and Classical Worlds*, Andromeda, Oxford, 2000, sections 1.23, 1.25 and 1.42.

29. C. Renfrew, *Before Civilization*, Jonathan Cape, London, 1973, 37–91, 120–129; E. MacKie, *The Megalith Builders*, Phaedon/Book Club Associates, London, 1977, 33–50; *The Times History of the World*, Times Books, London, 1999, 40–41.

30. D. Wilson, *Lost Worlds*, 428–440; E. Zangger, *The Flood from Heaven*, 59–106; P. James and N. Thorpe, *Ancient Mysteries*, 26–32.

31. I. Velikovsky, *Worlds in Collision*, Sphere Books, London, 1972, 148–150.

32. C. Beaumont, *The Mysterious Comet*, Rider, London, 1932, 221–226; P. Jordan, *The Atlantis Syndrome*, 271.

33. N. Zhirov, *Atlantis – Basic Problems*, 358; C. Beaumont, *Riddle of Prehistoric Britain*, Rider, London, 1946.

34. J.V. Luce, *The End of Atlantis*, 28–134; P. James, *The Sunken Kingdom*, 57–84; R. Castleden, *Atlantis Destroyed*, 1–147; P. Jordan, *The Atlantis Syndrome*, 33–39;

M. Magnusson, The ashes of Atlantis, in R. Sutcliffe (ed.), *Chronicle*, BBC Publications, London, 1978, 53–66.

35. J.V. Luce, *The End of Atlantis*, 34–43; P. James, *The Sunken Kingdom*, 58–63, 67; R. Castleden, *Atlantis Destroyed*, 15–26; J.W. Mavor, *Voyage to Atlantis*, Souvenir Press, 1969, Fontana/Collins, London, 1973, 51–54.

36. J.V. Luce, *The End of Atlantis*, 43–44, 76–77; P. James, *The Sunken Kingdom*, 63–66; R. Castleden, *Atlantis Destroyed*, 26–28.

37. S. Marinatos, The volcanic destruction of Minoan Crete, *Antiquity*, 13, 1939, 425–439.

38. J.V. Luce, *The End of Atlantis*, 86–96; R. Castleden, *Atlantis Destroyed*, 28–57; R. Ellis, *Imagining Atlantis*, 143–187; C. Doumas, *Santorini*, Ekdotike Athenon S.A., Athens, 1999, 14–59.

39. J.W. Mavor, *Voyage to Atlantis*, 23–43; P. James, *The Sunken Kingdom*, 66–67; R. Castleden, *Atlantis Destroyed*, 34–35.

40. A.G. Galanopoulos and E. Bacon, *Atlantis – The Truth Behind the Legend*, Nelson, London, 1969.

41. J.V. Luce, *The End of Atlantis*, 135–159; J.W. Mavor, *Voyage to Atlantis*, 119–265.

42. P. James, *The Sunken Kingdom*, 70–77; P. James and N. Thorpe, *Ancient Mysteries*, 28–31; A. Collins, *Gateway to Atlantis*, 37–40.

43. J.V. Luce, *The End of Atlantis*, 45–85; R. Castleden, *Atlantis Destroyed*, 8–13, 114–126; I. Wilson, *The Exodus Enigma*, Weidenfeld and Nicolson/ Book Club Associates, London, 1985, 86–100.

Chapter 14

1. A. Hallam, *Great Geological Controversies*, Oxford University Press, 2nd edn, 1989, 84; P.J. Bowler, *Evolution – The History of an Idea*, University of California Press, Berkeley, Los Angeles and London, 1989, 124.

2. S.M. Stanley, *Extinction*, Scientific American Books, New York, 1987, 1–19, 96–99, 133–161; G.R. McGhee Jr., Catastrophes in the history of life, in K.C. Allen and D.E.G. Briggs (eds.), *Evolution and the Fossil Record*, Belhaven Press, London, 1989, 26–50; D.M. Raup, *Extinction – Bad Genes or Bad Luck*, Norton, New York, 1991, 64–87.

3. A. Hallam, *Great Geological Controversies*, 184–186; A. Hoffman, Changing palaeontological views on mass extinction phenomena, in S.K. Donovan (ed.), *Mass Extinctions – Processes and Evidence*, Belhaven Press, London, 1989, 1–18; R. Huggett, *Catastrophism*, Verso, London and New York, 1997, 163–174.

4. H.T. Marshall, Ultra-violet and extinction, *American Naturalist*, 62, 1928, 165–187 (quotations from pages 166 and 183); E. Hennig, *Wege und Wesen der Paläontologie*, Berlin, 1932.

5. A. Hoffman, in *Mass Extinctions – Processes and Evidence*, 1–18 (particularly pages 3–4).

6. J. Huxley, *Evolution – The Modern Synthesis*, Allen and Unwin, London, 1942, 1963, 1974; E. Mayr, *The Growth of Biological Thought*, Harvard University Press, Cambridge, Mass., 1982, 620; P. Bowler, *Evolution – The History of an Idea*, University of California Press, Berkeley, Los Angeles and London, 1989, 351.

7. P. Bowler, *Evolution – The History of an Idea*, 351–352; E. von Däniken, *Chariots of the Gods*, Souvenir Press, London, 1969; E. von Däniken, *The Return of the Gods*, Element Books, Shaftesbury, Dorset, 1997.

8. A. Hoffman, in *Mass Extinctions – Processes and Evidence*, 1–18; D.M. Raup, *The Nemesis Affair*, Norton, New York, 1986, 38–39; C.C. Albritton, *Catastrophic Episodes in Earth History*, Chapman and Hall, London, 1989, 120–121, 175–176.

9. O.H. Schindewolf, Neokatastrophismus?, *Deutsche Geologisch Gesellschaft Zeitschrift*, 114, 1963, 430–445; O.H. Schindewolf, Neocatastrophism?, *Catastrophist Geology*, 2, 1977, 9–21.

10. H. Liniger, Über das Dinosauriersterben in der Provence, *Leben und Umwelt*, 18(2), 27–33, 1961 (quotation from page 32, translation by Claudia Stockenhüber).

11. D. Russell and W. Tucker, Supernovae and the extinction of the dinosaurs, *Nature*, 229, 1971, 533–534.

12. A.O. Kelly and F. Dachille, *Target Earth – The Role of Large Meteors in Earth Science*, Carlsbad, California, 1953; D. Steel, *Rogue Asteroids and Doomsday Comets*, Wiley, New York, 1995, 243–244; G.L. Verschuur, *Impact!*, Oxford University Press, 1996, 88–92; C. Hapgood, *The Path of the Pole*, Chilton Books, New York, 1970, Souvenir Press, London, 1999, 26; D. Steel, *Target Earth*, Quarto Publishing, London, 2000, 93.

13. M.W. de Laubenfels, Dinosaur extinction – one more hypothesis, *Journal of Paleontology*, 30, 1956, 207–218 (quotation from page 207).

14. P. Lancaster Brown, *Comets, Meteorites and Men*, Hale, London, 1973, 192–197; S. Welfare and J. Fairley, *Arthur C. Clarke's Mysterious World*, Fontana/Collins, London, 1982, 223–243; P. James and N. Thorpe, *Ancient Mysteries*, Ballantine, New York, 1999, 147–148.

15. G.L. Verschuur, *Impact!*, 53; V. Clube and B. Napier, *The Cosmic Serpent*, Faber and Faber, London, 1982, 72; J. Gribbin and M. Gribbin, *Fire on Earth*, Simon and Schuster, London, 1996, 97–98; C. Sagan and A. Druyan, *Comet*, Headline, London, 1997, 144–151.

16. C.C. Albritton, *Catastrophic Episodes in Earth History*, 62–65, 70–71; D. Steel, *Rogue Asteroids and Doomsday Comets*, 75–86; J.S. Lewis, *Rain of Iron and Ice*, Addison-Wesley, Reading, Mass., 1996, 30–34.

17. P. Lancaster Brown, *Comets, Meteorites and Men*, 188–190; J.K. Davies, *Cosmic Impact*, St. Martin's Press, New York, 1986, 18–22; F. Heide and F. Wlotzka, *Meteorites*, Springer-Verlag, Berlin, Heidelberg and New York, 1995, 34–38.

18. P. Lancaster Brown, *Comets, Meteorites and Men*, 190–192; C.C. Albritton, *Catastrophic Episodes in Earth History*, 69–92; D. Steel, *Rogue Asteroids and Doomsday Comets*, 86–87.

19. C.C. Albritton, *Catastrophic Episodes in Earth History*, 66–69, 81, 90–92; F. Heide and F. Wlotzka, *Meteorites*, 38–54; J.S. Lewis, *Rain of Iron and Ice*, 34–36.

20. A. Hallam, *Great Geological Controversies*, 184–185; R. Huggett, *Catastrophism*, 165–166; N. Eldredge, *Life Pulse*, Penguin Books, Harmondsworth, 1989, 205–208.

21. N.D. Newell, Paleontological gaps and geochronology, *Journal of Paleontology*, 36, 1962, 592–610 (quotation from page 592).

22. N.D. Newell, Crises in the history of life, *Scientific American*, 208(2), 1963, 76–92.

23. N.D. Newell, Revolutions in the history of life, *Geological Society of America Special Paper*, 89, 1967, 63–91.

24. A. Hallam, *Great Geological Controversies*, 135–162; S.M. Stanley, *Earth and Life Through Time*, Freeman, New York, 1986, 169–181; C. Officer and J. Page, *Tales of the Earth*, Oxford University Press, 1993, 9–11; H. Hellman, *Great Feuds in Science*, Wiley, New York, 1998, 141–158.

25. S.M. Stanley, *Earth and Life Through Time*, 181–185; A. Hallam, *Great Geological Controversies*, 162–167; *The Hutchinson Dictionary of Scientific Biography*, Helicon, Oxford, 1994, 328.

26. S.M. Stanley, *Earth and Life Through Time*, 185–189; A. Hallam, *Great Geological Controversies*, 168–178; W. Glen (ed.), *The Mass-Extinction Debates*, Stanford University Press, Stanford, California, 1994, 126–128.

27. T.S. Kuhn, *The Structure of Scientific Revolutions*, University of Chicago Press, Chicago, 1962.

28. W. Glen (ed.), *The Mass-Extinction Debates*, 72–79.

29. K. Popper, *The Logic of Scientific Discovery*, Hutchinson, London, 1959.

30. A. Hallam, *Great Geological Controversies*, 222–229; P.J. Bowler, *Evolution – The History of an Idea*, 14–20; E. Mayr, *The Growth of Biological Thought*, 3–6, 25–28, 856–858.

31. *The Oxford Dictionary of Quotations*, Oxford University Press, 5th edn, 1999, 577.

32. W. Glen (ed.), *The Mass-Extinction Debates*, 39–40.

33. W. Glen (ed.), *The Mass-Extinction Debates*, 75.

34. R. Huggett, *Catastrophism*, 167–169; R.T. Bakker, Tetrapod mass extinctions, in A. Hallam (ed.), *Patterns of Evolution as Illustrated by the Fossil Record*, Elsevier, Amsterdam, 1977, 439–468; S.M. Stanley, *Extinction*, Scientific American Books, New York, 1987, 33–36.

35. T.J.M. Schopf, Permo-Triassic extinctions – relation to sea-floor spreading, *Journal of Geology*, 82, 1974, 129–143; S.J. Gould, *Ever Since Darwin*, Burnett Books, New York, 1978, Penguin Books, Harmondsworth, 1980, Chapter 16, 134–138.

36. S.M. Stanley, *Extinction*, 96–101; N. Eldredge, *The Miner's Canary*, Virgin, London, 1992, 88–98; J. Hecht, *Vanishing Life*, Charles Scribner's Sons, New York, 1993, 83–90.

37. E.J. Öpik, On the catastrophic effects of collisions with celestial bodies, *Irish Astronomical Journal*, 5, 1958, 34–36.

38. R. Gallant, *Bombarded Earth*, John Baker, London, 1964, 39–125 (quotations from pages 119–120).

39. D.J. McLaren, Time, life and boundaries, *Journal of Paleontology*, 44, 1970, 801–815 (quotations from pages 801 and 812).

40. H.H. Nininger, Cataclysm and evolution, *Popular Astronomy*, 50, 1942, 270–272.

41. H.C. Urey, Cometary collisions and geological periods, *Nature*, 242, 1973, 32–33.

42. C.C. Albritton, *Catastrophic Episodes in Earth History*, 92–93; G.L. Verschuur, *Impact!*, 87–94; J. Gribbin and M. Gribbin, *Fire on Earth*, 10–11; D. Steel, *Target Earth*, 90–95.

43. F. Hoyle and N.C. Wickramasinghe, Comets, ice-ages and ecological catastrophes, *Astrophysics and Space Science*, 53, 1978, 523–526; F. Hoyle and N.C. Wickramasinghe, *Living Comets*, University College Cardiff Press, 1985, 111; C. Wickramasinghe, *Cosmic Dragons*, Souvenir Press, London, 2001, 100.

44. D.M. Raup, *The Nemesis Affair*, 39–45; D. Steel, *Rogue Asteroids and Doomsday Comets*, 71–72, 106–107; J.S. Lewis, *Rain of Iron and Ice*, 100–108; R. Huggett, *Catastrophism*, 124–127, 173–174; D. Steel, *Target Earth*, 90–97.

45. D.J. MacLaren, Time, life and boundaries, *Journal of Paleontology*, 44, 1970, 801–815 (quotations from page 812).

46. E. Mayr, *Populations, Species and Evolution*, Harvard University Press, Cambridge, Mass., 1970, 372–373.

47. J. Huxley, *Evolution – The Modern Synthesis*, 3rd edn, 560.

48. J. Huxley, *Evolution – The Modern Synthesis*, 3rd edn, 446.

49. T. Dobzhansky, F.J. Ayala, G.L. Stebbins and J.W. Valentine, *Evolution*, Freeman, San Francisco, 1977, 421.

50. T. Dobzhansky, F.J. Ayala, G.L. Stebbins and J.W. Valentine, *Evolution*, 426.

51. T. Dobzhansky, F.J. Ayala, G.L. Stebbins and J.W. Valentine, *Evolution*, 436.

52. G.G. Simpson, *How and Why*, Pergamon Press, Oxford, 1980, 35.

53. G.L. Stebbins, *Darwin to DNA, Molecules to Humanity*, Freeman, San Francisco, 1982, 27.

Chapter 15

1. *The Oxford English Reference Dictionary*, Oxford University Press, 2nd edn, 1996.

2. D. Ager, *The New Catastrophism*, Cambridge University Press, 1993, xix.

3. A. Woodcock and M. Davis, *Catastrophe Theory*, Penguin Books, Harmondsworth, 1980; R. Thom, *Mathematical Models of Morphogenesis*, Ellis Horwood, Chichester, 1983.

4. T. Hunkapiller, H. Huang, L. Hood and J.J. Campbell, The impact of modern genetics on evolutionary theory, in R. Milkman (ed.), *Perspectives on Evolution*, Sinauer Associates, Sunderland, Mass., 1982, 164–189; D.J. Futuyma, *Evolutionary Biology*, Sinauer Associates, Sunderland, Mass., 1986, 47–50, 68–71, 456–472; D.L. Nelson and M.M. Cox, *Lehninger's Principles of Biochemistry*, Worth, New York, 3rd edn, 2000, 908–915, 970–973, 991–996; J.M. Berg, J.L. Tymoczko and L. Stryer, *Biochemistry*, Freeman, New York, 5th edn, 2002, 117–142, 745–844, 867–896.

5. N. Eldredge, *Unfinished Synthesis*, Oxford University Press, 1985, 3–128; C.D. Rollo, *Phenotypes – Their Epigenetics, Evology and Evolution*, Chapman and Hall, London, 1994, 1–81; B. Goodwin, *How the Leopard Changed its Spots*, Weidenfeld and Nicolson, London, 1994, 1–39.

6. R. Dawkins, *The Selfish Gene*, Oxford University Press, 1976; R. Dawkins, *The Extended Phenotype*, Freeman, Oxford and San Francisco, 1982, 1–155.

7. S.M. Stanley, *Macroevolution*, Freeman, San Francisco, 1979, 11–212; S.J. Gould, *The Panda's Thumb*, Norton, New York, 1980, Penguin Books, Harmondsworth, 1983, Chapter 8, 72–78; J. Maynard Smith, *Evolutionary Genetics*, Oxford University Press, 1989, 273–303.

8. S.M. Stanley, *The New Evolutionary Timetable*, Basic Books, New York, 1981, 55, 102, 106.

9. E. Mayr, *Systematics and the Origin of Species*, Columbia University Press, New York, 1942, 225–226, 234–237.

10. E. Mayr, Change of genetic environment and evolution, in J. Huxley (ed.), *Evolution as a Process*, Allen and Unwin, London, 1954, 157–180.

11. J. Maynard Smith, *The Theory of Evolution*, Penguin Books, Harmondsworth, 1958, 240–244, 251–254; G. de Beer, *Evolution*, British Museum (Natural History), London, 3rd edn, 1964, 24, 67–68; N. Eldredge, *Time Frames*, Simon and Schuster, New York, 1985, 55.

12. N. Eldredge, *Time Frames*, 144.

13. N. Eldredge and S.J. Gould, Punctuated equilibria, an alternative to phyletic gradualism, in T.J.M. Schopf (ed.), *Models in Paleobiology*, Freeman, San Francisco, 1972, 82–115.

14. N. Eldredge, *Time Frames*, 57–122; N. Eldredge, The allopatric model and phylogeny in Paleozoic invertebrates, *Evolution*, 25, 1971, 156–167.

15. S.J. Gould, An evolutionary microcosm – Pleistocene and Recent history of the land snail P. (Poecilozonites) in Bermuda, *Bulletin of the Museum of Comparative Zoology*, 138, 1969, 407–531; S.J. Gould, Coincidence of climate and faunal fluctuations in Pleistocene Bermuda, *Science*, 168, 1970, 572–573.

16. S.J. Gould, in *Models in Paleobiology*, 82–115.

17. S.J. Gould, Is a new and general theory of evolution emerging?, *Paleobiology*, 6, 1980, 119–130.

18. D. King-Hele, *Erasmus Darwin*, Giles de la Mare, London, 1999, 300–301; D. Ospovat, *The Development of Darwin's Theory*, Cambridge University Press, 1981, 33–86; C. Darwin, *On the Origin of Species*, 6th edn, John Murray, London, 1872, Studio Editions, London, 1994, 279; S. Wright, Comments on the preliminary working papers of Eden and Waddington, in P.S. Moorehead and M.M. Kaplan (ed.), *Mathematical Challenges to the Neo-Darwinian Interpretation of Evolution*, Wistar Institute Press, Philadelphia, 1967, 117–120; S. Stanley, *Macroevolution*, 1–30.

19. E. Mayr, *Populations, Species and Evolution*, Harvard University Press, Cambridge, Mass., 1970, 373–374.

20. M.T. Ghiselin, A radical solution to the species problem, *Systematic Zoology*, 23, 1974, 536–544; D.L. Hull, Are species really individuals?, *Systematic Zoology*, 25, 1976, 174–191.

21. S.M. Stanley, *The New Evolutionary Timetable*, 77–78; N. Eldredge, *Time Frames*, 111–121; J.H. Schwartz, *Sudden Origins*, Wiley, New York, 1999, 310–330.

22. E. Mayr, *The Growth of Biological Thought*, Harvard University Press, Cambridge, Mass., 1982, 616–618.

23. N. Eldredge, *Time Frames*, 121, 141–142.

24. N. Eldredge, *Time Frames*, 120–122.

25. C.K. Yoon, Stephen Jay Gould , *The New York Times*, 21 May 2002, obituaries; S. Rose; S. Jones, Stephen Jay Gould, *The Guardian*, 22 May 2002, 20; R.A. Fortey, Stephen Jay Gould (1941–2002), *Science*, 296, 2002, 1984; D.E.G. Briggs, Stephen Jay Gould

(1941–2002), *Nature*, 417, 2002, 706; S.J. Gould, *The Panda's Thumb*, Chapter 18, 155–161.

26. S.J. Gould, *Ontogeny and Phylogeny*, Harvard University Press, Cambridge, Mass., 1977 (quotation from page 2).

27. S.J. Gould, *Paleobiology*, 6, 119–130.

28. S.M. Stanley, *Macroevolution*, 90–96, 118–122, 186–188, 200–202; S.M. Stanley, Trends, rates and patterns of evolution in the Bivalvia, in A. Hallam (ed.), *Patterns of Evolution As Illustrated by the Fossil Record*, Elsevier, Amsterdam, 1977, 209–250.

29. S.M. Stanley, *The New Evolutionary Timetable*, 111.

30. S.M. Stanley, *Macroevolution*, 181.

31. S.M. Stanley, *The New Evolutionary Timetable*, 182–186; S.J. Gould, Speciation and sorting as the source of evolutionary trends, in K.J. MacNamara (ed.), *Evolutionary Trends*, Belhaven Press, London, 1990, 3–27; N. Eldredge, *The Miner's Canary*, Virgin, London, 1992, 60–61.

32. S.M. Stanley, *Macroevolution*, 186–195; S.M. Stanley, *The New Evolutionary Timetable*, 180–202; S.M. Stanley, A theory of evolution above the species level, *Proceedings of the National Academy of Sciences*, 72, 1975, 646–650.

33. S.M. Stanley, *Macroevolution*, 55–56, 157–158; S.M. Stanley, *The New Evolutionary Timetable*, 127–130; S.J. O'Brien, The ancestry of the giant panda, *Scientific American*, 257(5), 1987, 82–87.

34. S.M. Stanley, *Macroevolution*, 42–47; S.M. Stanley, *The New Evolutionary Timetable*, 117–120; C. Tudge, All fish bright and beautiful, *New Scientist*, 8 February 1992, 50; R. Osborne and M. Benton, *The Viking Atlas of Evolution*, Viking, London, 1996, 60–61; M.L.J. Stiassny and A. Meyer, Cichlids of the Rift Lakes, *Scientific American*, 280(2), 1999, 44–49.

35. S.M. Stanley, *Macroevolution*, 164–169; S.M. Stanley, *The New Evolutionary Timetable*, 126; S.J. Gould, *Paleobiology*, 6, 119–130.

36. G.L. Bush, Sympatric host race formation and speciation in frugivorous flies of the genus *Rhagoletis*, *Evolution*, 23, 1969, 237–251; J.L. Feder, C.A. Chilcote and G.L. Bush, Genetic differentiation between sympatric host races of the apple maggot fly *Rhagoletis pomonella*, *Nature*, 336, 1988, 61–64; J.L. Feder and G.L. Bush, A field test of differential host-plant usage between the sibling species of *Rhagoletis pomonella* fruit flies and its consequences for sympatric models of speciation, *Evolution*, 43, 1989, 1813–1819; P.R. Grant and B.R. Grant, Adaptive radiation of Darwin's finches, *American Scientist*, 90, 2002, 130–139.

37. E. Mayr, *One Long Argument*, Penguin Books, Harmondsworth, 1991, 153–154.

38. C. Darwin, *On the Origin of Species, by Means of Natural Selection*, John Murray, London, 1859, Penguin Books, Harmondsworth, 1968, 171–172, 211, 224.

39. G. de Beer, *Evolution*, British Museum (Natural History), London, 3rd edn, 1964, 31; J. Huxley, *Evolution – The Modern Synthesis*, Allen and Unwin, London, 1942, 389.

40. S.J. Gould, Punctuated equilibrium – a different way of seeing, in J. Cherfas (ed.), *Darwin up to Date*, New Science Publications, London, 1982, 26–30; also *New Scientist*, 15 April 1982, 137–141.

Chapter 16

1. R. Lewin, *Bones of Contention*, Simon and Schuster, New York, 1987, Penguin Books, Harmondsworth, 1989, 85–104; R. Leakey and R. Lewin, *Origins Reconsidered*, Little, Brown & Co., London, 1992, 77–80; I. Tattersall, *The Fossil Trail*, Oxford University Press, 1995, 119–122.

2. A.C. Walker and P. Andrews, Reconstruction of the dental arcade of *Ramapithecus wickeri*, *Nature*, 244, 1973, 313–314; P. Andrews and I. Tekkaya, A revision of the Turkish Miocene hominoid, *Sivapithecus meteai*, *Palaeontology*, 23, 1980, 83–95; P. Andrews and J.E. Cronin, The relationships of *Sivapithecus* and *Ramapithecus* and the evolution of the orangutan, *Nature*, 297, 1982, 541–546.

3. D.R. Pilbeam, Recent finds and interpretation of Miocene hominoids, *Annual Review of Anthropology*, 8, 1979, 333–352; D.R. Pilbeam, Miocene hominoids and hominid origins, *American Journal of Physical Anthropology*, 52, 1980, 268; D.R. Pilbeam, New hominoid skull material from the Miocene of Pakistan, *Nature*, 295, 1982, 232–234.

4. S.L. Washburn, The evolution of man, *Scientific American*, 239(3), 1978, 146–154.

5. R. Lewin, *Bones of Contention*, 105–115; R. Leakey and R. Lewin, *Origins Reconsidered*, 75–78; I. Tattersall, *The Fossil Trail*, 122–126.

6. R. Lewin, *Bones of Contention*, 105–127; J. Gribbin and J. Cherfas, *The Monkey Puzzle*, The Bodley Head, London, 1982, Triad/Granada, London, 1993, 15–137; M.H. Brown, *The Search for Eve*, Harper and Row, New York, 1990, 41–57; J. Gribbin and J. Cherfas, *The First Chimpanzee*, Penguin Books, Harmondsworth, 2001, 88–123.

7. V.M. Sarich and A.C. Wilson, Quantitative immunochemistry and the evolution of the primate albumins, *Science*, 154, 1966, 1563–1566; V.M. Sarich and A.C. Wilson, Rates of albumin evolution in primates, *Proceedings of the National Academy of Sciences*, 58, 1967, 142–148; V.M. Sarich and A.C. Wilson, Immunological time scale for hominid evolution, *Science*, 158, 1967, 1200–1203.

8. R.E. Leakey and R. Lewin, *Origins*, Macdonald and Jane's, London, 1977, 27; B.A. Wood, *Human Evolution*, Chapman and Hall, London, 1978, 27; R.E. Leakey, *The Making of Mankind*, Michael Joseph, London, 1981, 48; S. Tomkins, *The Origins of Mankind*, Cambridge University Press, 1984, 41–44.

9. J.E. Pfeiffer, *The Emergence of Humankind*, Harper and Row, New York, 4th edn, 1985, 37.

10. M.H. Brown, *The Search for Eve*, 125–135; C.G. Sibley and J.E. Ahlquist, The phylogeny of the hominoid primates, as indicated by DNA–DNA hybridization, *Journal of Molecular Evolution*, 20, 1984, 2–15; J. Diamond, *The Rise and Fall of the Third Chimpanzee*, Radius, London, 1991, Vintage, London, 1992, 14–20; T. Palmer, *Controversy – Catastrophism and Evolution: The Ongoing Debate*, Kluwer/Plenum, New York, 1999, 217–223.

11. R. Britten, Rates of DNA sequence evolution differ between taxonomic groups, *Science*, 231, 1986, 2393–2398; W.-H. Li and M. Tanimura, The molecular clock runs more slowly in man than in apes and monkeys, *Nature*, 326, 1987, 93–96; R. Lewin, Molecular clocks run out of time, *New Scientist*, 10 February 1990, 38–41; J. Gribbin and J. Cherfas, *The First Chimpanzee*, 117–143.

12. R. Lewin, *Bones of Contention*, 110–127; I. Tattersall, *The Fossil Trail*, 121–126; D. Pilbeam, The descent of hominoids and hominids, *Scientific American*, 250(3), 1984, 60–69.

13. M.A. Edey and D.C. Johanson, *Blueprints*, Oxford University Press, 1990, 325–368; R. Lewin, *Human Evolution*, Blackwell, Oxford, 3rd edn, 1993, 84; R. Leakey, *The Origin of Humankind*, Weidenfeld and Nicolson, London, 1994, 6–9; C. Tudge, *The Day Before Yesterday*, Jonathan Cape, London, 1995, Pimlico, London, 1996, 184–185.

14. N. Eldredge and I. Tattersall, Evolutionary models, phylogenetic reconstruction, and another look at hominid phylogeny, in F.S. Szalay (ed.), *Approaches to Primate Paleobiology*, Karger, Basel, 1975, 218–242; N. Eldredge and I. Tattersall, *The Myths of Human Evolution*, Columbia University Press, New York, 1982, 119–159.

15. J.E. Cronin, N.T. Boaz, C.B. Stringer and Y. Rak, Tempo and mode in hominid evolution, *Nature*, 292, 1981, 113–122.

16. G.P. Rightmire, The tempo of change in the evolution of Mid-Pleistocene *Homo*, in E. Delson (ed.), *Ancestors – The Hard Evidence*, Liss, New York, 1985, 255–264; M.H. Wolpoff, Evolution in *Homo erectus* – the question of stasis, *Paleobiology*, 10, 1985, 389–406; G.P. Rightmire, *The Evolution of Homo erectus*, Cambridge University Press, 1990, 191–201.

17. S.J. Gould, *Ontogeny and Phylogeny*, Harvard University Press, Cambridge, Mass., 1977, 352–404; J.H. Schwartz, *Sudden Origins*, Wiley, New York, 1999, 126–161.

18. R.E.F. Leakey and A.C. Walker, *Australopithecus*, *Homo erectus* and the single-species hypothesis, *Nature*, 261, 1976, 572–574; C.L. Brace and A. Montagu, *Human Evolution*, Macmillan, New York, 1977, 292.

19. I. Tattersall, *The Fossil Trail*, 229–246; S.M. Stanley, *The New Evolutionary Timetable*, Basic Books, New York, 1981, 138–164; S.J. Gould, *Dinosaur in a Haystack*, Jonathan Cape, London, 1996, Chapter 11, 133–143.

20. B. Campbell, *Human Evolution*, Aldine, Chicago, 2nd edn, 1974, 89–94, 372; R. Lewin, *Bones of Contention*, Penguin Books, Harmondsworth, 1989, 86.

21. R. Leakey and R. Lewin, *Origins*, Macdonald and Jane's, London, 1977, 56; R. Leakey and R. Lewin, *Origins Reconsidered*, Little, Brown and Co., London, 1992, 78.

Chapter 17

1. J.H. Campbell, Autonomy in evolution, in R. Milkman (ed.), *Perspectives on Evolution*, Sinauer Associates, Sunderland, Mass., 1982, 190–201 (quotation from page 190).

2. C. Darwin, *On the Origins of Species by Natural Selection*, John Murray, London, 1859, Penguin Books, Harmondsworth, 1968, 69.

3. S.L. Miller, A production of amino acids under possible primitive Earth conditions, *Science*, 117, 1953, 528–529; S.L. Miller and H.C. Urey, Organic compound synthesis on the primitive Earth, *Science*, 130, 1959, 245–251.

4. G. Wald, The origin of life, *Scientific American*, 191(2), 1954, 44–52 (reproduced in *Molecules to Living Cells*, Scientific American/Freeman, San Francisco, 1980, 11–19); C.E. Folsome, *The Origin of Life*, Freeman, San Francisco, 1979, 140–144.

5. G.A. Dover, Molecular drive, a cohesive mode of species evolution, *Nature*, 299, 1982, 111–117; G.A. Dover, Molecular drive in multigene families – how biological

novelties arise, spread and are assimilated, *Trends in Genetics*, 2, 1986, 159–165; G. Dover, *Dear Mr Darwin*, Weidenfeld and Nicolson, London, 2000.

6. M. Kimura, *The Neutral Theory of Molecular Evolution*, Cambridge University Press, 1983; M. Kimura, Recent developments in the neutral theory viewed from the Wrightian tradition of theoretical population genetics, *Proceedings of the National Academy of Sciences, USA*, 88, 1991, 5969–5973.

7. T. Palmer, *Controversy – Catastrophism and Evolution: The Ongoing Debate*, Kluwer/Plenum, New York, 1999, 261–326.

8. R. Shapiro, *Origins*, Penguin Books, Harmondsworth, 1988; M.A. Edey and D.C. Johanson, *Blueprints*, Oxford University Press, 1990, 281–322; J. Horgan, In the beginning . . . , *Scientific American*, 264(2), 1991, 100–109.

9. C. Böhler, P.E. Nielsen and L.E. Orgel, Template switching between PNA and RNA oligonucleotides, *Nature*, 376, 1995, 578–581; P. Cohen, Let there be life, *New Scientist*, 6 July 1996, 22–27; I. Fry, *The Emergence of Life on Earth*, Free Association Books, London, 1999, 112–178; J. McFadden, *Quantum Evolution*, Flamingo, London, 2000, 90–98; M. Hagmann, Between a rock and a hard place, *Science*, 295, 2002, 2006–2007.

10. P. Radetsky, Life's crucible, *Earth*, 7(1), 1998, 34–41; M. Milstein, Cooking up a volcano, *Earth*, 7(2), 1998, 24–31; M. Kaplan, A fresh start – life may have begun not in the sea but in some warm little freshwater pond, *New Scientist*, 11 May 2002, 7; J.L. Bada and A. Lazcano, Some like it hot, but not the first biomolecules, *Science*, 296, 2002, 1982–1983; A. Coghlan, Power of three, *New Scientist*, 22 June 2002, 10; F. Hoyle and C. Wickramasinghe, *Our Place in the Cosmos*, Phoenix, London, 1996; I. Fry, *The Emergence of Life on Earth*, 112–178; J. McFadden, *Quantum Evolution*, 72–90; P. Davies, Survivors from Mars, *New Scientist*, 12 September 1998, 24–29; G. Schueller, Stuff of life, *New Scientist*, 12 September 1998, 30–35; D.H. Levy, *Comets – Creators and Destroyers*, Touchstone, New York, 1998, 57–70; C. Wickramasinghe, *Cosmic Dragons*, Souvenir Press, London, 2001, 28–44, 168–181; C. Wickramasinghe, The long road to panspermia, *Astronomy Now*, 16(4), 2002, 57–60; M. Garlick, Life and the Moon, *Astronomy Now*, 16(5), 2002, 28–30.

11. M. Ruse, *Darwinism Defended*, Addison-Wesley, Reading, Mass., 1982, 285–329; J. Cherfas (ed.), *Darwin up to Date*, New Science Publications, London, 1982, 7–11; M. Shermer, *Why People Believe Weird Things*, Freeman, New York, 1997, 125–172; J. Rennie, Fifteen answers to creationist nonsense, *Scientific American*, 287(1), 2002, 62–69.

12. R. Shapiro, *Origins*, 248–265; M. Shermer, *Why People Believe Weird Things*, 154–172; G.H. Gallup and F. Newport, Belief in paranormal phenomena among adult Americans, *Skeptical Inquirer*, 15(2), 1991, 137–147; C. Sagan and A. Druyan, *Shadows of Forgotten Ancestors*, Century, London, 1992, 423–425; H. Hellman, *Great Feuds in Science*, Wiley, New York, 1998, 92–103; R. Doyle, Down with evolution!, *Scientific American*, 286(3), 2002, 20.

13. Don't mention Darwin, *New Scientist*, 21 August 1999, 4; R. Doyle, *Scientific American*, 20; Evolution critics seek role for unseen hand in education, *Nature*, 416, 2002, 250.

14. For examples of consistent results being obtained by different methods see D.C. Johanson and M.E. Edey, *Lucy – The Beginnings of Humankind*, Granada, London, 1982, 191–211; C.B. Stringer and R. Grün, Time for the last Neanderthals, *Nature*, 351, 1991, 701–702; J.C. Cloué-Long, Z. Zichao, M. Guogan and D. Shaohua, The age of the Permian–Triassic boundary, *Earth and Planetary Science Letters*, 105, 1991, 182–190; G. Curtis, C. Swisher and R. Lewin, *Java Man*, Little, Brown & Co., London, 2000, 220–221.

15. D.C. Johanson and M.E. Edey, *Lucy – The Beginnings of Humankind*, 191–200; R. Lewin, *Human Evolution*, Blackwell, Malden, Mass. and Oxford, 4th edn, 1999, 33–34.

16. V. Courtillot, G. Feraud, H. Maluski, *et al.*, The Deccan flood basalts and the Cretaceous–Tertiary boundary, *Nature*, 333, 1988, 843–846; V. Courtillot, A volcanic eruption, *Scientific American*, 263(4), 1990, 53–60.

17. C. Renfrew, *Before Civilization*, Jonathan Cape, London, 1973, 48–83, 255–268; M. Baillie, *Exodus to Arthur*, Batsford, London, 1999, 14–30; R.E. Taylor, Fifty years of radiocarbon dating, *American Scientist*, 88(1), 2000, 60–67.

18. A.N. Beal, A bit creaky – tree rings, radiocarbon and ancient history, *Chronology and Catastrophism Review*, 13, 1991, 38–42; B. Newgrosh, Living with radiocarbon dates, *Journal of the Ancient Chronology Forum*, 5, 1992, 59–67; M. Baillie, *Exodus to Arthur*, 43; S.W. Manning, B. Cromer, P.I. Kuniholm and M.W. Newton, Anatolian tree rings and a new chronology for the East Mediterranean Bronze–Iron Ages, *Science*, 294, 2001, 2532–2535.

19. A. Koestler, *Janus – A Summing Up*, Pan Books, London, 1979, 165–226; R. Dawkins, *The Extended Phenotype*, Freeman, Oxford and San Francisco, 1982, 164–178.

20. R. Sheldrake, *A New Science of Life*, Blond and Briggs, London, 1981.

21. R. Sheldrake, *The Presence of the Past*, Collins, London, 1988.

22. S.W. Hawking, *A Brief History of Time*, Guild, London, 1988, 46–47; P. Davies, *The Cosmic Blueprint*, Unwin, London, 1989, 152–154; J. Gribbin, *In the Beginning*, Viking, London, 1993, 165–166; J.D. Barrow, Is nothing sacred?, *New Scientist*, 24 July 1999, 28–32.

23. See in particular R. Sheldrake, *The Rebirth of Nature*, Century, London, 1990.

24. R. Sheldrake, *A New Science of Life*, Revised edition, Paladin, London, 1987, 255–265; T. Palmer, Review of *The Presence of the Past*, *Chronology and Catastrophism Workshop*, 1989(1), 33–35; T. Palmer, Review of *The Rebirth of Nature*, *Chronology and Catastrophism Review*, 1991(1), 24–25.

25. H.M. Temin and S. Mizutani, RNA-dependent DNA polymrase in virions of Rous Sarcoma Virus, *Nature*, 226, 1970, 1211–1213; H.M. Temin, The DNA provirus hypothesis, *Science*, 192, 1976, 1075–1080.

26. E.J. Steele, *Somatic Selection and Adaptive Evolution*, Croom Helm, London, 1979; R.M. Gorczynski and E.J. Steele, Simultaneous yet independent inheritance of somatically acquired tolerance to two distinct H-2 antigenic haplotype determinants in mice, *Nature*, 289, 1981, 678–681; E.J. Steele, R.A. Lindley and R.V. Blanden, *Lamarck's Signature*, Perseus Books, Reading, Mass., 1998, 125–186; L. Brent, P. Chandler, W. Fierz, *et al.*, Further studies on supposed lamarckian inheritance of immunological tolerance, *Nature*, 295, 1982, 242–244.

27. R. Dawkins, *The Extended Phenotype*, 164–177; J. Cherfas (ed.), *Darwin up to Date*, New Science Publications, London, 1982, 56–61; P.J. Bowler, *Evolution – The History of an Idea*, University of California Press, Berkeley, Los Angeles and London, 1989, 340–341.

28. J. Cairns, J. Overbaugh and S. Miller, The origin of mutants, *Nature*, 335, 1988, 142–145; B.G. Hall, Adaptive evolution that requires multiple spontaneous mutations, *Genetics*, 120, 1988, 887–897; B.G. Hall, Selection-induced mutations occur in yeast, *Proceedings of the National Academy of Sciences, USA*, 89, 1992, 4300–4303.

29. J.P. Radicella, P.W. Park and M.S. Fox, Adaptive mutations in *Escherichia coli* – a role for conjugation, *Science*, 268, 1995, 418–420; E.R. Moxon and D.S. Thaler, the tinkerer's evolving tool box, *Nature*, 387, 1997, 659–662; R.V. Millar, Bacterial gene swapping in bacteria, *Scientific American*, 278(1), 1998, 47–51; M. Brookes, Day of the mutators, *New Scientist*, 14 February 1998, 38–42.

30. G. Nicolis and I. Prigogine, *Self-organization in Nonequilibrium Systems*, Wiley, New York, 1977; I. Progogine and I. Stengers, *Order out of Chaos*, Heinemann, London, 1984; I. Prigogine, Origins of complexity, in A.C. Fabian (ed.), *Origins*, Cambridge University Press, 1988, 69–88.

31. S.A. Kauffman, *The Origins of Order*, Oxford University Press, 1993.

32. S.A. Kauffman, *The Origins of Order*, xiii.

33. S.A. Kauffman, *The Origins of Order*, 173–235.

34. S.A. Kauffman, *The Origins of Order*, 235.

35. S.A. Kauffman, *The Origins of Order*, 262.

36. S.A. Kauffman, *The Origins of Order*, 409.

37. D.J. Pritchard, *Foundations of Developmental Genetics*, Taylor and Francis, London, 1986.

38. D.J. Pritchard, The missing chapter in evolution theory, *The Biologist*, 37, 1990, 149–152.

39. B. Goodwin, *How the Leopard Changed its Spots*, Weidenfeld and Nicolson, London, 1994.

40. B. Goodwin, *How the Leopard Changed its Spots*, x–xi.

41. B. Goodwin, *How the Leopard Changed its Spots*, xiii.

42. B. Goodwin, *How the Leopard Changed its Spots*, 73–103.

43. B. Goodwin, *How the Leopard Changed its Spots*, vii.

44. L. Margulis, Re-reading the origins of species, *The Times Higher Education Supplement*, 23 December 1994, 20.

45. M.M. Waldrop, *Complexity*, Simon and Schuster, New York, 1992, Penguin Books, Harmondsworth, 1992, 299.

46. J. Horgan, From complexity to perplexity, *Scientific American*, 272(6), 1995, 74–79.

Chapter 18

1. E. Mayr, *One Long Argument*, Penguin Books, Harmondsworth, 1991 (quotations froms pages 147 and 164).

2. D.C. Dennet, *Darwin's Dangerous Idea*, Penguin Books, Harmondsworth, 1995, 20.

3. E. Mayr, *What Evolution Is*, Weidenfeld and Nicolson, London, 2002 (quotation from page 272).

4. T. Hunkapillar, H. Huang, L. Hood and J.H. Campbell, in R. Milkman (ed.), *Perspectives on Evolution*, Sinauer Associates, Sunderland, Mass., 1982, 164–189; A. Coghlan, Fast and loose, *New Scientist*, 28 November 1998, 12; G. Vines, Hidden inheritance, *New Scientist*, 28 November 1998, 26–30; D.E. Ingber, The architecture of life, *Scientific American*, 278(1), 1998, 30–39; S. Jones, *Almost Like a Whale*, Doubleday, London, 1999; E.R. Moxon and C. Wills, DNA microsatellites: agents of evolution?, *Scientific American*, 280(1), 1999, 72–77; S.-H. Lee, K.K. Fu, J.N. Hui and J.M. Richman, Noggin and retinoic acid transforms the identity of avian facial prominences, *Nature*, 414, 2001, 909–912; R. Galant and S.B. Carroll, Evolution of a transcriptional repression domain in an insect *Hox* protein, *Nature*, 415, 2002, 910–913; Y. Shigetani, F. Sugahara, Y. Kawakami, *et al.*, Heterotopic shift of epithelial–mesenchymal interactions in vertebrate jaw evolution, *Science*, 292, 2002, 1316–1310; J.H. Schultz, *Sudden Origins*, Wiley, New York, 1999; J. McFadden, *Quantum Evolution*, HarperCollins, London, 2000; G. Dover, *Dear Mr Darwin*, Weidenfeld and Nicolson, London, 2000; E. Mayr, *What Evolution Is*, 106–114, 144–146; M. Pigliucci, Buffer zone, *Nature*, 417, 2002, 598–599; C. Queitsch, T.A. Sangster and S. Lindquist, Hsp90 as a capacitor of phenotypic variation, *Nature*, 417, 2002, 618–624; P. Callaerts, P.N. Lee, B. Hartmann, *et al.*, HOX genes in the sepiolid squid *Eupryma scalopes* – implications for the evolution of complex body plans, *Proceedings of the National Academy of Sciences, USA*, 99, 2002, 2088–2093; M. Lynch, Intron evolution as a population-genetic process, *Proceedings of the National Academy of Sciences, USA*, 99, 2002, 6118–6123; L. Wolpert, *Principles of Development*, Oxford University Press, 2nd edn, 2002, 493–519; S.J. Gould, *The Structure of Evolutionary Theory*, Harvard University Press, Cambridge, Mass., 2002, 595–1343.

5. P. Whitfield, *The Natural History of Evolution*, Marshall Editions, London, 1993, 195.

6. N. Shubin, C. Tabin and S.B. Carroll, Fossils, genes and the evolution of animal limbs, *Nature*, 388, 1997, 639–638 (quotation from page 639).

7. A. Kopp, I. Duncan and S.B. Carroll, Genetic control and evolution of sexually dimorphic characters in *Drosophila*, *Nature*, 408, 2000, 553–559.

8. C.J. Lowe and G.A. Wray, Radical alterations in the roles of homeobox genes during echinoderm evolution, *Nature*, 389, 1997, 718–721 (quotation from page 718); M. Kmita-Cunisse, F. Loosli, J. Bièrne and W.J. Gehring, Homeobox genes in the ribbonworm *Lineus sanguineus* – evolutionary implications, *Proceedings of the National Academy of Sciences, USA*, 95, 1998, 3030–3039 (quotation from page 3030).

9. M. Ronshaugen, N. McGinnis and W. McGinnis, *Hox* protein mutation and macroevolution of the insect body plan, *Nature*, 415, 2002, 914–917 (quotations from pages 914 and 917).

10. J.H. Schwartz, *Sudden Origins*, 320–379; J. McFadden, *Quantum Evolution*, 259–274; G. Dover, *Dear Mr Darwin*, 148–140; S.J. Gould, *The Structure of Evolutionary Theory*, 744–1024; S.J. Gould and N. Eldredge, Punctuated equilibrium comes of age, *Nature*, 366, 1993, 223–227; S.J. Gould, *Dinosaur in a Haystack*, Jonathan Cape, London, 1996, 123–144; W. Fontana and P. Schuster, Continuity of evolution – on the nature of transitions, *Science*, 280, 1998, 1451–1455; T.S. Kemp, *Fossils and*

Evolution, Oxford University Press, 1999, 129–187; S.J. Gould, The Lying Stones of Marrakech, Jonathan Cape, London, 2000, 333–346; P.R. Grant and B.R. Grant, Adaptive radiation of Darwin's finches, American Scientist, 90, 2002, 130–139; P.R. Grant and B.R. Grant, Unpredictable evolution in a 30-year study of Darwin's finches, Science, 296, 2002, 707–711.

11. G. Dover, Dear Mr Darwin, 118–129; E. Mayr, What Evolution Is, 174–187; N. Eldredge, Reinventing Darwin, Weidenfeld and Nicolson, London, 1995, 93–123; R.K. Butlin and T. Tregenza, Is speciation no accident?, Nature, 387, 1997, 551–552; W.D. McMillan, C.D. Jiggins and J. Mallett, What initiates speciation in passion-vine butterflies?, Proceedings of the National Academy of Sciences, USA, 94, 1997, 8628–8633.

12. R. Dawkins, Universal Darwinism, in D.S. Bendall (ed.), Evolution from Molecules to Men, Cambridge University Press, 1982, 403–425; R. Dawkins, Climbing Mount Improbable, Penguin Books, Harmondsworth, 1996, 64–96.

13. F. Hoyle, The Intelligent Universe, Michael Joseph, London, 1983, 19.

14. J. Maynard Smith, Evolutionary Genetics, Oxford University Press, 1989, 273.

15. J. Maynard Smith, Evolutionary Genetics, Oxford University Press, 2nd edn, 1998, 267.

16. M.-W. Ho, P. Saunders and S. Fox, A new paradign for evolution, New Scientist, 27 February 1986, 41–43 (quotation from page 43).

17. C.D. Rollo, Phenotypes: Their Epigenetics, Ecology and Evolution, Chapman and Hall, London, 1994.

18. C.D. Rollo, Phenotypes: Their Epigenetics, Ecology and Evolution, xi.

19. N. Eldredge, Unfinished Synthesis, Oxford University Press, 1985, 7.

20. N. Eldredge, Reinventing Darwin, 1–10.

21. N. Eldredge, Unfinished Synthesis, 214; S.J. Gould, The Structure of Evolutionary Theory, 595–1343 (particularly pages 595–743). See also D. Jablonski, Micro- and macroevolution: scale and hierarchy in evolutionary theory, Paleobiology, 26(4) (Supplement), 2000, 15–52.

22. R. Dawkins, River out of Eden, Weidenfeld and Nicolson, London, 1995, xi–xii.

23. R. Dawkins, River out of Eden, 77–83; D.E. Nilsson and S. Pelger, A pessimistic estimate of the time required for an eye to evolve, Proceedings of the Royal Society of London, B, 256, 1994, 53–58.

24. E. Mayr, One Long Argument, 143.

25. J.H. Campbell and J.W. Schopf (eds.), Creative Evolution?!, Jones and Bartlett, Boston, 1994, ix.

26. J.H. Campbell and J.W. Schopf (eds.), Creative Evolution?!, 100.

27. S. Jones, Almost Like a Whale, 125.

28. S. Jones, Almost Like a Whale, 318.

29. G. Dover, Dear Mr Darwin, xii–xiii.

Chapter 19

1. A. Roy (ed.), Oxford Illustrated Encyclopedia of the Universe, Oxford University Press, 1992, 151; N. Henbest, The Planets, Penguin Books, Harmondsworth, 1994, 180–191; D. McNab and J. Younger, The Planets, BBC Books, London, 1999, 20–23.

2. P.J. Bowler, *Evolution – The History of an Idea*, University of California Press, Berkeley, Los Angeles and London, 1989, 35–39; *The Hutchinson Dictionary of Scientific Biography*, Helicon, Oxford, 1994, 377–378, 410–411; C. Dwyer, Laplace – the supreme calculator, *Astronomy Now*, 12(5), 1998, 52–53; S.R. Taylor, *Destiny or Chance*, Cambridge University Press, 1998, 13–16.

3. C.J. Allègre and S.H. Schneider, The evolution of the Earth, *Scientific American*, 271(4), 1994, 44–51; I. Nicholson, D.A. Rothery and F. Taylor, Birth of the Solar System, *Astronomy Now*, 12(2), 1998, 41–50; T. Yulsman, From pebbles to planets, *Astronomy*, 26(2), 1998, 56–61; S.R. Taylor, *Destiny or Chance*, 49–64; D. McNab and J. Younger, *The Planets*, 25–29; G. Vanin, *Cosmic Phenomena*, Firefly Books, Buffalo, New York, 1999, 30–34.

4. J. Laskar, A numerical experiment on the chaotic behaviour of the Solar System, *Nature*, 338, 1989, 237–238; C. Murray, Is the Solar System stable?, *New Scientist*, 25 November 1989, 60–63; C. Murray, Is the Solar System stable?, in N. Hall (ed.), *The New Scientist Guide to Chaos*, Penguin Books, Harmondsworth, 1992, 96–107; S.R. Taylor, *Destiny or Chance*, 213; J.J. Lissauer, Chaotic motion in the Solar System, *Reviews of Modern Physics*, 71, 1999, 835–845.

5. A. Roy (ed.), *Oxford Illustrated Encyclopedia of the Universe*, 125; N. Henbest, *The Planets*, 173–179.

6. R. Malhotra, Migrating planets, *Scientific American*, 281(3), 1999, 46–53.

7. A. Roy (ed.), *Oxford Illustrated Encyclopedia of the Universe*, 107, 174; N. Henbest, *The Planets*, 159–171; H. Couper with N. Henbest, *The Planets*, Pan Books, London, 1985, 99–100; S.R. Taylor, *Destiny or Chance*, 85–87.

8. N. Henbest, *The Planets*, 143–157; J.K. Davies, *Cosmic Impact*, St Martin's Press, New York, 1986, 40–41; D.J. Jankowski and S.W. Squyres, Solid-state ice volcanism on the satellites of Uranus, *Science*, 241, 1988, 897–899; *The Far Planets*, Time-Life Books, Amsterdam, 1990, 116–118; S.R. Taylor, *Destiny or Chance*, 85–87; D. McNab and J. Younger, *The Planets*, 129–131.

9. N. Henbest, *The Planets*, 110–121, 127–133, 147–157, 168–171; Focus – Guide to the planets, *Astronomy Now*, 12(11), 1998, 52–58; D. Rothery, An unusual family, *Astronomy Now*, 14(11), 2000, 60–64.

10. J.K. Davies, *Cosmic Impact*, 23–26; C. Sagan and A. Druyan, *Shadows of Forgotten Ancestors*, Century, London, 1992, 11–17; J. Gribbin and M. Gribbin, *Fire on Earth*, Simon and Schuster, London, 1996, 61–65; I. Semeniuk, Neptune attacks!, *New Scientist*, 7 April 2001, 26–29; J. Man, *Comets, Meteors and Asteroids*, BBC Books, London, 2001, 8–17.

11. N. Henbest, *The Planets*, 127–133; *The Far Planets*, 96–97; P. Moore, *Stars and Planets*, Chancellor Press, London, 1992, 56–57; S.R. Taylor, *Destiny or Chance*, 83–85; D. McNab and J. Younger, *The Planets*, 126–128; J. Foust, Astronomers find 11 new moons orbiting Jupiter, *Astronomy Now*, 16(7), 2002, 7.

12. N. Henbest, *The Planets*, 110–121; *The Far Planets*, 60–68; P. Moore, *Stars and Planets*, 51–52; S.R. Taylor, *Destiny or Chance*, 76–83; J. Foust, *Astronomy Now*, 16(7), 2002, 7; 11 more moons of Jupiter discovered, *Astronomy and Space*, July 2002, 11; D. Tytell, Jupiter reclaims title as "King of the Moons", *Sky and Telescope*, 104(2), 2002, 26.

13. A. Roy (ed.), *Oxford Illustrated Encyclopedia of the Universe*, 37, 119; N. Henbest, *The Planets*, 94–95; P. Moore, *Stars and Planets*, 40–41; S.R. Taylor, *Destiny or Chance*, 59–61; D. McNab and J. Younger, *The Planets*, 43.

14. N. Henbest, *The Planets*, 57–63; C. Sagan and A. Druyan, *Shadows of Forgotten Ancestors*, 21–22; G.J. Taylor, The scientific legacy of Apollo, *Scientific American*, 271(1), 1994, 26–33; How to make Earth's Moon, *Astronomy*, 26(1), 1998, 24; P. Bond, How old is the Moon?, *Astronomy Now*, 12(1), 1998, 7; S.R. Taylor, *Destiny or Chance*, 147–161; D. McNab and J. Younger, *The Planets*, 39–72; N. English, The day the Moon was made, *Astronomy Now*, 15 (10), 2001, 67; M. Garlick, Life and the Moon, *Astronomy Now*, 16(5), 2002, 28–30.

15. A. Roy (ed.), *Oxford Illustrated Encyclopedia of the Universe*, 92, 96, 101–103; N. Henbest, *The Planets*, 22–29, 35–55; 58–60; P. Moore, *Stars and Planets*, 28–39; Focus – guide to the planets, *Astronomy Now*, 12(11), 1998, 45–51; S.R. Taylor, *Destiny or Chance*, 118–126, 161–167; D. McNab and J. Younger, *The Planets*, 88–99.

16. N. Henbest, *The Planets*, 48–49; D.H. Grinspoon, Venus unveiled, *Astronomy*, 25(5), 1997, 44–49; M.A. Bullock and D.H. Grinspoon, Global climate change on Venus, *Scientific American*, 280(3), 1999, 34–41; S.R. Taylor, *Destiny or Chance*, 127–134; D. McNab and J. Younger, *The Planets*, 82–88.

17. A. Roy (ed.), *Oxford Illustrated Encyclopedia of the Universe*, 9; N. Henbest, *The Planets*, 181–184; P. Moore, *Stars and Planets*, 44–45; S.R. Taylor, *Destiny or Chance*, 111–117; J. Man, *Comets, Meteors and Asteroids*, 30–37.

18. J.F. Veverka and R.W. Farquar, NEAR views of Mathilde, *Sky and Telescope*, 94(4), 1997, 30–32; R. Zimmerman, Ice cream sundaes and mashed potatoes, *Astronomy*, 27(2), 1999, 54–59; D. McNab and J. Younger, *The Planets*, 27; W. Harwood, NEAR orbits asteroid Eros, *Astronomy Now*, 14(4), 2000, 30–33; E. Asphaug, The small planets, *Scientific American*, 282(5), 2000, 28–37; J. Veverka, M. Robinson, P. Thomas, et al., NEAR at Eros – imaging and spectral results, *Science*, 289, 2000, 2088–2097; N. Henbest and H. Couper, *Extreme Universe*, Chaneel 4 Books, London, 2001, 68–84.

19. R.P. Binzel, M.A. Barucci and M. Fulchignoni, The origins of the asteroids, *Scientific American*, 265(4), 1991, 66–73; D. Steel, *Rogue Asteroids and Doomsday Comets*, Wiley, New York, 1995, 15–26, 187–192; G.L. Verschuur, *Impact!*, Oxford University Press, 1996, 44–45; M.E. Bailey, Sources and populations of Near-Earth Objects, in B. Peiser, T. Palmer and M.E. Bailey (eds.), *Natural Catastrophes During Bronze Age Civilisations*, Archaeopress, Oxford, 1998, 10–20; S.R. Taylor, *Destiny or Chance*, 112–114.

20. N. Henbest, *The Planets*, 184–186; D. Steel, *Rogue Asteroids and Doomsday Comets*, 27–28, 126–128, 267; M.E. Bailey, Recent results in cometary astronomy, *Vistas in Astronomy*, 39, 1995, 647–671; B. Napier, Comets, *Astronomy Now*, 11(3), 1997, 41–43; S.R. Taylor, *Destiny or Chance*, 103–104; G. Vanin, *Cosmic Phenomena*, 44–45; W.J. Altenhoff, K.M. Menten and F. Bertoldi, Size determination of the Centaur Chariklo from millimetre-wavelength bolometer observations, *Astronomy and Astrophysics*, 366, 2001, L9–L12; J. Man, *Comets, Meteors and Asteroids*, 37–39; Y.R. Fernandez, D.C. Jewitt and S.S. Sheppard, Thermal properties of Centaurs Asbolus and Chiron, *Astronomical Journal*, 123, 2002, 1050–1055; S.A. Stern and H.F. Levison, Toward a

planet paradigm, *Sky and Telescope*, 104(2), 2002, 42–46; *Unusual Minor Planets*, see http://cfa-www.harvard.edu/iau/lists/Unusual.html

21. D. Steel, *Rogue Asteroids and Doomsday Comets*, 128–129; J. Gribbin and M. Gribbin, *Fire on Earth*, 106–107, 120–126; P. Farinella and D.R. Davis, Short-period comets – primordial bodies or collisional fragments?, *Science*, 273, 1996, 938–941; A. Hollis, Kenneth Essex Edgeworth – a biographical note, *Journal of the British Astronomical Association*, 106(6), 1996, 354; W.M. Napier and S.V.M. Clube, Our cometary environment, *Reports on Progress in Physics*, 60, 1997, 293–343.

22. J.X. Luu and D.C. Jewitt, The Kuiper Belt, *Scientific American*, 274(5), 1996, 32–38; B. Napier, Comets, *Astronomy Now*, 11(3), 1997, 41–43; S.C. Tegler and W. Romanishin, Two distinct populations of Kuiper-belt objects, *Nature*, 392, 1998, 49–51; S.R. Taylor, *Destiny or Chance*, 93–94; G. Vanin, *Cosmic Phenomena*, 34–36; J.K. Beatty, Giants found lurking in the Kuiper belt, *Sky and Telescope*, 101(3), 2001, 26–27; J. Foust, Ceres dethroned as largest minor planet, *Astronomy Now*, 15(10), 2001, 35; P. Bond, Asteroids and KBOs go in pairs, *Astronomy Now*, 16(6), 2002, 10; Hubble finds Kuiper belt binaries, *Astronomy and Space*, June 2002, 10–11; C. Veillet, J.W. Parker, I. Griffin, *et al.*, The binary Kuiper-belt object 1998 WW$_{31}$, *Nature*, 416, 2002, 711–713; D. Tytell, Making binary asteroids in the Kuiper belt, *Sky and Telescope*, 104(1), 2002, 22; *Unusual Minor Planets*.

23. M. Kidger, And then there were eight?, *Astronomy Now*, 13(3), 1999, 17; R. Graham, Is Pluto a planet?, *Astronomy*, 27(7), 42–47; S.R. Taylor, *Destiny or Chance*, 99–102; J. Man, *Comets, Meteors and Asteroids*, 40–41; W.B. McKinnon, Out on the edge, *Nature*, 418, 2002, 152–155; S.A. Stern and H.F. Levison, Toward a planet paradigm, *Sky and Telescope*, 104(2), 2002, 42–46.

24. J. Luu, B.G. Marsden, D. Jewitt, *et al.*, A new dynamical class of objects in the outer Solar System, *Nature*, 387, 1997, 573–575; G.R. Stewart, The frontier beyond Neptune, *Nature*, 387, 1997, 658–659; M.J. Duncan and H.F. Levison, A disk of scattered icy objects and the origin of Jupiter-family comets, *Science*, 276, 1997, 1670–1672; M.J. Holman, A possible long-lived belt of objects between Uranus and Neptune, *Nature*, 387, 1997, 785–788; G. Vanin, *Cosmic Phenomena*, 36; *Unusual Minor Planets*.

25. N. Henbest, *The Planets*, 186–188; G. Vanin, *Cosmic Phenomena*, 36–37; J.H. Oort, The structure of the cloud of comets surrounding the Solar System, and a hypothesis concerning its origin, *Bulletin of the Astronomical Institutes of the Netherlands*, 11, 1950, 91–110; M.E. Bailey, S.V.M. Clube and W.M. Napier, *The Origin of Comets*, Pergamon Press, Oxford, 1990, 179–207; C. Sagan and A. Druyan, *Comet*, Headline, London, 1997, 201–211; S.R. Taylor, *Destiny or Chance*, 94; P.R. Weissman, The Oort cloud, *Scientific American*, 279(3), 1998, 62–67.

26. V. Clube and B. Napier, *The Cosmic Winter*, Basil Blackwell, Oxford, 1990, 136–137; M.E. Bailey, Recent results in comtary astronomy, *Vistas in Astronomy*, 39, 1995, 647–671; B. Napier, Comets, *Astronomy Now*, 11(3), 1997, 41–43.

27. H. Muir, Doom star may be heading our way, *New Scientist*, 24 May 1997, 17; Close stellar approach predicted, *Sky and Telescope*, 94(2), 1997, 17; P.R. Weissman, *Scientific American*, 279(3), 1998, 62–67; S.A. Stern, Seeking rogue comets, *Astronomy*,

25(2), 1997, 46–51; Z.Q. Zheng and M.J. Valtonen, On the probability that a comet from another Solar System will collide with the Earth, *Monthly Notices of the Royal Astronomical Society*, 304, 1999, 579–582; J. Fount, Possible comet belt found around distant star, *Astronomy Now*, 15(9), 2001, 7.

28. C. Sagan and A. Druyan, *Comet*, 94–107; J. Man, *Comets, Meteors and Asteroids*, 43–46; P. Farinella and D.R. Davis, Short-period comets – primordial bodies or collisional fragments?, *Science*, 273, 1996, 938–941; D.M. Levy, *Comets*, Touchstone, New York, 1998, 24–25; M.E. Bailey, Sources and populations of Near-Earth Objects, in *Natural Catastrophes During Bronze Age Civilisations*, 10–20; N. English, Cometary origins, *Astronomy Now*, 14(5), 2000, 61–63; H.F. Levison, A. Morbidelli, L. Dones, *et al.*, The mass disruption of Oort cloud comets, *Science*, 296, 2002, 2212–2215.

29. M.E. Bailey, Recent results in cometary astronomy, *Vistas in Astronomy*, 39, 1995, 647–671; W.M. Napier and S.V.M. Clube, Our cometary environment, *Reports on Progress in Physics*, 60, 1997, 293–343; D. Steel, *Rogue Asteroids and Doomsday Comets*, 25, 34–35; S.R. Taylor, *Destiny or Chance*, 91–99.

30. A. Roy (ed.), *Oxford Illustrated Encyclopedia of the Universe*, 84–85; D. Steel, *Rogue Asteroids and Doomsday Comets*, 77–78; C. Sagan and A. Druyan, *Comet*, 97–99.

31. G.L. Verschuur, *Impact*, 140–155; Guide to the planets, *Astronomy Now*, 12(11), 1998, 45–56; W.K. Hartmann, The great Solar System revision, *Astronomy*, 26(8), 40–45.

32. J. Gribbin and M. Gribbin, *Fire on Earth*, 64–65; V. Clube and B. Napier, *The Cosmic Winter*, 246–247, 255–259; V. Clube and B. Napier, *The Cosmic Serpent*, Faber and Faber, London, 1982, 54–55; C.R. Chapman and D. Morrison, *Cosmic Catastrophes*, Plenum Press, New York and London, 1989, 43–58; R.B. Stothers, Impacts and tectonism in Earth and Moon history of the past 3800 million years, *Earth, Moon and Planets*, 58, 1992, 145–152; E.M. Shoemaker, Long-term variations in the impact cratering record on Earth, in M.M. Grady, R. Hutchison, G.J.H. McCall and D.A. Rothery, *Meteorites – Flux with Time and Impact Effects*, Geological Society, London, Special Publication 140, 1998, 7–10.

33. V. Clube and B. Napier, *The Cosmic Serpent*, 15, 86–91; S.P. Parker and J.M. Pasachoff (eds.), *McGraw-Hill Encyclopedia of Astronomy*, McGraw-Hill, New York, 2nd edn, 1992, 252–270; F. Heide and F. Wlotzka, *Meteorites*, Springer-Verlag, Berlin, 1995, 133–134, 172–175, 180–186; B.J. Gladwin, J.A. Burns, M. Duncan, *et al.*, The exchange of impact ejecta between terrestrial planets, *Science*, 271, 1996, 1387–1392; D. Hughes, Meteorites, *Astronomy Now*, 11(11), 1997, 41–44; A.J.T. Jull, S. Cloudt and E. Cielaszyk, ^{14}C terrestrial ages of meteorites from Victoria Land, Antarctica, and the infall rates of meteorites, in *Meteorites – Flux with Time and Impact Effects*, 75–91; M. Zolensky, The flux of meteorites to Antarctica, in *Meteorites – Flux with Time and Impact Effects*, 93–104; M. Grady, Visitors from outer space, *Astronomy Now*, 11(11), 1997, 45–47; C. Seife, It looks like a bug . . . , *New Scientist*, 13 March 1999, 7; B. Livermore, Meteorites on ice, *Astronomy*, 27(7), 1999, 54–58; A. Longstaff, Does this piece of rock come from Mercury?, *Astronomy Now*, 16(6), 2002, 12; A. Yamaguchi, R.N. Clayton, T.K. Mayeda, *et al.*, A new source of basaltic meteorites inferred from Northwest Africa 011, *Science*, 296, 2002, 334–336.

34. W.B. McKenna, Vestal voyagers unveiled, *Nature*, 363, 1993, 211–212; R.P. Binzel and S. Xu, Chips off asteroid 4 Vesta – evidence for the parent body of basaltic achondrite meteorites, *Science*, 260, 1993, 186–191; Hubble finds big crater on Vesta, *Astronomy*, 25(12), 1997, 30–34; P.C. Thomas, E.P. Binzel, M.J. Gaffey, *et al.*, Impact excavation on asteroid 4 Vesta – Hubble Space Telescope results, *Science*, 277, 1997, 1492–1495.

35. P. Farinella, Chaotic routes between the asteroid belt and the Earth, *Meteorite*, May 1996, 8–10; C.R. Chapman, Mantles were battered to bits, *Nature*, 385, 1997, 293–295; S.G. Love and T.J. Ahrens, Origin of asteroid rotation rates in catastrophic impacts, *Nature*, 386, 1997, 154–156; P. Farinella, D. Vokrouhlicky and W.H. Hartmann, Meteorite delivery via Yarkovsky orbital drift, *Icarus*, 132, 1998, 378–387; F. Migliorini, P. Michel, A. Morbidelli, *et al.*, Origin of multililometre Earth- and Mars-crossing asteroids, *Science*, 281, 1998, 2022–2024; Asteroids have astronomers seeing double, *Astronomy and Space*, June 2002, 9; J.L. Margot, M.C. Nolan, L.A.M. Benner, *et al.*, Binary asteroids in the Near-Earth object population, *Science*, 296, 2002, 1445–1448; J.K. Beatty, Double trouble among near-Earth asteroids, *Sky and Telescope*, 104(1), 2002, 23; J. Hecht, Asteroid's family tree, *New Scientist*, 15 June 2002, 12; J. Foust, Asteroid's ancient breakup, *Astronomy Now*, 16(8), 2002, 15; D.C. Richardson, Rocks that go bump in the night, *Nature*, 417, 2002, 697–698; D. Nesvorny, W.F. Bottke, L. Dones and H.F. Levison, The recent breakup of an asteroid in the main-belt region, *Nature*, 417, 2002, 720–722.

36. P. Michel, P. Farinella and C. Froeschlé, The orbital evolution of the asteroid Eros and implications for collision with the Earth, *Nature*, 380, 1996, 689–691; P. Bond, Eros on collision course?, *Astronomy Now*, 10(6), 1996, 4; R.P. Binzel, Eros's extended family, *Nature*, 388, 1997, 516–517; J.S. Lewis, *Rain of Iron and Ice*, Addison-Wesley, Reading, Mass., 1996, 75–77.

37. M.E. Bailey, S.V.M. Clube and W.M. Napier, The origin of comets, *Vistas in Astronomy*, 29, 1986, 53–112; A. Theokas, The origin of comets, *New Scientist*, 11 February 1988, 42–45; M.E. Bailey, S.V.M. Clube and W.M. Napier, *The Origin of Comets*, Pergamon Press, Oxford, 1990, 249–259; M.E. Bailey, Comets and molecular clouds – the sink and the source, in R.A. James and T.J. Millar (eds.), *Molecular Clouds*, Cambridge University Press, 1991, 273–289; M. Bailey, Where have all the comets gone?, *Science*, 296, 2002, 2151–2153.

38. V. Clube and B. Napier, *The Cosmic Winter*, 131–136; D. Steel, *Rogue Asteroids and Doomsday Comets*, 96–99; G.L. Verschuur, *Impact!*, 128–133; W.M. Napier and S.V.M. Clube, A theory of terrestrial catastrophism, *Nature*, 282, 1979, 455–459.

39. A. Roy (ed.), *Oxford Illustrated Encyclopedia of the Universe*, 27–28; C. Sagan and A. Druyan, *Comet*, 108–143; G. Vanin, *Cosmic Phenomena*, 44–45; D.H. Levy, *The Quest for Comets*, Oxford University Press, 1995, 70–71; F.L. Whipple, A comet model, *Astrophysical Journal*, 111, 1950, 375–394; 113, 1951, 464–474.

40. N. Henbest, *The Planets*, 186–191; C. Sagan and A. Druyan, *Comet*, 328–336; S.R. Taylor, *Destiny or Chance*, 95–96; D. McNab and J. Younger, *The Planets*, 35, 157; J. Man, *Meteors, Comets and Asteroids*, 60–63; R. Reinhard, The Giotto encounter with Comet Halley, *Nature*, 321, 1986, 313–318; H. Balsiger, H. Fechtig and J. Geisl, A close look

at Halley's Comet, *Scientific American*, 259(3), 1988, 96–103; G. Schwehm, A night to remember, *Astronomy Now*, 15(9), 2001, 22–23; J.K. Beatty, The super-black heart of Comet Borrelly, *Sky and Telescope*, 103(3), 2002, 22; R. Talcott, Comet Borrelly's dark nature, *Astronomy*, 30(4), 2002, 42–45; T. Clarke, Space probe shows comet sense, *Nature Science Update*, see http://www.nature.com/nsu/011220/011220-5.html; Deep Space 1 finds comet has hot, dry surface, *Astronomy and Space*, May 2002, 11; L.A. Soderblom, T.L. Becker, G. Bennett, *et al.*, *Science*, 296, 2002, 1087–1091.

41. V. Clube and B. Napier, *The Cosmic Winter*, 140–141; J. Gribbin and M. Gribbin, *Fire on Earth*, 106–113; C. Sagan and A. Druyan, *Comet*, 141; E.M. Shoemaker, Asteroid and comet bombardment of the Earth, *Annual Review of Earth and Planetary Science*, 11, 1983, 461–494; W.S. Weed, Chasing a comet, *Astronomy*, 29(9), 2001, 32–37.

42. J. Davies, Asteroids – the comet connection, *New Scientist*, 20/27 December 1984, 46–48; J. Mason, Asteroids, dead comets and meteor streams, *New Scientist*, 10 December 1988, 34–38; R.P. Binzel, S. Xu, S.J. Bus and E. Bowell, Origins for the near-Earth asteroids, *Science*, 257, 1992, 779–782; A. Caradini, E. Capaccione, M.T. Capria, *et al.*, Transition between comets and asteroids – from the Kuiper-belt to NEO objects, *Icarus*, 129, 1997, 337–347; S.R. Taylor, *Destiny or Chance*, 96; G. Vanin, *Cosmic Phenomena*, 37–44, 90–91; M.E. Bailey and V.V. Emel'Yanenko, Cometary capture and the nature of impactors, in *Meteorites – Flux with Time and Impact Effects*, 11–17.

43. V. Clube and B. Napier, *The Cosmic Winter*, 147–154; D. Steel, *Rogue Asteroids and Doomsday Comets*, 132–136, 242; G. Vanin, *Cosmic Phenomena*, 75–91; D.J. Steel, The limitations of NEO-uniformitarianism, *Earth, Moon and Planets*, 72, 1996, 279–292; D.J. Steel, D.J. Asher, W.M. Napier and S.V.M. Clube, Are impacts correlated in time?, in T. Gehrels (ed.), *Hazards due to Comets and Asteroids*, University of Arizona Press, Tucson, 1994, 463–477; M.E. Bailey, S.V.M. Clube, G. Hahn, *et al.*, Hazards due to giant comets – climate and short-term catastrophism, in *Hazards due to Comets and Asteroids*, 479–533; G.W. Kronk, Halley's Comet crumbs, *Sky and Telescope*, 103(5), 2002, 92–93; D. Steel, Chasing a comet's tail, *The Guardian*, 27 June 2002, Online section, 10.

44. C.R. Chapman and D. Morrison, *Cosmic Catastrophes*, 61–77; G.L. Verschuur, *Impact!*, 44–49; J.S. Lewis, *Rain of Iron and Ice*, 75–90, 138–149; N. Henbest and H. Couper, *Extreme Universe*, 68–93; D.B. Carlisle, *Dinosaurs, Diamonds, and Things from Outer Space*, Stanford University Press, Stanford, Calif., 1995, 121–124; M.E. Bailey, Sources and populations of Near-Earth Objects, in *Natural Catastrophes During Bronze Age Civilisations*, 10–20; W.K. Hartmann, The great Solar System revision, *Astronomy*, 26(8), 1998, 40–45; A. Atkinson, *Impact Earth*, Virgin, London, 1999, xxviii–xxxiii, 3, 21–22; H. Muir, How dust sent asteroids on crash course for the Sun, *New Scientist*, 9 February 2002, 17; C. Ryback, Tearing apart asteroid families, *Astronomy*, 30(3), 2002, 22.

Chapter 20

1. D.M. Raup, *Extinction – Bad Genes or Bad Luck*, Norton, New York, 1991, 195–199; D. Steel, *Rogue Asteroids and Doomsday Comets*, Wiley, New York, 1995, 15–31;

G.L. Verschuur, *Impact!*, Oxford University Press, 1996, 156–168; J. Gribbin and M. Gribbin, *Fire on Earth*, Simon and Schuster, London, 1996, 228–231; J.S. Lewis, *Rain of Iron and Ice*, Addison-Wesley, Reading, Mass., 1996, 201–205; B. McGuire, *Apocalypse*, Cassell, London, 1999, 207–209; E. Bryant, *Tsunami – The Underated Hazard*, Cambridge University Press, 2001, 231–233.

2. E.M. Shoemaker, Asteroid and comet bombardment of the Earth, *Annual Review of Earth and Planetary Science*, 11, 1983, 461–494; C.R. Chapman and D. Morrison, Impacts on the Earth by asteroids and comets – assessing the hazard, *Nature*, 367, 1994, 33–40; D. Morrison, C.R. Chapman and P. Slovic, The impact hazard, in T. Gehrels (ed.), *Hazards due to Comets and Asteroids*, University of Arizona Press, Tucson, 1994, 59–91; D. Morrison, Target Earth, *Astronomy*, 23(10), 1995, 34–41.

3. D. Steel, *Rogue Asteroids and Doomsday Comets*, 8, 30–38, 56–59; G.L. Verschuur, *Impact!*, 156–168; J.S. Lewis, *Rain of Iron and Ice*, 201–205; B. McGuire, *Apocalypse*, 207–209; E. Bryant, *Tsunami – The Underated Hazard*, 229–233; C.R. Chapman and D. Morrison, *Cosmic Catastrophes*, Plenum Press, New York and London, 1989, 74–77, 275–287; V. Clube and B. Napier, *The Cosmic Winter*, Basil Blackwell, Oxford, 1990, 244–248; D.W. Hughes, The mass distribution of crater-producing bodies, in M.M. Grady, R. Hutchison, G.J.H. McCall and D.A. Rothery, *Meteorites – Flux with Time and Impact Effects*, Geological Society, London, Special Publication 140, 1998, 31–42; R.A.F. Grieve, Extraterrestrial impacts on Earth, in *Meteorites – Flux with Time and Impact Effects*, 105–131.

4. C. Sagan and A. Druyan, *Comet*, Headline, London, 1997, 257–261; E.M. Shoemaker, P.R. Weisman and C.S. Shoemaker, The flux of periodic comets near Earth, in *Hazards due to Comets and Asteroids*, 313–335; D. Morrison, The contemporary hazard of cometary impacts, in P.J. Thomas, C.F. Chyba and C.P. McKay (eds.), *Comets and the Origin and Evolution of Life*, Springer-Verlag, New York, 1997, 243–258; D.H. Levy, *Comets – Creators and Destroyers*, Touchstone, New York, 1998, 19–33, 155–156.

5. E.M. Shoemaker, *Annual Review of Earth and Planetary Science*, 11, 1983, 461–494 (particularly pages 468–469); C. Sagan and A. Druyan, *Comet*, 262–266; E.M. Shoemaker, Long-term variations in the impact cratering rate on Earth, in *Meteorites – Flux with Time and Impact Effects*, 7–10.

6. R.A. Grieve, Terrestrial impact structures, *Annual Review of Earth and Planetary Science*, 15, 1987, 247–270; R.A.F. Grieve, Impact cratering on the Earth, *Scientific American*, 262(4), 1990, 44–51; R.A.F. Grieve and E.M. Shoemaker, The record of past impacts on Earth, in *Hazards due to Comets and Asteroids*, 417–462; R. Osborne and D. Tarling, *The Viking Historical Atlas of the Earth*, Viking, London, 1995, 20–21; R.A.F. Grieve and I.J. Personen, Terrestrial impact craters, *Earth, Moon and Planets*, 72, 1996, 357–376; R.A.F. Grieve, Extraterrestrial impact events – the record in the rocks and the stratigraphic column, *Palaeogeography, Palaeoclimate, Palaeoecology*, 132, 1997, 5–23; R.A.F. Grieve, Extraterrestrial impacts on Earth, in *Meteorites – Flux with Time and Impact Effects*, 105–131.

7. G.L. Verschuur, *Impact!*, 17–31; J. Gribbin and M. Gribbin, *Fire on Earth*, 28–32; D.H. Levy, *Comets – Creators and Destroyers*, 93–97; A. Atkinson, *Impact Earth*, Virgin, London, 1999, 16–19; J. Man, *Comets, Meteors and Asteroids*, BBC Books, London, 2001,

81–83; N. Henbest and H. Couper, *Extreme Universe*, Channel 4 Books, London, 2001, 96–117; P. Hadfield, Destroyer of worlds, *New Scientist*, 26 January 2002, 11.

8. D.M. Raup, *Extinction – Bad Genes or Bad Luck?*, 156–160; R.A.F. Grieve, Taget Earth – evidence for large-scale impact craters, *Annals of the New York Academy of Science*, 822, 1997, 161–181; R.A.F. Grieve, Extraterrestrial impact events, *Palaeogeography, Palaeoclimatology and Palaeoecology*, 132, 1997, 5–23.

9. D. Steel, *Rogue Asteroids and Doomsday Comets*, 81–92; G.L. Verschuur, *Impact!*, 41–42; D.H. Levy, *Comets – Creators and Destroyers*, 119–121; J. Man, *Asteroids, Meteors and Comets*, 25–27; C.C. Albritton, *Catastrophic Episodes in Earth History*, Chapman and Hall, London, 1989, 63–69; D.A. Kring, Calamity at Meteor Crater, *Sky and Telescope*, 98(5), 1999, 48–53.

10. J.S. Lewis, *Rain of Iron and Ice*, 21–22, 88; C.C. Albritton, *Catastrophic Episodes in Earth History*, 71–72; F. Heide and F. Wlotzka, *Meteorites*, Springer-Verlag, Berlin, 1995, 103–105; P.H. Schulz and R.E. Lianza, Recent grazing impacts on the Earth recorded in the Rio Cuarto crater field, Argentina, *Nature*, 355, 1992, 234–237; H.L. Melosh, Traces of an unusual impact, *Science*, 296, 2002, 1037–1038; P.A. Bland, C.R. de Souza Filho, A.J.T. Jull, *et al.*, A possible tektite strewn field in the Argentinian Pampa, *Science*, 296, 2002, 1109–1111; P. Bond, Argentina's mystery craters, *Astronomy Now*, 16(7), 2002, 10.

11. J.S. Lewis, *Rain of Iron and Ice*, 35–36; C.C. Albritton, *Catastrophic Episodes in Earth History*, 75–76; J. Man, *Comets, Meteors and Asteroids*, 24–25; C. Koeberl, Identification of meteoritic components in impactites, in *Meteorites – Flux with Time and Impact Effects*, 133–153; J.C. Wynn and E.M. Shoemaker, Secrets of the Wabar craters, *Sky and Telescope*, 94(5), 1997, 44–49; J.C. Wynn and E.M. Shoemaker, The day the sands caught fire, *Scientific American*, 279(5), 1998, 36–43; B.E. Schaeffer, Meteors that changed the world, *Sky and Telescope*, 96(6), 1998, 68–75; A. Raukas, R. Tiirmaa, E. Kaup and K. Kimmel, The age of the Ilumetsa metorite craters in southeast Estonia, *Meteoritics and Planetary Science*, 36, 2001, 1507–1514; M.C.L. Rocca-Mendoza, A Wabar-like site in eastern Uruguay?, *Meteoritics and Planetary Science*, 36(9), Supplement, 2001, A176.

12. G. Wright, The riddle of the sands, *New Scientist*, 10 July 1999, 42–45.

13. J.S. Lewis, *Rain of Iron and Ice*, 179; A. Atkinson, *Impact Earth*, 80–82.

14. D. Steel, *Rogue Asteroids and Doomsday Comets*, 92, 175; F. Heide and F. Wlotzke, *Meteorites*, 215.

15. C.C. Albritton, *Catastrophic Episodes in Earth History*, 70; F. Heide and F. Wlotzka, *Meteorites*, 6–7, 19–21; J. Gribbin and M. Gribbin, *Fire on Earth*, 56–57; R.A. Gallant, Sikhote–Aline 50 years later, *Sky and Telescope*, 93(2), 1997.

16. J. Baxter and T. Atkins, *The Fire Came By*, Futura, London, 1977; J. Stoneley, *Tunguska – Cauldron of Hell*, W.H. Allen, London, 1977; S. Welfare and J. Fairley, *Arthur C. Clarke's Mysterious World*, Collins, London, 1980, 153–167; R. Story, *Guardians of the Universe?*, New English Library, London, 1980, 95–100.

17. F. Heide and F. Wlotzka, *Meteorites*, 55–58; J. Gribbin and M. Gribbin, *Fire on Earth*, 43–59; J. Man, *Comets, Meteors and Asteroids*, 77–78; C. Chyba, P. Thomas and K. Zahnle, The 1908 Tunguska explosion – atmospheric detonation of a stony asteroid, *Nature*,

361, 1993, 40–44; J.E. Lynne and M. Tauber, Origin of the Tunguska event, *Nature*, 375, 1995, 638–639; Proceedings of the Tunguska 96 Workshop (Bologna), *Planetary and Space Science*, 46, 1998, 125–340; D. Morrison, Target Earth, *Astronomy*, 30(2), 2002, 46–51.

18. S. Welfare and J. Fairley, *Arthur C. Clarke's Mysterious World*, 153–167 (see comments at end of chapter 9); J.S. Lewis, *Rain of Iron and Ice*, 3–8, 51–54; V. Clube and B. Napier, *The Cosmic Winter*, Basil Blackwell, Oxford, 1990, 155–159; P. James and N. Thorpe, *Ancient Mysteries*, Ballantine, New York, 1999, 147–149.

19. C.C. Albritton, *Catastrophic Episodes in Earth History*, 76–78; D. Steel, *Rogue Asteroids and Doomsday Comets*, 44; D.M. Raup, *Bad Genes or Bad Luck*, 160–161.

20. V. Clube and B. Napier, *The Cosmic Winter*, 244–245; J.S. Lewis, *Rain of Iron and Ice*, 54; A. Atkinson, *Impact Earth*, 76–82; R.M. Schoch, *Voices of the Rocks*, Thorsons, London, 1999, 186–194; E. Bryant, *Tsunami – The Underated Hazard*, Cambridge University Press, 2001, 229–243.

21. G.L. Verschuur, *Impact!*, 108–112; J. Hecht, Asteroid 'airburst' may have devastated New Zealand, *New Scientist*, 5 October 1991, 19; D. Steel, *Rogue Asteroids and Doomsday Comets*, 44–45; E. Bryant, *Tsunami – The Underated Hazard*, 251–252 .

22. V. Clube and B. Napier, *The Cosmic Winter*, 159–161; D. Steel, *Rogue Asteroids and Doomsday Comets*, 135; J.S. Lewis, *Rain of Iron and Ice*, 50–51; I. Seymour, Moonstruck, *Astronomy Now*, 12(6), 1998, 58.

23. D. Steel, *Rogue Asteroids and Doomsday Comets*, 190–191, 203; G.L. Verschuur, *Impact!*, 88, 116; J.S. Lewis, *Rain of Iron and Ice*, 106; J. Gribbin and M. Gribbin, *Fire on Earth*, 98–99; D.W. Cox and J.H. Chestek, *Doomsday Asteroid*, Prometheus Books, Amherst, New York, 1998, 51–61.

24. C. Milmo, Close encounter prompts call for asteroid watch, *The Independent*, 8 January 2002, 5; 2001 YB₅'s close flyby, *Astronomy and Space*, March 2002, 10–11; Close call for planet Earth, *Astronomy*, 30(4), 2002, 26; B.J. Peiser (moderator), *Cambridge Conference Network* (see archive at http://abob.libs.uga.edu/bobk/cccmenu.html).

25. A. Atkinson, *Impact Earth*, xiv–xix, 142–147; D.K. Yeomans and P.W. Chodas, Predicting close approaches of asteroids and comets to Earth, in *Hazards due to Comets and Asteroids*, 241–258; D.L. Rabinowitz, E. Bowell, E.M. Shoemaker and K. Muininen, The population of Earth-crossing asteroids, in *Hazards due to Comets and Asteroids*, 285–312; S.J. Ostro, R.S. Hudson, R.F. Jurgens, *et al.*, Radar images of asteroid 4179 Toutatis, *Science*, 270, 1995, 80–83; J. Scotti, The tale of an asteroid, *Sky and Telescope*, 96(1), 1998, 30–34; S. Clark, Asteroid prediction sparks scare, *Astronomy Now*, 12(5), 1998, 15–16.

26. J. Hecht, World's end, *New Scientist*, 29 May 1999, 14; Phew, we're safe, *New Scientist*, 24 July 1999, 25; Radar pushes limits of asteroid prediction, *Asteroid and Space*, May 2002, 10; R.A. Kerr, Celestial billiards threaten hit in 2880, *Science*, 296, 2002, 27; J.D. Giorgini, S.J. Ostro, L.A.M. Benner, *et al.*, Asteroid 1950 DA's encounter with the Earth in 2880, *Science*, 296, 2002, 132–136; J. Foust, Distant threat, *Astronomy Now*, 16(6), 2002, 10; J.K. Beatty, "Threatening" asteroid aids planetary prognosticators, *Sky and Telescope*, 104(1), 2002, 24; D.L. Chandler, Impact threat comes

with silver lining, Astronomy, 30(7), 2002, 22; B. Peiser (mod.), Cambridge Conference Network.

27. D. Steel, Rogue Asteroids and Doomsday Comets, 122–123; J. Gribbin and M. Gribbin, Fire on Earth, 136–137; J.S. Lewis, Rain of Iron and Ice, 41–48; G. Vanin, Cosmic Phenomena, Firefly Books, Buffalo, New York, 1999, 70–75, 78, 107–113.

28. J. Rao, Will the Leonid storms continue?, Sky and Telescope, 99(6), 2000, 31–37; D. Fischer, The Leonids' rising stars, Sky and Telescope, 99(6), 2000, 38–40; P. Bond, A Leonid on the Moon?, Astronomy Now, 14(1), 2000, 6; A. Graps and A. Juhász, Dusty phenomena in the Solar System, Sky and Telescope, 101(1), 2001, 57–63; Lunar Leonids strike again, Sky and Telescope, 103(3), 2002, 106.

29. V. Clube and B. Napier, The Cosmic Winter, 141–142; A. Atkinson, Impact Earth, 152.

30. D. Steel, Rogue Asteroids and Doomsday Comets, 203; J.S. Lewis, Rain of Iron and Ice, 80; R.A. Kerr, Earth gains a retinue of mini-asteroids, Science, 258, 1992, 403.

31. D.H. Levy, Comets – Creators and Destroyers, 30–32; R. Naeye, Cosmic rain of minor comets, Astronomy, 25(9), 1997, 24–26; The small comet hypothesis, Geophysics Research Letters, 24, 1997, 3105–3124.

32. E.M. Shoemaker, Annual Review of Earth and Planetary Science, 11, 1983, 461–494; D. Steel, Rogue Asteroids and Doomsday Comets, 30–32; D.H. Levy, The Quest for Comets, Oxford University Press, 1995, 202–205; F. Saunders, The sound of one rock falling, Discover 23(2), 2002, 18.

33. D. Steel, Rogue Asteroids and Doomsday Comets, 175; J. Gribbin and M. Gribbin, Fire on Earth, 57–58; J.S. Lewis, Rain of Iron and Ice, 86–87; G. Vanin, Cosmic Phenomena, 74.

34. J. Gribbin and M. Gribbin, Fire on Earth, 58; R. Matthews, It was only a meteor, Mr President, New Scientist, 2 April 1994, 4.

35. A. Atkinson, Impact Earth, 86–87; W.W. Gibbs, The search for Greenland's mysterious meteor, Scientific American, 279(5), 1998, 44–51; B.J. Peiser (mod.), Cambridge Conference Network: R.M. Schoch, Voices of the Rocks, Thorsons, London, 1999, 184.

36. A. Atkinson, Impact Earth, 70–71; Meteorite fragments found following Texas fireball, Sky and Telescope, 9596), 1998, 17; H. Povenmire, New Mexico's sky is falling, Astronomy, 26(11), 1998, 30–32; Target Turkmenistan, Sky and Telescope, 96(6), 1998, 20; A narrow meteorite miss, Sky and Telescope, 97(2), 1999, 24; T. Yulsman, In search of fresh fall, Astronomy, 29(3), 2001, 48–53; B.J. Peiser (mod.), Cambridge Conference Network.

37. New Zealand meteorite, Astronomy and Space, September 1999, 16; New Irish meteorite, Astronomy and Space, July 2000, 18–19; B.J. Peiser (mod.), Cambridge Conference Network.

38. P. Bond, Yukon meteorite excites scientists, Astronomy Now, 14(5), 2000, 6; Yukon meteorite bonanza, Sky and Telescope, 99(6), 2000, 22; J.K. Beatty and D. Tytell, Daylight meteor lights the northeast, Sky and Telescope, 102(5), 2001, 23; N. English, UK Spaceguard Centre approved, Astronomy Now, 15(10), 2001, 8; N. Bone, UK fireballs during autumn, Astronomy Now, 16(1), 2002, 10; B.J. Peiser (mod.), Cambridge Conference Network.

39. D. Steel, Rogue Asteroids and Doomsday Comets, 29–30, 191; D.H. Levy, The Quest for Comets, 148–149; G.L. Verschuur, Impact!, 88, 196–203; D. Morrison, Target Earth,

Astronomy, 23(10), 1995, 34–41; D.W. Cox and J.H. Chestek, *Doomsday Asteroid*, 51–61, 97; Spacewatch snares long-lost asteroid 719 Albert, *Sky and Telescope*, 100(2), 2000, 22–23; Finding long lost asteroid Albert, *Astronomy Now*, 14(7), 2000, 29; D. Rabinowitz, E. Helin, K. Lawrence and S. Pravdo, A reduced estimate of the number of kilometre-sized near-Earth asteroids, *Nature*, 403, 2000, 165–166; J.S. Stuart, A near-Earth asteroid population estimate from the LINEAR survey, *Science*, 294, 2001, 1691–1693; J.K. Beatty, Killer asteroid count rises, *Sky and Telescope*, 103(5), 2002, 24; B.J. Peiser (mod.), *Cambridge Conference Network*.

40. A. Atkinson, *Impact Earth*, 3, 108–109; T.B. Spahr, C.W. Hergenrother, S.M. Larsen, et al., The discovery and physical characteristics of 1996 JA_1; A close call . . . and the search goes on, *Amateur Astronomy and Earth Sciences*, 1(8), 1996, 8–9; J. Hecht, Asteroid buzzes Earth from 'blind spot', NewScientist.com 15 March 2002 (see http://www.newscientist.com/news/news.jsp?id=ns99992052); T. Radford, Asteroid size of football pitch just misses Earth, *The Guardian*, 21 June 2002, 5; R. Uhlig, Earth has 'close shave' from large asteroid, *The Daily Telegraph*, 21 June 2002, 1; M. Henderson, Asteroid just missed us, *The Times*, 21 June, 2002, 3; D. Morrison, Two recent asteroid 'near misses', *Asteroid and Comet Impact Hazard*, 24 June 2002 (see http://impact.arc.nasa.gov/); P. Bond, Our brush with asteroid 2002 MN, *Astronomy Now*, 16(8), 2002, 8; B.J. Peiser (mod.), *Cambridge Conference Network*.

41. D.H. Levy, *Comets – Creators and Destroyers*, 148–150; G. Vanin, *Cosmic Phenomena*, 49–53; G. Hurst, Hyakutake spectacular, *Astronomy Now*, 10(6), 1996, 49–51; E.I. Aguirre, A great comet visits Earth, *Sky and Telescope*, 91(6), 1996, 20–23.

42. D. Steel, *Rogue Asteroids and Doomsday Comets*, 46–53; J. Gribbin and M. Gribbin, *Fire on Earth*, 228–233; G.L. Verschuur, *Impact!*, 163–166; D. Morrison, *Astronomy*, 23(10), 1995, 34–41.

43. G.L. Verschuur, *Impact!*, 208–209; C. Sagan and A. Druyan, *Comet*, 179–180; D.H. Levy, *The Quest for Comets*, 81–83, 207, 223; G. Vanin, *Cosmic Phenomena*, 46.

44. D. Steel, *Rogue Asteroids and Doomsday Comets*, 7–12; G.L. Verschuur, *Impact!*, 116–119; J. Gribbin and M. Gribbin, *Fire on Earth*, 224–226; D.H. Levy, *The Quest for Comets*, 1–12.

45. D. Steel, *Rogue Asteroids and Doomsday Comets*, 247–259; D.H. Levy, *Comets – Creators and Destroyers*, 197–213; J. Man, *Comets, Meteors and Asteroids*, 86–87; Articles: Comet Shoemaker–Levy 9, *Science*, 267, 1995, 1277–1323; D.H. Levy, E.M. Shoemaker and C.S. Shoemaker, Comet Shoemaker–Levy 9 meets Jupiter, *Scientific American*, 273(2), 1995, 68–75.

46. G.L. Verschuur, *Impact!*, 59–62; J. Gribbin and M. Gribbin, *Fire on Earth*, 141; C. Sagan and A. Druyan, *Comet*, 248–250; D.H. Levy, *The Quest for Comets*, 73; B.G. Marsden, The sun-grazing comet group, *Astronomical Journal*, 72, 1967, 1170–1183; Death of a comet, *Astronomy Now*, 14(9), 2000, 21–22; I. Toth, Impact-triggered breakup of comet c/1999 S4 (LINEAR) – identification of the closest intersecting orbits of other small bodies with its orbit, *Astronomy and Astrophysics*, 368, 2001, L25-L28; H.A. Weaver, Z. Sekanina, I. Toth, et al., HST and VLT investigation of the fragments of Comet C/1999 S4(LINEAR), *Science*, 292, 2001, 1329–1333; M. Bailey, Where have

all the comets gone?, *Science*, 296, 2002, 2151–2153; H.F. Levison, A. Morbidelli, L. Dones, *et al.*, The mass disruption of Oort cloud comets, *Science*, 296, 2002, 2212–2215; R. Massey, Where have the comets gone?, *Astronomy Now*, 16(8), 2002, 10.

47. D. Steel, *Rogue Asteroids and Doomsday Comets*, 257–259; J. Gribbin and M. Gribbin, *Fire on Earth*, 132–133; J.S. Lewis, *Rain of Iron and Ice*, 149.

48. A. Atkinson, *Impact Earth*, 75–80; D.H. Levy, *Comets – Creators and Destroyers*, 141–152; G. Vanin, *Cosmic Phenomena*, 53–67; D.J. Eicher, Here comes Hale–Bopp, *Astronomy*, 24(2), 1996, 68–73; D.W. Hughes, Farewell great comet Hale–Bopp, *Modern Astronomer*, 1(5), 1997, 41–45; N. English, K2 – the comet that everyone missed, *Astronomy Now*, 14(7), 2000, 5; R. Graham, A century of comets, *Astronomy*, 28(8), 2000, 58–62.

49. D. Steel, *Rogue Asteroids and Doomsday Comets*, 46–73; J.S. Lewis, *Rain of Iron and Ice*, 103–112; A. Atkinson, *Impact Earth*, 3–15; N. Henbest and H. Couper, *Extreme Universe*, 113–114; D.J. McLaren and W.D. Goodfellow, Geological and biological consequences of giant impacts, *Annual Review of Earth and Planetary Science*, 18, 1990, 123–171; O.B. Toon, K. Zahnle, R.P. Turco and C. Covey, Environmental perturbations caused by asteroid impacts, in *Hazards due to Comets and Asteroids*, 791–826.

50. C.R. Chapman and D. Morrison, *Cosmic Catastrophes*, 113–123; G.L. Verschuur, *Impact!*, 10–11, 189; C. Sagan and A. Druyan, *Comet*, 286–288; R.P. Turco, O.B. Toon, T.P. Ackerman, *et al.*, Nuclear winter – global consequences of multiple nuclear explosions, *Science*, 222, 1983, 1283–1292; R.P. Turco, O.B. Toon, T.P. Ackerman, *et al.*, Nuclear winter – physics and physical mechanisms, *Annual Review of Earth and Planetary Science*, 19, 1991, 383–422.

51. J. Gribbin and M. Gribbin, *Fire on Earth*, 32–38; C. Emiliani, E.B. Kraus and E.M. Shoemaker, Sudden death at the end of the Mesozoic, *Earth and Planetary Science Letters*, 55, 1985, 317–334; C.C. Albritton, *Catastrophic Episodes in Earth History*, Chapman and Hall, London, 1989, 108–119; W. Glen (ed.), *The Mass-Extinction Debates*, Stanford University Press, Stanford, Calif., 1994, 18–22; R.Cowen, The day the dinosaurs died, *Astronomy*, 24(4), 1996, 34–41.

52. D.H. Levy, *Comets – Creators and Destroyers*, 217–229; E. Bryant, *Tsunami – The Underated Hazard*, 228–262; R. Huggett, *Cataclysms and Earth History*, Oxford University Press, 1989, 171–186; G.L. Verschuur, Impact hazards – truth and consequences, *Sky and Telescope*, 95(6), 1998, 26–34; C. Frankel, *The End of the Dinosaurs*, Cambridge University Press, 1999, 110–137.

53. C.C. Albritton, *Catastrophic Episodes in Earth History*, 120–122; K.D. Terry and W.H. Tucker, Biological effects of supernovae, *Science*, 159, 1968, 421–423; A. Roy (ed.), *Oxford Illustrated Encyclopedia of the Universe*, Oxford University Press, 1992, 165–166; S. Blair, The end of the world is . . . weird, *Focus*, December 1996, 56–62; W. Alvarez, *T. rex and the Crater of Doom*, Princeton University Press, Princeton, New Jersey, 1997, 71–73; R. Irion, Exploding stars tell all, *Astronomy*, 26(11), 1998, 50–55; Gaseous collision begins in supernova 1987A, *Astronomy*, 95(1), 1998, 18; G. Schilling, X-ray "gamma-ray bursts", *Sky and Telescope*, 104(2), 2002, 20.

54. J. Ellis and D.N. Schramm, Could a nearby supernova have caused a mass extinction?, *Proceedings of the National Academy of Sciences, USA*, 92, 1995, 235–238;

B. Holmes, Did deathrays destroy species?, *New Scientist*, 14 January 1995, 15; J. Hecht, Death star – a supernova takes the blame for Earth's worst catastrophe, *New Scientist*, 4 April 1998, 20; N. Henbest and H. Couper, *Extreme Universe*, 147–151; P. Bond, Gamma rays from star deaths, *Astronomy Now*, 16(7), 2002, 10.

55. C.C. Albritton, *Catastrophic Episodes in Earth History*, 122–124; J. Gribbin and M. Gribbin, *Fire on Earth*, 192; G.C. Reid, J.R. McAfee and P.J. Crutzen, Effects of intense stratospheric ionisation events, *Nature*, 275, 1978, 489–492; Souped-up supernova, *Astronomy*, 26(11), 1998, 26; P. Bond, Powerful gamma-ray flare blasts Earth, *Astronomy Now*, 12(11), 1998, 4; M. Chown, Life force, *New Scientist*, 15 December 2001, 10; J. Roth, Gamma-ray bursts next door, *Sky and Telescope*, 103(2), 2002, 20.

56. P. Parsons, Did neutron stars bomb species?, *New Scientist*, 14 September 1996, 15; P. Bond, Did radiation wipe out the dinosaurs?, *Astronomy Now*, 11(2), 1997, 5; J. Kanipe, Dark matter blamed for mass extinctions on Earth, *New Scientist*, 11 January 1997, 14; P.J. Leonard and J.T. Bonnell, Gamma-ray bursts of doom, *Sky and Telescope*, 95(2), 1998, 28–34; Stupendous explosion challenges theory, *Astronomy*, 26(8), 1998, 18–20; Muon this, *Astronomy*, 26(11), 1998, 38; Gamma-ray bursts, *Astronomy and Space*, May 2002, 8; J.N. Reeves, D. Watson, J.P. Osborne, *et al.*, The signature of supernova ejecta in the X-ray afterglow of the gamma-ray burst 011211, *Nature*, 416, 2002, 512–515; N. Henbest and H. Couper, *Extreme Universe*, 140–165.

57. R. Huggett, *Cataclysms and Earth History*, 165.

58. C. Chapman and D. Morrison, *Cosmic Catastrophes*, 185–196 (quotation from page 185).

59. V. Clube and B. Napier, *The Cosmic Serpent*, Faber and Faber, London, 1982, 257.

60. V. Clube and B. Napier, *The Cosmic Winter*, 125.

61. G.L. Verschuur, *Impact!*, 96.

62. D. Steel, *Rogue Asteroids and Doomsday Comets*, 155.

63. J.E. Strickling, *Origins – Today's Science, Tomorrow's Myth*, Peripheral Vision, Norcross, Georgia, 1996, 173.

64. W. Thornhill, New physics supports planetary catastrophism, *Chronology and Catastrophism Review*, 1999:2, 11–15 (quotation from page 11).

65. W. Glen (ed.), *The Mass-Extinction Debates*, 152–169; J. Gribbin and M. Gribbin, *Fire on Earth*, 119–143; G.L. Verschuur, *Impact!*, 132–139; J.S. Lewis, *Rain of Iron and Ice*, 144–146; A. Atkinson, *Impact Earth*, 35–39.

66. V. Clube and B. Napier, *The Cosmic Winter*, 136–163, 225–232, 255–273; D. Steel, *Rogue Asteroids and Doomsday Comets*, 109–136; D.J. Asher and S.V.M. Clube, An extraterrestrial influence during the current glacial-interglacial, *Quarterly Journal of the Royal Astronomical Society*, 34, 1993, 481–511; D.J. Asher and S.V.M. Clube, Towards a dynamical history of proto-Encke, *Celestial Mechanics and Dynamical Astronomy*, 69, 1998, 149–170; M.E. Bailey, Sources and populations of Near-Earth Objects, in B.J. Peiser, T. Palmer and M.E. Bailey (eds.), *Natural Catastrophes During Bronze Age Civilisations*, Archaeopress, Oxford, 1998, 10–20; W.M. Napier, Cometary catastrophes, cosmic dust and ecological disasters in historical times, in *Natural Catastrophes During Bronze Age Civilisations*, 21–32.

67. D. Steel, *Rogue Asteroids and Doomsday Comets*, 124–125; G.L. Verschuur, *Impact!*, 136–139; D.I. Steel, D.J. Asher, W.M. Napier and S.V.M. Clube, Are impacts correlated in time?, in *Hazards due to Comets and Asteroids*, 463–477; B. Peiser, Cosmic uncertainty, *Astronomy Now*, 14(3), 2000, 53–55.

68. J. Gribbin and M. Gribbin, *Fire on Earth*, 119–126; M.E. Bailey, S.V.M. Clube, G. Hahn, *et al.*, Hazards due to giant comets, in *Hazards due to Comets and Asteroids*, 479–533; M.E. Bailey, Rapid dynamical evolution of giant comet Chiron, *Nature*, 348, 1990, 132–136; M.E. Bailey, Recent results in cometary astronomy, *Vistas in Astronomy*, 39, 1995, 647–671; W.M. Napier and S.V.M. Clube, Our cometary environment, *Reports on Progress in Physics*, 60, 1997, 293–343.

69. D. Steel, *Rogue Asteroids and Doomsday Comets*, 207–220; J. Gribbin and M. Gribbin, *Fire on Earth*, 223–243; G.L. Verschuur, *Impact!*, 191–211; D. Morrison, Target Earth, *Astronomy*, 23(10), 1995, 34–41.

70. A.C. Clarke, *Rendezvous with Rama*, Gollancz, London, 1973, chapter 1.

71. Latest sensors bring new asteroids and a comet into focus, *Amateur Astronomy and Earth Sciences*, 2(2), 1997, 12–13; D. Steel, Panning the sky for asteroids, *Amateur Astronomy and Earth Sciences*, 2(2), 1997, 52–53, 75; K.J. Lawrence, Near-Earth asteroid tracking (NEAT) program, *Annals of the New York Academy of Sciences*, 822, 1997, 6–25; D. Steel, Meteoroid orbits – implications for near-Earth object search programs, *Annals of the New York Academy of Sciences*, 822, 1997, 31–51; D.W. Cox and J.H. Chestek, *Doomsday Asteroid*, 95–114; D. Steel, *Target Earth*, Time-Life Books/Quarto Publishing, London, 2000, 100–135; D. Morrison, Target Earth, *Astronomy*, 30(2), 2002, 46–51.

72. D. Steel, *Rogue Asteroids and Doomsday Comets*, 221–240; G.H. Canavan, J.C. Solem and J.D.G. Rather, Near-Earth object intervention workshop, in *Hazards due to Comets and Asteroids*, 93–125; J. Hecht, Pentagon hot shots take aim at asteroids, *New Scientist*, 23 March 1996, 12; D. Hughes, Waiting for the sky to fall, *New Scientist*, 6 April 1996, 42–43.

73. A.W. Harris, G.H. Canavan, C. Sagan and S.J. Ostro, The deflection dilemma, in *Hazards due to Comets and Asteroids*, 1145–1155; C. Sagan and S.J. Ostro, Dangers of asteroid deflection, *Nature*, 368, 1994, 501; P.E. Tyler, Chinese seek atom option to fend off asteroids, *Amateur Astronomy and Earth Sciences*, 2(2), 1997, 12.

74. D. Steel, *Rogue Asteroids and Doomsday Comets*, 229–230; J.S. Lewis, *Rain of Iron and Ice*, 206–222; T.J. Ahrens and A.W. Harris, Deflection and fragmentation of near-Earth asteroids, in *Hazards due to Comets and Asteroids*, 897–927; T.J. Ahrens and A.W. Harris, Deflection and fragmentation of near-Earth asteroids, *Nature*, 360, 1992, 429–433.

75. D.W. Cox and J.H. Chestek, *Doomsday Asteroid*, 147–166; A. Atkinson, *Impact Earth*, 90–129; H.J. Melosh and I.V. Nemchinov, Solar asteroid diversion, *Nature*, 366, 1993, 21–22; D.K. Hill, Gathering airs schemes for averting asteroid doom, *Science*, 268, 1995, 1562–1563; H.J. Melosh and E.V. Ryan, Asteroids – shattered, but not dispersed, *Icarus*, 129, 1997, 562–564; A.W. Harris, Making and braking asteroids, *Nature*, 393, 1998, 418–419; E. Asphaug, S.J. Ostro, R.S. Hudson, *et al.*, Disruption of kilometre-sized asteroids by energetic collisions, *Nature*, 393, 1998,

437–440; J.N. Spitale, Asteroid hazard mitigation using the Yarkovsky effect, *Science*, 296, 2002, 77; D.L. Chandler, Impact threat comes with silver lining, *Astronomy*, 30(7), 2002, 22.

76. A. Atkinson, *Impact Earth*, 129–133.

77. D. Steel, *Target Earth*, 100–153; J. Man, *Comets, Meteors and Asteroids*, 90–93; N. Henbest and H. Couper, *Extreme Universe*, 120–130; S. Clark, Asteroid prediction sparks scare, *Astronomy Now*, 12(5), 1998, 15–16; B. McGuire, *Apocalypse*, Cassell, London, 1999, 201–246; C. Tickell, Space invaders, *Geographical*, 72(12), 2000, 14–17; D. Steel, Asteroid impacts – task force calls for action, *Astronomy Now*, 14(11), 2000, 20–22; T. Ryan, UK task force report on potentially hazardous near-Earth objects published, *Astronomy and Space*, February 2001, 14–15; H. Muir, Target Earth, *New Scientist*, 3 March 2001, 40–44; N. English, UK Spaceguard Centre approved, *Astronomy Now*, 15(10), 2001, 8; D. Morrison, Target Earth, *Astronomy*, 30(2), 2002, 46–51; Leicester hosts NEO Centre, *Astronomy Now*, 16(2), 2002, 10; E. Samuel, Incoming! To deflect an asteroid, choose your shot carefully, *New Scientist*, 16 February 2002, 11; A. Harris and D. Morrison, Progress in the Spaceguard Survey, *Astronomy and Space*, July 2002, 16–17.

78. C. Officer and J. Page, *Tales of the Earth*, 32–58; B. McGuire, *Apocalypse*, 155–200; A. Scarth, *Savage Earth*, HarperCollins, London, 2001, 74–79, L. O'Hanlon, Quakin' all over, *New Scientist*, 7 April 2001, 30–33; K. Segupta, Afghan quake leaves 2,000 dead and 30,000 homeless, *The Independent*, 27 March 2002, 1; S. Salahuddin, Tremors continue as death toll rises, *The Guardian*, 28 March 2002, 17.

79. C. Officer and J. Page, *Tales of the Earth*, 32–58; B. McGuire, *Apocalypse*, 109–200; A. Scarth, *Savage Earth*, 36–37, 56–57, 70–73; 84–85; D.A. Hardy and J. Murray, *The Fires Within*, Dragon's World, Limpsfield, Surrey, 1991, 32–36; R. O'Neill, *Natural Disasters*, Parragon, Bristol, 1998, 20–21, 44–49; A. Zaremba, A survivor's tale, *Newsweek*, 29 January 2001, 52; I. MacKinnon, Nothing between earth and sky, *Newsweek*, 5 February 2001, 30–31; E. Bryant, *Tsunami – The Underated Hazard*, 13–24, 60–66, 98–100, 135–177, 187–188.

80. C. Officer and J. Page, *Tales of the Earth*, 32–58; R. Osborne and D. Tarling, *The Viking Historical Atlas of the Earth*, 162–165; B. McGuire, *Apocalypse*, 155–200; A. Scarth, *Savage Earth*, 64–65, 80–83, 162–169; R. O'Neill, *Natural Disasters*, 18–19, 22–29; R. Bonson and R. Platt, *Disaster!*, Dorling Kindersley, London, 1997, 18–19; K.P.N. Shuker, A sense of disaster, in A. Grayson (ed.), *Equinox: The Earth*, Channel 4 Books, London, 2000, 177–211. See also D. Appell, Danger signals, *New Scientist*, 15 June 2002, 12; S. Uyeda, M. Hayakawe, T. Nagao, *et al.*, Electrical and magnetic phenomena observed before the volcano-seismic activity in 2000 in the Iza Island region, Japan, *Proceedings of the National Academy of Sciences, USA*, 99, 2002, 7352–7355.

81. D.A. Hardy and J. Murray, *The Fires Within*, 14–15, 97–105, 144–157; C. Officer and J. Page, *Tales of the Earth*, 3–31; B. McGuire, *Apocalypse*, 57–108; A. Scarth, *Savage Earth*, 138–151, 180–181; R. O'Neill, *Natural Disasters*, 34–35, 38–39, 62; R.M. Schoch, *Voices of the Rocks*, Thorsons, London, 1999, 130–133; H. Sigurdsson (ed.), *Encyclopedia of Volcanoes*, Academic Press, San Diego, Calif., 2000, 1302–1305;

O. Bowcott, Thousands flee as river of molen rock engulfs villages in Congo, *The Guardian*, 19 January 2002, 2; Caught out again, *New Scientist*, 2 February 2002, 3; T. Clarke, Seismic rumbling foretold Congo eruption, *Nature*, 415, 2002, 353; J.Z. de Boer and D.T. Sanders, *Volcanoes in Human History*, Princeton University Press, Princeton, New Jersey, 2002, 74–107, 186–208; B. Mason, Pinatubo dust is still a killer, *New Scientist*, 13 July 2002, 7.

82. D.A. Hardy and J. Murray, *The Fires Within*, 118–121, 125–128; C. Officer and J. Page, *Tales of the Earth*, 3–31; R. Osborne and D. Tarling, *The Viking Historical Atlas of the Earth*, 160–161; B. McGuire, *Apocalypse*, 57–108; A. Scarth, *Savage Earth*, 114–137; R. O'Neill, *Natural Disasters*, 42–43; H. Sigurdsson (ed.), *Encyclopedia of Volcanoes*, 1310–1311; J.Z. de Boer and D.T. Sanders, *Volcanoes in Human History*, 228–249.

83. V. Clube and B. Napier, *The Cosmic Serpent*, 76; D.A. Hardy and J. Murray, *The Fires Within*, 9–11, 96–97; C. Officer and J. Page, *Tales of the Earth*, 3–31; B. McGuire, *Apocalypse*, 57–108, 126–130; A. Scarth, *Savage Earth*, 152–153; R. O'Neill, *Natural Disasters*, 32–33, 36–37; J.S. Lewis, *Rain of Iron and Ice*, 99, 151–152, 204; H. Sigurdsson (ed.), *Encyclopedia of Volcanoes*, 1306–1308; E. Bryant, *Tsunami – The Underrated Hazard*, 213–227; J.Z. de Boer, *Volcanoes in Human History*, 47–73, 157–185; J.V. Luce, *The End of Atlantis*, Paladin, London, 1970, 45–50, 60–72; R. Castleden, *Atlantis Destroyed*, Routledge, London and New York, 1998, 114–126.

84. C.C. Albritton, *Catastrophic Episodes in Earth History*, 150–151; D.A. Hardy and J. Murray, *The Fires Within*, 82–83; C. Officer and J. Page, *Tales of the Earth*, 3–31; B. McGuire, *Apocalypse*, 57–108; A. Scarth, *Savage Earth*, 110–111, 143; R. O'Neill, *Natural Disasters*, 62–63; R.M. Schoch, *Voices of the Rocks*, 130–133; J.Z. de Boer and D.T. Sanders, *Volcanoes in Human History*, 108–156.

85. C. Officer and J. Page, *Tales of the Earth*, 3–31; B. McGuire, *Apocalypse*, 57–108; R.M. Schoch, *Voices of the Rocks*, 130–133; M.R. Voorhies, Ancient ashfall creates a Pompeii of prehistoric animals, *National Geographic*, 159(1), 1981, 66–75; Plumes of gold, *Earth*, 6(6), 1997, 13–14; Ashfall Fossil Beds State Historical Park website, http://ngp.ngpc.state.ne.us/parks/ashfall.html; Horizon – Supervolcanoes, shown on BBC Television 3 February 2000 (see http://www.bbc.co.uk/horizon/supervolcanoes_script.shtml); Will we be wiped out by a super-eruption?, *New Scientist*, 20 July 2002, 23.

86. V. Courtillot, G. Ferraud H. Maluski, *et al.*, The Deccan flood basalts and the Cretaceous/Tertiary boundary, *Nature*, 333, 1988, 843–846; M.R. Rampino and R.B. Stothers, Flood basalt volcanism during the past 250 million years, *Science*, 241, 1988, 663–668; R.L. Larson, The Mid-Cretaceous superplume episode, *Scientific American*, 272(2), 1995, 66–70; L.J. Penvenne, Turning up the heat, *New Scientist*, 16 December 1995, 26–30.

87. C.C. Albritton, *Catastrophic Episodes in Earth History*, 151–155; B. McGuire, *Apocalypse*, 58–61, 86; M.R. Rampino, S. Self and R.B. Stothers, Volcanic winters, *Annual Review of Earth and Planetary Science*, 16, 1988, 73–99.

88. C. Officer and J. Page, *Tales of the Earth*, 140–147; R.S. White and D.P. McKenzie, Volcanism at rifts, *Scientific American*, 261(1), 1989, 44–55; V.E. Courtillot, A volcanic eruption, *Scientific American*, 263(4), 1990, 53–60; A.P.M. Vaughan, Circum Pacific

mid-Cretaceous deformation and uplift – a superplume related event, *Geology*, 23, 1995, 491–494.

89. V. Clube and B. Napier, *The Cosmic Serpent*, 127–129; D. Steel, *Rogue Asteroids and Doomsday Comets*, 104–106; M.R. Rampino, Impact cratering and flood basalt volcanism, *Nature*, 327, 1987, 468; M.R. Rampino, B.M. Haggerty and T.C. Pagano, A unified theory of impact crises and mass extinctions, *Annals of the New York Academy of Science*, 822, 1997, 403–431; R.W. Carlson, Do continents part passively, or do they need a shove?, *Science*, 278, 1997, 240–241.

90. J. Gribbin and M. Gribbin, *Fire on Earth*, 190–194; M. Rampino, Dinsosaurs, comets and volcanoes, *New Scientist*, 18 February 1989, 54–58; M.R. Rampino and K. Caldeira, Major episodes of geologic change – correlation, time structure and possible causes, *Earth and Planetary Science Letters*, 114, 1993, 215–227.

91. L. Kaufman, Why the Ark is sinking, in L. Kaufman and K. Mallory (eds.), *The Last Extinction*, MIT Press, Cambridge, Mass., 1987, 1–41; N. Eldredge, *The Miner's Canary*, Virgin, London, 1992, 200–207; S.A. Kauffman, *The Origins of Order*, Oxford University Press, 238, 263–270; R. Leakey and R. Lewin, *The Sixth Extinction*, Weidenfeld and Nicolson, London, 1996, 223–254; R. Lewin, Critical mass – complexity theory may explain mass extinctions better than asteroids, *New Scientist*, 23 August 1997, 7; S. Jain and S. Krishna, Large extinctions in an evolutionary model – the role of innovation and keystone species, *Proceedings of the National Academy of Sciences, USA*, 99, 2002, 2055–2060.

Chapter 21

1. J. Smit and J. Hertogen, An extraterrestrial event at the Cretaceous–Tertiary boundary, *Nature*, 285, 1980, 198–200; S. M. Stanley, *Extinction*, Scientific American Books, New York, 1987, 122–154; D.J. McLaren and W.D. Goodfellow, Geological and biological consequences of giant impacts, *Annual Review of Earth and Planetary Science*, 18, 1990, 123–171; D.M. Raup, *Extinction – Bad Genes or Bad Luck*, Norton, New York and London, 1991, 64–70; D.B. Carlisle, *Dinosaurs, Diamonds, and Things from Outer Space*, Stanford University Press, Stanford, Calif., 1995, 1–64; R. Osborne and D. Tarling, *The Viking Historical Atlas of the Earth*, Viking, London, 1995, 106–109, 120–121.

2. D.M. Raup, *The Nemesis Affair*, Norton, New York and London, 1986, 61–64; R. Muller, *Nemesis*, Guild Publishing, London, 1989, 18–48; W. Alvarez, *T. rex and the Crater of Doom*, Princeton University Press, Princeton, New Jersey, 1997, 59–68.

3. D.M. Raup, *The Nemesis Affair*, 64–71; R. Muller, *Nemesis*, 49–72; W. Alvarez, *T. rex and the Crater of Doom*, 68–81.

4. L.W. Alvarez, W. Alvarez, F. Asaro and H.V. Michel, Extraterrstrial cause for the Cretaceous –Tertiary extinction, *Science*, 208, 1980, 1095–1108 (quotation from page 1095).

5. K.J. Hsü, Terrestrial catastrophe caused by cometary impact at the end of the Cretaceous, *Nature*, 285, 1980, 201–203.

6. K.J. Hsü, J.A. McKenzie and Q.X. He, Terminal Cretaceous environmental environmental and evolutionary changes, *Geological Society of America Special Paper*, 190, 1982, 317–328; M. Allaby and J. Lovelock, *The Great Extinction*, Secker and Warburg,

London, 1983, 25–26; K.J. Hsü, Geochemical markers of impacts and their effects on environments, in H.G. Holland and A.F. Trendall (eds.), *Patterns of Change in Earth Evolution*, Springer-Verlag, Berlin, 1984, 63–74.

7. D.M. Raup, *The Nemesis Affair*, 66–74, 102; R. Muller, *Nemesis*, 70–82; V. Clube and B. Napier, *The Cosmic Serpent*, Faber and Faber, London, 1982, 111–118; J.M. Wilford, *The Riddle of the Dinosaur*, Vintage Books, New York, 1987, 261–271; A. Hallam, *Great Geological Controversies*, Oxford University Press, 2nd edn, 1989, 186–193; C.C. Albritton, *Catastrophic Episodes in Earth History*, Chapman and Hall, London, 1989, 113–114.

8. S.J. Gould, *Hen's Teeth and Horse's Toes*, Norton, New York and London, 1983, 320–331 (quotations from page 322).

9. F.T. Kyte, Z. Zhou and J.T. Wasson, Siderophile-enriched sediments from the Cretaceous–Tertiary boundary, *Nature*, 288, 1980, 651–656; R. Ganapathy, A major meteorite impact on the Earth 65 million years ago, *Science*, 209, 1980, 921–923.

10. J. Smit and J. Hertogen, *Nature*, 285, 1980, 198–200.

11. F.T. Kyte, Z. Zhou and J.T. Wasson, *Nature*, 288, 1980, 651–656.

12. C.J. Orth, J.S. Gilmore, J.D. Knight, *et al.*, An iridium abundance anomaly at the palynological Cretaceous–Tertiary boundary in northern New Mexico, *Science*, 214, 1981, 1341–1343; J.S. Gilmore, J.D. Knight, C.J. Orth, *et al.*, Trace element patterns at a non-marine Cretaceous–Tertiary boundary, *Nature*, 307, 1984, 224–228; C.L. Pillmore, R.H. Tschudy, C.J. Orth, *et al.*, Geologic framework of non-marine Cretaceous–Tertiary boundary sites, Raton Basin, New Mexico and Colorado, *Science*, 223, 1984, 1180–1183.

13. J. Smit and G. Klaver, Sanidine spherules at the Cretaceous–Tertiary boundary indicate a large impact event, *Nature*, 292, 1981, 47–49.

14. S.M. Stanley, *Extinction*, 155–160; A. Hallam, *Great Geological Controversies*, 190–191; C.C. Albritton, *Catastrophic Episodes in Earth History*, 110–113.

15. R. Muller, *Nemesis*, 81–82; M. Rampino, A non-catastrophist explanation for the iridium anomaly at the Cretaceous–Tertiary boundary, *Geological Society of America Special Paper*, 190, 1982, 455–460.

16. D.M. Raup, *The Nemesis Affair*, 75–87; R. Muller, *Nemesis*, 72–81; W. Alvarez, *T. rex and the Crater of Doom*, 86–89; L.W. Alvarez, Experimental evidence that an asteroid impact led to the extinction of many species 65 million years ago, *Proceedings of the National Academy of Sciences, USA*, 80, 1983, 627–642.

17. D.M. Raup, *The Nemesis Affair*, 81; A. Hallam, *Great Geological Controversies*, 194; J.M. Luck and K.K. Turekian, Osmium-187/osmium-186 ratios in manganese nodules and the Cretaceous–Tertiary boundary, *Science*, 222, 1983, 613–615; W. Alvarez and F. Asaro, An extraterrestrial impact, *Scientific American*, 263(4), 1990, 44–52.

18. W. Alvarez, *T. rex and the Crater of Doom*, 92–95; A. Montanari, R.L. Hay, W. Alvarez, *et al.*, Spheroids at the Cretaceous–Tertiary boundary are altered impact droplets of basaltic composition, *Geology*, 11, 1983, 668–671; J. Smit and A.J.T. Romain, A sequence of events across the Cretaceous–Tertiary boundary, *Earth and Planetary Science Letters*, 74, 1985, 155–170; W. Alvarez, Towards a theory of impact crises, *Eos*, 67, 1986, 649–658.

19. B.F. Bohor, E.E. Foord, P.J. Modreski and D.M. Triplehorn, Mineralogical evidence for an impact event at the Cretaceous–Tertiary boundary, *Science*, 224, 1984, 867–869; B.F. Bohor, P.J. Modreski and E.E. Foord, Shocked quartz in the Cretacous–Tretiary boundary clays – evidence for a global distribution, *Science*, 236, 1987, 705–709; W. Alvarez and F. Asaro, *Scientific American*, 263(4), 1990, 44–52; J. Hecht, *Vanishing Life*, Charles Scribner's Sons, New York, 1993, 60–61.

20. D.M. Raup, *The Nemesis Affair*, 82–87; W.S. Wolbach, R.S. Lewis and E. Anders, Cretaceous exticntion – evidence for wildfires and search for meteoritic material, *Science*, 230, 1985, 167–170; W.S. Wolbach, I. Gilmour, E. Anders, *et al.*, Global fire at the Cretaceous–Tertiary boundary, *Nature*, 334, 1988, 665–669; I. Gilmour, W.S. Wolbach and E. Anders, Major wildfires at the Cretaceous–Tertiary boundary, in S.V.M. Clube (ed.), *Catastrophes and Evolution*, Cambridge University Press, 1989, 195–213.

21. S.M. Stanley, *Extinction*, 145–154; K.J. Hsü, Q. He, J.A. Mackenzie, *et al.*, Mass mortality and its environmental and evolutionary consequences, *Science*, 216, 1982, 249–256; S. Conway Morris, *The History in our Bones*, BBC Educational Developments, London, 1996, 10–11.

22. M. Allaby and J. Lovelock, *The Great Extinction*, 83–85; D.M. Raup, *The Nemesis Affair*, 94; L.W. Alvarez, *Proceedings of the National Academy of Sciences, USA*, 80, 1983, 627–642; S. Lamb and B. Sington, *Earth Story*, BBC Books, London, 1998, 55–57, 112.

23. W. Alvarez and F. Asaro, *Scientific American*, 263(4), 1990, 44–52; W. Alvarez, *T. rex and the Crater of Doom*, 90–98.

24. M. Allaby and J. Lovelock, *The Great Extinction*, 92–123; W. Alvarez and F. Asaro, *Scientific American*, 263(4), 1990, 44–52; D. Norman, *Dinosaur!*, Boxtree Books, London, 1991, 152–153.

25. G.S. Paul, *Predatory Dinosaurs of the World*, Simon and Schuster, New York, 1988; *Dinosaur Encyclopedia*, Bloomsbury Books, London, 1994; D.E. Fastovsky and D.B. Weishampel, *The Evolution and Extinction of the Dinosaurs*, Cambridge University Press, 1996; D. Norman, *The Illustrated Encyclopedia of Dinosaurs*, Greenwich Editions, London, 1998; *The Ultimate Book of Dinosaurs*, Parragon, Bath, 2001; I. Cranfield (ed.), *The Illustrated Directory of Dinosaurs and other Prehistoric Creatures*, Greenwich Editions, London, 2001, 16–19.

26. D.B. Carlisle, *Dinosaurs, Diamonds, and Things from Outer Space*, 41–46; D.A. Russell, The mass extinctions of the Late Mesozoic, *Scientific American*, 246(1), 1982, 48–55.

27. J.B. Pollock, O.B. Toon, T.P Ackerman, *et al.*, Environmental effects of an impact-generated dust-cloud, *Science*, 219, 1983, 287–289; P.M. Sheehan and T.A. Hansen, Detritis feeding as a buffer to extinction at the end of the Cretaceous, *Geology*, 14, 1986, 868–879; W. Glen (ed.), *The Mass-Extinction Debates*, Stanford University Press, Stanford, Calif., 1994, 18–25.

28. J.N. Wilford, *The Riddle of the Dinosaur*, 251–262; D. Norman, *Dinosaur!*, 144–151; W. Alvarez, *T. rex and the Crater of Doom*, 54–58; D.E. Fastovsky and D.B. Weishampel, *The Evolution and Extinction of the Dinosaurs*, 420–426.

29. A. Hallam, *Great Geological Controversies*, 184–215; C.C. Albritton, *Catastrophic Episodes in Earth History*, 158–162; R.T. Bakker, Tetrapod mass extinctions, in

A. Hallam (ed.), *Patterns of Evolution as Illustrated by the Fossil Record*, Elsevier, Amsterdam, 1977, 439–468; W. Alvarez and F. Asaro, *Scientific American*, 263(4), 1990, 44–52.

30. C.C. Albritton, *Catastrophic Episodes in Earth History*, 151–156; D.I. Axelrod and H.P. Bailey, Cretaceous dinosaur extinction, *Evolution*, 22, 1968, 595–611; D.I. Axelrod, Role of volcanism in climate and evolution, *Geological Society of America Special Paper*, 185, 1981, 1–59.

31. W. Glen (ed.), *The Mass Extinction Debates*, 29–33; P.R. Vogt, Evidence for local synchronism in mantle plume convection, and possible significance for geology, *Nature*, 240, 1972, 338–342; D.M. McLean, Deccan volcanism and the Cretaceous–Tertiary transition scenario – a unifying causal mechanism, *Museum of Canada, Syllogeus*, 39, 1982, 143–144; V.E. Courtillot, J. Besse, D. Vandamme, *et al.*, Deccan flood basalts at the Cretaceous–Tertiary boundary, *Earth and Planetary Science Letters*, 80, 1986, 361–374; V.E. Courtillot, A volcanic eruption, *Scientific American*, 263(4), 1990, 53–60; D.M. McLean, Impact winter in the global K/T extinctions – no definitive evidence, in J.S. Levine (ed.), *Global Biomass Burning*, MIT Press, Cambridge, Mass., 1991, 493–503; V. Courtillot, *Evolutionary Catastrophes*, Cambridge University Press, 1999, 45–72.

32. D.M. Raup, *The Nemesis Affair*, 25; W. Alvarez, *T. rex and the Crater of Doom*, 82–85; L.T. Silver and P.H. Schultz (eds.), *Geological Implications of Impacts of Large Asteroids and Comets*, *Geological Society of America Special Paper*, 190, 1982.

33. V. Clube and B. Napier, *The Cosmic Serpent*, 111–118; M. Allaby and J. Lovelock, *The Great Extinction*, 19–21; A. Milne, *The Fate of the Dinosaurs*, Prism Press, Bridport, Dorset, 1991, 166–169; C. Frankel, *The End of the Dinosaurs*, Cambridge University Press, 1999, 19–31.

34. D.M. Raup, *The Nemesis Affair*, 66–74; R. Muller, *Nemesis*, 69–85; D.M. McLaren, Impacts that changed the course of evolution, *New Scientist*, 24 November 1983, 588–592.

35. W.A. Clemens, J.D. Archibald and L.J. Hickey, Out with a whimper, not a bang, *Paleobiology*, 7, 1981, 293–298; J.D. Archibald, Late Cretaceous extinctions, *American Scientist*, 70, 1982, 377–385.

36. W. Alvarez, *T. rex and the Crater of Doom*, 86–89; P. Ward, *The End of Evolution*, Weidenfeld and Nicolson, London, 1995, 147–152; S.J. Gould, *Dinosaur in a Haystack*, Jonathan Cape, London, 1996, 147–158.

37. P.W. Signor and J.H. Lipps, Sampling bias, gradual extinction patterns and catastrophes in the fossil record, *Geological Society of America Special Paper*, 190, 1982, 291–296.

38. K. Perch-Nielsen, J.A. McKenzie and Q. He, Biostratigraphy and isotope stratigraphy and the "catastrophic" extinctions of calcareous nannoplankton at the Cretaceous–Tertiary boundary, *Geological Society of America Special Paper*, 190, 1982, 353–372; T. Birkelund and E. Hakansson, The terminal Cretaceous extinction in the Boreal shelf seas – a multicausal event, *Geological Society of America Special Paper*, 190, 1982, 373–384; T.J.M. Schopf, Extinction of the dinosaurs – a 1982 understanding, *Geological Society of America Special Paper*, 190, 1982, 415–422; W. Alvarez,

E.G. Kauffman, F. Surlyk, *et al.*, Impact theory of mass extinctions and the invertebrate fossil record, *Science*, 223, 1984, 1135–1141.

39. J.N. Wilford, *The Riddle of the Dinosaur*, 278–288; D.A. Russell, The gradual decline of the dinosaurs – fact or fallacy?, *Nature*, 307, 1984, 360–361.

40. B. Halstead, Dinosaurs survive one more disaster, *New Scientist*, 99, 1 September 1983, 633.

41. L.B. Halstead and J. Halstead, *Dinosaurs*, Blandford Press, Poole, Dorset, 1981, 160–165.

42. L.B. Halstead and J. Halstead, *Dinosaurs*, 162.

43. L.B. Halstead and J. Halstead, *Dinosaurs*, 161; D.A. Russell, The mass extinctions of the Late Mesozoic, *Scientific American*, 246(1), 1982, 48–55.

44. S.M. Stanley, *Extinction*, 137.

45. L.B. Halstead and J. Halstead, *Dinosaurs*, 162.

46. L.B. Halstead and J. Halstead, *Dinosaurs*, 162–165.

47. S.M. Stanley, *Extinction*, 168–171; D.M. McLaren and W.D. Goodfellow, *Annual Review of Earth and Planetary Sciences*, 18, 1990, 123–171; S.V. Margolis, J.F. Mount, E. Doehne, *et al.*, The Cretaceous-Tertiary boundary carbon and oxygen isotope stratigraphy, diagenesis, and paleoceanography, *Paleoceanography*, 2, 1987, 361–378; J.C. Zachos and M.A. Arthur, Geochemical evidence for suppression of pelagic marine productivity at the Cretaceous–Tertiary boundary, *Nature*, 337, 1989, 61–64.

48. D.M. Raup, *The Nemesis Affair*, 91–94; R. Muller, *Nemesis*, 83–85; A. Milne, *The Fate of the Dinosaurs*, 169–174; W. Alvarez, *T. rex and the Crater of Doom*, 98–100; C. Frankel, *The End of the Dinosaurs*, 37–47.

49. C.B. Officer and C.L. Drake, The Cretaceous–Tertiary transition, *Science*, 219, 1983, 1383–1390 (quotation from page 1390).

50. W. Alvarez, L.W. Alvarez, F. Asaro and H.V. Michel, The end of the Cretaceous – sharp boundary or gradual transition, *Science*, 223, 1984, 1183–1186; F. Surlyk and M.B. Johansen, End-Cretaceous brachiopod extinctions in the chalk of Denmark, *Science*, 223, 1984, 1174–1177; G.D. Leahy, M.D. Spoon and G.J. Retallack, Linking impacts and plant extinctions, *Nature*, 318, 1985, 318; J.R. Richardson, Brachiopods, *Scientific American*, 255(3), 1986, 100–106.

51. C.B. Officer and C.L. Drake, Terminal Cretaceous environmental events, *Science*, 227, 1985, 1161–1167; J. Smit, F.T. Kyte and B.M. French; C.B. Officer and C.L. Drake, Cretaceous–Tertiary extinctions – alternative models, *Science*, 230, 1985, 1292–1295.

52. L. Dingus, Effects of stratigraphic completeness on interpretation of extinction rates across the K–T boundary, *Paleobiology*, 10, 1984, 420–438.

53. W. Glen (ed.), *The Mass-Extinction Debates*, 29–33; R. Ganapathy, The Tunguska explosion of 1908 – discovery of meteoritic debris near the explosion site and at the south pole, *Science*, 220, 1983, 1158–1161; W.H. Zoller, J.R. Parrington and J.M. Phelan Kotra, Iridium enrichment in airborne particles from Kilauea Volcano, January 1983, *Science*, 222, 1983, 1118–1120.

54. J. Maddox, Extinction by catastrophe?, *Nature*, 308, 1984, 685.

55. A. Hallam, *Great Geological Controversies*, 211–212.

56. C.C. Albritton, *Catastrophic Episodes in Earth History*, 173–175; A. Hallam, *Great Geological Controversies*, 210–211; A. Hoffman and M.H. Nitecki, Reception of the asteroid hypothesis of terminal Cretaceous extinctions, *Geology*, 13, 1985, 884–887.

57. S.M. Stanley, *Extinction*, 163.

58. S.J. Gould, *Hen's Teeth and Horse's Toes*, 322–331 (quotation from page 329).

59. R. Muller, *Nemesis*, 112–113; C.C. Albritton, *Catastrophic Episodes in Earth History*, 115; A. Hallam, *Great Geological Controversies*, 204; M. Redfern, Giant comet spells slow death for the dinosaurs, *New Scientist*, 20/27 December 1984, 27; W. Glen (ed.), *The Mass-Extinction Debates*, 60–61, 70–71, 167; V. Clube and B. Napier, *The Cosmic Winter*, Basil Blackwell, Oxford, 1990, 222–228.

60. A. Hallam, *Great Geological Controversies*, 204–214; W. Glen (ed.), *The Mass-Extinction Debates*, 7–38.

61. D.J. McLaren, Abrupt extinctions, in D.K. Elliott (ed.), *Dynamics of Extinction*, Wiley, New York, 1986, 37–46 (quotation from page 43).

Chapter 22

1. J.N. Wilford, *The Riddle of the Dinosaur*, Vintage Books, New York, 1987, 263–297; S.M. Stanley, *Extinction*, Scientific American Books, New York, 1987, 163–171; W. Alvarez and F. Asaro, An extraterrestrial impact, *Scientific American*, 263(4), 1990, 42–52; V.E. Courtillot, A volcanic eruption, *Scientific American*, 263(4), 1990, 53–60.

2. S.M. Stanley, *Extinction*, x.

3. C.B. Officer and C.L. Drake, Terminal Cretaceous environmental events, *Science*, 227, 1985, 1161–1167.

4. W. Alvarez, Towards a theory of impact crises, *Eos*, 67, 1986, 649–658; R. Muller, *Nemesis*, Guild Publishing, London, 1988, 84–85.

5. S.M. Stanley, *Extinction*, 142–145.

6. A. Hoffman, Changing palaeontological views on mass extinction phenomena, in S.K. Donovan (ed.), *Mass Extinctions – Processes and Evidence*, Belhaven Press, London, 1989, 1–18; D.J. McLaren and W.D. Goodfellow, Geological and biological consequences of giant impacts, *Annual Review of Earth and Planetary Sciences*, 18, 1990, 123–171; M. Lindiger and G. Keller, Stable isotope stratigraphy across the Cretaceous/Tertiary boundary in Tunisia – evidence for a multiple extinction mechanism, *Geological Society of America Abstracts with Programs*, 19, 1987, 747.

7. R. Lewin, Extinctions and the history of life, *Science*, 221, 1983, 935–937; P. Ward, The extinction of the ammonites, *Scientific American*, 249(4), 1983, 114–124; P. Ward, J. Wiedmann and J.F. Mount, Maastrichtian molluscan biostratigraphy and extinction patterns in a K–T boundary section exposed at Zumaya, Spain, *Geology*, 14, 1986, 899–903.

8. A. Hallam, *Great Geological Controversies*, Oxford University Press, 2nd edn, 1989, 206–207; P.D. Ward, *On Methuselah's Trail*, Freeman, New York, 1992, 117–123; P. Ward, *The End of Evolution*, Weidenfeld and Nicolson, London, 1995, 148–154; S.J. Gould, *Dinosaur in a Haystack*, Jonathan Cape, London, 1996, 153–158.

9. C.B. Officer and C.L. Drake, The Cretaceous–Tertiary transition, *Science*, 219, 1983, 1383–1390.

10. W.A. Clemens, Evolution of the vertebrate fauna during the Cretaceous–Tertiary transitions, in G.K. Elliott (ed.), *Dynamics of Extinction*, Wiley, New York, 1986, 63–85; C.C. Albritton, *Catastrophic Episodes in Earth History*, Chapman and Hall, London, 1989, 139.

11. J.E. Fassett, Dinosaurs in the San Juan Basin, New Mexico, may have survived the event that resulted in creation of an iridium-enriched zone near the Cretaceous–Tertiary boundary, *Geological Society of America Special Paper*, 190, 1982, 435–447; J.E. Fassett, S.G. Lucas and F.M. O'Neill, Dinosaurs, pollen and spores, and the age of the Ojo Alamo Sandstone, San Juan Basin, New Mexico, *Geological Society of America Special Paper*, 209, 1987, 17–34.

12. J.N. Wilford, *The Riddle of the Dinosaur*, 278–297; S.M. Stanley, *Extinction*, 160–162; C.C. Albritton, *Catastrophic Episodes in Earth History*, 137–140, 164–165; J.D. Archibald, *Dinosaur Extinction and the End of an Era*, Columbia University Press, 1996, 5–11, 29–51.

13. L.M. Van Valen, Catastrophes, expectation and the evidence, *Paleobiology*, 10, 1984, 121–137; L.M. Van Valen, The case against impact extinctions, *Nature*, 311, 1984, 17–18; L.M. Van Valen, Iridium, impacts and para-volcanism, *Nature*, 316, 1985, 396.

14. L.M. Van Valen and R.E. Sloan, Ecology and the extinction of the dinosaurs, *Evolutionary Theory*, 2, 1977, 37–64; J. Smit and S. van der Kaars, Terminal Cretaceous extinctions in the Hell Creek area, Montana, compatible with catastrophic extinction, *Science*, 223, 1984, 1177–1179.

15. R.E. Sloan, J.K. Rigby, L.M. Van Valen and D. Gabriel, Gradual dinosaur extinction and simultaneous ungulate radiation in the Hell Creek formation, *Science*, 232, 1986, 629–633.

16. G.J. Retallack, G.D. Leahy, P.M. Sheehan, *et al.*, Cretaceous–Tertiary dinosaur extinctions, *Science*, 234, 1986, 1170–1172; G.J. Retallack, G.D. Leahy and M.D. Spoon, Evidence from paleosols for ecosystem changes across the Cretacous–Tertiary boundary in eastern Montana, *Geology*, 15, 1987, 1090–1093; S. Argast, J.O. Farlow, R.M. Gabet and D.L. Brinkman, Transport-induced abrasion of fossil reptilian teeth – implications for existence of Tertiary dinosaurs in the Hell Creek formation, Montana, *Geology*, 15, 1987, 927–930; J.K. Rigby, K.R. Newman, J. Smit, *et al.*, Dinosaurs from the Paleocene part of the Hell Creek formation, McCone County, Montana, *Palaios*, 2, 1987, 296–302; J.G. Easton, J.I. Kirkland and K. Doi, Evidence of re-worked Cretaceous fossils and their bearing on the existence of Tertiary dinosaurs, *Palaios*, 4, 1989, 281–286; D.L. Lofgren, C.I. Hotton and A.C. Runkel, Re-working of Cretaceous dinosaurs into Paleocene channel deposits, upper Hell Creek formation, Montana, *Geology*, 18, 1990, 874–877.

17. P. Ward, *The End of Evolution*, 156–158; R. Kerr, Dinosaurs and friends snuffled out, *Science*, 251, 1991, 160–162; W. Glen (ed.), *The Mass-Extinction Debates*, Stanford University Press, Stanford, Calif., 1994, 22–25.

18. R. Bakker, *The Dinosaur Heresies*, Penguin Books, Harmondsworth, 1986, 15–392.

19. D. Norman, *Dinosaur!*, Boxtree Books, London, 1991, 25, 79–80, 90–93, 177; T. Gardom with A. Milner, *The Natural History Museum Book of Dinosaurs*, Virgin Books, London, 1993, 13, 75–77; D.E. Fastovsky and D.B. Weishampel, *The Evolution and*

Extinction of the Dinosaurs, Cambridge University Press, 1996, 325–358; R. Fortey, *Life – An Unauthorised Biography*, Flamingo, London, 1998, 250–261; T. Haines, *Walking with Dinosaurs*, BBC Books, London, 1999, 12–13, 104, 188–189; *The Ultimate Book of Dinosaurs*, Parragon, Bath, 2000, 62–63; J.R. Hutchinson and M. Garcia, *Tyrannosaurus* was not a fast mover, *Nature*, 415, 2002, 1018–1021.

20. R. Bakker, *The Dinosaur Heresies*, 406–444; T. Haines, *Walking with Dinosaurs*, 246–249; R. Osborne and M. Benton, *The Viking Atlas of Evolution*, Viking, London, 1996, 24–25, 72–73, 90–91; R. Redfern, *Origins*, Cassell, London, 2000, 192–197.

21. P. Dodson, Counting dinosaurs – how many kinds were there?, *Proceedings of the National Academy of Sciences, USA*, 87, 1990, 7608–7612.

22. W. Glen (ed.), *The Mass-Extinction Debates*, 232–234; P. Dodson, Maastrichtian dinosaurs, *Geological Society of America Abstracts*, San Diego Meeting, 1991, 184.

23. S.J. Gould, *Dinosaur in a Haystack*, 156–158; P.M. Sheehan, D.E. Fastovsky, R.G. Hoffman, *et al.*, Sudden extinction of the dinosaurs – latest Cretaceous, Upper Great Plains, USA, *Science*, 254, 1991, 835–839; P.M. Sheehan and D.E. Fastovsky, Major extinction of land-dwelling vertebrates at the Cretaceous–Tertiary boundary, eastern Montana, *Geology*, 20, 1992, 556–560; P.M. Sheehan and D.A. Russell, Faunal change following the Cretaceous–Tertiary impact – using paleontological data to assess the hazards of impacts, in T. Gehrels (ed.), *Hazards due to Comets and Asteroids*, University of Arizona Press, Tucson, 1994, 879–896; D.E. Fastovsky and D.B. Weishampel, *The Evolution and Extinction of the Dinosaurs*, 417–419.

24. W.A. Clemens, J.D. Archibald, P.M. Sheehan, *et al.*, Dinosaur diversity and extinction, *Science*, 256, 1992, 159–161; S.H. Hurlbert and D. Archibald, No statistical support for sudden (or gradual) extinction of the dinosaurs, *Geology*, 23, 1995, 881–884; P.M. Sheehan, D.E. Fastovsky, R.G. Hoffman, *et al.*, No statistical support for sudden (or gradual) extinction of the dinosaurs – comment and reply, *Geology*, 24, 1996, 957–959.

25. S.M. Stanley, *Extinction*, 155–160; W. Glen (ed.), *The Mass-Extinction Debates*, 22–25; G.D. Leahy, M.D. Spoon and G.J. Retallack, Linking impacts and plant extinctions, *Nature*, 318, 1985, 318; R.H. Tschudy and B.D. Tschudy, Extinction and survival of plant life following the Cretaceous–Tertiary boundary event, Western Interior, North America, *Geology*, 14, 1986, 667–670;

26. G.R. Upchurch, Terrestrial environmental changes and extinction patterns at the Cretaceous–Tertiary boundary, North America, in K. Donovan (ed.), *Mass Extinctions – Processes and Evidence*, 195–216; J.A. Wolfe and G.R. Upchurch, vegetation, climatic and floral changes at the Cretaceous–Tertiary boundary, *Nature*, 324, 1986, 148–151; J.A. Wolfe, Late Cretaceous–Cenozoic history of deciduousness and the terminal Cretaceous event, *Paleobiology*, 13, 1987, 215–226; R.A. Spicer, Plants at the Cretaceous–Tertiary boundary, in W.G. Chaloner and A. Hallam (eds.), *Evolution and Extinction*, Cambridge University Press, 1989, 51–65.

27. J.A. Wolfe, Palaeobotanical evidence for a June "impact winter" at the Cretaceous/Tertiary boundary, *Nature*, 352, 1991, 421–423.

28. C.C. Albritton, *Catastrophic Episodes in Earth History*, 117–119; W. Glen (ed.), *The Mass-Extinction Debates*, 18–22; P.J. Crutzen, Acid rain at the K/T boundary, *Nature*, 330,

1987, 108–109; R.G. Prinn and B. Fegley, Bolide impacts, acid rain, and biospheric traumas at the Cretaceous–Tertiary boundary, *Earth and Planetary Science Letters*, 83, 1987, 1–15.

29. N.L. Carter, C.B. Officer, C.A. Chesner and W.I. Rose, Dynamic deformation of volcanic ejecta from the Toba caldera – possible relevance to the Cretaceous/Tertiary boundary phenomena, *Geology*, 14, 1986, 380–383.

30. S.M. Stanley, *Extinction*, 164–165; W. Glen (ed.), *The Mass-Extinction Debates*, 29–33; F. Langenhorst, A. Deutsch, D. Stöffler and U. Hornemann, Effect of temperature on shock metamorphism of single-crystal quartz, *Nature*, 356, 1992, 507–509; W. Alvarez, *T. rex and the Crater of Doom*, Princeton University Press, Princeton, New Jersey, 1997, 96–98.

31. R.A. Kerr, Asteroid impact gets more support, *Science*, 236, 1987, 666–668.

32. C.B. Officer, A. Hallam, C.L. Drake and J.D. Devine, Late Cretaceous and paroxysmal K–T extinctions, *Nature*, 326, 1987, 143–149.

33. A. Hallam, End-Cretaceous mass extinction events – argument for terrestrial causation, *Science*, 238, 1987, 1237–1242.

34. V.E. Courtillot, *Scientific American*, 263(4), 1990, 53–60 (quotation from page 60).

35. A. Hallam, *Great Geological Controversies*, 197–198; W. Alvarez, *T. rex and the Crater of Doom*, 98–100; V. Courtillot, G. Féraud, H. Maluski, *et al.*, The Deccan flood basalts and the Cretaceous–Tertiary boundary, *Nature*, 333, 1988, 843–846; M. Rampino, Dinosaurs, comets and volcanoes, *New Scientist*, 18 February 1989, 54–58; C. Frankel, *The End of the Dinosaurs*, Cambridge University Press, 1999, 40–45, 54–57, 115–117; V. Courtillot, *Evolutionary Catastrophes*, Cambridge University Press, 1999, 45–72.

36. D.B. Carlisle, *Dinosaurs, Diamonds, and Things from Outer Space*, Stanford University Press, Stanford, Calif., 1995, 46, 102–104.

37. V. Courtillot, G. Féraud, H. Maluski, *et al.*, *Nature*, 333, 1988, 843–846; V.E. Courtillot, *Scientific American*, 263(4), 1990, 53–60; R.S. White and D.P. McKenzie, Volcanism at rifts, *Scientific American*, 261(1), 1989, 44–55; N. Henbest, Geologist hit back at impact theory of extinctions, *New Scientist*, 29 April 1989, 34–35; V. Courtillot, *Evolutionary Catastrophes*, 73–87.

38. W. Glen (ed.), *The Mass-Extinction Debates*, 30–31; V.E. Courtillot, *Scientific American*, 263(4), 1990, 53–60; C. Officer and J. Page, *Tales of the Earth*, Oxford University Press, 1993, 146.

39. A. Hallam, *Great Geological Controversies*, 196–198; V.E. Courtillot, *Scientific American*, 263(4), 1990, 53–60; R.B. Stothers, J.A. Wolfe, S. Self and M.R. Rampino, Basaltic fissure eruptions, plume heights and atmospheric aerosols, *Geophysics Research Letters*, 13, 1986, 725–728; M.R. Rampino, S. Self and R.B. Stothers, Volcanic winters, *Annual Review of Earth and Planetary Sciences*, 16, 1988, 73–99; A. Grayson, *Equinox: The Earth*, Channel 4 Books, London, 2000, 49–80.

40. A. Hoffman, in *Mass Extinctions – Processes and Evidence*, 1–18; W. Alvarez and F. Asaro, *Scientific American*, 263(4), 1990, 44–52; D.J. McLaren and W.D. Goodfellow, *Annual Review of Earth and Planetary Sciences*, 18, 1990, 123–171; W. Alvarez, *T. rex and the Crater of Doom*, 107–109; C. Frankel, *The End of the Dinosaurs*, 79–82.

41. J. Bourgeois, T.A. Hansen, P.L. Wilberg and E.G. Kauffman, A tsunami deposit at the Cretaceous–Tertiary boundary in Texas, *Science*, 241, 1988, 567–570.

42. M.R. Owen and M.H. Anders, Evidence from cathodoluminescence for non-volcanic origin of shocked quartz at the Cretaceous/Tertiary boundary, *Nature*, 334, 1988, 145–147.

43. G.I. Bekov, V.S. Letokhov, V.N. Radaev, *et al.*, Rhodium distribution at the Cretaceous–Tertiary boundary analysed by ultrasensitive laser photoionization, *Nature*, 332, 1988, 146–148.

44. S.M. Stanley, *Extinction*, 165–167; W. Alvarez, *T. rex and the Crater of Doom*, 89–98; C. Frankel, *The End of the Dinosaurs*, 61–78; R.A. Kerr, Searching land and sea for the dinosaur killer, *Science*, 237, 1987, 856–857.

45. W. Alvarez, *T. rex and the Crater of Doom*, 104–105; B.M. French, Impact event at the Cretaceous–Tertiary boundary – a possible site, *Science*, 226, 1984, 353; C. Joyce, Fragmentary theory of dinosaur extinctions, *New Scientist*, 11 February 1988, 27; R.A.F. Grieve, Manson structure implicated, *Nature*, 340, 1989, 438–429.

46. C. Officer and J. Page, *Tales of the Earth*, 147.

47. A. Hallam, *Great Geological Controversies*, Oxford University Press, 1983, 107.

48. A. Hallam, *Great Geological Controversies*, Oxford University Press, 2nd edn, 1989, 213–214.

49. W. Alvarez, *T. rex and the Crater of Doom*, 97–98; R.A. Kerr, Snowbird II – clues to the Earth's impact history, *Science*, 242, 1988, 1380–1382; J. Hecht, Evolving theories for old extinctions, *New Scientist*, 12 November 1988, 28–30; V.L. Sharpton and P.D. Ward (eds.), *Global Catastrophes in Earth History*, Geological Society of America Special Paper, 247, 1990.

50. R.A. Kerr, Huge impact is favored K–T boundary killer, *Science*, 242, 1988, 865–867.

51. W. Glen (ed.), *The Mass-Extinction Debates*, 9–12; J.D. O'Keefe and T. Ahrens, Impact production of CO_2 by the Cretaceous/Tertiary extinction bolide and the resultant heating of the Earth, *Nature*, 338, 1989, 247–249; H.J. Melosh, N.M. Schneider, K.J. Zahnle and D. Latham, Ignition of global wildfires at the Cretaceous/Tertiary boundary, *Nature*, 343, 1990, 251–254; J. Beard, Did nickel poisening finish off the dinosaurs? *New Scientist* 19 May 1990, 31; W. Alvarez, F. Asaro and A. Montanari, Iridium profile for 10 million years across the Cretaceous–Tertiary boundary at Gubbio (Italy), *Science*, 250, 1990, 1700–1702; W.B. McKinnon, Killer acid at the K/T boundary, *Nature*, 357, 1992, 15–16; I. Gilmour, Geochemistry of carbon in terrestrial impact processes, in M.M. Grady, R. Hutchison, G.J.H. McCall and D.A. Rothery, *Meteorites – Flux with Time and Impact Effects*, Geological Society, London, Special Publication 140, 1998, 205–216.

52. D.B. Carlisle, *Dinosaurs, Diamonds, and Things from Outer Space*, 98; J.F. McHone, R.A. Nieman, C.F. Lewis and A.M. Yates, Stishovite at the Cretaceous–Tertiary boundary, Raton, New Mexico, *Science*, 243, 1989, 1182–1184.

53. D.B. Carlisle, *Dinosaurs, Diamonds, and Things from Outer Space*, 135–145; M. Zhao and J.L. Bada, Extraterrestrial amino acids in Cretaceous/Tertiary boundary sediments

at Stevns Klint, Denmark, *Nature*, 339, 1989, 463–465; J.R. Cronin, Amino acids and bolide impacts, *Nature*, 339, 1989, 423–424; K. Zahnle and D. Grinspoon, Comet dust as a source of amino acids at the Cretaceous/Tertiary boundary, *Nature*, 348, 1990, 157–160; C.F. Chyba, Meteorites – extraterrestrial amino acids and terrestrial life, *Nature*, 348, 1990, 113–114.

54. D.B. Carlisle, *Dinosaurs, Diamonds, and Things from Outer Space*, 131–135; D.B. Carlisle and D.R. Braman, Nanometre-size diamonds in the Cretaceous/Tertiary boundary clay of Alberta, *Nature*, 352, 1991, 708–709; D.B. Carlisle, Diamonds at the K/T boundary, *Nature*, 357, 1992, 119–120.

55. D.B. Carlisle, *Dinosaurs, Diamonds, and Things from Outer Space*, 127–156.

56. D.B. Carlisle, *Dinosaurs, Diamonds, and Things from Outer Space*, 157–200.

57. A.R. Hildebrand and W.V. Boynton, Proximal Cretaceous–Tertiary boundary impact deposits in the Caribbean, *Science*, 248, 1990, 843–846; B.F. Bohor and R. Seitz, Cuban K/T catastrophe, *Nature*, 344, 1990, 593.

58. A.R. Hildbrand and W.V. Boynton, *Science*, 248, 1990, 843–846; J. Hecht, Did a Caribbean impact wipe out the dinosaurs?, *New Scientist*, 2 June 1990, 30; J. Smit, Where did it happen?, *Nature*, 349, 1991, 461–462; F.J-M.R. Maurrasse and G. Sen, Impacts, tsunamis, and the Haitian Cretaceous–Tertiary boundary layer, *Science*, 252, 1991, 1690–1693; H. Sigurdsson, S. D'Hondt, M.A. Arthur, *et al.*, Glass from the Cretaceous/Tertiary boundary in Haiti, *Nature*, 349, 1991, 482–487; H. Sigurdsson, S. D'Hondt and S. Carey, The impact of the Cretaceous/Tertiary bolide on evaporite terrane and generation of major sulfuric acid aerosol, *Earth and Planetary Science Letters*, 109, 1992, 543–559; J.D. Blum and C.P. Chamberlain, Oxygen isotope constraints on the origin of impact glasses from the Cretaceous–Tertiary boundary, *Science*, 257, 1992, 1104–1107.

59. W. Alvarez, *T. rex and the Crater of Doom*, 106–129; C. Frankel, *The End of the Dinosaurs*, 79–110.

60. J. Smit, A. Montanari, N.H.M. Swinburne, *et al.*, Tektite-bearing, deep water clastic unit at the Cretaceous–Tertiary boundary in northeastern Mexico, *Geology*, 20, 1992, 99–103; D.A. Kring and W.V. Boynton, Petrogenesis of an augite-bearing melt rock in the Chicxulub structure and its relationship to K/T impact spherules in Haiti, *Nature*, 358, 1992, 141–144; R.A. Kerr, Extinction by a one-two comet punch, *Science*, 255, 1992, 160–161; A.R. Hildebrand, G.T. Penfield, D.A. Kring, *et al.*, Chicxulub crater – a possible Cretaceous/Tertiary boundary impact crater on the Yucatán Peninsula, Mexico, *Geology*, 19, 1992, 867–871.

61. C. Officer and J. Page, *Tales of the Earth*, 146.

62. J.B. Lyons and C.B. Officer, Mineralogy and petrology of the Haiti Cretaceous/Tertiary section, *Earth and Planetary Science Letters*, 109, 1992, 205–224; M.A. Iturralde-Vinent, A short note on the Cuban late Maastrichtian megaturbidite (an impact-derived deposit?), *Earth and Planetary Science Letters*, 109, 1992, 225–228; C. Jehanno, D. Boclet, L. Forget, *et al.*, The Cretaceous–Tertiary boundary at Beloc, Haiti – no evidence for an impact in the Caribbean Area, *Earth and Planetary Science Letters*, 109, 1992, 229–241; H. Montgomery, E. Pessagno, K. Soegaard, *et al.*, Misconceptions concerning the Cretaceous/Tertiary boundary at the Brazos River, Falls County,

Texas, *Earth and Planetary Science Letters*, 109, 1992, 593–600; C. Officer, Victims of volcanoes, *New Scientist*, 20 February 1993, 34–38; A. Meyerhoff, J.B. Lyons and C.B. Officer, Chicxulub structure – a volcanic sequence of Late Cretaceous age, *Geology* 22, 1994, 3–4.

63. C. Officer and J. Page, *The Great Dinosaur Extinction Controversy*, Addison-Wesley, Reading, Mass., 1996, xi.

64. C. Officer and J. Page, *The Great Dinosaur Extinction Controversy*, 178.

65. C. Officer and J. Page, *The Great Dinosaur Extinction Controversy*, 182–187.

66. J.D. Archibald, *Dinosaur Extinction and the End of an Era*, 205–206.

67. V. Courtillot, *Evolutionary Catastrophes*, 105–106, 140–143 (quotations from pages 140 and 143).

68. R. Rocchia, D. Boclet, P. Bonté, *et al.*, The Cretaceous–Tertiary boundary at Gubbio revisited, *Earth and Planetary Science Letters*, 99, 1990, 206–219; C. Koeberl, Chicxulub crater, Yucatan – tektites, impact glasses, and the geochemistry of target rocks and breccias, *Geology*, 21, 1993, 211–214; K.R. Johnson, Extinctions at the antipodes, *Nature*, 366, 1993, 511–512; W. Stinnesbeck, J.M. Barbarin, G. Keller, *et al.*, Deposition of channel deposits near the Cretaceous–Tertiary boundary in northeastern Mexico – catastrophic or normal sedimentary deposits?, *Geology*, 21, 1993, 797–800; D.H. Elliott, P.R.A. Askin, F.T. Kyte and W.J. Zinsmeister, Iridium and dinocysts at the Cretaceous–Tertiary boundary on Seymour Island, Antarctica, *Geology*, 22, 1994, 675–678; N. MacLeod and G. Keller, Comparative biogeographic analysis of planktonic foraminiferal survivorship across the Cretaceous–Tertiary (K/T) boundary, *Paleobiology*, 20, 1994, 143–171; C.R. Marshall, Distinguishing between sudden and gradual extinctions in the fossil record, *Geology*, 23, 1995, 731–734; W.C. Ward, G. Keller, W. Stinnesbeck and T. Adatte, Yucatán subsurface stratigraphy, *Geology*, 23, 1995, 873–876; F. Longoria and M.A. Gamper, Planktonic formaminiferal faunas across the Cretaceous–Tertiary succession of Mexico, *Geology*, 23, 329–332, 1995; E. Barrera, Global environmental changes preceding the Cretaceous–Tertiary boundary, *Geology*, 22, 1994, 877–880; J.D. Archibald, *Dinosaur Extinction and the End of an Era*, 116–127; D.E. Fastovsky and D.B. Weishampel, *The Evolution and Extinction of the Dinosaurs*, 420–427; B. Galbrun, Did the European dinosaurs disappear before the K–T event?, *Earth and Planetary Science Letters*, 148, 1997, 569–579; G. Keller, T. Adatte, C. Hollis, *et al.*, The Cretaceous/Tertiary boundary event in Ecuador, *Marine Micropaleontology*, 31, 1997, 97-133; What really killed the dinosaurs?, *New Scientist*, 16 August 1997, 23–27; N. MacLeod, Impacts and marine invertebrate extinctions, in *Meteorites – Flux with Time and Impact Effects*, 217–246; A.C. Milner, Timing and causes of vertebrate extinction at the K–T boundary, in *Meteorites – Flux with Time and Impact Effects*, 247–257; G.C. Cadee, Mass mortalities and mass extinctions, *Lethaia*, 32, 1999, 318–320; S.A. Jafar, A magnitude induced model for mass extinctions in the geologic record, *Neues Jahrbuch fur Geologie und Palaontologie-Abhandlungen*, 217, 2000, 161–197; C.H. Jeffery, Heart urchins at the Cretaceous/Tertiary boundary – a tale of two clades, *Paleobiology*, 27, 2001, 140–158; C.J. Tsujita, Multiple causes of catastrophic events, *Canadian Journal of Earth Sciences*, 38, 2001, 271–292; J. Hecht, The end of the world, *New Scientist*, 2 February

2002, 13; K.O. Pope, Impact dust not the cause of the Cretaceous–Tertiary mass extinctions, *Geology*, 30, 2002, 99–102; R.A. Kerr, No 'darkness at noon' to do in the dinosaurs, *Science*, 295, 2002, 1445–1446; K. Ravilious, Killer blow, *New Scientist*, 4 May 2002, 29–31; E. Dobb, What wiped out the dinosaurs?, *Discover*, 23(6), 2002, 35–43.

69. J. Smit, A. Montanari, N.H.M. Swinburne, *et al.*, Tektite-bearing deep-water clastic unit at the Cretaceous–Tertiary boundary in northeastern Mexico, *Geology*, 20, 1992, 99–103; W. Alvarez, J. Smit, W. Lowrie, *et al.*, Proximal impact deposits at the Cretaceous–Tertiary boundary in the Gulf of Mexico, *Geology*, 20, 1992, 697–700; N. Swinburne, It came from outer space, *New Scientist*, 20 February 1993, 28–32; R. Cowen, The day the dinosaurs died, *Astronomy*, 24(4), 1996, 34–41; L.C. Ivany and R.J. Salawitch, Carbon isotopic evidence for biomass burning at the K–T boundary, *Geology*, 21, 1993, 487–490; T.E. Krogh, S.L. Kamo, V.L. Sharpton, *et al.*, U–Pb ages of single shocked zircons linking distal K/T ejecta to the Chicxulub crater, *Nature*, 366, 1993, 731–734; J. Smit, Extinctions at the Cretaceous–Tertiary boundary – the link to the Chicxulub impact, in *Hazards due to Comets and Asteroids*, 859–878; B.C. Shuraytz, V.L. Sharpton and L.E. Marín, Petrology of impact-melt rocks at the Chicxulub multi-ring basin, Yucatan, Mexico, *Geology*, 22, 1994, 868–872; R. Coccioni and S. Galeotti, K–T boundary extinction – geologically instantaneous or gradual event? *Geology*, 22, 1994, 779–782; D. Kring, A.R. Hildebrand and W.V. Boynton, Provenance of mineral phases in the Cretaceous–Tertiary boundary sediments exposed on the southern peninsula of Haiti, *Earth and Planetary Science Letters*, 128, 1994, 629–641; D. Heymann, L.P. Felipe Chibante, R.R. Brooks, *et al.*, Fullerenes in the Cretaceous–Tertiary boundary layer, *Science*, 265, 1994, 645–647; J.J. Pospichal, Calcareous nannofossils at the K–T boundary, El Kef – no evidence for stepwise, gradual or sequential extinctions, *Geology*, 22, 1994, 99–102; A.R. Hildebrand, M. Pilkington, M. Connors, *et al.*, Size and structure of the Chicxulub crater revealed by horizontal gravity gradients and cenotes, *Nature*, 376, 1995, 415–417; W. Alvarez, P. Claeys and S.W. Kieffer, Emplacement of Cretaceous–Tertiary boundary shocked quartz from Chicxulub crater, *Science*, 269, 1995, 930–935; F.T. Kyte and J.A. Bostwick, Magnesioferrite spinel in Cretaceous–Tertiary boundary sediments of the Pacific basin – remnants of hot, early ejecta from the Chicxulub impact?, *Earth and Planetary Science Letters*, 132, 1995, 113–127; T. Meisel, U. Krähenbühl and M.A. Nazarov, Combined osmium and strontium isotopic study of the Cretaceous–Tertiary boundary at Sumbar, Turkmenistan, *Geology*, 23, 1995, 313–316; S.L. Kamo and T.E. Krogh, Chicxulub crater source for shocked zircon crystals from the Cretaceous–Tertiary boundary layer, Saskatchewan, *Geology*, 23, 1995, 281–284; B.C. Schuraytz, D.J. Lindstrom, L.E. Marín, *et al.*, Iridium metal in Chicxulub impact melt, *Science*, 271, 1996, 1573–1576; J.J. Pospichal, Calcareous nannofossils and clastic sediments at the Cretaceous–Tertiary boundary, northeastern Mexico, *Geology*, 24, 1996, 255–258, K.O. Pope, A.C. Ocampo, G.C. Kinsland and R. Smith, Surface expression of the Chicxulub crater, *Geology*, 24, 1996, 527–530; S. D'Hondt, J. King and C. Gibson, Oscillatory marine response to the Cretaceous–Tertiary impact, *Geology*, 24, 1996, 611–614; P.H. Schultz and S. D'Hondt,

Cretaceous–Tertiary (Chicxulub) impact angle and its consequences, *Geology*, 24, 1996, 963–967; H.B. Vanhof and J. Smit, High-resolution late Maastrichtian-early Danian oceanic $^{87}Sr/^{86}Sr$ record, *Geology*, 25, 1997, 347–350; V.L. Sharpton and L.E. Marín, The Cretaceous–Tertiary impact crater and the cosmic projectile that produced it, *Annals of the New York Academy of Sciences*, 822, 1997, 353–380; J. Morgan, M. Warner, *et al.*, Size and morphology of the Chicxulub impact crater, *Nature*, 390, 1997, 472–476; A. Shukolyukov and G.W. Lugmair, Isotopic evidence for the Cretaceous–Tertiary impactor and its type, *Science*, 282, 1998, 927–929; A.R. Hildebrand, M. Pilkington, C. Ortiz-Aleman, *et al.*, Mapping Chicxulub crater structure with gravity and seismic reflection data, in *Meteorites – Flux with Time and Impact Effects*, 155–176; I. Gilmour, Geochemistry of carbon in terrestrial impact processes, in *Meteorites – Flux with Time and Impact Effects*, 205–216; Big splash, *New Scientist*, 6 November 1999, 27; R.D. Norris, B.T. Huber and J. Self Trail, Synchroneity of the K–T oceanic mass extinction and meteorite impact, *Geology*, 27, 1999, 419–422; K. Kaiho and M.A. Lamolda, Catastrophic extinction of planktonic foraminifera at the Cretaceous–Tertiary boundary evidenced by stable isotopes and foraminiferal abundance at Caravaca, Spain, *Geology*, 27, 1999, 355–358; J. Morgan and M. Warner, Chicxulub – the third dimension of a multi-ring impact basin, *Geology*, 27, 1999, 407–410; J.A. Arz, I. Arenillas, E. Molina, *et al.*, Planktonic foraminiferal stability in the Upper Maastrichtian and the catastrophic mass extinction at the Cretaceous–Tertiary boundary at Caravaca (Spain), *Revista Geologica de Chile*, 27, 2000, 27–47; D. Zaghbib-Turki, N. Karoui-Yaakoub, R. Rocchia, *et al.*, Characteristic events record of the K/T boundary in the Elles section (Tunisia), *Comptes Rendus de l'Academie des Sciences, Serie II*, 331, 2000, 141–149; I. Arenillas, J.A. Arz, E. Molina, *et al.*, The Cretaceous/Paleogene boundary at Ain Settara, Tunisia – sudden catastrophic mass extinction in planktic foraminifera, *Journal of Foraminiferal Research*, 30, 2000, 202–218; S. Mukhopadhyay, K.A. Farley and A. Montanari, A short duration of the Cretaceous–Tertiary boundary event – evidence from extraterrestrial helium-3, *Science*, 291, 2001, 1952–1955; S.V. Jeffers, J.P. Manley, M.E. Bailey and D.J. Asher, Near-Earth object velocity distributions and consequences for the Chicxulub impactor, *Monthly Notices of the Royal Astronomical Society*, 327(1), 2001, 126–132; V. Vajda, J.I. Raine and C.J. Hollis, Indication of global deforestation at the Cretaceous–Tertiary boundary by New Zealand fern spike, *Science*, 292, 2001, 1700–1702; J. Melosh, Deep down at Chicxulub, *Nature*, 414, 2001, 861–862; B.O. Dressler and W.U. Reimold, Terrestrial impact melt rocks and glasses, *Earth Science Reviews*, 56, 2001, 205–284; K. Ravilious, Killer blow, *New Scientist*, 4 May 2002, 29–31; C.C. Labandeira, K.R. Johnson and P. Wilf, Impact of the terminal Creatceous event on plant–insect associations, *Proceedings of the National Academy of Sciences, USA*, 99, 2002, 2061–2066; J. Randerson, Dead as a dino, *New Scientist*, 15 June 2002, 13; D.J. Beerling, B.H. Lomax, D.L. Royer, *et al.*, An atmospheric pCO_2 reconstruction across the Cretaceous–Tertiary boundary from leaf megafossils, *Proceedings of the National Academy of Sciences, USA*, 99, 2002, 7836–7840.

70. W. Alvarez, *T. rex and the Crater of Doom*, 126.

71. W. Alvarez, *T. rex and the Crater of Doom*, 114; V. Courtillot, *Evolutionary Catastrophes*, 129–132; C. Frankel, *The End of the Dinosaurs*, 99; G. Ryder, D. Fastovsky and S. Gartner (eds.), *The Cretaceous–Tertiary Event and Other Catastrophes in Earth History*, Geological Society of America Special Paper 307, 1996.

72. W. Alvarez, *T. rex and the Crater of Doom*, 94, 132; C. Frankel, *The End of the Dinosaurs*, 111–135; J. Gribbin and M. Gribbin, *Fire on Earth*, Simon and Schuster, London, 1996, 33–37; K.O. Pope, K.H. Baines, A.C. Ocampo and B.A. Ivanov, Energy, volatile production, and climatic effect of the Cretaceous/Tertiary boundary impact, *Journal of Geophysics Research – Planets*, 102, 1997, 21,645–21,664; Death by sulphur, *New Scientist*, 6 September 1997, 23.

73. M. Day, Hell on Earth, *New Scientist*, 20 November 1999, 5.

74. D.B. Carlisle, *Dinosaurs, Diamonds, and Things from Outer Space*, 102–104; J. Gribbin and M. Gribbin, *Fire on Earth*, 21, 193–194; M.R. Rampino, Impact cratering and flood basalt volcanism, *Nature*, 327, 1987, 468; D. Steel, *Rogue Asteroids and Doomsday Comets*, Wiley, New York, 1995, 105.

75. R.A. Kerr, Extinction by a one-two comet punch, *Science*, 255, 1992, 160–161; J. Hecht, *Vanishing Life*, Charles Scribner's Sons, New York, 1993, 73–82.

76. H. Leroux, R. Rocchia, L. Forget, *et al.*, The K/T boundary at Beloc (Haiti) – compared stratigraphic distributions of the boundary markers, *Earth and Planetary Science Letters*, 131, 1995, 255–268; G. Keller, J.G. Lopez-Oliva, W. Stinnesbeck and T. Adatte, Age, stratigraphy, and deposition of near-K/T siliciclastic deposits in Mexico – relation to bolide impact?, *Geological Society of America Bulletin*, 109, 1997, 410–428; E.A.M. Koutsoukos, An extraterrestrial impact in the early Danian – a secondary K/T boundary event?, *Terra Nova*, 10(2), 1998, 68–73.

77. M.J. Kunk, G.A. Izett, R.A. Haugerud and J.F. Sutter, ^{40}Ar-^{39}Ar dating of the Manson impact structure – a Cretaceous–Tertiary boundary crater candidate, *Science*, 244, 1989, 1565–1568; G.A. Izett, G.B. Dalrymple and L.W. Snee, ^{40}Ar/^{39}Ar age of Cretaceous–Tertiary boundary tektites from Haiti, *Science*, 252, 1991, 1539–1542; C.C. Swisher, J.M. Grajales-Nishimura, A. Montanari, *et al.*, Coeval ^{40}Ar/^{39}Ar ages of 65.0 million years ago from Chicxulub crater melt rock and Cretaceous–Tertiary boundary tektites, *Science*, 257, 1992, 954–958; R.A. Kerr, Huge impact tied to mass extinction, *Science*, 257, 1992, 878–880; L.V. Sharpton, G.B. Dalrymple, L.E. Marín, *et al.*, New links between the Chicxulub impact structure and the Cretaceous–Tertiary boundary, *Nature*, 359, 1992, 819–821.

78. G.A. Izett, W.A. Cobban, J.D. Obradovitch and M.J. Kunk, The Manson impact structure – ^{40}Ar/^{39}Ar age and its distal impact ejecta in the Pierre Shale in southeastern South Dakota, *Science*, 262, 1993, 729–732; G.A. Izett, W.A. Cobban, G.B. Dalrymple, *et al.*, Ar-40/Ar-39 age of the Manson impact structure, Iowa, and corrrelative impact ejecta in the New Creek member of the Pierre Shale (Upper Cretaceous), South Dakota and Nebraska, *Geological Society of America Bulletin*, 110, 1998, 361–376.

79. F.L. Sutherland, Volcanism around K/T boundary time – its rôle in an impact scenario for the K/T boundary events, *Earth Science Reviews*, 36, 1994, 1–26; R.A.F. Grieve, Target Earth – evidence for large-scale impact events, *Annals of the New York Academy of Sciences*, 822, 1997, 319–352; A. Montanari, A.Campo Bagatin and P.Farinella, Earth

cratering record and impact energy flux over the last 150 Ma, *Planetary and Space Science*, 46, 1998, 271–281; G.A. Izzett, V.L. Masaitis, E.M. Shoemaker, *et al.*, *LPI Contributions*, 825, 1994, 55; E.K. Jesserbereger and W.U. Reimold, A Late Cretaceous 40Ar-39Ar age for the Lappajärvi impact crater, Finland, *Journal of Geophysics*, 48, 1980, 57–59; C. Koeberl, V. Sharpton, T.M. Harrison, *et al.*, The Kara/Ust-Kara twin structure – a large-scale impact event in the Late Cretaceous, in V.L. Sharpton and P.D. Ward (eds.), *Global Catastrophes in Earth History*, Geological Society of America, Special Paper 247, 1990, 233–238.

80. W. Glen (ed.), *The Mass-Extinction Debates*, 39–91.

81. W. Glen (ed.), *The Mass-Extinction Debates*, 91.

82. W. Glen (ed.), *The Mass-Extinction Debates*, 230–236.

83. D. Archibald, Were dinosaurs born losers?, *New Scientist*, 13 February 1993, 24–27; J.D. Archibald, *Dinosaur Extinction and the End of an Era*, 198–206 (quotation from page 204).

84. E. Dobb, What wiped out the dinosaurs?, *Discover*, 23(6), 2002, 35–43.

85. M. Benton, Dinosaur summer, in S.J. Gould (ed.), *The Book of Life*, Ebury Hutchinson, London, 1993, 126–167 (quotation from page 167).

86. A. Hallam and P.B. Wignall, *Mass Extinctions and their Aftermath*, Oxford University Press, 1997, 219.

87. D.B. Carlisle, *Dinosaurs, Diamonds, and Things from Outer Space*, 103–104; V. Clube and B. Napier, *The Cosmic Serpent*, Faber and Faber, London, 1982, 126–129; M.R. Rampino and K. Caldeira, Major episodes of geologic change – correlations, time structure and possible causes, *Earth and Planetary Science Letters*, 114, 1993, 215–227; I. Liritzis, Cyclicity in terrestrial upheavals during the Phanerozoic Eon, *Quarterly Journal of the Royal Astronomical Society*, 34, 1993, 251–260; R.E. Plotnick and J.J. Sepkoski, A multiplicative multifractal model for originations and extinctions, *Paleobiology*, 27, 2001, 126–139.

88. A. Charig, *A New Look at the Dinosaurs*, British Museum (Natural History), London, 1979, 149–151 (quotation from page 191).

89. L.B. Halstead, *Hunting the Past*, Hamish Hamilton, London, 1982, 146–147 (quotation from page 147).

90. T. Gardom with A. Milner, *The Natural History Museum Book of Dinosaurs*, 78–79.

91. R. Fortey, *Life – An Unauthorised Biography*, 195–196.

Chapter 23

1. P.C. Sereno, The evolution of dinosaurs, *Science*, 284, 1999, 2137–2147; G. Lawton, Here be monsters, *New Scientist*, 23 September 2000, 22–26; J.B. Smith, M.C. Lamanna, K.J. Lacovara, *et al.*, Giant sauropod dinosaur from an Upper Cretaceous mangrove deposit in Egypt, *Science*, 292, 2001, 1704–1706.

2. F.H. Pough, J.B. Heiser and W.N. McFarland, *Vertebrate Life*, Macmillan, New York, 3rd edn, 1989, 779–791; M.J. Benton, *The Rise of the Mammals*, Crescent Books, New York, 1991, 16–61; C. Janis, Victors by default, in S.J. Gould (ed.), *The Book of Life*, Ebury Hutchinson, London, 1993, 168–217; W. Alvarez, *T. rex and the Crater of Doom*, Princeton University Press, Princeton, New Jersey, 1997, 15–17, 130–131; R. Fortey,

Life – An Unauthorised Biography, Flamingo, London, 1997, 301–317; C. Frankel, *The End of the Dinosaurs*, Cambridge University Press, 1999, 135–140; T. Haines, *Walking with Beasts*, BBC Books, London, 2001, 6–21; J.D. Archibald, A.O. Averianov and E.G. Ekdale, Late Cretaceous relatives of rabbits, rodents, and other extant eutherian mammals, *Nature*, 414, 2001, 62–65; E. Stokstad, 'Fantastic' fossil helps narrow data gap, *Science*, 296, 2002, 637–638; A. Weil, Upwards and onwards, *Nature*, 416, 2002, 798–799; Q. Ji, Z.-X. Luo, C.-X. Yuan, *et al.*, The earliest known eutherian mammal, *Nature*, 416, 2002, 816–822.

3. M. Benton, *The Rise of the Mammals*, Crescent Books, New York, 1991, 39.

4. N. Eldredge, *The Miner's Canary*, Virgin Books, London, 1991, 102.

5. C.J. Avers, *Process and Pattern in Evolution*, Oxford University Press, 1989, 455–456.

6. M. Ridley, *Evolution*, Blackwell Scientific Publications, Oxford, 1993, 615.

7. J.N. Wilford, *The Riddle of the Dinosaur*, Vintage Books, New York, 1987, 317–322; D. Norman, *Dinosaur!*, Boxtree, London, 1991, 187.

8. T. Haines, *Walking with Dinosaurs*, BBC Books, London, 1999, 280–281.

9. D.M. Raup, Large-body impact and extinction in the Phanerozoic, *Paleobiology*, 18, 1992, 80–88 (quotation from page 82).

10. M.J. Benton, Reptiles, in K.J. McNamara (ed.), *Evolutionary Trends*, Belhaven Press, London, 1990, 279–300.

11. R.D. Norris, Biased extinction and evolutionary trends, *Paleobiology*, 17, 1991, 388–399.

12. R. Fortey, *Trilobite!*, HarperCollins, London, 2000, 144.

13. C. Darwin, *On the Origin of Species by Means of Natural Selection*, John Murray, London, 1859, Penguin Books, Harmondsworth, 1968, 133.

14. C. Darwin, *On the Origin of Species by Means of Natural Selection*, 172.

15. C. Darwin, *On the Origin of Species by Means of Natural Selection*, 321–322.

16. C. Darwin, *On the Origin of Species by Means of Natural Selection*, 324–325.

17. C. Darwin, *On the Origin of Species by Means of Natural Selection*, 325.

18. S.J. Gould, Speciation and sorting as the source of evolutionary trends, or "things are seldom what they seem", in *Evolutionary Trends*, 3–27 (quotations from page 17).

19. F.J. Ayala, Darwin's revolution, in J.H. Campbell and J.W. Schopf (eds.), *Creative Evolution?!*, Jones and Bartlett, Boston, 1994, 1–17 (particularly on page 16).

20. D.M. Raup, *Extinction – Bad Genes or Bad Luck?*, Norton, New York and London, 1991, 187.

21. N. Eldredge, *The Miner's Canary*, 45–47.

22. N. Eldredge, *The Miner's Canary*, 80.

23. J.J. Sepkoski, Some implications of mass extinction for the evolution of complex life, in M.D. Papagiannis (ed.), *The Search for Extraterrestrial Life*, Reidel, Dordrecht, 1985, 223–232; D. Jablonski, Evolutionary consequences of mass extinction, in D.M. Raup and D. Jablonski (eds.), *Patterns and Processes in the History of Life*, Springer-Verlag, Berlin, 1986, 313–329; D. Norman, *Prehistoric Life*, Boxtree, London, 1994, 180–182; G.R. McGhee, *The Late Devonian Mass Extinction – The Frasnian–Fammenian Crisis*, Columbia University Press, New York, 1996, 246–254; S.J. Gould, *Dinosaur in a Haystack*, Jonathan Cape, London, 1996, 147–158; S.J. Gould, *Life's Grandeur*,

Jonathan Cape, London, 1996, 135–146; R. Osborne and M. Benton, *The Viking Atlas of Evolution*, Penguin Books, London, 1996, 66–75; R. Fortey, *Life – An Unauthorised Biography*, 174–300; R. Fortey, *Trilobite!*, 114–115, 176–181; S. Lamb and D. Sington, *Earth Story*, BBC Books, London, 1998, 175–177, 189–190; V. Courtillot, *Evolutionary Catastrophes*, Cambridge University Press, 1999, 144–156; E. Mayr, *What Evolution Is*, Weidenfeld and Nicolson, London, 2002, 201–203 (quotation from page 203).

24. S.L. Rutherford and S. Lindquist, Hsp90 as a capacitor for morphological evolution, *Nature*, 396, 1998, 336–342; M. Pigliucci, Buffer zone, *Nature*, 417, 2002, 598–599; C. Queitsch, A. Sangster and S. Lindquist, Hsp90 as a capacitor of phenotypic variation, *Nature*, 417, 2002, 618–624.

25. M. Benton, Introduction – life and time, in *The Book of Life*, 22–35 (quotations from page 25).

26. P. Whitfield, *The Natural History of Evolution*, Marshall Editions, London, 1993, 186.

27. P. Whitfield, *The Natural History of Evolution*, 7.

28. S. Lamb and D. Sington, *Earth Story*, 177.

29. S. Jones, *Almost Like a Whale*, Doubleday, London, 1999, 235–239.

Chapter 24

1. C.J. Orth, Geochemistry of the bio-event horizons, in S.K. Donovan (ed.), *Mass Extinctions – Processes and Evidence*, Belhaven Press, London, 1989, 37–72 (quotation from page 67).

2. L.W. Alvarez, Experimental evidence that an asteroid impact led to the extinction of many species 65 million years ago, *Proceedings of the National Academy of Sciences, USA*, 80, 1983, 627–642; D.B. Carlisle, *Dinosaurs, Diamonds, and Things from Outer Space*, Stanford University Press, Stanford, Calif., 1995, 128.

3. C.J. Orth, in *Mass Extinctions – Processes and Evidence*, 37–72; R.A.F. Grieve, The record of impact on Earth, *Geological Society of America Special Paper*, 190, 1982, 25–37.

4. N.H. Sleep, K.J. Zahnle, J.F. Kasting and H.J. Morowitz, Annihilation of ecosystems by large asteroid impacts on the early Earth, *Nature*, 342, 1989, 139–142; D.R. Lowe, G.R. Byerly, F. Asaro and F.J. Kyte, Geological and geochemical record of 3400-million-year-old terrestrial meteorite impacts, *Science*, 245, 1989, 959–962; C. Sagan and A. Druyan, *Shadows of Forgotten Ancestors*, Century, London, 1992, 22–27; P.K. Strother, Pre-Metazoan life, in K.C. Allen and D.E.G. Briggs (eds.), *Evolution and the Fossil Record*, Belhaven Press, London, 1989, 51–72; S.J. Gould, The evolution of life on the Earth, *Scientific American*, 271(4), 1994, 63–69; J. Gribbin and M. Gribbin, *Fire on Earth*, Simon and Schuster, London, 1996, 79–82; R. Fortey, *Life – An Unauthorised Biography*, Flamingo, London, 1997, 31–76; P. Stanley, Impact! A matter of life and death?, *Astronomy Now*, 11(3), 1997, 44–45; D.H. Levy, *Comets – Creators and Destroyers*, Touchstone, New York, 1998, 47–67; G. Cooper, N. Kimmich, W. Belisle, *et al.*, Carbonaceous meteorites as a source of sugar-related organic compounds for the early Earth, *Nature*, 414, 2001, 879–883; N. English, The carbon drenched Universe, *Astronomy Now*, 16(4), 2002, 50–53; E. Pierazzo, The stuff of life, *Astronomy Now*, 16(4), 2002, 54–56; C. Wickramasinghe, The long road to panspermia, *Astronomy*

Chapter 24

Now, 16(4), 2002, 57–60; C.M. Fedo and M.J. Whitehouse, Metasomatic origin of quartz–pyroxene rock, Akilia, Greenland, and implications for Earth's earliest life, Science, 296, 2002, 1448–1452; E. Samuel, Total meltdown – asteroids bombarded the Earth, and then came life, New Scientist, 9 March 2002, 7; J. Glausiuz, When life was hell, Discover, 23(3), 2002, 10; Building block of life found in deep space, New Scientist, 20 July 2002, 23.

5. M.A.S. McMenamin, The origin and radiation of the early metazoa, in Evolution and the Fossil Record, 73–98; L. Margulis, Symbiosis in Cell Evolution, Freeman, New York, 2nd edn, 1993; D. Norman, Prehistoric Life, Boxtree, London, 1994, 20–29; S.J. Gould, The evolution of life on Earth, Scientific American, 271(4), 1994, 63–69; R.J. Horodyski and L.P. Knauth, Life on land in the Precambrian, Science, 263, 1994, 494–498; R. Osborne and D. Tarling, The Viking Historical Atlas of the Earth, Viking, London, 1995, 44–45; M. Brasier, D. Green and G. Shields, Ediacaran sponge spicule clusters from southwestern Mongolia and the origins of the Cambrian fauna, Geology, 25, 1997, 303–306; R. Mestel, Kimberellas's slippers, Earth, 6(5), 1997, 24–29; B. Holmes, When we were worms, New Scientist, 18 October 1997, 30–35; R. Monastersky, Life grows up, National Geographic 193(4), 1998, 100–115; B. Daviss, Cast out of Eden, New Scientist, 16 May 1998, 26–30; S. Lamb and D. Sington, Earth Story, BBC Books, London, 1998, 177–189; H. Hartman and A. Fedorov, The origin of the eukaryotic cell – a genomic investigation, Proceedings of the National Academy of Sciences, USA, 99, 2002, 1420–1423.

6. S.M. Stanley, Extinction, Scientific American Books, New York, 1987, 49–54; D. Dathe, Fundamentals of Historical Geology, Brown, Dubuque, Iowa, 1993, 60–87; M.D. Brasier, On mass extinctions and faunal turnover near the end of the Precambrian, in Mass Extinctions – Processes and Evidence, 73–88; R. Osborne and D. Tarling, The Viking Historical Atlas of the Earth, 46–47; A.H. Knoll and M.R. Walter, Latest Proterozoic stratigraphy and Earth history, Nature, 356, 1992, 673–678; A. Latham and R. Riding, Fossil evidence for the location of the Precambrian/Cambrian boundary in Morocco, Nature, 344, 1990, 752–754; A.Y. Rozanov and E. Landinz, Precambrian–Cambrian boundary global stratotype ratified and a new perspective of Cambrian time – comment and reply, Geology, 23, 1995, 285–286; J.L. Kirshvink, R.L. Ripperdan and D.A. Evans, Evidence for a large-scale reorganization of Early Cambrian continental masses by inertial interchange true polar wander, Science, 277, 1997, 541–545; J. Hecht, Twist of fate, New Scientist, 2 August 1997, 15; D. Pendick, Tipsy-turvy Earth, Astronomy, 25(12), 1997, 26–27; J. Hecht, Tilt-a-world, Earth, 7(3), 1998, 34–37; G. Walker, Snowball Earth, New Scientist, 6 November 1999, 28–33; R. Redfern, Origins, Cassell, London, 2000, 52–73; P.E. Hoffman and D.P. Schrag, Snowball Earth, Scientific American, 282(1), 2000, 50–57; S. Simpson, Triggering a snowball, Scientific American, 285(3), 2001, 14–15; R. Adler, Melting away, New Scientist, 15 December 2001, 15.

7. C.J. Orth, in Mass Extinctions – Processes and Evidence, 37–72; M.A. Nazarov, L.D. Basukova, G.M. Kolesov and A.S. Alekseev, Iridium abundances in the Precambrian–Cambrian boundary deposits and sedimentary rocks of Russian platform, Abstracts, Lunar and Planetary Science Conference, 1983; K.J. Hsü, H. Oberhänsli, J.Y. Gao, et al.,

451

'Strangelove Ocean' before the Cambrian explosion, *Nature*, 316, 1985, 809–811; M. Margaritz, W.T. Holser and J.L. Kirschvink, Carbon-isotope events across the Precambrian–Cambrian boundary on the Siberian platform, *Nature*, 320, 1986, 258–259; D.J. McLaren and W.D. Goodfellow, Geological and biological consequences of giant impacts, *Annual Review of Earth and Planetary Sciences*, 18, 1990, 123–171; M.R. Rampino and B.M. Haggerty, Extraterrestrial impacts and mass extinctions of life, in T. Gehrels (ed.), *Hazards due to Comets and Asteroids*, University of Arizona Press, Tucson, 1994, 827–857; C. Frankel, *The End of the Dinosaurs*, Cambridge University Press, 1999, 145–146.

8. D.B. Carlisle, *Dinosaurs, Diamonds, and Things from Outer Space*, 186; C.C. Albritton, *Catastrophic Episodes in Earth History*, Chapman and Hall, London, 1989, 90–92; R.A.F. Grieve and E.M. Shoemaker, The record of past impacts on Earth, in *Hazards due to Comets and Asteroids*, 417–462; J.G. Sprag and L.M. Thompson, Friction melt distribution in a multi-ring impact basin, *Nature*, 373, 1995, 130–132; L. Becker, J.R. Poreda and J.L. Bada, Extraterrestrial helium trapped in fullerenes in the Sudbury impact structure, *Science*, 272, 1996, 249–252; P. Joreau, B.M. French and J.-C. Doukhan, A TEM investigation of shock metamorphism in quartz from the Sudbury impact structure (Canada), *Earth and Planetary Science Letters*, 138, 1996, 137–143; D.E. Moser, Dating the shock wave and thermal imprint of the giant Vredefort impact, South Africa, *Geology*, 25, 1997, 7–10; J.G. Spray, Localized shock- and friction-induced melting in response to hypervelocity impact, in M.M. Grady, R.Hutchison, G.J.H. McCall and D.A. Rothery (eds.), *Meteorites – Flux with Time and Impact Effects*, Geological Society, London, Special Publication 140, 1998, 195–204; B. Chadwick, P. Claeys and B. Simonson, New evidence for a large palaeoproterozoic impact, *Journal of the Geological Society*, 158, 2001, 331–340; L. Becker, Repeated blows, *Scientific American*, 286(3), 2002, 62–69.

9. R.A.F. Grieve and E.M. Shoemaker, in *Hazards due to Comets and Asteroids*, 417–462; M.R. Rampino and B.M. Haggerty, in *Hazards due to Comets and Asteroids*, 827–857; V.A. Gostin, P.W. Haines, R.J.F. Jenkins, *et al.*, Impact ejecta horizon within the Late Precambrian shales, Adelaide geosyncline, South Australia, *Science*, 233, 1986, 198–200; V.A. Gostin, R.R. Keays and M.W. Wallace, Iridium anomaly from the Acraman impact ejecta horizon, *Nature*, 340, 1989, 542–544; G.E. Williams, The Acraman impact structure – source of ejecta in Late Precambrian shales, South Australia, *Science*, 233, 1986, 200–203; R.B. Hargraves, C.E. Cullicott, K.S. Deffeyes, *et al.*, Shatter cones and shocked rocks in southwestern Montana – the Beaverhead impact structure, *Geology*, 18, 832–834, 1990; W. Alvarez, *T. rex and the Crater of Doom*, Princeton University Press, Princeton, New Jersey, 1997, 140–141.

10. S.M. Stanley, An ecological theory for the sudden origin of multicellular life in the Late Precambrian, *Proceedings of the National Academy of Sciences, USA*, 70, 1973, 1486–1489; S.J. Gould, *Ever Since Darwin*, Penguin Books, Harmondsworth, 1980, 126–133; A. Hallam and P.B. Wignall, *Mass Extinctions and their Aftermath*, Oxford University Press, 1997, 24–34; T.S. Kemp, *Fossils and Evolution*, Oxford University Press, 1999, 204–205.

11. D. Norman, *Prehistoric Life*, 30–55; R. Osborne and D. Tarling, *The Viking Historical Atlas of the Earth*, 50–55; R. Fortey, *Life – An Unauthorised Biography*, 98–121; S. Lamb and D. Sington, *Earth Story*, 185–189; S.J. Gould, *Wonderful Life*, Hutchinson Radius, London, 1989, 53–60; 107–207; J.J. Sepkoski, Foundations – life in the oceans, in S.J. Gould (ed.), *The Book of Life*, Ebury Hutchinson, London, 1993, 36–63; S. Conway Morris, *The Crucible of Creation*, Oxford University Press, 1998, 27–115; R. Fortey, *Trilobite!*, HarperCollins, London, 2000, 114–138.

12. S. Conway Morris, *The Crucible of Creation*, 116–137; S. Conway Morris, Burgess Shale faunas and the Cambrian explosion, *Science*, 246, 1989, 339–346; S. Conway Morris, Palaeontology's hidden agenda, *New Scientist*, 11 August 1990, 38–42; D.E.G. Briggs, D.H. Erwin and F.J. Collier, *The Fossils of the Burgess Shale*, Smithsonian Institution Press, Washington (1994); M. Reid, Ghosts of the Burgess Shale, *Earth*, September 1992, 38–45; T. Beardsley, Weird wonders, *Scientific American*, 266(6), 1992, 12–14; J.S. Levinton, The big bang of animal evolution, *Scientific American*, 267(5), 1992, 52–59; L. Ramsköld and Hou Xianguang, New early Cambrian animal and onychophoran affinities of enigmatic metazoans, *Nature*, 351, 1991, 225–228; Jun-yuan Chen, L. Ramsköld and Gui-qing Zhou, Evidence for monophyly and arthropod affinity of Cambrian giant predators, *Science*, 264, 1994, 1304–1308.

13. S.J. Gould, *Wonderful Life*, 98–100.

14. R. Fortey, *Life – An Unauthorised Biography*, 108–111; N.J. Butterfield, A reassessment of the enigmatic Burgess Shale fossil *Wiwaxia corrugata* (Matthew) and its relationship to the polychaete *Canadia spinosa* Walcott, *Paleobiology*, 16, 1990, 287–303; N.J. Butterfield, Burgess Shale-type fossils from a lower Cambrian shallow-shelf sequence in northwestern Canada, *Nature*, 369, 1994, 477–479; D.E.G. Briggs, R.A. Fortey and M.A. Wills, Morphological disparity in the Cambrian, *Science*, 256, 1992, 1670–1673; M. Foote and S.J. Gould; M.S.Y. Lee; D.E.G. Briggs, R.A. Fortey and M.A. Wills, Cambrian and Recent morphological disparity, *Science*, 258, 1992, 1816–1818.

15. S. Conway Morris, *The Crucible of Creation*, 185–206.

16. S. Bengtson, Oddballs from the Cambrian start to get even, *Nature*, 351, 1991, 184–185.

17. S.M. Stanley, *Extinction*, 54–63; T.S. Kemp, *Fossils and Evolution*, 205; N. Eldredge, *The Miner's Canary*, Virgin Books, London, 1991, 69–77; A.R. Palmer, The biomere problem – evolution of an idea, *Journal of Paleontology*, 58, 1984, 599–611; G.R. McGhee, Catastrophes in the history of life, in *Evolution and the Fossil Record*, 26–50; J. Hecht, The biggest mass extinction of them all?, *New Scientist*, 1 August 1992, 14; T. Beardsley, Cambrian jolt, *Scientific American*, 267(6), 1992, 20–21; J. Hecht, *Vanishing Life*, Charles Scribner's Sons, New York, 1993, 102–104; A.Y. Zhuravlev and R.A. Wood, Anoxia as the cause of the mid-Early Cambrian (Botomian) extinction event, *Geology*, 24, 1996, 311–314; A. Hallam and P.B. Wignall, *Mass Extinctions and their Aftermath*, 31–38.

18. N. Eldredge, *Life Pulse*, Penguin Books, Harmondsworth, 1989, 62–66.

19. C.J. Orth, in *Mass Extinctions – Processes and Evidence*, 37–72; S.R. Westrop, Trilobite mass extinctions near the Cambrian–Ordovician boundary in North America, in

Mass Extinctions – Processes and Evidence, 89–103; C.J. Orth, J.D. Knight, R.L. Quintana and A.R.Palmer, A search for iridium abundance anomalies at two Late Cambrian biomere boundaries in Western Utah, *Science*, 223, 1984, 163–165; N. Eldredge, *The Miner's Canary*, 74–77; S.R. Westrop and M.B. Cuggy, Comparative paleoecology of Cambrian trilobite extinctions, *Journal of Paleontology*, 73, 1999, 337–354.

20. S.R. Westrop, Macroevolutionary implications of mass extinctions – evidence from an Upper Cambrian stage boundary, *Paleobiology*, 15, 1989, 46–52 (quotation from page 46). See also J.M. Adrain, S.R. Westrop, B.D.E. Chatterton and L. Ramskold, Silurian trilobite alpha diversity and the end-Ordovician mass extinction, *Paleobiology*, 26, 2000, 625–646.

21. S.J. Gould, *Wonderful Life*, 188.

22. S. Conway Morris, *The Crucible of Creation*, 199–205.

23. N. Eldredge, *Life Pulse*, 79–119; D. Norman, *Prehistoric Life*, 56–68; J.J. Sepkoski, in *The Book of Life*, 36–63; R. Fortey, *Life – An Unauthorised Biography*, 122–156; S. Lamb and D. Sington, *Earth Story*, 188–189; D.M. Raup, *Extinction – Bad Genes or Bad Luck?*, Norton, New York and London, 1991, 64–87.

24. R. Fortey, *Life – An Unauthorised Biography*, 154–156 (quotation from page 155).

25. S.M. Stanley, *Extinction*, 65–75; J. Hecht, *Vanishing Life*, 102–105; N. Eldredge, *The Mine's Canary*, 79–85; R. Osborne and D. Tarling, *The Viking Historical Atlas of the Earth*, 56–57; R. Fortey, *Trilobite!*, 174–178; P.J. Brenchley, The Late Ordovician extinction, in *Mass Extinctions – Processes and Evidence*, 104–132; A. Hallam and P.B. Wignall, *Mass Extinctions and their Aftermath*, 39–61; T.S. Kemp, *Fossils and Evidence*, 205–207; R. Redfern, *Origins*, 76–83; S. Bowler, Three steps on the road to extinction, *New Scientist*, 8 September 1990, 37; R.A. Berner, Paleo-CO_2 and climate, *Nature*, 358, 1992, 114; I.W.D. Dalziel, Earth before Pangea, *Scientific American*, 272(1), 1995, 38–43; S.R. Connolly and A.I. Miller, Global Ordovician faunal transitions in the marine benthos – proximate causes, *Paleobiology*, 27, 2001, 779–795; S.R. Connolly and A.I. Miller, Global Ordovician faunal transitions in the marine benthos – ultimate causes, *Paleobiology*, 28, 2002, 26–40.

26. C.J. Orth, in *Mass Extinctions – Processes and Evidence*, 37–72; S.K. Donovan, Palaeontological criteria for the recognition of mass extinctions, in *Mass Extinctions – Processes and Evidence*, 19–36; D.J. McLaren and W.D. Goodfellow, *Annual Review of Earth and Planetary Sciences*, 18, 1990, 123–171; P. Wilde, W.B.N. Berry, M.S. Quinby-Hunt, *et al.*, Iridium abundances across the Ordovician–Silurian stratotype, *Science*, 233, 1986, 339–341; M.R. Rampino and B.M. Haggerty, in *Hazards due to Comets and Asteroids*, 827–857; W.D. Goodfellow, G.S. Nowlan, A.D. McCracken, *et al.*, Geochemical anomalies near the Ordovician–Silurian boundary, North Yukon Territory, Canada, *Historical Biology*, 6, 1992, 1–23; K. Wang, C.J. Orth, M. Attrep, *et al.*, The great Latest Ordovician on the South China plate – chemostratigraphic studies of the Ordovician-Silurian boundary interval on the Yangtze platform, *Palaeogeography, Palaeoclimatology, Palaeoecology*, 104, 1993, 61–79; C. Frankel, *The End of the Dinosaurs*, 147–148.

27. N. Eldredge, *Life Pulse*, 95–138; M. Benton, The rise of the fishes, in *The Book of Life*, 54–77; D. Norman, *Prehistoric Life*, 63–99; R. Fortey, *Life – An Unauthorised*

Biography, 157–190; S. Lamb and D. Sington, *Earth Story*, 189–193; R. Redfern, *Origins*, 118–141.

28. G.R. McGhee, in *Evolution and the Fossil Record*, 26–50; D.J. McLaren, Time, life and boundaries, *Journal of Paleontology*, 44, 1970, 801–815; D.J. McLaren, Impacts that changed the course of evolution, *New Scientist*, 24 November 1983, 588–592; D.M. Raup, *Extinction – Bad Genes or Bad Luck?*, 64–87; N. Eldredge, *The Miner's Canary*, 85–88.

29. S.M. Stanley, *Extinction*, 75–89; G.R. McGhee, The Frasnian–Famennian Extinction event, in *Mass Extinctions – Processes and Evidence*, 133–151; R. Osborne and D. Tarling, *The Viking Historical Atlas of the Earth*, 62–63; A. Hallam and P.B. Wignall, *Mass Extinctions and their Aftermath*, 62–91; M.V. Caputo and J.C. Crowell, Migration of glacial centers across Gondwana during the Paleozoic, *Bulletin of the Geological Society of America*, 96, 1985, 1020–1036; G.R. McGhee, The Late devonian extinction crisis – evidence for abrupt ecosystem collapse, *Paleobiology*, 14, 1988, 250–257; M.M. Joachimski and W. Buggisch, anoxic events in the Late Frasnian – causes of the Frasnian-Famennian faunal crisis?, *Geology*, 21, 1993, 675–678; V. Courtillot, *Evolutionary Catastrophes*, Cambridge University Press, 1999, 95–96.

30. S.K. Donovan, in *Mass Extinctions – Processes and Evidence*, 19–36; C.J. Orth, in *Mass Extinctions – Processes and Evidence*, 37–72; M.R. Rampino and B.M. Haggerty, in *Hazards due to Comets and Asteroids*, 827–857; G.R. McGhee, J.S. Gilmore, C.J. Orth and E. Olsen, No geochemical evidence for an asteroid impact at late Devonian mass extinction horizon, *Nature*, 308, 1984, 629–631; P.E. Playford, D.J. McLaren, C.J. Orth, *et al.*, Iridium anomaly in the Upper Devonian of the Canning Basin, Western Australia, *Science*, 226, 1984, 437–439; D.J. McLaren, Ammonoids and extinctions, *Nature*, 313, 1985, 12–13; D.J. McLaren, Mass extinction and iridium anomaly in the Upper Devonian of Western Australia, *Geology*, 13, 1985, 170–172; G.R. McGhee, C.J. Orth, L.R. Quintana, *et al.*, Late Devonian 'Kellwasser Event' mass-extinction horizon in Germany – no geochemical evidence for a large-body impact, *Geology*, 14, 1986, 776–779.

31. J. Hecht, *Vanishing Life*, 105–107; W. Alvarez, *T. rex and the Crater of Doom*, 141; C. Frankel, *The End of the Dinosaurs*, 148–152; T.S. Kemp, *Fossils and Evolution*, 207–209; G.R. McGhee, *The Late Devonian Mass Extinction – The Frasnian–Famennian Crisis*, Columbia University Press, New York, 1996, 166–245; K. Wang, C.J. Orth, M. Attrep, *et al.*, Geochemical evidence for a catastrophic biotic event at the Frasnian/Famennian boundary in south China, *Geology*, 19, 1991, 776–779; K. Wang, Glassy microspherules (microtektites) from an Upper Devonian limestone, *Science*, 256, 1992, 1547–1550; H. Leroux, J.E. Warme and J.-C. Doukhas, Shocked quartz in the Alamo breccia, southern Nevada – evidence for a Devonian impact event, *Geology*, 23, 1995, 1003–1006; J.Hecht, At last. . . the truth about Area 51, *New Scientist*, 1 November 1997, 13; P. Claeys, J-G. Casier and S.V. Margolis, Microtektites and mass extinctions – evidence for a Late Devonian asteroid impact, *Science*, 257, 1992, 1102–1104; P. Claeys and J.G. Casier, Microtektite-like impact glass associated with the Frasnian-Famennian boundary mass extinction, *Earth and Planetary Science*

Letters, 122, 1994, 303–315; P. Copper, Great Phanerozoic crises – the mass extinction series, *Paleobiology*, 22, 1996, 568–572; J.E. Warme and H.C. Kuehner, Anatomy of an impact – the Devonian catastrophic Alamo impact Breccia of southern Nevada, *International Geology Review*, 40, 1998, 189–216; L. Becker, *Scientific American*, 286(3), 2002, 62–69.

32. K. Wang, H.H.J. Geldsetzer, W.D. Goodfellow and H.R. Krouse, Carbon and sulfur isotope anomalies across the Frasnian–Famennian extinction boundary, Alberta, Canada, *Geology*, 24, 1996, 187–191.

33. G.R. McGhee, *The Late Devonian Mass Extinction – The Frasnian–Famennian Crisis*, 202. See also G.R. McGhee, The 'multiple impact hypothesis' for mass extinctions – a comparison of the Late Devonian and the Late Eocene, *Palaeogeography, Palaeoclimatology, Palaeoecology*, 176, 2001, 47–58.

34. K. Wang, M. Attrep and C.J. Orth, Global iridium anomaly, mass extinction and redox change at the Devonian–Carboniferous boundary, *Geology*, 21, 1993, 1071–1074; J. Hecht, Earth rocked by shattered asteroids, *New Scientist*, 30 March 1996, 15; Multiple impact on Earth by asteroid or comet string, *Astronomy*, 24(4), 1996, 26; Prehistoric asteroid impact detected in Chad, *Amateur Astronomy and Earth Sciences*, April/May 1996, 6; R. Kerr, Impact craters all in a row?, *Science*, 272, 1996, 33; B. Napier, Comets, *Astronomy Now*, 11(3), 1997, 41–43; D. Miallier, S. Sanzelle, C. Falgueres, *et al.*, TL and ESR of quartz from the astrobleme of Aorounga (Sahara of Chad), *Quaternary Science Reviews*, 16, 1997, 265–274; M.L. Caplan and R.M. Bustin, Devonian–Carboniferous Hangenberg mass extinction event, widespread organic-rich mudrock and anoxia – causes and consequences, *Palaeogeography, Palaeoclimatology, Palaeoecology*, 148, 1999, 187–207; A. Hallam and P.B. Wignall, *Mass Extinctions and their Aftermath*, 62–93; A.J. Mory, R.P. Iasky, A.Y. Glikson and F. Pirajno, Woodleigh, Carnarvon Basin, Western Australia – a new 120 km diameter impact structure, *Earth and Planetary Science Letters*, 177, 2000, 119–128; P. Hadfield, Destroyer of worlds, *New Scientist*, 26 January 2002, 11; L. Becker, *Scientific American*, 286(3), 2002, 62–69.

35. N. Eldredge, *Life Pulse*, 121–153; M. Benton, Four feet on the ground, in *The Book of Life*, 78–125 (particularly pages 78–108); D. Norman, *Prehistoric Life*, 98–119; R. Fortey, *Life – An Unauthorised Biography*, 191–241; S. Lamb and D. Sington, *Earth Story*, 190–197; M. Benton, *The Reign of the Reptiles*, Crescent Books, New York, 1990, 38–61.

36. S.M. Stanley, *Extinction*, 93–107; W.D. Maxwell, The end Permian mass extinction, in *Mass Extinction – Processes and Evidence*, 152–173; D.M. Raup, *Extinction – Bad Genes or Bad Luck?*, 64–87; N. Eldredge, *The Miner's Canary*, 88–98; J. Hecht, *Vanishing Life*, 83–92; C. Frankel, *The End of the Dinosaurs*, 152–155; R. Redfern, *Origins*, 118–141; D.H. Erwin, Carboniferous–Triassic gastropod diversity patterns and the Permo-Triassic mass extinctions, *Paleobiology*, 16, 1990, 187–203; M. Haag and F. Heller, Late Permian to Early Triassic magnetostratigraphy, *Earth and Planetary Science Letters*, 107, 1991, 42–54; J.C. Claoué-Long, Z. Zichao, M. Guogan and D. Shaohua, The age of the Permian–Triassic boundary, *Earth and Planetary Science Letters*, 105, 1991, 182–190; D.H. Erwin, The Permo-Triassic extinction, *Nature*, 367, 1994, 231–235; D.H. Erwin, *The Great Paleozoic Crisis – Life and Death in the Permian*, Columbia

University Press, New York, 1995; D. Jablonski, Causes and consequences of mass extinctions – a comparative approach, in D.K. Elliott (ed.), *Dynamics of Extinction*, Wiley, New York, 1986, 183–229; P. Ward, *The End of Evolution*, Weidenfeld and Nicolson, London, 1995, 44–90; R. Osborne and M. Benton, *The Viking Atlas of Evolution*, Viking, London, 1996, 70–71; A. Grayson, *Equinox: The Earth*, Channel 4 Books, London, 2000, 19–48.

37. J.F. Thackeray, N.J. van der Merwe, J.A. Lee-Thorp, *et al.*, Changes in carbon isotope ratios in the late Permian recorded in therapsid tooth apatite, *Nature*, 347, 1990, 751–753; P. Wignall, The day the world nearly died, *New Scientist*, 25 January 1992, 51–55; K. Faure, M.J. de Wit and J.P. Willis, Late Permian global coal hiatus linked to ^{13}C-depleted CO_2 flux into atmosphere during the final consolidation of Pangea, *Geology*, 23, 1995, 507–510; D.R. Gröcke; K. Faure, M.J. de Wit and J.P. Willis, Late Permian global coal hiatus linked to ^{13}C-depleted CO_2 flux into the atmosphere during the final consolidation of Pangea – comment and reply, *Geology*, 24, 1996, 473–474; J.B. Graham, R. Dudley, N.M. Aguilar and C. Gans, Implications of the late Palaeozoic oxygen pulse for physiology and evolution, *Nature*, 375, 1995, 117–120; A.H. Knoll, R.K. Bambach, D.E. Caulfield and J.P. Grotzinger, Comparative earth history and the Late Permian mass extinction, *Science*, 273, 1996, 452–453; P.B. Wignall and R.J. Twitchett, Oceanic anoxia and the End Permian mass extinction, *Science*, 272, 1996, 1155–1158; D. Pendick, Four disasters that shaped the world, *Earth*, 6(1), 1997, 26–35; J. Gall, L. Grauvogel-Stamm, A. Nel and F. Papier, The Permian mass extinction and the Triassic recovery, *Comptes Rendus de l'Academie des Sciences Serie II Fascicule A – Sciences de la Terre et des Planetes*, 326, 1998, 1–12; R. Osborne and D. Tarling, *The Viking Historical Atlas of the Earth*, 80–87; A. Hallam and P.B. Wignall, *Mass Extinctions and their Aftermath*, 94–141; T.S. Kemp, *Fossils and Evolution*, 209–211; R.A. Berner, Examination of hypotheses for the Permo-Triassic boundary extinction by carbon cycle modelling, *Proceedings of the National Academy of Sciences, USA*, 99, 2002, 4172–4177.

38. D. McLaren, Impacts that changed the course of evolution, *New Scientist*, 24 November 1983, 588–592.

39. C.J. Orth, in *Mass Extinctions – Processes and Evidence*, 37–72; Y.-Y. Sun, Z.-F. Chai, S.-L. Ma, *et al.*, The discovery of iridium anomaly in the Permian–Triassic boundary clay in Changxing, Zhijing, China, and its significance, in G. Tu (ed.), *Developments in Geoscience*, Science Press, Beijing, 1984, 235–245; Y.-Y. Sun, Z.-F. Chai, S.-L. Ma, *et al.*, Discovery of anomalies of platinum group elements at the Permian–Triassic boundary in Changxing, Zhejiang, China and their significance, *Abstracts, 27th International Geological Congress*, 1985; D.-Y. Xu, S.-L. Ma, Z.-F. Chai, *et al.*, Abundance variation of iridium and trace elements at the Permian/Triassic boundary at Shangsi in China, *Nature*, 314, 1985, 154–156; F. Asaro, H.V. Michel, W. Alvarez, *et al.*, Geochemical study of the Permian–Triassic boundary in Meishan, Tangshan and Watchapo Mountain sections in the Peoples Republic of China, *Abstracts, 27th International Geological Congress*, 1985; D.L. Clark, C.-Y. Wang, C.J. Orth and J.S. Gilmore, Conodont survival and low iridium abundances across the Permian–Triassic boundary in south China, *Science*, 233, 1986, 984–986; D.M. Raup, Mass extinctions – a commentary, *Paleontology*, 30, 1987, 1–13; L. Zhou and

F.T. Kyte, The Permian–Triassic boundary event – a geochemical study of three Chinese sections, *Earth and Planetary Science Letters*, 90, 1988, 411–421.

40. D.J. McLaren and W.D. Goodfellow, *Annual Review of Earth and Planetary Sciences*, 18, 1990, 123–171 (quotation from page 143).

41. K. Wang, H.H.J. Geldsetzer and H.R. Krouse, Permian-Triassic extinction – organic δ^{13}C evidence from British Columbia, Canada, *Geology*, 22, 1994, 580–584; R.A. Kerr, Growth, death and climate featured in Salt Lake City, *Science*, 278, 1997, 1017–1018; R.A. Kerr, Biggest extinction looks catastrophic, *Science*, 280, 1998, 1007; S.A. Bowring, D.H. Erwin, Y.G. Yin, *et al.*, U/Pb Zircon geochronology and tempo of the end-Permian mass extinction, *Science*, 280, 1998, 1039–1045; S.M. Stanley and X. Yang, A double mass extinction at the end of the Paleozoic era, *Science*, 266, 1994, 1340–1344; I.H. Campbell, G.K. Czamanske, V.A. Fedorenko, *et al.*, Synchronism of the Siberian Traps and the Permian–Triassic boundary, *Science*, 258, 1992, 1760–1763; R.A. Kerr, A volcanic crisis for early life, *Science*, 270, 1995, 27–28; R. Osborne and M. Benton, *The Viking Atlas of Evolution*, 70–71; V. Courtillot, *Evolutionary Catastrophes*, 81–92; T.S. Kemp, *Fossils and Evolution*, 209–211; Blasts from the past, *New Scientist*, 7 July 2001, 10; R.J. Twitchett, C.V. Looy, R. Morante, *et al.*, Rapid and synchronous collapse of marine and terrestrial ecosystems during the end-Permian biotic crisis, *Geology*, 29, 2001, 351–354; P.R. Renne, Flood basalts – bigger and badder, *Science*, 296, 2002, 1812–1813; M.K. Reichow, A.D. Saunders, R.V. White, *et al.*, ^{40}Ar/^{39}Ar data on basalts from the West Siberian basin – Siberian flood basalt province doubled, *Science*, 296, 2002, 1846–1849.

42. D.H. Erwin, The Permo-Triassic extinction, *Nature*, 367, 1994, 231–235.

43. M.R. Rampino and B.M. Haggerty, in *Hazards due to Comets and Asteroids*, 827–857; C. Frankel, *The End of the Dinosaurs*, 152–155; X. Dao-Yi and Y. Zeng, Carbon isotope and iridium event markers near the Permian/Triassic boundary in the Meishan section, Zhejiang Province, China, *Palaeogeography, Palaeoclimatology, Palaeoecology*, 104, 1993, 171–176; R.A. Kerr, A shocking view of the Permian–Triassic, *Science*, 274, 1996, 1080; Y.G. Jin, Y. Wang, W. Wang, *et al.*, Pattern of marine mass extinction near the Permian–Triassic boundary in South China, *Science*, 289, 2000, 432–436; K. Kaiho, Y. Kajiwara, T. Nakano, *et al.*, End-Permian catastrophe by a bolide impact – evidence of a gigantic release of sulfur from the mantle, *Geology*, 29, 2001, 815–818; L. Becker, R.J. Poreda, A.G. Hunt, *et al.*, Impact event at the Permian–Triassic boundary – evidence from extraterrestrial noble gases in fullerenes, *Science*, 291, 2001, 1530–1533.

44. R.A.F. Grieve and E.M. Shoemaker, in *Hazards due to Comets and Asteroids*, 417–462; W.N. Engelhardt, S.K. Matthai and J. Walzebuck, The Araguinha impact crater, Brazil, *Meteoritics*, 27, 1992, 442–457; W. Masero, G. Fischer and P.A. Schnegg, Electrical conductivity and crustal deformation from magnetotelluric results in the region of the Araguinha impact, Brazil, *Physics of the Earth and Planetary Interiors*, 101, 1997, 271–289.

45. M.R. Rampino and B.M. Haggerty, in *Hazards due to Comets and Asteroids*, 827–857; J. Gribbin and M. Gribbin, *Fire on Earth*, 82; C. Frankel, *The End of the Dinosaurs*, 155; M.R. Rampino, A large Late Permian impact structure from the Falkland

Plateau, *Eos*, 73, 1992, 136; D. Steel, *Rogue Asteroids and Doomsday Comets*, Wiley, New York, 1995, 104–106; L. Becker, *Scientific American*, 286(3), 2002, 62–69.

46. N. Eldredge, *Life Pulse*, 155–174; C. Pellant, *the Pocket Guide to Fossils*, Parkgate Books, London, 1992, 128–155; R. Osborne and M. Benton, *The Viking Atlas of Evolution*, 73.

47. N. Eldredge, *Life Pulse*, 171.

48. S.M. Stanley, *Extinction*, 109–115; M. Benton, *The Reign of the Reptiles*, 62–88, 108–120; M. Benton, Four feet on the ground, in *The Book of Life*, 78–125 (particularly pages 110–125); D. Norman, *Prehistoric Life*, 113–155; R. Redfern, *Origins*, 140–163; J.J. Flynn and A.R. Wyss, Madagascar's Mesozoic secrets, *Scientific American*, 286(2), 2002, 42–51.

49. T. Kemp, The reptiles that became mammals, *New Scientist*, 4 March 1982, 581–584 (quotations from page 584); also published in J. Cherfas (ed.), *Darwin up to Date*, New Science Publications, London, 1982, 31–34 (quotations from page 34); T.S. Kemp, *Fossils and Evolution*, 227–234 (quotation from page 234).

50. M. Benton, *The Reign of the Reptiles*, 62–81 (quotation from p. 81); M. Benton, in *The Book of Life*, 114–125; N. Eldredge, *The Miner's Canary*, 95–104; D. Norman, *Dinosaur!*, Boxtree, London, 1991, 94–99; M.J. Benton, Late Triassic extinctions and the origin of the dinosaurs, *Science*, 260, 1993, 769–770.

51. N. Eldredge, *Life Pulse*, 153.

52. M. Benton, *The Reign of the Reptiles*, 80–107, 135–137; D. Norman, *Dinosaur!*, 94–146; M. Benton, Dinosaur summer, in *The Book of Life*, 126–167; D.E. Fastovsky and W.B. Weishampel, *The Evolution and Extinction of the Dinosaurs*, Cambridge University Press, 1996, 293–321; D. Norman, *The Illustrated Encyclopedia of Dinosaurs*, Greenwich Editions, London, 1998, 191–195; P. Shipman, *Taking Wing*, Weidenfeld and Nicolson, London, 1998, 47–279; *The Ultimate Book of Dinosaurs*, Parragon, Bath, 2000, 60–61; E. Stokstad, New fossil fills gap in bird evolution, *Science*, 291, 2001, 225; E. Stokstad, Exquisite Chinese fossils add new pages to book of life, *Science*, 291, 2001, 232–236; J. Hecht, Dinos of a feather, *New Scientist*, 9 March 2002, 11; M. Norell, Qiang Ji, Keqin Gao, *et al.*, 'Modern' feathers on a non-avian dinosaur, *Nature*, 416, 2002, 36–37.

53. S.M. Stanley, *Extinction*, 115–117; D.M. Raup, *Extinction – Bad Genes or Bad Luck?*, 64–87; J. Hecht, *Vanishing Life*, 93–101; P. Ward, *The End of Evolution*, 91–106; C. Frankel, *The End of the Dinosaurs*, 156–158.

54. A.L.A. Johnson and M.J. Simms, The timing and cause of Late Triassic marine invertebrate extinctions – evidence from scallops and crinoids, in *Mass Extinctions – Processes and Evidence*, 174–194; D.J. McLaren and W.D. Goodfellow, *Annual Review of Earth and Planetary Sciences*, 18, 1990, 123–171; M.R. Rampino and B.M. Haggerty, in *Hazards due to Comets and Asteroids*, 827–857; M.J. Benton, More than one event in the Late Triassic mass extinction, *Nature*, 321, 1986, 857–861; K. Padian; P.E. Olsen, N.H. Shubin and M.H. Anders, Triassic–Jurassic extinctions, *Science*, 241, 1988, 1358–1360; M. Rampino and R.B. Stothers, Flood basalt volcanism during the past 250 Myr, *Science*, 241, 1988, 663–668; A. Hallam, The end-Triassic mass extinction event, *Geological Society of America Special Paper*, 247, 1990, 577–583; R. Morante and A. Hallam, Organic carbon isotope record across the Triassic-Jurassic boundary

of Austria and its bearing on the cause of the mass extinction, *Geology*, 24, 1996, 391–394; A. Hallam and P.B. Wignall, *Mass Extinctions and their Aftermath*, 142–160; V. Courtillot, *Evolutionary Catastrophes*, 93–94; T.S. Kemp, *Fossils and Evolution*, 211–212; P.E. Olsen, Giant lava flows, mass extinctions, and mantle plumes, *Science*, 284, 1999, 604–605; A. Marzoli, P.R. Renne, E.M. Piccirillo, *et al.*, Extensive 200-million-year-old continental flood basalts of the Central Atlantic Magmatic Province, *Science*, 284, 1999, 616–618; J.C. McElwain, D.J. Beerling and F.T. Woodward, Fossil plants and global warming at the Triassic–Jurassic boundary, *Science*, 285, 1999, 1386–1390; L.H. Tanner, J.F. Hubert, B.P. Coffey and D.P. McInerney, Stability of atmospheric CO_2 level across the Triassic-Jurassic boundary, *Nature*, 411, 2001, 675–677; D. Beerling, CO_2 and the end-Triassic mass extinction, *Nature*, 415, 2002, 386–387; G.L. Retallack, L.H. Tanner, Triassic–Jurassic atmospheric CO_2 spike, *Nature*, 415, 2002, 387–388.

55. M.R. Rampino and B.M. Haggerty, in *Hazards due to Comets and Asteroids*, 827–857; R.E. Weems, The 'terminal Triassic catastrophic extinction event' in perspective, *Palaeogeography, Palaeoclimatology, Palaeoecology*, 94, 1992, 1–29; C.A. McRoberts and C.R. Newton, Selective extinction among end-Triassic European bivalves, *Geology*, 23, 1995, 102–104; J. Hecht, Did a comet end the Triassic period?, *New Scientist*, 24 November 1990, 20; D.M. Bice, C.R. Newton, S. McCauley, *et al.*, Shocked quartz at the Triassic–Jurassic boundary in Italy, *Science*, 255, 1992, 443–446; P.D. Ward, J.W. Haggard, E.S. Carter, *et al.*, Sudden productivity collapse associated with the Triassic–Jurassic boundary mass extinction, *Science*, 292, 2001, 1148–1151; J. Hecht, Jurassic spark, *New Scientist*, 25 May 2002, 18; R.A. Kerr, Did an impact trigger the dinosaurs' rise?, *Science*, 296, 2002, 1215–1216; P.E. Olsen, D.V. Kent, H.-D. Sues, *et al.*, Ascent of dinosaurs linked to an iridium anomaly at the Triassic–Jurassic boundary, *Science*, 296, 2002, 1305–1307.

56. R.A.F. Grieve and E.M. Shoemaker, in *Hazards due to Comets and Asteroids*, 417–462; W.M. Napier, Galactic periodicity and the geological record, in *Meteorites – Flux with Time and Impact Effects*, 19–29; R.A.F. Grieve, Extraterrestrial impacts on Earth, in *Meteorites – Flux with Time and Impact Effects*, 105–131; C. Koeberl, Identification of meteoritic components in impactites, in *Meteorites – Flux with Time and Impact Effects*, 133–153; J.P. Hodych and G.R. Dunning, Did the Manicouagan impact trigger end-of-Triassic mass extinction?, *Geology*, 20, 1992, 51–54; J.G. Spray, S.P. Kelley and D.B. Rowley, Evidence for a late Triassic multiple impact event on Earth, *Nature*, 392, 1998, 171–173; C. Frankel, *The End of the Dinosaurs*, 156–158, 165–166.

57. M.R. Rampino and R.B. Stothers, *Science*, 241, 1988, 663–668; C.T.S. Little and M.J. Benton, Early Jurassic mass extinctions – a global long-term event, *Geology*, 23, 1995, 495–498; V. Courtillot, *Evolutionary Catastrophes*, 83.

58. C.J. Orth, in *Mass Extinctions – Processes and Evidence*, 37–72; M.R. Rampino and B.M. Haggerty, in *Hazards due to Comets and Asteroids*, 827–857; W. Brochwicz-Lewinski, A. Gasiewicz, G. Melendez, *et al.*, A possible Middle/Upper Jurassic boundary event, *Abstracts, 27th International Geology Congress*, 1985; A. Hallam and P.B. Wignall, *Mass Extinctions and their Aftermath*, 161–166.

59. V.A. Zhakarov, A.S. Lapukhov and O.V. Shenfil, Iridium anomaly at the Jurassic–Cretaceous boundary in northern Siberia, *Russian Journal of Geology and Geophysics*, 34, 1993, 83–90; S.T. Gudlaugsson, Large impact crater in the Barents Sea, *Geology*, 21, 1993, 291–294; H. Dypvik, S.T. Gudlaugsson, F. Tsikalas, *et al.*, Mjølnir structure – an impact crater in the Barents Sea, *Geology*, 24, 1996, 779–782; J. Hecht, Big blast brought in the Cretaceous, *New Scientist*, 29 March 1997, 21; R.J. Hart, M.A.G. Andreoli, M. Tredoux, *et al.*, Late Jurassic age for the Morokweng impact structure, southern Africa, *Earth and Planetary Science Letters*, 147, 1997, 25–35; W.B. McKinnon, Extreme cratering, *Science*, 276, 1997, 1346–1348; H. Dypvik and R.E. Ferrell, Clay mineral alteration associated with a meteorite impact in the marine environment (Barents Sea), *Clay Minerals*, 33, 1998, 51–64; R.E.F. Grieve, Extraterrestrial impacts on Earth, in *Meteorites – Flux with Time and Impact Effects*, 105–131; C. Frankel, *The End of the Dinosaurs*, 158–161; V. Courtillot, *Evolutionary Catastrophes*, 95; I. McDonald, M.A.G. Andreoli, R.J. Hart and M. Tredoux, Platinum-group elements in the Morokweng impact structure, South Africa – evidence for the impact of a large ordinary chondrite projectile at the Jurassic–Cretaceous boundary, *Geochimica et Cosmochimica Acta*, 65, 2001, 299–309; A. Hallam and P.B. Wignall, *Mass Extinctions and their Aftermath*, 166–168.

60. C.J. Orth, in *Mass Extinctions – Processes and Evidence*, 37–72; M.R. Rampino and B.M. Haggerty, in *Hazards due to Asteroids and Comets*, 827–857; C.J. Orth, A. Attrep, X.-Y. Mao, *et al.*, Iridium abundance maxima in the upper Cenomanian extinction interval, *Geophysical Research Letters*, 15, 1988, 346–349; M.A. Arthur, W.E. Dean and L.M. Pratt, Geochemical and climatic effects of increased marine organic carbon burial at the Cenomanian/Turonian boundary, *Nature*, 335, 1988, 714–717; C.J. Orth, M. Attrep, L.R. Quintana, *et al.*, Elemental abundance anomalies in the late Cenomanian extinction interval – a search for the source(s), *Earth and Planetary Science Letters*, 117, 1993, 189–204; A. Banerjee and G. Boyajian, Changing biologic selectivity of extinction in the Foraminifer over the past 150 m.y., *Geology*, 24, 1996, 607–610; J.G. Eaton, J.I. Kirkland, J.H. Hutchinson, *et al.*, Nonmarine extinctions across the Cenomanian–Turonian boundary, southwestern Utah, with a comparison to the Cretaceous–Tertiary extinction events, *Geological Society of America Bulletin*, 109, 1997, 560–567; A. Hallam and P.B. Wignall, *Mass Extinctions and their Aftermath*, 171–183; V. Courtillot, *Evolutionary Catastrophes*, 95; A.B. Smith, A.S. Gale and N.E.A. Monks, Sea-level change and rock-record bias in the Cretaceous, *Paleobiology*, 27, 2001, 241–253.

61. R. Osborne and M. Benton, *The Viking Atlas of Evolution*, 72.

62. G.R. McGhee, in *Evolution and the Fossil Record*, 26–50; D.M. Raup, *Extinction – Bad Genes or Bad Luck?*, 64–87; M.J. Benton, in *The Book of Life*, 162–167; N. Eldredge, *The Miner's Canary*, 99–124; J. Hecht, *Vanishing Life*, 52–59; W. Alvarez, *T. rex and the Crater of Doom*, 15–17; A. Hallam and P.B. Wignall, *Mass Extinctions and their Aftermath*, 184–222; C. Frankel, *The End of the Dinosaurs*, 135–140; T.S. Kemp, *Fossils and Evolution*, 212–213; L. Becker, *Scientific American*, 286(3), 2002, 62–69.

63. D. Norman, *Prehistoric Life*, 182.

64. T. Haines, *Walking with Beasts*, BBC Books, London, 2001, 8–11.

65. S.M. Stanley, *Extinction*, 173–186; C. Janis, Victors by default, in *The Book of Life*, 168–217 (particularly pages 182–192); C. Tudge, *The Day before Yesterday*, Pimlico, London, 1996, 64; I.C. Sloan, J.C.G. Walker, T.C. Moore, *et al.*, Possible methane-induced polar warming in the early Eocene, *Nature*, 357, 1992, 320–322; R.A. Kerr, When climate twitches, evolution takes great leaps, *Science*, 257, 1992, 1622–1624; D.G. Greenwood and S.L. Wing, Eocene continental climates and latitudinal temperature gradients, *Geology*, 23, 1995, 1044–1048; G.J. Jordan; S.L. Wing and D. Greenwood, Eocene continental climates and latitudinal temperature gradients – comment and reply, *Geology*, 24, 1996, 1054–1055; J.V. Browning, K.G. Miller and D.K. Pak, Global implications of lower to middle Eocene sequence boundaries on the New Jersey coastal plane – the icehouse cometh, *Geology*, 24, 1996, 639–642; J. Hecht, Scorched Earth, *New Scientist*, 8 November 1997, 20; F. Pearce, Wind of change, *New Scientist*, 2 May 1998, 34–37; P.N. Pearson, P.W. Ditchfield, J. Singano, *et al.*, Warm tropical sea surface temperatures in the Late Cretaceous and Eocene epochs, *Nature*, 413, 2001, 481–487; C. Beard, East of Eden at the Paleocene/Eocene boundary, *Science*, 295, 2002, 2028–2029; G.J. Bowen, W.C. Clyde, P.L. Koch, *et al.*, Mammalian dispersal at the Paleocene/Eocene boundary, *Science*, 295, 2002, 2062–2065; V. Courtillot, *Evolutionary Catastrophes*, 95–99; T. Haines, *Walking with Beasts*, 14.

66. S.M. Stanley, *Extinction*, 186–189; A. Hallam and P.B. Wignall, *Mass Extinctions and their Aftermath*, 223–226; R. Redfern, *Origins*, 216–228; L. Diester-Haas and R. Zahn, Eocene–Oligocene transition in the Southern Ocean – history of water mass circulation and biological productivity, *Geology*, 24, 1996, 163–166; B. Livermore, Life behind the cold wall, *Earth*, 6(6), 1997, 50–56.

67. L.B. Halstead, *The Evolution of the Mammals*, Eurobook/Book Club Associates, London, 1979, 42.

68. S.M. Stanley, *Extinction*, 173–189; J.C. Zachos, J.R. Breza and S.W. Wise, Early Oligocene ice-sheet expansion on Antarctica, *Geology*, 20, 1992, 569–573; M.R. Rampino and R.B. Stothers, *Science*, 241, 663–668, 1988; M.R. Rampino and K. Caldeira, Major episodes of geological change – correlations, time structure and possible causes, *Earth and Planetary Science Letters*, 114, 1993, 215–217; J.P. Kennet, C. von der Borch, P.A. Baker, *et al.*, Palaeotectonic implications of increased late Eocene – early Oligocene volcanism from South Pacific DSDP sites, *Nature*, 316, 1985, 507–511; V. Courtillot, *Evolutionary Catastrophes*, 81, 96, 142.

69. S.M. Stanley, *Extinction*, 175–186; F.H. Pough, J.B. Heiser and W.N. McFarland, *Vertebrate Life*, Macmillan, New York, 3rd edn, 1989, 779–791; C. Janis, in *The Book of Life*, 185–196.

70. R.A.F. Grieve and E.M. Shoemaker, in *Hazards due to Comets and Asteroids*, 417–462; I. Gilmour, Geochemistry of carbon in terrestrial impact processes, in *Meteorites – Flux with Time and Impact Effects*, 205–216; R.A.F. Grieve, Target Earth, *Annals of the New York Academy of Sciences*, 822, 1997, 319–352; A. Montanari, A. Campo Bagatin and P. Farinella, Earth cratering record and impact energy flux in the last 150 Ma, *Planetary and Space Science*, 46, 1998, 271–281; A.K. Clymes, D.M. Bice and A. Montanari, Shocked quartz from the late Eocene – impact evidence from Massignano, Italy, *Geology*. 24, 1996, 483–486; F. Langenhorst and A.K. Clymes, Characteristics of shocked

quartz in the late Eocene impact ejecta from Massignano (Ancona, Italy), *Geology*, 24, 1996, 487–490; A. Montanari, F. Asaro, H.V. Michel and J.P. Kennett, Iridium anomalies of Late Eocene age at Massignano, Italy, and ODP Site 689B (Maud Rise, Antarctica), *Palaeos*, 8, 1993, 420–438; D. Stöffler and P. Claeys, Earth rocked by combination punch, *Nature*, 388, 1997, 332–333; R. Bottomley, R. Grieve, D. York and V. Masaitis, The age of the Popigai impact event and its relation to events at the Eocene–Oligocene boundary, *Nature*, 388, 1997, 365–368; O. Pierrard, E. Robin, R. Rocchia and A. Montanari, Extraterrestrial Ni-rich spinel in upper Eocene sediments from Massignano, Italy, *Geology*, 26, 1998, 307–310; V.L. Masaitis, Popigai crater – origin and distribution of diamond-bearing impactites, *Meteoritics and Planetary Science*, 33, 1998, 349–359; C. Frankel, *The End of the Dinosaurs*, 162–164.

71. G. Keller, S. D'Hondt and T.L. Vallier, Multiple microtektite horizons in Upper Eocene marine sediments, *Science*, 221, 1983, 150–152; A. Sanfilippo, W.R. Riedel, B.P. Glass and F.T. Kyte, Late Eocene microtektites and radiolarian extinctions on Barbados, *Nature*, 314, 1985, 613–615; G. Keller, S.L. D'Hondt, C.J. Orth, *et al.*, Late Eocene impact microspherules – stratigraphy, age and geochemistry, *Meteoritics*, 22, 1987, 25–60; B.P. Glass and J. Wu, Coesite and shocked quartz discovered in the Australasian and N. American microtektite layers, *Geology*, 21, 1993, 435–438; M.R. Rampino and B.M. Haggerty, in *Hazards due to Comets and Asteroids*, 827–857; C.W. Poag, D.S. Powars, L.J. Poppe and R.B. Mixon, Meteoroid mayhem in Ole Virginny – source of the North American tektite strewn field, *Geology*, 22, 1994, 691–694; W.B. McKinnon, Extreme cratering, *Science*, 276, 1997, 1346–1348; C.W. Poag, The Chesapeake Bay bolide impact, *Sedimentary Geology*, 108, 1997, 45–90; K.A. Farley, A. Montanari, E.M. Shoemaker and C.S. Shoemaker, Geochemical evidence for a comet shower in the Late Eocene, *Science*, 280, 1998, 1250–1253; C. Koeberl, C.W. Poag, W.V. Reimold and D. Brandt, Impact origin of the Chesapeake Bay structure and the source of the North American tektites, *Science*, 271, 1996, 1263–1266; C. Frankel, *The End of the Dinosaurs*, Cambridge University Press, 1999, 162–164.

72. S.M. Stanley, *Extinction*, 186–189; D.J. McLaren and W.D. Goodfellow, *Annual Review of Earth and Planetary Sciences*, 18, 1990, 123–171; D.R. Prothero, North American mammalian diversity and Eocene–Oligocene extinctions, *Paleobiology*, 11, 1985, 389–405; D.R. Prothero, Mid-Oligocene extinction event in North American land animals, *Science*, 229, 1985, 550–551; D.R. Prothero, Stepwise extinctions and climatic decline during the later Eocene and Oligocene, in *Mass Extinctions – Processes and Evidence*, 217–234; D.R. Prothero, The Late Eocene–Oligocene extinctions, *Annual Review of Earth and Planetary Sciences*, 22, 1994, 145–165; D.R. Prothero and W.A. Berggren (eds.), *Eocene–Oligocene Climatic and Biotic Evolution*, Princeton University Press, New Jersey, 1992; D.R. Prothero, *The Eocene–Oligocene Transition*, Columbia University Press, New York, 1994; D.R. Prothero and T.H. Heaton, Faunal stability during the Early Oligocene climatic crash, *Palaeogeography, Palaeoclimatology and Palaeoecology*, 27, 1996, 257–283; A. Hallam and P.B. Wignall, *Mass Extinctions and their Aftermath*, 227–234; V. Courtillot, *Evolutionary Catastrophes*, 96; T. Haines, *Walking with Beasts*, 60–99.

73. D.R. Prothero, in *Mass Extinctions – Processes and Evidence*, 228–229.

74. R. Ganapathy, Evidence for a major meteorite impact on the Earth 34 million years ago – implications for Eocene extinctions, *Science*, 216, 1982, 885–886; W. Alvarez, F. Asaro, H.V. Michel and L.W. Alvarez, Iridium anomaly approximately synchronous with terminal Eocene extinctions, *Science*, 216, 1982, 886–888.

75. P. Hut, W. Alvarez, W.P. Elder, *et al.*, Comet showers as a cause of mass extinctions, *Nature*, 329, 1987, 118–125; P.R. Weissman, The Oort Cloud, *Scientific American*, 279(3), 1998, 62–67. See also G.R. McGhee, The 'multiple impact hypothesis' for mass extinctions, *Palaeogeography, Palaeoclimatology and Palaeoecology*, 176, 2001, 47–58.

76. V. Clube and B. Napier, *The Cosmic Winter*, Basil Blackwell, Oxford, 1990, 229.

77. S.M. Stanley, *Extinction*, 182; C. Janis, in *The Book of Life*, 185–196; T. Haines, *Walking with Beasts*, 14; R. Lewin, *Human Evolution*, Blackwell Science, Malden, Mass. and Oxford, 4th edn, 1999, 21–25; L.C. Ivany, W.P. Patterson and K.C. Lehman, Cooler winters as a possible cause of mass extinctions at the Eocene/Oligocene boundary, *Nature*, 407, 2000, 887–890.

78. T. Haines, *Walking with Beasts*, 65.

79. M.J. Benton, *The Rise of the Mammals*, 86; R. Osborne and M. Benton, *The Viking Atlas of Evolution*, 82–83; C. Tudge, *The Day before Yesterday*, 42–46; R. Redfern, *Origins* 216–228.

Chapter 25

1. A.G. Fischer and M.A. Arthur, Secular variations in the pelagic realm, *Society of Economic Paleontologists and Mineralogists Special Publications*, 25, 1977, 19–50; D.M. Raup, *The Nemesis Affair*, Norton, New York and London, 1986, 107–129; S.M. Stanley, *Extinction*, Scientific American Books, New York, 1987, 210–216; C.C. Albritton, *Catastrophic Episodes in Earth History*, Chapman and Hall, London, 1989, 99–106; W. Glen (ed.), *The Mass-Extinction Debates*, Stanford University Press, Stanford, Calif., 1994, 25–29.

2. J.J. Sepkoski, A factor analytic description of the Phanerozoic marine fossil record, *Paleobiology*, 7, 1981, 36–53; J.J. Sepkoski, A kinetic model of Phanerozoic taxonomic diversity – post-Paleozoic families and mass extinctions, *Paleobiology*, 10, 1984, 246–267; J.J. Sepkoski, Mass extinctions in the Phanerozoic oceans – a review, *Geological Society of America Special Paper*, 190, 1982, 283–289; J.J. Sepkoski, *A Compendium of Fossil Marine Families*, Milwaukee Public Museum Contributions in Biology and Geology, 51, 1982; J.J. Sepkoski, What I did with my research career – or how research on biodiversity yielded data on extinction, in *The Mass-Extinction Debates*, 132–144.

3. D.M. Raup and J.J. Sepkoski, Mass extinctions in the marine fossil record, *Science*, 215, 1982, 1501–1503; D.M. Raup, Extinction in the geologic past, in D.E. Osterbrock and P.H. Raven (eds.), *Origins and Extinction*, Yale University Press, New Haven and London, 1988, 109–119; D.M. Raup, *Extinction – Bad Genes or Bad Luck?*, Norton, New York and London, 1991, 64–74; E. Mayr, *What Evolution Is*, Weidenfeld and Nicolson, London, 2002, 201–203.

4. D.M. Raup and J.J. Sepkoski, Periodicity of extinctions in the geologic past, *Proceedings of the National Academy of Sciences*, USA, 81, 1984, 801–805; J.J. Sepkoski and

D.M. Raup, Periodicity in marine extinction events, in D.K. Elliott (ed.), *Dynamics of Evolution*, Wiley, New York, 1986, 3–36; D.M. Raup, Mass extinctions – a commentary, *Paleontology*, 30, 1987, 1–13; W.B. Harland, A.V. Cox, R.G. Llewellyn, *et al.*, *A Geologic Time Scale*, Cambridge University Press, 1982.

5. D.M. Raup and J.J. Sepkoski, *Proceedings of the National Academy of Sciences, USA*, 81, 1984, 801–805 (quotation from page 805).

6. D.M. Raup, *The Nemesis Affair*, 126–145; R. Muller, *Nemesis*, Guild Publishing, London, 1988, 89–116; A. Hallam, *Great Geological Controversies*, Oxford University Press, 2nd edn, 1989, 200–204; W. Alvarez, *T. rex and the Crater of Doom*, Princeton University Press, Princeton, New Jersey, 1997, 100–101.

7. W.M. Napier and S.V.M. Clube, A theory of terrestrial catastrophism, *Nature*, 282, 1979, 455–459; S.V.M. Clube and W.M. Napier, The role of episodic bombardment in geophysics, *Earth and Planetary Science Letters*, 57, 1982, 251–262; V. Clube and B. Napier, *The Cosmic Serpent*, Faber and Faber, 1982, 30–41.

8. J.G. Hills, Comet showers and the steady-state infall of comets from the Oort cloud, *Astronomical Journal*, 86, 1981, 1730–1740; P.R. Weissman, The Oort Cloud, *Scientific American*, 279(3), 1998, 62–67.

9. M. Davis, P. Hut and R.A. Muller, Extinction of species by periodic comet showers, *Nature*, 308, 1984, 715–717; M. Davis, P. Hut and R.A. Muller, Terrestrial catastrophism – Nemesis or galaxy?, *Nature*, 313, 1985, 503; R.A. Muller, P. Hut, M. Davis and W. Alvarez, Cometary showers and unseen solar companions, *Nature*, 312, 1984, 380–381; R. Muller, *Nemesis*, 89–185.

10. D.P. Whitmire and A.A. Jackson, Are periodic mass extinctions driven by a distant solar companion?, *Nature*, 308, 1984, 713–715; D.P. Whitmire and A.A. Jackson, Cometary showers and unseen solar companions, *Nature*, 312, 1984, 381.

11. S.J. Gould, *The Flamingo's Smile*, Norton, New York, 1985, Penguin Books, Harmondsworth, 1986, 438–450.

12. M.E. Bailey, Nemesis for Nemesis?, *Nature*, 311, 1984, 602–603; S.V.M. Clube and W.M. Napier, Terrestrial catastrophism – Nemesis or galaxy?, *Nature*, 311, 1984, 635–636; J.G. Hills, Dynamical constraints on the mass and perihelion distance of Nemesis and the stability of its orbit, *Nature*, 311, 1984, 636–638; P. Hut, How stable is an astronomical clock that can trigger mass extinctions on Earth?, *Nature*, 311, 1984, 638–640; M.V. Torbett and R. Smoluchowski, Orbital stability of the unseen solar companion linked to periodic extinction events, *Nature*, 311, 1984, 641–642; P.R. Weissman, Cometary showers and unseen solar companions, *Nature*, 312, 1984, 380; R.A. Kerr, Periodic extinctions and impacts challenged, *Science*, 227, 1985, 1451–1452.

13. D.M. Raup, *Extinction – Bad Genes or Bad Luck?*, 164–165; D.B. Carlisle, *Dinosaurs, Diamonds, and Things from Outer Space*, Stanford University Press, Stanford, Calif., 1995, 111–113; J. Gribbin and M. Gribbin, *Fire on Earth*, Simon and Schuster, London, 1996, 182–186; G.L. Verschuur, *Impact!*. Oxford University Press, 1996, 130–131; W. Alvarex, *T. rex and the Crater of Doom*, 100–101; V. Courtillot, *Evolutionary Catastrophes*, Cambridge University Press, 1999, 102–103; C. Frankel, *The End of the Dinosaurs*,

Cambridge University Press, 1999, 170–171; D. Grossman, One disaster after another, *Scientific American*, 284(2), 2001, 21–22.

14. M.R. Rampino and R.B. Stothers, Terrestrial mass extinctions, cometary impacts and the Sun's motion perpendicular to the galactic plane, *Nature*, 308, 1984, 709–712; R.B. Stothers, Mass extinctions and missing matter, *Nature*, 311, 1984, 17.

15. J.A. Kitchell and D. Pena, Periodicity of extinctions in the geologic past – determinist versus stochastic explanations, *Science*, 226, 1984, 689–692.

16. R.D. Schwartz and P.B. James, Periodic mass extinctions and the Sun's oscillation about the galactic plane, *Nature*, 308, 1984, 712–713.

17. S.V.M. Clube and W.M. Napier, Terrestrial catastrophism – Nemesis or galaxy?, *Nature*, 313, 1984, 503; A. Theokas, The origin of comets, *New Scientist*, 11 February 1988, 42–45.

18. M.R. Rampino and R.B. Stothers, Geological rhythms and cometary impacts, *Science*, 226, 1984, 1427–1431; M.R. Rampino and R.B. Stothers, Terrestrial mass extinctions and galactic plane crossings, *Nature*, 313, 1985, 159–160; R.B. Stothers, Terrestrial record of the Solar System's oscillation about the galactic plane, *Nature*, 317, 1985, 338–341.

19. S.M. Stigler, Terrestrial mass extinctions and galactic plane crossings, *Nature*, 313, 1985, 159; P. Thaddeus and G.A. Chanan, Cometary impacts, molecular clouds and the motion of the Sun perpendicular to the galactic plane, *Nature*, 314, 1985, 73–75; J.N. Bahcall and S. Bahcall, The Sun's motion perpendicular to the galactic plane, *Nature*, 316, 1985, 706–708; R.A. Kerr, Periodic extinctions and impacts challenged, *Science*, 227, 1985, 1451–1452.

20. V. Clube and B. Napier, *The Cosmic Winter*, 207–216; S.V.M. Clube and W.M. Napier, Catastrophism is still viable, *Nature*, 318, 1985, 238; W.M. Napier and S.V.M. Clube, Our cometary environment, *Reports on Progress in Physics*, 60, 1997, 293–343; E.M. Leitch and G. Vasisht, Mass extinctions and the Sun's encounters with spiral arms, *New Astronomy*, 3, 1997, 51–56; P. Parsons, Danger zone – spiral arms, *Astronomy*, 26(6), 1998, 22–24; W.M. Napier, Galactic periodicity and the geological record, in M.M. Grady, R. Hutchison, G.J.H. McCall and D.A. Rothery, *Meteorites – Flux with Time and Impact Effects*, Geological Society, London, Special Publication 140, 1998, 19–29.

21. H. Couper, *The Planets*, Pan Books, London, 1985, 104–112; P. Moore, *New Guide to the Planets*, Sidgwick and Jackson, London, 1993, 180–193; D.P. Whitmire and J.J. Matese, Periodic comet showers and planet X, *Nature*, 313, 1985, 36–38, 744.

22. W. Glen (ed.), *The Mass-Extinction Debates*, 27; D.B. Carlisle, *Dinosaurs, Diamonds, and Things from Outer Space*, 115–117; R.A. Kerr, *Science*, 227, 1985, 1451–1452; N. Henbest, Say goodbye to the tenth planet, *New Scientist*, 30 November 1991, 21; R. Matthews, Planet X – going, going . . . but not quite gone, *Science*, 254, 1991, 1454–1455; N. Henbest, *The Planets*, Penguin Books, Harmondsworth, 1994, 197–199; J. Gribbin, Galactic tides bring showers of comets to Earth, *New Scientist*, 14 December 1996, 15; C. Seife, Do comets get a nudge from the galaxy?, *Science*, 274, 1996, 920; S.R. Taylor, *Destiny or Chance*, Cambridge University Press, 1998, 62–63; J. Hecht,

Then there were ten, *New Scientist*, 16 October 1999, 5; P. Bond, Is planet X there?, *Astronomy Now*, 14(6), 2000, 56–58.

23. W.A. Alvarez and R.A. Muller, Evidence from crater ages for periodic impacts on Earth, *Nature*, 308, 1984, 718–720; M.R. Rampino and R.B. Stothers, *Nature*, 308, 1984, 709–712.

24. P.R. Weissman, Terrestrial impactors at geological boundary events – comets or asteroids?, *Nature*, 314, 1985, 517–518.

25. J. Gribbin and M. Gribbin, *Fire on Earth*, 107–111; G.L. Verschuur, *Impact!*, 53–54; V. Clube and B. Napier, *The Cosmic Winter*, Basil Blackwell, Oxford, 1980, 133; C. Sagan and A. Druyan, *Comet*, Headline, London, 1997, 141; M.E. Bailey and V.V. Emel'Yanenko, Cometary capture and the nature of impactors, in *Meteorites – Flux with Time and Impact Effects*, 11–17.

26. C.C. Albritton, *Catastrophic Episodes in Earth History*, 105–106; A.G. Fischer, The two Phanerozoic supercycles, in W.A. Berggren and J.A. van Couvering (eds.), *Catastrophes and Earth History*, Princeton University Press, Princeton, New Jersey, 1984, 129–150; T.S. Kemp, *Fossils and Evolution*, Oxford University Press, 1999, 214–215.

27. A. Hallam, *Great Geological Controversies*, 30–64, 216–226.

28. A. Hallam, Concepts of change, *Nature*, 310, 1984, 435.

29. A. Hallam, *Nature*, 310, 1984, 435; A. Hallam, *Great Geological Controversies*, 200–204; A. Hallam, The causes of mass extinctions, *Nature*, 308, 1984, 686–687; A. Hallam, Pre-quaternary sea-level changes, *Annual Review of Earth and Planetary Sciences*, 12, 1984, 205–243; A. Hallam, Asteroids and extinction – no cause for concern, *New Scientist*, 8 November 1984, 30–33; A. Hallam, The Pliensbachian and Tithonian extinction events, *Nature*, 319, 1986, 765–768; A. Hallam, The case for sea-level change as a dominant causal factor in mass extinction of marine invertebrates, in W.G. Chaloner and A. Hallam (eds.), *Evolution and Extinction*, Cambridge University Press, 1989, 197–216.

30. R.A.F. Grieve, V.L. Sharpton, A.K. Goodacre and J.B. Garvin, A perspective on the evidence for periodic cometary impacts on Earth, *Earth and Planetary Science Letters*, 76, 1985/86, 1–9; R.A.F. Grieve, Hypervelocity impact cratering, in S.V.M. Clube (ed.), *Catastrophes and Evolution – Astronomical Foundations*, Cambridge University Press, 1989, 57–79.

31. A. Hoffman, Patterns of family extinction depend on definition and geological timescale, *Nature*, 315, 1985, 659–662; A. Hoffman, Periodicity of extinctions, *Science*, 230, 1985, 8; M.J. Benton, Interpretations of mass extinction, *Nature*, 314, 1985, 496–497; J. Maddox, Periodic extinctions undermined, *Nature*, 315, 1985, 627; C. Patterson and A.B. Smith; J.J. Sepkoski, Is the periodicity of extinctions a taxonomic artifact?, *Nature*, 330, 1987, 248–252; J.F. Quinn, On the statistical detection of cycles in extinctions in the marine fossil record, *Paleobiology*, 13, 1987, 465–478; H.R. Shaw, The periodic structure of the natural record and non-linear dynamics, *Eos*, 68, 1987, 1651–1665; A.J. Boucot, Periodic extinctions within the Cenozoic, *Nature*, 331, 1988, 395–396; A. Hoffman, Mass extinctions – the view of a sceptic, *Journal of the Geological Society, London*, 146, 1989, 21–35; A. Hoffman, *Arguments on Evolution – a Paleontologists Perspective*, Oxford University Press, 1989;

A. Hoffman, What, if anything, are mass extinctions?, in *Evolution and Extinction*, 13–22; C.R.C. Paul, Patterns of evolution and extinction in invertebrates, in K.C. Allen and D.E.G. Briggs (eds.), *Evolution and the Fossil Record*, Belhaven Press, London, 1989, 99–121.

32. D.M. Raup and J.J. Sepkoski, Periodic extinction of families and genera, *Science*, 231, 1986, 833–836; J.J. Sepkoski, Phanerozoic overview of mass extinctions, in D.M. Raup and D. Jablonski (eds.), *Patterns and Processes in the History of Life*, Springer-Verlag, Berlin, 1986, 277–295; J.S. Trefil and D.M. Raup, Numerical simulations and the problem of periodicity in the cratering record, *Earth and Planetary Science Letters*, 82, 1987, 159–164.

33. D.M. Raup, Magnetic reversals and mass extinctions, *Nature*, 314, 1985, 341–343; T.M. Lutz, The magnetic reversal record is not periodic, *Nature*, 317, 1985, 404–407; D.M. Raup, Rise and fall of periodicity, *Nature*, 317, 1985, 384–385.

34. P.C. Pal and K.M. Creer, Geomagnetic reversal spurts and episodes of extraterrestrial catastrophism, *Nature*, 320, 1986, 148–150; S.M. Stigler, R.B. Stothers, P.L. McFadden, Aperiodicity of magnetic reversals, *Nature*, 330, 1987, 26–27; A. Mazaud and C. Laj, The 15 m.y. geomagnetic reversal periodicity – a quantitative test, *Earth and Planetary Science Letters*, 107, 1991, 689–696; M.R. Rampino and B.M. Haggerty; M.J. Benton, Mass extinctions and periodicity, *Science*, 269, 1995, 617–619.

35. M.R. Rampino and R.B. Stothers, Flood basalt volcanism during the past 250 Myr, *Science*, 241, 1988, 663–668.

36. D.M. Raup and J.J. Sepkoski, *Science*, 231, 1986, 833–836 (quotation from page 836).

37. W.T. Fox, Harmonic analysis of periodic extinctions, *Paleobiology*, 13, 1987, 257–271; D.M. Raup and J.J. Sepkoski; S.M. Stigler and M.J. Wagner, Testing for periodicity of extinction, *Science*, 241, 1988, 94–99; D.M. Raup and G.E. Boyajian, Patterns of generic extinction in the fossil record, *Paleobiology*, 14, 1988, 109–125; K.C. Allen and D.E.G. Briggs (eds.), *Evolution and the Fossil Record*, 35–37.

38. A. Hallam and P.B. Wignall, *Mass Extinctions and their Aftermath*, Oxford University Press, 1997, 4–12 (quotation from page 11).

39. D. Jablonski, Background and mass extinctions, *Science*, 231, 1986, 129–133; J.J. Sepkoski, Environmental trends in extinction during the Paleozoic, *Science*, 235, 1987, 64–66; D.M. Raup, A kill curve for Phanerozoic marine species, *Paleobiology*, 17, 1991, 37–48; D.M. Raup, Large-body impact and extinction in the Phanerozoic, *Paleobiology*, 18, 1992, 80–88; D. Jablonski, The biology of mass extinctions, in *Evolution and Extinction*, 117–128; D. Jablonski and D.M. Raup, Selectivity of end-Cretaceous marine bivalve extinctions, *Science*, 268, 1995, 389–391; K. Roy, The roles of mass extinction and biotic interaction in large-scale replacements – a re-examination using the fossil record of stromboidean gastropods, *Paleobiology*, 22, 1996, 436–452; D. Jablonski, Lessons from the past – evolutionary impacts of mass extinctions, *Proceedings of the National Academy of Sciences, USA*, 98, 2001, 5393–5398; D. Jablonski, Survival without recovery, *Proceedings of the National Academy of Sciences, USA*, 99, 2002, 8139–8144; T.S. Kemp, *Fossils and Evolution*, 188–202.

40. M.L. McKinney, Mass extinction patterns of marine invertebrate groups and some implications for a causal phenomenon, *Paleobiology*, 11, 1985, 227–233;

M.L. McKinney, Taxonomic selectivity and continuous variation in mass and background extinctions of marine taxa, *Nature*, 325, 1987, 143–145; M.L. McKinney and D. Frederick, Extinction and population dynamics – new methods and evidence from Paleogene foraminifera, *Geology*, 20, 1992, 343–346; M.L. Droser, D.J. Bottjer and P.M. Sheehan, Evaluating the ecological architecture of major events in the Phanerozoic history of marine invertebrate life, *Geology*, 25, 1997, 167–170; D. Jablonski, Extinctions – a paleontological perspective, *Science*, 253, 1991, 754–757; N.L. Gilinsky and I.J. Good, Probabilities of origination, persistence, and extinction of families of marine invertebrate life, *Paleobiology*, 17, 1991, 145–166; M.C. Rhodes and C.W. Thayer, Mass extinctions – ecological selectivity and primary production, *Geology*, 19, 1991, 877–880; A.E. Hubbard and N.L. Gilinsky, Mass extinctions as statistical phenomena – an examination of the evidence using χ^2 tests and bootstrapping, *Paleobiology*, 18, 1992, 148–160.

41. D. Jablonski, The future of the fossil record, *Science*, 284, 1999, 2114–2116 (quotation from page 2114).

42. M.J. Benton, Mass extinction among non-marine tetrapods, *Nature*, 316, 1985, 811–814; M.J. Benton, Patterns of evolution and extinction in vertebrates, in *Evolution and the Fossil Record*, 218–241; M.J. Benton, Mass extinctions among tetrapods and the quality of the fossil record, in *Evolution and Extinction*, 129–146.

43. M.J. Benton, Diversification and extinction in the history of life, *Science*, 268, 1995, 52–58. See also D. Hewzulla, M.C. Boulter, M.J. Benton and J.M. Halley, Evolutionary patterns from mass originations and mass extinctions, *Philosophical Transactions of The Royal Society of London*, B, 354, 1999, 463–469.

44. V. Courtillot and Y. Gaudemer, Effects of mass extinctions on biodiversity, *Nature*, 381, 1996, 146–148.

45. V. Courtillot, *Evolutionary Catastrophes*, 9–12; M.J. Benton, M.A. Wills and R. Hitchin, Quality of the fossil record through time, *Nature*, 403, 2000, 534–537.

46. M.J. Benton, *Science*, 268, 1995, 52–58.

47. D.B. Carlisle, *Dinosaurs, Diamonds, and Things from Outer Space*, 8.

48. S.M. Stanley, Delayed recovery and the spacing of mass extinctions, *Paleobiology*, 16, 1990, 401–414.

49. W. Glen (ed.), *The Mass-Extinction Debates*, 25–29; D.B. Carlisle, *Dinosaurs, Diamonds, and Things from Outer Space*, 108–117; A.W. Wolfendale and D.A. Wilkinson, Periodic mass extinctions – some astronomical difficulties, in *Catastrophes and Evolution – Astronomical Foundations*, 231–239; C.R. Chapman and D. Morrison, Plenum Press, New York, 1989, 97–109; J.S. Lewis, *Rain of Iron and Ice*, Addison-Wesley, Reading, Mass., 1996, 143–144.

50. D.M. Raup, *Extinction – Bad Genes or Bad Luck?*, 165.

51. S.J. Gould, Taxonomy of death, *Nature*, 313, 1995, 505–506.

52. D.M. Raup, *Extinctions – Bad Genes or Bad Luck?*, 165–171; D.M. Raup, Extinctions – bad genes or bad luck?, *Acta Geologica Hispanica*, 16, 1981, 25–33; D.M. Raup, Extinctions – bad genes or bad luck?, *New Scientist*, 14 September 1991, 46–49; D.M. Raup, *Paleobiology*, 17, 1991, 37–48; D.M. Raup, *Paleobiology*, 18, 1992, 80–88.

53. A. Montanari, A. Campo Bagatin and P. Farinella, Earth cratering record and impact energy flux in the last 150 Ma, *Planetary and Space Science*, 46, 1998, 271–281. (First presented at the Tuguska96 conference in Bologna.)

54. D.B. Carlisle, *Dinosaurs, Diamonds, and Things from Outer Space*, 184–200; W. Alvarez, T. Hansen, P. Hut, *et al*., Uniformitarians and the response of Earth scientists to the theory of impact crises, in *Catastrophes and Evolution – Astronomical Foundations*, 13–24.

55. P. Hut, W. Alvarez, W.P. Elder, *et al*., Comet showers as a cause of mass extinctions, *Nature*, 329, 1987, 118–126.

56. V. Clube and B. Napier, *The Cosmic Winter*, 131–154; D. Steel, *Rogue Asteroids and Doomsday Comets*, Wiley, New York, 1995, 109–136; J. Gribbin and M. Gribbin, *Fire on Earth*, 119–143; W.M. Napier and S.V.M. Clube, Our cometary environment, *Reports on Progress in Physics*, 60, 1997, 293–343; S.V.M. Clube, The catastrophic role of giant comets, in *Catastrophes and Evolution – Astronomical Foundations*, 81–112; B. Napier, Comets, *Astronomy Now*, 11(3), 1997, 41–43; M.E. Bailey, Sources and populations of near-Earth objects, in B.J. Peiser, T. Palmer and M.E. Bailey, *Natural Catastrophes during Bronze Age Civilisations*, Archaeopress, Oxford, 1998, 10–20.

57. L.M. Van Valen, Catastrophes, expectations and the evidence, *Paleobiology*, 10, 1984, 121–137; L.M. Van Valen, The case against impact extinctions, *Nature*, 311, 1984, 17–18; L.M. Van Valen, Iridium, impacts and para-volcanism, *Nature*, 316, 1985, 396; A. Hallam, *Nature*, 308, 1984, 686–687; A. Hallam, in *Evolution and Extinction*, 197–216; N. MacLeod, Impacts and marine invertebrate extinctions, in *Meteorites – Flux with Time and Impact Effects*, 217–246.

58. A. Hallam, *Nature*, 310, 1984, 435.

59. A. Hallam, Radiations and extinctions in relation to environmental change in the marine Lower Jurassic of northwest Europe, *Paleobiology*, 13, 1987, 152–168; A. Hallam, The end-Triassic mass extinction event, *Geological Society of America Special Paper*, 247, 1990, 577–583; R. Morante and A. Hallam, Organic carbon isotope record across the Triassic–Jurassic boundary of Austria and its bearing on the cause of the mass extinction, *Geology*, 24, 1996, 391–394; A. Hallam, Catastrophism and geology, in *Catastrophes and Evolution – Astronomical Foundations*, 25–55; P. Wignall, The day the world nearly died, *New Scientist*, 25 January 1992, 51–55; R.A. Berner, Drying, O_2 and mass extinction, *Nature*, 340, 1989, 603–604; T.J. Crowley and G.R. North, Abrupt climate change and extinction events in Earth history, *Science*, 240, 1988, 996–1002; P.B. Wignall and R.J. Twitchett, Oceanic anoxia and the End Permian mass extinction, *Science*, 272, 1996, 1155–1158; A. Hallam, Mass extinctions in Phanerozoic time, in *Meteorites – Flux with Time and Impact Effects*, 259–274.

60. A. Hallam and P.B. Wignall, *Mass Extinctions and their Aftermath*, 245.

61. A. Hallam and P.B. Wignall, *Mass Extinctions and their Aftermath*, 251–252.

62. V. Clube and B. Napier, *The Cosmic Serpent*, 122–130; V. Clube and B. Napier, *The Cosmic Winter*, 219–236; W.M. Napier, Galactic periodicity and the geological record, in *Meteorites – Flux with Time and Impact Effects*, 19–29; W.M. Napier, Cometary catastrophes, cosmic dust and ecological disasters in historical times – the astronomical framework, in *Natural Catastrophes during Bronze Age Civilisations*, 21–32; S.V.M. Clube,

The problem of historical catastrophism, in *Natural Catastrophes during Bronze Age Civilisations*, 232–249.

63. V. Clube and B. Napier, *The Cosmic Serpent*, 128–129.

64. R.S. White and D.P. McKenzie, Volcanism at rifts, *Scientific American*, 261(1), 1989, 44–55 (quotation from page 55).

65. V.E. Courtillot, A volcanic eruption, *Scientific American*, 263(4), 1990, 53–60.

66. M.R. Rampino and K. Caldeira, Major episodes of geologic change – correlations, time structure and possible causes, *Earth and Planetary Science Letters*, 114, 1993, 215–227.

67. I. Liritzis, Cyclicity in terrestrial upheavals during the Phanerozoic Eon, *Quarterly Journal of the Royal Astronomical Society*, 34, 1993, 251–260.

68. D.B. Carlisle, *Dinosaurs, Diamonds, and Things from Outer Space*, 102–104; D. Steel, *Rogue Asteroids and Doomsday Comets*, 100–107; M. Gribbin and M. Gribbin, *Fire on Earth*, 186–195; M.R. Rampino and R.B. Stothers, Geological rhythms and cometary impacts, *Science*, 226, 1984, 1427–1431; M.R. Rampino and R.B. Stothers, Flood basalt volcanism during the past 250 Myr, *Science*, 241, 1988, 663–668; M.R. Rampino, Impact cratering and flood basalt volcanism, *Nature*, 327, 1987, 468; M. Rampino, Dinosaurs, comets and volcanoes, *New Scientist*, 18 February 1989, 54–58; M.R. Rampino, Shiva versus Gaia: cosmic effects on the long-term evolution of the biosphere, in S.H. Schneider and P. Boston (eds.), *Scientists on Gaia*, MIT Press, Cambridge, Mass., 1991, 382–391.

69. W.M. Napier, Galactic periodicity and the geological record, in *Meteorites – Flux with Time and Impact Effects*, 19–29; M.R. Rampino, The Shiva hypothesis: impacts, mass extinctions and the galaxy, *Earth, Moon and Planets*, 72, 1996, 441–460; M.R. Rampino, B.M. Haggerty and T.C. Pagano, A unified theory of impact crises and mass extinctions, *Annals of the New York Academy of Sciences*, 822, 1997, 403–431; M.R. Rampino, The Shiva hypothesis: impacts, mass extinctions and the galaxy, *Planetary Report* (Planetary Society, Pasadena, Calif.), 18(1), 1998, 6–11.

70. V. Courtillot, *Evolutionary Catastrophes*, 88–118, 135–143; A. Prokoph, A.D. Fowler and R.T. Patterson, Periodically forced self-organization in the long-term evolution of planktic foraminifera, *Canadian Journal of Earth Sciences*, 38, 2001, 293–308.

Chapter 26

1. M.J. Benton, *The Rise of the Mammals*, Crescent Books, New York, 1991, 120–122; P. Andrews and C. Stringer, The primates' progress, in S.J. Gould (ed.), *The Book of Life*, Ebury Hutchinson, London, 1993, 218–251 (particularly pages 218–229); D. Norman, *Prehistoric Life*, Boxtree, London, 1994, 210–216; P. Ward, *The End of Evolution*, Weidenfeld and Nicolson, London, 1995, 177–182; P. Jordan, *Early Man*, Sutton, Stroud, Gloucestershire, 1999, 1–9; R. Lewin, *Principles of Human Evolution*, Blackwell Science, Malden, Mass, and Oxford, 1998, 49–63, 201–212; R. Lewin, *Human Evolution*, Blackwell Science, Malden, Mass. and Oxford, 1999, 21–24, 89–92; T. Haines, *Walking with Beasts*, BBC Books, London, 2001, 14, 100–141.

2. C.C. Albritton, *Catastrophic Episodes in Earth History*, Chapman and Hall, London, 1989, 86–88; R.A.F. Grieve and E.M. Shoemaker, The record of past impacts on

Earth, in T. Gehrels (ed.), *Hazards due to Comets and Asteroids*, University of Arizona Press, Tucson, 1994, 417–462; R.M. Hough, I. Gilmour, C.T. Pillinger, *et al.*, Diamond and silicon carbide in impact melt rock from the Ries impact crater, *Nature*, 378, 1995, 41–44; E. von Engelhardt, Suevite breccia of the Ries impact crater, Germany, *Meteoritics and Planetary Science*, 32, 1997, 545–554; A. Deutsch, M. Ostermann and V.L. Masaitis, Geochemistry and neodymium-strontium isotope signature of tektite-like objects from Siberia, *Meteoritics and Planetary Science*, 32, 1997, 679–686; R.A.F. Grieve, in M.M. Grady, R. Hutchison, G.J.H. McCall and D.A. Rothery, *Meteorites – Flux with Time and Impact Effects*, Geological Society, London, Special Publication 140, 1998, 105–131; I. Gilmour, Geochemistry of carbon in terrestrial impact processes, in *Meteorites – Flux with Time and Impact Effects*, 205–216.

3. M. Pickford, Geochronology of the Hominoidea, in J.G. Else and P.C. Lee (eds.), *Primate Evolution*, Cambridge University Press, 1986, 123–128; C. Janis, Victors by default, in *The Book of Life*, 168–217 (particularly pages 192–206); C. Tudge, *The Day before Yesterday*, Pimlico, London, 1995, 42–74; R. Osborne and D. Tarling, *The Viking Historical Atlas of the Earth*, Viking, London, 1995, 130–131, 140–141; S. Lamb and D. Sington, *Earth Story*, BBC Books, London, 1998, 199–202; R. Redfern, *Origins*, Cassell, London, 2000, 227–232.

4. M.R. Rampino and R.B. Stothers, Flood basalt volcanism during the past 250 million years, *Science*, 241, 1988, 663–668; P. Lipman, Chasing the volcano, *Earth*, 6(6), 1997, 32–39; V. Courtillot, *Evolutionary Catastrophes*, Cambridge University Press, 1999, 139.

5. R. Osborne and D. Tarling, *The Viking Historical Atlas of the Earth*, 138–139; S. Lamb and D. Sington, *Earth Story*, 112; V. Courtillot, *Evolutionary Catastrophes*, 95; R. Redfern, *Origins*, 220–227; S.M. Stanley, *Extinction*, Scientific American Books, New York, 1987, 191–196; B. Livermore, Life behind the cold wall, *Earth*, 6(6), 1997, 50–56.

6. R.A.F. Grieve and E.M. Shoemaker, in *Hazards due to Comets and Asteroids*, 417–462; M.R. Rampino and B.M. Haggerty, Extraterrestrial impacts and mass extinctions of life, in *Hazards due to Comets and Asteroids*, 827–857; M.R. Voorhies, Ancient ashfall creates a Pompeii of prehistoric animals, *National Geographic*, 159(1), 1981, 66–75; D.M. Raup and J.J. Sepkoski, Periodicity of extinctions in the geologic past, *Proceedings of the National Academy of Sciences, USA*, 81, 1984, 801–805; F. Asaro, H.V. Michel. W. Alvarez and L.W. Alvarez, Impacts and multiple iridium anomalies, *Eos*, 69, 1988, 301; C.J. Orth, Geochemistry of the bio-event horizon, in S.K. Donovan (ed.), *Mass Extinctions – Processes and Evidence*, Belhaven Press, London, 1989, 37–72; D.M. Raup, *Extinction – Bad Genes or Bad Luck*, Norton, New York and London, 1991, 173; R. Lewin, *Human Evolution*, 21–25; Ashfall Fossil Beds State Historical Park website, http://ngp.ngpc.state.ne.us/parks/ashfall.html.

7. A. Smith, *The Great Rift*, Sterling, New York, 1989, 15–59; Y. Coppens, East side story – the origin of humankind, *Scientific American*, 270(5), 1994, 62–69; R. Leakey, *The Origin of Humankind*, Weidenfeld and Nicolson, London, 1994, 15–16; R. Caird, *Ape Man*, Boxtree, London, 1994, 55–57.

8. M.J. Benton, *The Rise of the Mammals*, 122–123; P. Andrews and C. Stringer, in *The Book of Life*, 229–234; D. Norman, *Prehistoric Life*, 216–219; G.D. Brown, *Human Evolution*, Brown, Dubuque, Iowa, 1995, 78–95; P. Andrews and D. Pilbeam, The nature of the evidence, *Nature*, 379, 1996, 123–124; R. Lewin, *Principles of Human Evolution*, 205–212; R. Lewin, *Human Evolution*, 89–92; C. Zimmer, Kenyan skeleton shakes ape family tree, *Science*, 285, 1999, 1335–1337; S. Ward, B. Brown, A. Hill, *et al.*, Equatorius – a new hominoid genus from the Middle Miocene of Kenya, *Science*, 285, 1999, 1382–1386; J. Hecht, Earliest ancestor unearthed, *New Scientist*, 13 July 2002, 6; A. Gibbons, First member of human family uncovered, *Science*, 297, 2002, 171–173; B. Wood, Hominid revelations from Chad, *Nature*, 418, 2002, 133–135; M. Brunet, F. Guy, D. Pilbeam, *et al.*, A new hominid from the Upper Miocene of Chad, Central Africa, *Nature*, 418, 2002, 145–151; P. Vignaud, P. Duringer, H.T. Mackaye, *et al.*, Geology and paleontology of the Upper Miocene Toros-Menalla hominid locality, Chad, *Nature*, 418, 2002, 152–155.

9. M. Pickford, in *Primate Evolution*, 123–128; C.K. Brain, *The Hunters or the Hunted?*, University of Chicago Press, Chicago and London, 1981; R. Leakey, *The Origin of Humankind*, 21–22, 61–73; G.D. Brown, *Human Evolution*, 96–97, 116–119; C. Tudge, *The Day before Yesterday*, 194–201; S. Armstrong, Taung child fell from the sky, *New Scientist*, 9 September 1995, 7; L.R. Berger and R.J. Clarke, Bird of prey involvement in the collection of the Taung child fauna, *Journal of Human Evolution*, 29, 1995, 275–299; C. Stringer and R. McKie, *African Exodus*, Jonathan Cape, London, 1996, 20–21; R. McKie, *Ape Man*, BBC Books, London, 2000, 48–49; L.R. Berger and B. Hilton-Barber, *In the Footsteps of Eve*, National Geographic Adventure Press, Washington, D.C., 2000, 151–163.

10. R. Lewin, *Human Evolution*, 21–24, 93–98; R. McKie, *Ape Man*, 18–23; P. Jordan, *Early Man*, 9–18, 23–28; L.R. Berger and B. Hilton-Barber, *In the Footsteps of Eve*, 21–26; A.C. Hardy, Was man more aquatic in the past?, *New Scientist*, 17 March 1960, 642–645; E. Morgan, *The Descent of Woman*, Souvenir Press, London, 1972; E. Morgan, *The Aquatic Ape*, Souvenir Press, London, 1982; E. Morgan, *The Aquatic Ape Hypothesis*, Souvenir Press, London, 1997; K. Douglas, Taking the plunge, *New Scientist*, 25 November 2000, 28–33; G. WoldeGabriel, G. Heiken, T.D. White, *et al.*, Volcanism, tectonism, sedimentation and the paleoanthropological record in the Ethiopian Rift System, in F.W. McCoy and G. Heiken (eds.), *Volcanic Hazards and Disasters in Human Antiquity*, Geological Society of America Special Paper, 345, 2000, 83–99; J. Gribbin and J. Cherfas, *The First Chimpanzee*, Penguin Books, Harmondsworth, 2001, 182–204.

11. Y. Coppens, *Scientific American*, 270(5), 1994, 62–69; R. Leakey, *The Origin of Humankind*, 12–20; R. Caird, *Ape Man*, 52–65; R. Lewin, *Principles of Human Evolution*, 211–228; R. Lewin, *Human Evolution*, 91–98; G.D. Brown, *Human Evolution*, 96–102; P. Jordan, *Early Man*, 9–18; R. McKie, *Ape Man*, 22–23; N.T. Boaz, First steps into the human dawn, *Earth*, March 1992, 36–43; C.O. Lovejoy, Modelling human origins, in D.T. Rasmussen (ed.), *The Origins and Evolution of Humans and Humanness*, Jones and Bartlett, Boston, 1993, 1–28; C. Stringer and R. McKie, *African Exodus*, Jonathan Cape, London, 1996, 9–18.

12. C. Stringer and R. McKie, *African Exodus*, 13–15; R. Lewin, *Principles of Human Evolution*, 253–258; R. Lewin, *Human Evolution*, 107–109; P. Jordan, *Early Man*, 18–20; R. McKie, *Ape Man*, 23–25; L.R. Berger and B. Hilton-Barber, *In the Footsteps of Eve*, 31–33, 191–196, 220–222; T.D. White, G. Suwa and B. Asfaw, *A. ramidus*, a new species of early hominid from Aramis, Ethiopia, *Nature*, 371, 1994, 306–312; M.G. Leakey, C.S. Feibel, I. McDougal and A. Walker, New four-million-year-old hominid species from Kanapoi and Allia Bay, Kenya, *Nature*, 376, 1995, 565–571; J.H. Schwartz, *Sudden Origins*, Wiley, New York, 1999, 15–18; L.C. Aiello and M. Collard, Our newest oldest ancestor?, *Nature*, 410, 2001, 526–527; H. Gee, Return to the planet of the apes, *Nature*, 412, 2001, 131–132; Y. Haile-Selassie, Late Miocene hominids from the Middle Awash, Ethiopia, *Nature*, 412, 2001, 178–181; P. Shipman, Hunting the first hominid, *American Scientist*, 90(1), 2002, 25–27; J. Hecht, *New Scientist*, 13 July 2002, 6; A. Gibbons, *Science*, 297, 2002, 171–173; B. Wood, *Nature*, 418, 2002, 133–136; M. Brunet, F. Guy, D. Pilbeam, *et al.*, *Nature*, 418, 2002, 145–151; M. Lemonick and A. Dorfman, Father of us all?, *Time*, 160(4), 2002, 54–61; F. Guterl, All in the family, *Newsweek*, 22/29 July 2002, 68–71.

13. G.D. Brown, *Human Evolution*, 125–128; P. Jordan, *Early Man*, 21–22; R. Lewin, *Principles of Human Evolution*, 242–258; R. Lewin, *Human Evolution*, 104–109; R. McKie, *Ape Man*, 10–35; T. Haines, *Walking with Beasts*, 142–181; M.D. Leakey and R.L. Hay, Pliocene footprints in the Laetoli beds at Laetoli, northern Tanzania, *Nature*, 278, 1979, 317–323; D.C. Johanson and M. Edey, *Lucy – The Beginnings of Humankind*, Granada, London, 1981, 154–256; J.D. Clark, B. Asfaw, G. Assefa, *et al.*, Palaeoanthropological discoveries in the Middle Awash Valley, Ethiopia, *Nature*, 307, 1984, 423–428; N. Agnew and M. Demas, Preserving the Laetoli footprints, *Scientific American*, 279(3), 1998, 26–37.

14. R. Moore, *Man, Time and Fossils*, Jonathan Cape, London, 1962, 374.

15. R. Leakey, *The Origin of Humankind*, 34–36; R.L. Susman and J. Stern, The locomotor behavior of *Australopithecus afarensis*, *American Journal of Physical Anthropology*, 60, 1983, 279–317; D.C. Johanson and J. Shreeve, *Lucy's Child*, Penguin Books, London, 1990, 193–201; P. Shipman, These ears were made for walking, *New Scientist*, 30 July 1994, 26–29.

16. R. Lewin, *Principles of Human Evolution*, 274–278; J.H. Schwartz, *Sudden Origins*, 20–27; L.R. Berger and B. Hilton-Barber, *In the Footsteps of Eve*, 226–257; J. Gribbin and J. Cherfas, *The First Chimpanzee*, 43; R.J. Clarke and P.V. Tobias, Sterkfontein Member 2 foot bones of the oldest South African hominid, *Science*, 269, 1995, 521–524; R. Lewin, Little Foot stumbles into the crossfire, *New Scientist*, 5 August 1995, 14; J.F. McKee, P.V. Tobias and R.J. Clarke, Faunal evidence and Sterkfontein Member 2 foot bones of early hominid, *Science*, 271, 1996, 1301–1302; A.D. Smith, Vital clue in evolution riddle, *The Guardian*, 10 December 1998, 20; H. Gee, The face of Cinderella, *Nature*, 396, 1998, 521; R.J. Clarke, First ever discovery of a well-preserved skull and associated skeleton of *Australopithecus*, *South African Journal of Science*, 94, 1998, 460–463; D.E. Lieberman, Another face in our family tree, *Nature*, 410, 2001, 419–420; M.G. Leakey, F. Spoor, F.H. Brown, *et al.*, New hominin genus from eastern Africa shows diverse middle Pliocene lineages, *Nature*, 410,

2001, 433–440; K.E. Lange, Meet Kenya Man, *National Geographic*, 200(4), 2001, 84–89.

17. R. Caird, *Ape Man*, 56–65; G.D. Brown, *Human Evolution*, 100–102; C. Stringer and R. McKie, *African Exodus*, 18–19; R. Lewin, *Principles of Human Evolution*, 215–229; R. Lewin, *Human Evolution*, 93–98; P. Jordan, *Early Man*, 23–28; R. McKie, *Ape Man*, 22–23; L.R. Berger and B. Hilton-Barber, *In the Footsteps of Eve*, 25–29; P. Shipman, *New Scientist*, 30 July 1994, 26–29; A. Walker and P. Shipman, *The Wisdom of Bones*, Weidenfeld and Nicolson, London, 1996, 199–207; R. Gore, The first steps, *National Geographic*, 191(2), 1997, 72–99; I. Tattersall, *Becoming Human*, Harcourt Brace, New York, 1998, 109–124; M. Walker, Walk this way, *New Scientist*, 16 October 1999, 16.

18. G.D. Brown, *Human Evolution*, 129–132; R. Lewin, *Principles of Human Evolution*, 263–282; *Human Evolution*, 112–118; P. Jordan, *Early Man*, 23–25; R. McKie, *Ape Man*, 41–49; L.R. Berger and B. Hilton-Barber, *In the Footsteps of Eve*, 36–38, 164–187, 225, 305; R. Fortey, *Life – An Unauthorised Biography*, Flamingo, London, 1998, 337–345; H.M. McHenry and L.R. Berger, Apelike body proportions in *Australopithecus africanus* and their implications for the origin of *Homo*, *American Journal of Physical Anthropology*, 22 (suppl.), 1996, 163–164; H.M. McHenry and L.R. Berger, Body proportions in *Australopithecus afarensis* and *A. Africanus* and the origin of the genus *Homo*, *Journal of Human Evolution*, 35, 1998, 1–22; D. Falk, Hominid brain evolution – looks can be deceiving, *Science*, 280, 1998, 1714; G.C. Conroy, G.W. Weber, H. Seidler, *et al.*, Endocranial capacity in an early hominid cranium from Sterkfontein, South Africa, *Science*, 280, 1998, 1730–1731; E. Culotta, A new human ancestor?, *Science*, 284, 1999, 572–573; J. de Heinzelin, J.D. Clark, T. White, *et al.*, Environment and behavior of 2.5-million-year-old Bouri hominids, *Science*, 284, 1999, 625–629; B. Asfaw, T. White, O. Lovejoy, *et al.*, *A. garhi*, a new species of early hominid from Ethiopia, *Science*, 284, 1999, 629–633; D.E. Lieberman, *Nature*, 410, 2001, 419–420.

19. E. Samuel, Lethal supernova, *New Scientist*, 19 January 2002, 17; A. Longstaff, Impact hazard overated?, *Astronomy Now*, 16(3), 2002, 13; R. McKie, How star blasts forged mankind, *The Observer*, 17 February 2002, 14.

20. P. Andrews and C. Stringer, in *The Book of Life*, 235–239; R. Leakey, *The Origin of Humankind*, 22–26; R. Caird, *Ape Man*, 16–17, 56, 121–124; G.D. Brown, *Human Evolution*, 132–137; R. Lewin, *Principles of Human Evolution*, 57, 263–282; R. Lewin, *Human Evolution*, 24, 112–118; R. McKie, *Ape Man*, 45–67; T.G. Bromage and F. Schrenk, Biogeographic and climatic basis for a narrative of early hominid evolution, *Journal of Human Evolution*, 28, 1995, 109–114; J.R. Marlow, C.B. Lange, G. Wefer and A. Rossell-Melé, Upwelling intensification as part of the Pliocene–Pleistocene climate transition, *Science*, 290, 2000, 2288–2291; R. McKie, How Ice Ages increased our brainpower, *The Observer*, 14 April 2002, 16; W. Calvin, *Brain for All Seasons*, Chicago University Press, Chicago and London, 2002.

21. D.C. Johanson and J. Shreeve, *Lucy's Child*, 137–210; R. Leakey, *The Origin of Humankind*, 26–29, 36–41; R. Caird, *Ape Man*, 81–85; G.D. Brown, *Human Evolution*, 138–151; P. Jordan, *Early Man*, 29–30, 34–36; R. McKie, *Ape Man*, 57–61; D.C. Johanson, F.T. Masao, G.G. Eck, *et al.*, New partial skeleton of *H. habilis* from Olduvai Gorge, Tanzania, *Nature*, 327, 1987, 205–209.

22. R. Lewin, *Principles of Human Evolution*, 283–296; R. Lewin, *Human Evolution*, 119–124; C.B. Stringer, The credibility of *Homo habilis*, in B. Wood, L. Martin and P. Andrews (eds.), *Major Topics in Primate and Human Evolution*, Cambridge University Press, 1986, 266–294; C.P. Groves, *A Theory of Human and Primate Evolution*, Clarendon Press, Oxford, 1989, 213–321; R. Leakey and R. Lewin, *Origins Reconsidered*, Little, Brown & Co, London, 1992, 97–120; A. Bilsborough, *Human Evolution*, Blackie, London, 1992, 109–135 (particularly pages 124–130), 224–234; I. Tattersall, Once we were not alone, *Scientific American*, 282(1), 2000, 38–44; S.H. Ambrose, Paleolithic technology and human evolution, *Science*, 291, 2001, 1748–1753.

23. M.A. Edey and D.C. Johanson, *Blueprints*, Oxford University Press, 1990, 325–368; D. Johanson and J. Shreeve, *Lucy's Child*, 102–134, 245–280; R. Leakey and R. Lewin, *Origins Reconsidered*, 121–134; R. Leakey, *The Origin of Humankind*, 30–36; R. Fortey, *Life – An Unauthorised Biography*, 335–349; R. Lewin, *Principles of Human Evolution*, 297–307; R. Lewin, *Human Evolution*, 125–129; R. McKie, *Ape Man*, 50–51; L.R. Berger and B. Hilton-Barber, *In the Footsteps of Eve*, 201–205; H. Hellman, *Great Feuds in Science*, Wiley, New York, 1998, 159–176; L. Berger, The dawn of humans – redrawing our family tree?, *National Geographic*, 194(2), 1998, 90–99; D.E. Liebermann, *Nature*, 410, 2001, 419–420; L.C. Aiello and M. Collard, *Nature*, 410, 2001, 526–527; K.E. Lamge, *National Geographic*, 200(4), 2001, 84–89.

24. J. Gribbin and J. Cherfas, *The Monkey Puzzle*, Triad/Granada, London, 1983, 176–185; S. Easteal and G. Herbert, Molecular evidence from the nuclear genome for the time frame of human evolution, *Journal of Molecular Evolution*, 44 (supplement 1), 1997, S121–S132; J. Gribbin and J. Cherfas, *The First Chimpanzee*, 204–294.

25. D.C. Johanson and M.A. Edey, *Lucy – The Beginnings of Humankind*, 283–292; M.A. Edey and D.C. Johanson, *Blueprints*, 353; R. Leakey, *The Origin of Humankind*, 33; J. Diamond, *The Rise and Fall of the Third Chimpanzee*, Vintage, London, 1991, 30.

26. P. Andrews and C. Stringer, in *The Book of Life*, 239–243; R. Leakey, *The Origin of Humankind*, 43–58; G.D. Brown, *Human Evolution*, 153–176; P. Jordan, *Early Man*, 38–55; R. Lewin, *Principles of Human Evolution*, 325–341; R. Lewin, *Human Evolution*, 138–154; R. McKie, *Ape Man*, 70–115; G.P. Rightmire, *The Evolution of Homo erectus*, Cambridge University Press, 1990, 10–179; I. Tattersall, *The Fossil Trail*, Oxford University Press, 1995, 171–195; P.G. Bahn, Further back down under, *Nature*, 383, 1996, 577–578; M.J. Morwood, P.B. O'Sullivan, F. Aziz and A. Raza, Fission-track ages of stone tools and fossils on the east Indonesian island of Flores, *Nature*, 392, 1998, 173–176; G. Curtis, C. Swisher and R. Lewin, *Java Man*, Little, Brown & Co, London, 2000, 215–235; A. Gibbons, Chinese stone tools reveal high-tech *Homo erectus*, *Science*, 297, 2000, 1566; Hou Yamei, R. Potts, Yuan Bayin, *et al.*, Mid-Pleistocene Acheulean-like stone technology of the Bose Basin, South China, *Science*, 287, 2000, 1622–1626; J. Hecht, 'Dumb' hominids made it out of Africa, *New Scientist*, 13 July 2002, 21; M. Balter and A. Gibbons, Were 'little people' the first to venture out of Africa?, *Science*, 297, 2002, 26–27; A. Vekua, D. Lordkipanidze, G.P. Rightmire, *et al.*, A new skull of early *Homo* from Dmanisi, Georgia, *Science*, 297, 2002, 85–89.

27. R. Leakey and R. Lewin, *Origins Reconsidered*, 26–64; I. Tattersall, *The Fossil Trail*, 187–198; A. Walker and P. Shipman, *The Wisdom of Bones*, 7–26, 145–241; R.A. Walker and R.E. Leakey (eds.), *The Nariokotome* Homo erectus *skeleton*, Harvard University Press, Cambridge, Mass. and Springer-Verlag, Berlin, 1993.

28. R.A. Walker and P. Shipman, *The Wisdom of Bones*, 237.

29. S.M. Stanley, *Extinction*, 196–203; M. Gribbin and J. Gribbin, *Being Human*, Dent, London, 1993, 98–140; C. Stringer and C. Gamble, *In Search of the Neanderthals*, Thames and Hudson, London, 1993, 39–50; C. Charles, Cool tropical punch of the ice ages, *Nature*, 385, 1997, 681–683; R.S. Webb, D.H. Rind, S.J. Lehman, *et al.*, Influence of ocean heat transport on the climate in the last glacial maximum, *Nature*, 385, 1997, 695–699; J. Fleischman, Tropics on ice, *Earth*, 6(5), 1997, 39–45; S. Lehman, Climate change – sudden end of an interglacial, *Nature*, 390, 1997, 117–119; M.E. Raymo, K. Ganley, S. Carter, *et al.*, Millennial-scale climate instability during the early Pleistocene epoch, *Nature*, 392, 1998, 699–702; S. Lamb and D. Sington, *Earth Story*, 141–171; R. Redfern, *Origins*, 264–271; T. Haines, *Walking with Beasts*, 182–219.

30. P. Rightmire, *The Evolution of* Homo erectus, 204–233; P. Andrews and C. Stringer, in *The Book of Life*, 243–247; R. Leakey, *The Origin of Humankind*, 79–83; G.D. Brown, *Human Evolution*, 164–201; P. Jordan, *Early Man*, 55–83; R. Lewin, *Principles of Human Evolution*, 365–411; R. Lewin, *Human Evolution*, 156–175; R. McKie, *Ape Man*, 118–167; M.R. Roberts, C.B. Stringer and S.A. Parfitt, A hominid tibia from Middle Pleistocene sediments at Boxgrove, UK, *Nature*, 369, 1994, 311–313; E. Carbonell, J.M. Bermúdez de Castro, J.L. Arsuaga, *et al.*, Low Pleistocene hominids and artifacts from Atapuerca-TD6 (Spain), *Science*, 269, 1995, 826–830; M. Pitts and M. Roberts, *Fairweather Eden*, Century, London, 1997; H. Thieme, Lower Palaeolithic hunting spears from Germany, *Nature*, 385, 1997, 807–810; J.L. Arsuaga, The first Europeans, *Discovering Archaeology*, 2(5), 2000, 48–65; M. Batter, In search of the first Europeans, *Science*, 291, 2001, 1722–1723.

31. C. Stringer and C. Gamble, *In Search of the Neanderthals*; E. Trinkaus and P. Shipman, *The Neandertals*, Jonathan Cape, London, 1993; J. Shreeve, *The Neanderthal Enigma*, Morrow, New York, 1995; P. Jordan, *Neanderthal*, Sutton Publishing, Stroud, Gloucestershire, 1999; D. Palmer, *Neanderthal*, Channel 4 Books, London, 2000.

32. M.H. Brown, *The Search for Eve*, Harper and Row, New York, 1990; R. Leakey and R. Lewin, *Origins Reconsidered*, 203–236; R. Leakey, *The Origin of Humankind*, 79–99; G.D. Brown, *Human Evolution*, 224–241; C. Stringer and R. McKie, *African Exodus*, 44–183; R. Lewin, *Principles of Human Evolution*, 385–442; R. Lewin, *Human Evolution*, 164–188; R. McKie, *Ape Man*, 170–187; G. Curtis, C. Swisher and R. Lewin, *Java Man*, 183–200.

33. M.H. Wolpoff, Wu Xinzhan and A. Thorne, Modern Homo sapiens origins – a general theory of hominid evolution involving the fossil evidence from East Asia, in F. Smith and F. Spencer (eds.), *The Origins of Modern Humans*, Liss, New York, 1984, 411–483; P. Shipman, On the origin of races, *New Scientist*, 16 January 1993, 34–37; A.G. Thorne and M.H. Wolpoff, The multiregional evolution of humans, *Scientific American*, 266(4), 1992, 28–33; M. Wolpoff and R. Caspari, *Race and Human Evolution*,

Simon and Schuster, New York, 1996; B. Wood, Skulls and crossed bones, *New Scientist*, 22 February 1997, 42–43; W. Howells, *Mankind in the Making*, Penguin Books, Harmondsworth, 1967, 276–288.

34. W. Howells, *Mankind in the Making*, 279; J.S. Jones, The origin of *H. sapiens* – the genetic evidence, in *Major Topics in Primate and Human Evolution*, 317–330; C.B. Stringer and P. Andrews, Genetic and fossil evidence for the origin of modern humans, *Science*, 239, 1988, 1263–1268; H. Valladas, J.L. Reyss, J.L. Joron, *et al.*, Thermoluminescence dating of Mousterian 'Proto-Cro-Magnon' remains from Israel and the origin of modern man, *Nature*, 331, 1988, 614–616.

35. R.L. Cann, M. Stoneking and A.C. Wilson, Mitochondrial DNA and human evolution, *Nature*, 325, 1987, 31–36; L.L. Cavalli-Sforza, A. Piazza, P. Menozzi and J. Mountain, Reconstruction of human evolution – bringing together genetic, archaeological and linguistic data, *Proceedings of the National Academy of Sciences, USA*, 85, 1988, 6002–6006; L. Vigilant, M. Stoneking, H. Harpending, *et al.*, African populations and the evolution of human mitochondrial DNA, *Science*, 253, 1991, 1503–1507; A.C. Wilson and R.L. Cann, The recent African genesis of humans, *Scientific American*, 266(4), 1992, 22–27; R. Flanagan, Out of Africa, *Earth*, 5(1), 1996, 26–35; M. Wolpoff and A. Thorne, The case against Eve, *New Scientist*, 22 June 1991, 37–41; A. Gibbons, Mitochondrial Eve – wounded but not dead yet, *Science*, 257, 1992, 873–875; R.L. Dorit, H. Akashi and W. Gilbert, Absence of polymorphism at the ZFY locus, *Science*, 268, 1995, 1183–1185; B. Wood, *Ecce Homo* – behold mankind, *Nature*, 390, 1997, 120–121; S.L. Smith and F.B. Harrold, A paradigm's worth of difference? Understanding the impasse over modern human origins, *Yearbook of Physical Anthropology*, 40, 1997, 113–138; A. Gibbons, Calibrating the mitochondrial clock, *Science*, 279, 1998, 28–29; M. Day, All about Eve, *New Scientist*, 13 March 1999, 4; S.Pääbo, Human evolution, *Trends in Biochemical Sciences*, 24(12), 1999, M13–M16; P.A. Underhill, P. Shen, A.A. Lin, *et al.*, Y chromosome sequence variation and the history of human populations, *Nature Genetics*, 26, 2000, 358–361; M.H. Wolpoff, J. Hawks, D.W. Frayer and K. Hunley, Modern human ancestry at the peripheries – a test of the replacement theory, *Science*, 291, 2001, 293–297; G.J. Adcock, E.S. Dennis, S. Easteal, *et al.*, Mitochondrial DNA sequences in ancient Australians – implications for modern human origins, *Proceedings of the National Academy of Sciences, USA*, 98, 2001, 537–542; M.P.H. Stumpf and D.S. Goldstein, Genealogical and evolutionary inference with the human Y chromosome, *Science*, 291, 2001, 1738–1742; R.L. Cann, Genetic clues to dispersal in human populations, *Science*, 291, 2001, 1742–1748; C. Soares, Talking heads, *New Scientist*, 14 April 2001, 26–29; Yuehai Ke, Bing Su, Xiufeng Song, *et al.*, African origin of modern humans in East Asia – a tale of 12,000 Y chromosomes, *Science*, 292, 2001, 1151–1153; B. Sykes, *The Seven Daughters of Eve*, Bantam, London, 2001, 32–51, 185–194, 276–277.

36. C. Stringer and C. Gamble, *In Search of the Neanderthals*, 195–219; I. Tattersall, *Becoming Human*, 136–181; P. Jordan, *Neanderthal*, 190–214; D. Palmer, *Neanderthal*, 199–213; C.B. Stringer and R. Grün, Time for the last Neanderthals, *Nature*, 351, 1991, 701–702; A. Gibbons, Did Neandertals lose an evolutionary "arms" race?,

Science, 272, 1996, 1586–1587; A. Gibbons, The riddle of coexistence, *Science*, 291, 2001, 1725–1729.

37. Li Tianyuan and D.A. Etler, New Middle Pleistocene hominid crania from Yunxian in China, *Nature*, 357, 1992, 404–407; A. Gibbons, An about-face for modern human origins, *Science*, 256, 1992, 1521; P.E. Ross, East of Eden, *Scientific American*, 267(2), 1992, 14.

38. R. Lewin, *Principles of Human Evolution*, 395–397; R. Lewin, *Human Evolution*, 167–168; Chen Tiemei, Yang Quan and Wu En, Antiquity of *Homo sapiens* in China, *Nature*, 368, 1994, 55–56; C. Soares, *New Scientist*, 14 April 2001, 26–29.

39. D.M. Waddle, Matrix correlation tests support a single origin for modern humans, *Nature*, 368, 1994, 452–454; L.W. Konigsberg, A. Kramer, S.M. Donnelly, *et al.*, Modern human origins, *Nature*, 372, 1994, 228–229.

40. C. Stringer and R. McKie, *African Exodus*, 1–8, 143–183; I. Tattersall, *The Fossil Trail*, 213–246; R. Lewin, *Human Evolution*, 164–175; R. McKie, *Ape Man*, 173–187; L.R. Berger and B. Hilton-Barber, *In the Footsteps of Eve*, 1–15, 276–301; D. Roberts and L.R. Berger, 117k year old human footprints from Langebaan Lagoon, South Africa, *South African Journal of Science*, 93, 1997, 349–380; G. Bräuer, Y. Yokoyana, C. Falguères and E. Mbua, Modern human origins backdated, *Nature*, 386, 1997, 337–338; W.U. Reimold, D. Brandt, R. de Jong and J. Hancox, *Tswaing Meteorite Crater*, Council for Geoscience of South Africa, 1999; M. Frost, Impact in Africa, *Astronomy Now*, 16(1), 2002, 71–72.

41. B. McGuire, *Apocalypse*, Cassell, London, 1999, 84.

42. S.H. Ambrose, Late Pleistocene human population bottlenecks, volcanic winter and differentiation of modern humans, *Journal of Human Evolution*, 34, 1998, 623–651; *Horizon – Supervolcanoes*, a film shown on BBC Television, 3 February 2000 (see http://www.bbc.co.uk/horizon/supervolcanoes_script.shtml); B. Sykes, *The Seven Daughters of Eve*, 195–286; M.R. Rampino and S.H. Ambrose, Volcanic winter in the Garden of Eden – the Toba supereruption and the Late Pleistocene human population crash, in F.W. McCoy and G. Heiken, *Volcanic Hazards and Disasters in Human Antiquity*, 71–82; Will we be wiped out by a super-eruption?, *New Scientist*, 20 July 2002, 23.

43. C. Stringer and C. Gamble, *In Search of the Neanderthals*, 73–122; R. Lewin, *Human Evolution*, 156–162; P. Jordan, *Neanderthal*, 65–113, R. McKie, *Ape Man*, 146–163; D. Palmer, *Neanderthal*, 38–101; G. Curtis, C. Swisher and R. Lewin, *Java Man*, 194–200, 223–235; S.J. Gould, *Leonardo's Mountain of Clams and the Diet of Worms*, Vintage, London, 1999, 197–212; R.L. Cann, Tangled genetic routes, *Nature*, 416, 2002, 32–33; A.R. Templeton, Out of Africa again and again, *Nature*, 416, 2002, 45–51.

44. C. Stringer and C. Gamble, *In Search of the Neanderthals*, 123–194; R. Lewin, *Principles of Human Evolution*, 385–508; R. Lewin, *Human Evolution*, 164–220; P. Jordan, *Neanderthal*, 171–227; R. McKie, *Ape Man*, 162–209; D. Palmer, *Neanderthal*, 150–213; *The Cassell Atlas of World History, Volume 1 – The Ancient and Classical Worlds*, Andromeda, Oxford, 2000, section 1.02; S. Oppenheimer, The first exodus, *Geographical*, 74(7), 2002, 32–35.

45. S.M. Stanley, *The New Evolutionary Timetable*, Basic Books, New York, 1981, 138–164; N. Eldredge and I. Tattersall, *The Myths of Human Evolution*, Columbia University Press, New York, 1982, 119–159; C.J. Avers, *Process and Pattern in Evolution*, Oxford University Press, 1989, 508–546; I. Tattersall, *The Fossil Trail*, 229–246; A. Walker and P. Shipman, *The Wisdom of Bones*, 164–185; R. Lewin, *Human Evolution*, 190–194; L.R. Berger and B. Hilton-Barber, *In the Footsteps of Eve*, 164–211.

46. P. Jordan, *Early Man*, 36–43; R. Lewin, *Principles of Human Evolution*, 325–341; R. Lewin, *Human Evolution*, 138–144; R. McKie, *Ape Man*, 73; P. Jordan, *Neanderthal*, 125–128; I. Tattersall, *Scientific American*, 282(1), 2000, 38–44; A. Gibbons, African skull points to one human ancestor, *Science*, 295, 2002, 2192–2193; B. Asfaw, W.H. Gilbert, Y. Beyene, *et al.*, Remains of *Homo erectus* from Bouri, Middle Awash, Ethiopia, *Nature*, 416, 2002, 317–320.

47. A. Walker and P. Shipman, *The Wisdom of Bones*, 239.

48. N. Eldredge, *The Miner's Canary*, Virgin Books, London, 1992, 176.

49. C. Groves, *A Theory of Human and Primate Evolution*, 60.

50. C. Groves, *A Theory of Human and Primate Evolution*, 317.

51. J.H. Schwartz, *Sudden Origins*, 377–379 (quotation from page 377).

52. G.P. Rightmire, *The Evolution of Homo erectus*, 204–237; I. Tattersall, *The Fossil Trail*, 229–246; R. Lewin, *Principles of Human Evolution*, 385–411; R. Lewin, *Human Evolution*, 164–175; P. Jordan, *Neanderthal*, 114–129; I. Tattersall, *Scientific American*, 282(1), 2000, 38–44; D. Palmer, *Neanderthal*, 153–167; M.S. Ponce de León and C.P.E. Zollikofer, Neanderthal cranial ontology and its implications for late hominid diversity, *Nature*, 412, 2001, 534–538.

53. R. Gore, The dawn of humans – the first Europeans, *National Geographic*, 192(1), 1997, 96–113; G.P. Rightmire, Deep roots for the Neanderthals, *Nature*, 389, 1997, 917–918; R. Lewin, *Human Evolution*, 169–170; D. Palmer, *Neanderthal*, 162; L.R. Berger and B. Hilton-Barber, *In the Footsteps of Eve*, 286–288; J.L. Arsuaga, *Discovering Archaeology*, 2(5), 2000, 48–65.

54. I. Tattersall, *Scientific American*, 282(1), 2000, 38–44 (quotation from page 43). See also I. Tattersall, *Becoming Human*, 181–187.

55. C. Stringer and R. McKie, *African Exodus*, 153.

56. T. Lindahl, Facts and artifacts of ancient DNA, *Cell*, 90, 1997, 1–3; M. Krings, A. Stone, R.W. Schmitz *et al.*, Neandertal DNA sequences and the origin of modern humans, *Cell*, 90, 1997, 19–30; R. Lewin, Distant cousins, *New Scientist* 19 July 1997, 5; P. Kahn and A. Gibbons, DNA from an extinct human, *Science*, 227, 1997, 176–178; R. Ward and C. Stringer, A molecular handle on the Neanderthals, *Nature*, 388, 1997, 225–226; R. Lewin, Back from the dead, *New Scientist*, 18 October 1997, 42–43; K. Wong, Ancestral quandry, *Scientific American*, 278(1), 1998, 19–20; R. Lewin, *Principles of Human Evolution*, 382; R. Lewin, *Human Evolution*, 162; D. Palmer, *Neanderthal*, 206–211; R. McKie, *Ape Man*, 183–187.

57. P. Jordan, *Neanderthal*, 214; D. Palmer, *Neanderthal*, 199–206; R. McKie, *Ape Man*, 186–187; B. Sykes, *The Seven Daughters of Eve*, 116–130; K. Wright, Neanderthals like us, *Discover*, 23(3), 2002, 26–27.

Chapter 27

1. S.M. Stanley, *Earth and Life through Time*, Freeman, New York, 1986, 556–578, 589–614; C. Tudge, *The Day before Yesterday*, Pimlico, London, 1996, 45–49; R. Flanagan, Sea change, *Earth*, 7(1), 1998, 42–47; R. Lewin, *Human Evolution*, Blackwell Science, Malden, Mass. and Oxford, 1999, 21–25; R. Redfern, *Origins*, Cassell, London, 2000, 216–247; M.A. Cane and P. Molnar, Closing of the Indonesian seaway as a precursor to East African aridification around 3–4 million years ago, *Nature*, 411, 2001, 157–162.

2. V. Clube and B. Napier, *The Cosmic Winter*, Basil Blackwell, Oxford, 1990, 144–146; M.E. Bailey, S.V.M. Clube, G. Hahn, *et al.*, Hazards due to giant comets, in T. Gehrels (ed.), *Hazards due to Comets and Asteroids*, University of Arizona Press, Tucson, 1994, 479–535; R.A.F. Grieve and E.M. Shoemaker, The record of past impacts on Earth, in *Hazards due to Comets and Asteroids*, 417–462; M.E. Bailey, Recent results in cometary astronomy – implications for the ancient sky, *Vistas in Astronomy*, 39, 1995, 647–671; J. Gribbin and M. Gribbin, *Fire on Earth*, Simon and Schuster, London, 1996, 145–165; M.E. Bailey, Sources and populations of near-Earth objects, in B.J. Peiser, T. Palmer and M.E. Bailey (eds.), *Natural Catastrophes during Bronze Age Civilisations*, Archaeopress, Oxford, 1998, 10–20; B. Napier, temporal variation of the zodiacal dust cloud, *Monthly Notices of the Royal Astronomical Society*, 321, 2001, 463–470.

3. F.T. Kyte, L. Zhou and J.T. Wasson, New evidence on the size and possible effects of a Late Pliocene oceanic asteroid impact, *Science*, 241, 1988, 63–65; R. Gersonde, F.T. Kyte, U. Bleil, *et al.*, Geological record of the late Pliocene impact of the Eltanin asteroid in the Southern Ocean, *Nature*, 350, 1997, 357–363; J. Smit, The big splash, *Nature*, 390, 1997, 340–341; J. Hecht, A bigger splash, *New Scientist*, 1 November 1997, 11; P. Bond, Deep sea crater studied, *Astronomy Now*, 12(1), 1998, 6; C. Frankel, *The End of the Dinosaurs*, Cambridge University Press, 1999, 164; A. Atkinson, *Impact Earth*, Virgin, London, 1999, 60–61; E. Bryant, *Tsunami – The Underated Hazard*, Cambridge University Press, 2001, 249; B. Peiser (moderator), *Cambridge Conference Network* (see http://abob.libs.uga.edu/bobk/cccmenu.html).

4. R.A.F. Grieve and E.M. Shoemaker, in *Hazards due to Comets and Asteroids*, 417–462; B.P. Glass, D.V. Kent, D.A. Schneider and L. Tauxe, Ivory coast microtektite strewn field – description and relation to the Jaramillo geomagnetic event, *Earth and Planetary Science Letters*, 107, 1991, 182–196; R.A.F. Grieve, Extraterrestrial impacts on Earth, in M.M. Grady, R. Hutchison, G.J.H. McCall and D.A. Rothery, *Meteorites – Flux with Time and Impact Effects*, Geological Society, London, Special Publication 140, 1998, 105–131; C. Koeeberl, Identification of meteoritic components in impactites, in *Meteorites – Flux with Time and Impact Effects*, 133–153; B. Peiser (moderator), *Cambridge Conference Network*; M. Paine, Source of the Australasian tektites?, *Meteorite*, February 2001; M. Paine, Signs of devastating asteroid impact in Indochina 800,000 years ago, see http://www4.tpg.com.au/users/tps-seti/paine-indochina.pdf.

5. S.M. Stanley, *Earth and Life through Time*, 565–578; N. Roberts, Pleistocene environments in time and space, in R. Foley (ed.), *Hominid Evolution and Community Ecology*, Academic Press, London, 1984, 25–51; M.J. Benton, *The Rise of the Mammals*, Crescent Books, New York, 1991, 124–130; C. Stringer and C. Gamble, *In Search*

of the Neanderthals, Thames and Hudson, London, 1993, 39–50; D. Dathe, *Fundamentals of Historical Geology*, Brown, Dubuque, Iowa, 1993, 189–195; C. Whitlock and P.J. Bartlein, Vegetation and climate change in northwest America during the past 125 kyr, *Nature*, 388, 1997, 57–61; M.E. Raymo, K. Ganley, S. Carter, *et al.*, Millennial-scale climate instability during the early Pleistocene epoch, *Nature*, 392, 1998, 699–702; S. Lamb and D. Sington, *Earth Story*, BBC Books, London, 1998, 142–157.

6. M. Milankovitch, *Durch ferne Welten und Zeiten*, Koehler and Amalang, Leipzig, 1936; J. Imbrie and K.P. Imbrie, *Ice Ages – Solving the Mystery*, Macmillan, London, 1979; A. Hallam, *Great Geological Controversies*, Oxford University Press, 2nd edn, 1989, 102; K.D. Bennett, Milankovitch cycles and their effects on species in ecological and evolutionary time, *Paleobiology*, 16, 1990, 11–21; N. Eldredge, *The Miner's Canary*, Virgin, London, 1992, 191–192; M. Gribbin and J. Gribbin, *Being Human*, Dent, London, 1993, 126–136; C. Tudge, *The Day before Yesterday*, 67–70; S. Lamb and D. Sington, *Earth Story*, 159–163.

7. C. Officer and J. Page, *Tales of the Earth*, Oxford University Press, 1993, 119–120; I.J. Winograd, B.J. Szabo, T.B. Coplen and A.C. Riggs, A 250,000 year climatic record from Great Basin vein calcite – implications for Milankovitch theory, *Science*, 242, 1988, 1275–1280; New calculations rock theory of ice ages, *New Scientist*, 7 January 1989, 31; Devil's Hole discovery hots up row over ice ages, *New Scientist*, 21 November 1992, 15; W.S. Broeker, Upset for Milankovitch theory, *Nature*, 359, 1992, 779–780; R.A. Kerr, A revisionist timetable for the ice ages, *Science*, 258, 1992, 221–222; I.J. Winograd, T.B. Coplen, J.M. Landwehr, *et al.*, Continuous 500,000-year climate record from vein calcite in Devil's Hole, Nevada, *Science*, 258, 1992, 255–260; K.R. Ludwig, K.R. Simmons, B.J. Szabo, *et al.*, Mass-spectrometric ^{230}Th–^{234}U–^{238}U dating of the Devil's Hole calcite vein, *Science*, 258, 1992, 284–287; C.K. Yoon, Devil's Hole – The Ice Ages, *Earth*, 2(3), 1993, 20–23; R.L. Edwards, H. Cheng, M.T. Murrell and S.J. Goldstein, Protactinium-231 dating of carbonates by thermal ionization mass spectrometry – implications for Quaternary climate change, *Science*, 276, 1997, 782–786; P.C. Tzedakis, V. Andrieu, J.-L. de Beaulieu, *et al.*, Comparison of terrestrial and marine records of changing climate of the last 500,000 years, *Earth and Planetary Science Letters*, 150, 1997, 171–176; R.M. Schoch, *Voices of the Rocks*, Thorsons, London, 2000, 145–148.

8. F. Hoyle and N.C. Wickramasinghe, Comets, ice-ages and ecological catastrophes, *Astrophysics and Space Science*, 53, 1978, 523–526; F. Hoyle, *Ice*, New English Library, London, 1981, 131–183; F. Hoyle, On the causes of Ice Ages, *Earth, Moon and Planets*, 31, 1984, 229–248; E. Spedicato, *Apollo Objects, Atlantis and the Deluge – A Catastrophical Scenario for the End of the Last Glaciation*, Instituto Universitario di Bergamo, 1990, 4–30; F.T. Kyte, L. Zhou and J.T. Wasson, *Science*, 241, 1988, 63–65; J. Gribbin and M. Gribbin, *Fire on Earth*, 145–154; R.M. Schoch, *Voices of the Rocks*, 197–199; F. Hoyle and C. Wickramasinghe, On cosmic impacts and the cause of Ice Ages, *Cambridge Conference Network Essay*, 12 July 1999; C. Wickramasinghe, *Cosmic Dragons*, Souvenir Press, London, 2001, 157–167; F. Hoyle and C. Wickramasinghe, Cometary impacts and Ice Ages, *Astrophysics and Space Science*, 275, 2001, 367–376.

9. V. Clube and B. Napier, *The Cosmic Winter*, 145–154, 259–272; D.J. Asher and S.V.M. Clube, An extraterrestrial influence during the current glacial–interglacial, *Quarterly Journal of the Royal Astronomical Society*, 34, 1993, 481–511; D.I. Steel, D.J. Asher, W.M. Napier and S.V.M. Clube, Are impacts correlated in time?, in *Hazards due to Comets and Asteroids*, 463–478; J. Gribbin and M. Gribbin, *Fire on Earth*, 155–165; D.J. Asher and S.V.M. Clube, Towards a dynamical history of proto-Encke, *Celestial Mechanics and Dynamical Astronomy*, 69, 1998, 149–170; W.M. Napier, Cometary catastrophes, cosmic dust and ecological disasters in historical times, in *Natural Catastrophes during Bronze Age Civilisations*, 21–32; A. Atkinson, *Impact Earth*, 34–37; R.M. Schoch, *Voices of the Rocks*, 199–206.

10. V. Ramaswamy, Explosive start to last ice age, *Nature*, 359, 1992, 14; M.R. Rampino and S. Self, Volcanic winter and accelerated glaciation following the Toba super-eruption, *Nature*, 359, 1992, 50–52; M.R. Rampino and K. Caldeira, Major episodes of geological change – correlations, time structure and possible causes, *Earth and Planetary Science Letters*, 114, 1993, 215–227; V. Courtillot, *Evolutionary Catastrophes*, Cambridge University Press, 1999, 62–63; B. McGuire, *Apocalypse*, Cassell, London, 1999, 54–55, 80–86.

11. S.M. Stanley, *Earth and Life through Time*, 576–578; S.M. Stanley, *Extinction*, Scientific American Books, New York, 1987, 33–36, 72–75, 216.

12. C. Tudge, *The Day before Yesterday*, 49–67; S. Lamb and D. Sington, *Earth Story*, 163–165; V. Courtillot, *Evolutionary Catastrophes*, 144–146; R.M. Schoch, *Voices of the Rocks*, 141–148.

13. J. Gribbin and M. Gribbin, *Fire on Earth*, 145–165; C. Tudge, *The Day before Yesterday*, 27–72; S. Lamb and D. Sington, *Earth Story*, 141–171; R.M. Schoch, *Voices of the Rocks*, 141–148; R. Redfern, *Origins*, 240–261; W.S. Broecker and G.H. Denton, What drives glacial cycles? *Scientific American*, 262(1), 1990, 43–50; J. Gribbin, The end of the ice ages?, *New Scientist*, 17 June 1989, 48–52; R.A. Kerr, New way to switch Earth between hot and cold, *Science*, 243, 1989, 480; Did methane curb ice ages?, *New Scientist*, 25 May 1991, 24; C.D. Charles and R.G. Fairbanks, Evidence from Southern Ocean sediments for the effect of North Atlantic deep-water flux on climate, *Nature*, 355, 1992, 416–419; P. Williamson and J. Gribbin, How plankton changed the climate, *New Scientist*, 16 March 1991, 48–52; F. Pearce, Ice Ages – the peat bog connection, *New Scientist*, 3 December 1994, 18; W.R. Kuhn, Avoiding a permanent ice age, *Nature*, 359, 1992, 196–197; K. Caldeira and J.K. Kasting, Susceptibility of the early Earth to irreversible glaciation caused by carbon dioxide clouds, *Nature*, 359, 1992, 226–228; N. Eyles, Earth's glacial record and its tectonic setting, *Earth Science Reviews*, 35, 1993, 1–248; S.C. Clemens and R. Tiedermann, Eccentricity forcing of Pliocene – Early Pleistocene climate revealed in a marine oxygen isotope record, *Nature*, 385, 1997, 801–804; G.M. Filippelli, Intensification of the Asian monsoon and a chemical weathering event in the late Miocene – implications for a late Neogene climate change, *Geology*, 25, 1997, 27–30; J.F. Dutton and E.J. Barron, Miocene to present vegetation changes – a possible piece of the Cenozoic cooling puzzle, *Geology*, 25, 1997, 39–41; G. Ramstein, F. Fluteau, J. Besse and S. Joussaume, Effect of orogeny, plate motion and land-sea distribution on Eurasian climate change

over the past 30 million years, Nature, 386, 1997, 788–795; Possible ice age trigger, Astronomy Now, 11(7), 1997, 6; R.A. Kerr, Upstart ice age theory gets attentive but chilly hearing, Science, 277, 1997, 183–184; R.A. Muller and G.J. MacDonald, Glacial cycles and astronomical forcing, Science, 277, 1997, 215–218; G. Henderson, Deep freeze, New Scientist, 14 February 1998, 28–32; R.A. Kerr, A dusty ice age trigger looks too weak, Science, 280, 1998, 828–829; S.J. Kortenkamp and S.F. Dermill, A 100,000-year periodicity in the accretion rate of interplanetary dust, Science, 280, 1998, 874–876); R.B. Alley and M.L. Bender, Greenland ice cores – frozen in time, Scientific American, 278(2), 1998, 66–71; R.G. Rothwell, J. Thomson and G. Köhler, Low-sea-level emplacement of a very large Late Pleistocene 'megaturbidite' in the western Mediterranean Sea, Nature, 392, 1998, 377–380; M. Rossignol-Strick, M. Paterne, F.C. Bassinot, et al., An unusual mid-Pleistocene monsoon period over Africa and Asia, Nature, 392, 1998, 269–272; K. Taylor, Rapid climate change, American Scientist, 87(4), 1999, 320–327; D.M. Sigman and E.A. Boyle, Glacial/interglacial variations in atmospheric carbon dioxide, Nature, 407, 2000, 859–869; S. Simpson, Triggering a snowball, Scientific American, 285(3), 2001, 14–15; T.R. Naish, K.J. Woolfe, P.J. Barrell, et al., Nature, 413, 2001, 719–723; M.J. Siegert, J.A. Dowdeswell, J.-I. Svendsen and A. Elverhøi, The Eurasian Arctic during the last Ice Age, American Scientist, 90(1), 2002, 32–39; P.U. Clark, N.G. Pisias, T.F. Stocker and A.J. Weaver, The role of the thermohaline circulation in abrupt climate change, Nature, 415, 2002, 863–869; J. Hecht, Sun struck, New Scientist, 15 June 2002, 6.

14. F. Hoyle, Ice, 92–155; F. Hoyle and C. Wickramasinghe, On cosmic impacts and the causes of Ice Ages, Cambridge Conference Network Essay, 12 July 1999; F. Hoyle and C. Wickramasinghe, Cometary impacts and Ice Ages, Astrophysics and Space Science, 275, 2001, 367–376; D.R. Easterling, G.A. Meehl, C. Parmesan, et al., Climate extremes – observations, modelling and impacts, Science, 289, 2000, 2068–2074; F. Pearce, On the brink, New Scientist, 2 February 2002, 18; F. Saunders, Chaotic warnings from the last Ice Age, Discover, 23(6), 2002, 14.

15. S.M. Stanley, Earth and Life through Time, 565–570; D. Dathe, Fundamentals of Historical Geology, 189–195; S. Lamb and D. Sington, Earth Story, 152–155; R. Redfern, Origins, 250–251; A. Lister and P. Bahn, Mammoths, Macmillan, New York, 1994, 26–27; R. Osborne and M. Benton, The Viking Atlas of Evolution, Penguin Books, London, 1996, 114–115; M.J. Siegert, J.A. Dowdeswell, J.-I. Svendsen and A. Elverhøi, American Scientist, 90(1), 2002, 32–39.

16. S.M. Stanley, Extinction, 2–3; M.J. Benton, The Rise of the Mammals, 124–130; N. Eldredge, The Miner's Canary, 192–195; A. Lister and P. Bahn, Mammoths, 37–61, 145–156; F. Hitching, The World Atlas of Mysteries, Pan Books, London, 1979, 52–54; R. Osborne and M. Benton, The Viking Atlas of Evolution, 98–99; N.K. Vereschagin and G.F. Baryshnikov, Quaternary mammalian extinctions in northern Eurasia, in P.S. Martin and R.G. Klein (eds.), Quaternary Extinctions, University of Arizona Press, Tucson, 1984, 483–516; D.A. Young, The Biblical Flood, Paternoster Press, Carlisle, 1995, 204–209.

17. D.S. Allan and J.B. Delair, When the Earth Nearly Died, Gateway Books, Bath, 1995, 169–238.

18. D.S. Allan and J.B. Delair, *When the Earth Nearly Died*, 207–211; A. Roy (ed.), *The Oxford Illustrated Encyclopedia of the Universe*, Oxford University Press, 1992, 180. (Note that this supernova is separate from another supernova in the Vela constellation which appears to have exploded around 700 years ago – see B.Aschenbach, Discovery of a young nearby supernova remnant, *Nature*, 396, 1998, 141–142; R. Matthews, On ice – Antarctica yields clues to a 'lost' supernova, *New Scientist*, 18 September 1999, 7.)

19. D.S. Allan and J.B. Delair, *When the Earth Nearly Died*, 281–289.

20. D.S. Allan and J.B. Delair, *When the Earth Nearly Died*, 44–60, 241–323; D.S. Allan and J.B. Delair, Scientific evidence for a major world catastrophe about 11,500 years ago, *Chronology and Catastrophism Review*, 17, 1995, 41–48; D.S. Allan, An unexplained Arctic catastrophe, *Chronology and Catastrophism Review*, 2001:2, 3–7; R. Gallant, *Bombarded Earth*, John Baker, London, 1964, 115–117, 144, 148–149.

21. R.M. Schoch, *Voices of the Rocks*, 155–158; R. Huggett, *Cataclysms and Earth History*, Oxford University Press, 1989, 162–164; P. James and N. Thorpe, *Ancient Mysteries*, Ballantine Books, New York, 1999, 58–64, 71–76.

22. C.H. Hapgood, *The Path of the Pole*, Chilton Book Company, Philadelphia, New York and London, 1970 (reprinted, with the omission of the final two chapters, by Souvenir Press, London, 2001), 89–124, 155–184; C.H. Hapgood. *Maps of the Ancient Sea Kings*, revised edition, Turnstone Books, London, 1979, 174–178; R. Flem-Ath and R. Flem-Ath, *When the Sky Fell*, Weidenfeld and Nicolson, London, 1995, 82–87; R. Flem-Ath and C. Wilson, *The Atlantis Blueprint*, Little, Brown and Co., London, 2000, 36–38.

23. S.M. Stanley, *Earth and Life through Time*, 571–572; S. Lamb and D. Sington, *Earth Story*, 147–151; P. Jordan, *Neanderthal*, Sutton Publishing, Stroud, Gloucs., 1999, 65–67.

24. S. Lamb and D. Sington, *Earth Story*, 147–151; C. Stringer and C. Gamble, *In Search of the Neanderthals*, Thames and Hudson, London, 1993, 39–50; R. Lewin, *Principles of Human Evolution*, Blackwell Science, Malden, Mass. and Oxford, 1998, 54–59; S.C. Porter and A. Zhisheng, Correlation between climate events in the North Atlantic and China during the last glaciation, *Nature*, 375, 1995, 305–308.

25. A. Lister and P. Bahn, *Mammoths*, 134–135; S. Lamb and D. Sington, *Earth Story*, 152–155; P.D. Ward, *The Call of Distant Mammoths*, Copernicus, New York, 1997, 198–201; D. Palmer, *Neanderthal*, Channel 4 Books, London, 2000, 42–43.

26. J.Jouzel, Ice cores north and south, *Nature*, 372, 1994, 612–613; M. Bender, T. Sowers, M.-L. Dickerson, *et al.*, Climate correlations between Greenland and Antarctica during the past 100,000 years, *Nature*, 372, 1994, 663–666; T. Sowers and M. Bender, Climate records covering the last deglaciation, *Science*, 269, 1995, 210–214; E.J. Steig, E.J. Brook, J.W.C. White, *et al.*, Synchronous climate changes in Antarctica and the North Atlantic, *Science*, 282, 1998, 92–95; D.W. Oppo, J.F. McManus and J.L. Cullen, Abrupt climate change events 500,000 to 340,000 years ago, *Science*, 279, 1998, 1335–1338; P.A. Baker, Trans-Atlantic climate connections, *Science*, 296, 2002, 67–68; T.C. Johnson, E.T. Brown, J. McManus, *et al.*, A high-resolution paleoclimate record spanning the past 25,000 years in Southern East Africa, *Science*, 296, 2002, 113–114, 131–132.

27. S.M. Stanley, *Earth and Life through Time*, 181–191; S. Lamb and D. Sington, *Earth Story*, 78–81; R. Redfern, *Origins*, 14–33.

28. L. Krishtalka, The Pleistocene ways of death, *Nature*, 312, 1984, 225–226; P. Martin, Prehistoric overkill – the global model, in *Quaternary Extinctions*, 354–403; S.M. Stanley, *Extinction*, Scientific American Books, New York, 1987, 196–207; A.D. Barnosky, The late Pleistocene event as a paradigm for widespread mammal extinction, in S.K. Donovan (ed.), *Mass Extinctions – Processes and Evidence*, Belhaven Press, London, 1989, 235–254; M.J. Benton, *The Rise of the Mammals*, 131–137; C. Janis, Victors by default, in S.J. Gould (ed.), *The Book of Life*, Ebury Hutchinson, London, 1993, 169–217 (particularly pages 206–217); A. Lister and P. Bahn, *Mammoths*, 119–139; P. Ward, *The End of Evolution*, Weidenfeld and Nicolson, London, 1995, 194–212; P.D. Ward, *The Call of Distant Mammoths*, 119–165.

29. A. Lister and P. Bahn, *Mammoths*, 134–135; S. Lamb and D. Sington, *Earth Story*, 152–155, 165–171; D. Palmer, *Neanderthal*, 42–43; E.J. Steig, E.J. Brook, J.W.C. White, et al., *Science*, 282, 1998, 92–95.

30. S.D. Webb, Ten million years of mammal extinctions in North America, in *Quaternary Extinctions*, 189–210; R. Gruhn and A.L. Bryan, The record of Pleistocene megafaunal extinction at Taima-taima, northern Venezuela, in *Quaternary Extinctions*, 128–137; M.J. Benton, *The Rise of the Mammals*, 135–136; R. Osborne and M. Benton, *The Viking Atlas of Evolution*, 114–115; P.D. Ward, *The Call of Distant Mammoths*, 119–165.

31. N.K. Vereshchagin and G.F. Baryshnikov, Quaternary mammalian extinctions in northern Eurasia, in *Quaternary Extinctions*, 483–516; Liu Tung-sheng and Li Xing-guo, Mammoths in China, in *Quaternary Extinctions*, 517–527; R.G. Klein, Mammalian extinctions and Stone Age people in Africa, in *Quaternary Extinctions*, 553–573; P. Murray, Extinctions downunder, in *Quaternary Extinctions*, 600– 628; P. Ward, *The End of Evolution*, 194–212.

32. B.M. Gilbert and L.D. Martin, Late Pleistocene fossils of Natural Trap Cave, Wyoming, and the climatic model of extinction, in *Quaternary Extinctions*, 138–147; D.R. Horton, Red kangaroos – last of the Australian megafauna, in *Quaternary Extinctions*, 639–680; R.A. Kiltie, Seasonality, gestation time, and large mammalian extinctions, in *Quaternary Extinctions*, 299–314; A. Lister and P. Bahn, *Mammoths*, 120–139; N. Roberts, *The Holocene – An Environmental History*, Blackwell, Oxford, 2nd edn, 1998, 55–86; D. Palmer, Resurrecting the mammoth, in A. Grayson (ed.), *Equinox: The Earth*, Channel 4 Books, London, 2000, 81–113; T. Haines, *Walking with Beasts*, BBC Books, London, 2001, 220–255.

33. R. Osborne and M. Benton, *The Viking Atlas of Evolution*, 116–117; R. Lewin, *Principles of Human Evolution*, 485–497; P. Bahn, Dating the first American, *New Scientist*, 20 July 1991, 26–28; J. Horgan, Early arrivals, *Scientific American*, 266(2), 1992, 8–9; D.J. Meltzer, Monte Verde and the Pleistocene peopling of the Americas, *Science*, 276, 1997, 754–755; S.L. Bonatto and F.M. Salzano, A single and early migration for the peopling of the Americas supported by mitochondrial DNA sequence data, *Proceedings of the National Academy of Sciences, USA*, 94, 1997, 1866–1871; V. Morrell, Kennewick Man – more bones to pick, *Science*, 279, 1998,

25–26; S. Levy, Death by fire, *New Scientist*, 1 May 1999, 38–43; M. Parfit, Hunt for the first Americans, *National Geographic*, 198(6), 2000, 40–67; E. Marshall, Pre-Clovis sites fight for acceptance, *Science*, 291, 2001, 1730–1732; *The Cassell Atlas of World History, Volume 1 – The Ancient and Classical Worlds*, Andromeda, Oxford, 2000, section 1.02.

34. P.S. Martin, Catastrophic extinctions and late Pleistocene blitzkrieg, in M. Nitecki (ed.), *Extinctions*, University of Chicago Press, Chicago and London, 1984, 153–189; P.S. Martin, Prehistoric overkill – the global model, in *Quaternary Extinctions*, 354–403; L.G. Marshall, Who killed cock robin?, in *Quaternary Extinctions*, 785–806; P.S. Martin, Refuting late Pleistocene extinction models, in D.K. Elliott (ed.), *Dynamics of Extinction*, Wiley, New York, 1986, 107–130; A.D. Barnovsky, The late Pleistocene event as a paradigm for widespread mammalian extinction, in *Mass Extinctions – Processes and Evidence*, 235–254; A. Lister and P. Bahn, *Mammoths*, 119–139; M.W. Beck, On discerning the cause of the Late Pleistocene megafaunal extinctions, *Paleobiology*, 22, 1996, 91–103; P.D. Ward, *The Call of Distant Mammoths*, 119–201; K. Wong, Mammoth kill, *Scientific American*, 284(2), 2001, 15; M. Miller, Mammoth mystery, *New Scientist*, 5 May 2001, 32–35; A. Gill and A. West, *Extinct*, Channel 4 Books, London, 2001, 28–63.

35. P.D. Ward, *The Call of Distant Mammoths*, 201.

36. D.J. Asher and S.V.M. Clube, *Quarterly Journal of the Royal Astronomical Society*, 34, 1993, 481–511; J. Gribbin and M. Gribbin, *Fire on Earth*, 145–165; R.M. Schoch, *Voices of the Rocks*, 197–206.

37. R.M. Schoch, *Voices of the Rocks*, 172–174, 194–197; F. Barbiero, On the possibility of very rapid shifts of the poles, 1997, http://www.unibg.it/dmsia/dynamics/poles.html.

38. E.P. Izokh, Australo-Asian tektites and a global disaster of about 10,000 years B.P., caused by collision of the Earth with a comet, *Geologiya i Geofizika*, 38, 1997, 628–660 (in Russian); B.J. Peiser (moderator), *Cambridge Conference Network* (see archive at http://abob.libs.uga.edu/bobk/cccmenu.html).

39. A. Tollmann and E. Tollmann, *Und die Sintflut gab es doch*, Droemer-Knaur, Munich, 1993; E. Kristan-Tollmann and A. Tollmann, The youngest big impact on Earth deduced from geological and historical evidence, *Terra Nova*, 6, 1994, 209–217; E. Bryant, *Tsunami – The Underated Hazard*, 250–251.

40. E.M. Shoemaker and H.R. Uhlherr, Stratigraphic relations of australites in the Port Campbell embayment, Victoria, *Meteoritics and Planetary Science*, 34, 1999, 369–384; S.R. Taylor, The Australasian tektite paradox, *Meteoritics and Planetary Science*, 34, 1999, 311.

41. W.S. Broecker and G.H. Denton, The role of ocean–atmosphere reorganizations in glacial cycles, *Geochimica et Cosmochima Acta*, 53, 1989, 2465–2501; W.S. Broecker and G.H. Denton, What drives glacial cycles?, *Scientific American*, 262(1), 1990, 42–50; K. Taylor, *American Scientist*, 87(4), 1999, 320–327; S. Lamb and D. Sington, *Earth Story*, 165–171; F. Pearce, Coat of many colours, *New Scientist*, 10 November 2001, 36–39; P.U. Clark, N.G. Pisias, T.F. Stocker and A.J. Weaver, The role of the thermohaline circulation in abrupt climate change, *Nature*, 415, 2002, 863–869.

42. R. Huggett, *Cataclysms and Earth History*, 149–159; S.J. Gould, *The Panda's Thumb*, Penguin Books, Harmondsworth, 1983, 162–169; C. Officer and J. Page, *Tales of the Earth*, Oxford University Press, 1993, 73–75.

43. J.B. Delair, Planet in crisis, *Chronology and Catastrophism Review*, 1997:2, 4–11.

44. S. Lamb and D. Sington, *Earth Story*, 165–171; N. Roberts, *The Holocene – An Environmental History*, 87–126; R.M. Schoch, *Voices of the Rocks*, 147–148; P. James and N. Thorpe, *Ancient Mysteries*, 9–10; P. Blanchon and J. Shaw, Reef drowing during the last deglaciation – evidence for catastrophic sea-level rise and ice-sheet collapse, *Geology*, 23, 1995, 4–8; J.W. Beck, J. Récy, F. Taylor, *et al.*, Abrupt change in early Holocene tropical sea surface temperature derived from coral records, *Nature*, 385, 1997, 705–707; E. Baard, F. Rostek and C.Sonzogni, Interhemispheric synchrony of the last deglaciation inferred from alkenone palaeothermometry, *Nature*, 385, 1997, 707–710; K. Hughes, J.T. Overpeck, L.C. Peterson and S. Trumbone, Rapid climate change in the tropical Atlantic region during the last deglaciation, *Nature*, 380, 1996, 51–54; J.E. Smith, M.J. Risk, H.P. Schwarcz and T.A. McConnaughey, Rapid climate change in the North Atlantic during the Younger Dryas recorded by deep-sea corals, *Nature*, 386, 1997, 818–820; G.A. Zielinski and C.R. Mershon, Paleoenvironmental implications of the insoluble microparticle record in the GISP2 (Greenland) ice core during the rapidly changing climate of the Pleistocene–Holocene transition, *Geological Society of America Bulletin*, 109, 1997, 547–559; W.K. Stevens, If climate changes, it may change quickly, *The New York Times*, Science section, 27 January 1998; R.B. Alley, Icing the North Atlantic, *Nature*, 392, 1998, 335–337; A.M. McCabe and P.V. Clark, Ice-sheet variability around the North Atlantic Ocean during the last deglaciation, *Nature*, 392, 1998, 373–377; J.P. Severinghaus and E.J. Brook, Abrupt climate change at the end of the last glacial period inferred from trapped air in polar ice, *Science*, 286, 1999, 930–934; K. Taylor, *American Scientist*, 87(4), 1999, 320–327.

45. R. Mestel, Noah's Flood, *New Scientist*, 4 October 1997, 24–27; D. McInnes, And the waters prevailed, *Earth*, August 1998, 46–54; W. Ryan and W. Pitman, *Noah's Flood – The New Scientific Discoveries about the Event that Changed History*, Simon and Schuster, New York, 1999; I. Wilson, *Before the Flood*, Orion, London, 2001; R.D. Ballard, Black Sea mysteries, *National Geographic*, 199(5), 2001, 52–69; J. Hecht, Flood hypothesis seems to hold no water, *New Scientist*, 4 May 2002, 13; A.I. Aksu, R.N. Hiscott, P.J. Mudie, *et al.*, Persistent Holocene outflow from the Black Sea to the Eastern Mediterranean contradicts Noah's Flood hypothesis, *GSA Today*, 12(5), 2002, 4–10.

Chapter 28

1. C. Berlitz, *The Bermuda Triangle*, Souvenir Press, London, 1975, 123–140.

2. L.D. Kusche, *The Bermuda Triangle – Mystery Solved*, New English Library, London, 1975.

3. C. Berlitz, *Atlantis – The Lost Continent Revealed*, Fontana/Collins, London, 1985 (examples from pages 78–79, 100–101 and 149). (Previously published by MacMillan, London, 1984.)

4. C. Wilson, *From Atlantis to the Sphinx*, Virgin Books, London, 1997, 164; S. Gonzalez, A. Pastrana, C. Siebe and G. Deller, Timing of the prehistoric eruption of Xitle

Volcano and the abandonment of Cuicuilco Pyramid, southern basin of Mexico, *Geological Society, London, Special Publications*, 171, 2000, 205–224.

5. C. Berlitz, *Atlantis – The Lost Continent Revealed*, 94–99.

6. P. James, *The Sunken Kingdom*, Jonathan Cape, London, 1995, 52–56; P. James and N. Thorpe, *Ancient Mysteries*, Ballantine Books, New York, 1999, 598–604; K.L. Feder, *Frauds, Myths and Mysteries*, Mayfield, Mountain View, Calif., 3rd edn, 2000, 179–182; P. Jordan, *The Atlantis Syndrome*, Sutton Publishing, Stroud, Gloucs., 2001, 99–103; M. McKusick and E.A. Shinn, Bahamian Atlantis reconsidered, *Nature*, 287, 1980, 11–12.

7. C. Berlitz, *Atlantis – The Lost Continent Revealed*, 120–153.

8. C. Berlitz, *Atlantis – The Lost Continent Revealed*, 55.

9. A. Collins, *Gateway to Atlantis*, Headline, London, 2000, 50, 171; *The Cassell Atlas of World History, Volume 1 – The Ancient and Classical Worlds*, Andromeda, Oxford, 2000, section 1.23.

10. Z. Kukal, Atlantis in the light of modern research, *Earth Science Reviews*, 21, 1984, 1–224 (quotation from page 192).

11. Z. Kukal, *Earth Science Reviews*, 21, 1984, 1–224 (see pages 63–70).

12. P. James, *The Sunken Kingdom*, 37–40; A. Collins, *Gateway to Atlantis*, 294–300; C. Emiliani, S. Gartner, B. Lidz, *et al.*, Paleoclimatological analysis of Late Quaternary cores from the northeastern Gulf of Mexico, *Science*, 189, 1975, 1083–1088; R.M. Schoch, *Voices of the Rocks*, Thorsons, London, 1999, 140–148.

13. Z. Kukal, *Earth Science Reviews*, 21, 1984, 1–224 (see pages 161–169); C. Officer and J. Page, *Tales of the Earth*, Oxford University Press, 1993, 78–81; S. Lamb and D. Sington, *Earth Story*, BBC Books, London, 1998, 156–157; R. Redfern, *Origins*, Cassell, London, 2000, 278–286.

14. Z. Kukal, *Earth Science Reviews*, 21, 1984, 1–224 (see pages 73–76); P. James and N. Thorpe, *Ancient Mysteries*, 16–26; R.M. Schoch, *Voices of the Rocks*, 120–127; K.L. Feder, *Frauds, Myths and Mysteries*, 159–183; P. Jordan, *The Atlantis Syndrome*, 51–52; M. Eddy, Gran Canaria – A 'Fortunate Island', *Popular Archaeology*, March 1986, 15–21.

15. Z. Kukal, *Earth Science Reviews*, 21, 1984, 1–224 (see pages 18–20); K.L. Feder, *Frauds, Myths and Mysteries*, 179–180; D.H. Trump, *The Prehistory of the Mediterranean*, Penguin Books, Harmondsworth, 1981, 7–202; *The Times Archaeology of the World*, Times Books, London, 2001, 77–87, 104–115, 144–145; M. Andrews, *The Birth of Europe*, BBC Books, London, 1991, 34–55, 76–113; E. Zangger, *The Flood from Heaven*, Book Club Associates/Sidgwick and Jackson, 1992, 63–77, 119–131; N. Roberts, *The Holocene – An Environmental History*, Blackwell, Oxford, 2nd edn, 1998, 127–158; *The Times History of the World*, Times Books, London, 1999, 38–41; *The Cassell Atlas of World History, Volume 1 – The Ancient and Classical Worlds*, section 1.03; T.R. Martin, *Ancient Greece – From Prehistoric to Hellenistic Times*, Yale Nota Bene, New Haven and London, 2000, 1–69.

16. K.L. Feder, *Frauds, Myths and Mysteries*, 1–14; K.L. Feder, Irrationality and popular archaeology, *American Antiquity*, 49, 1984, 525–541.

17. K.L. Feder, *Frauds, Myths and Mysteries*, 1–14, 159–183; P. James and N. Thorpe, *Ancient Mysteries*, 16–26, 58–76; R.M. Schoch, *Voices of the Rocks*, 79–127.

18. R.M. Schoch, *Voices of the Rocks*, 113–120; K.L. Feder, *Frauds, Myths and Mysteries*, 184–215; P. Jordan, *The Atlantis Syndrome*, 250–256; E. von Däniken, *Chariots of the Gods?*, Souvenir Press, London, 1969, Z. Sitchin, *The Twelfth Planet*, Avon Books, New York, 1976; Z. Sitchin, *Genesis Revisited*, Avon Books, New York, 1990; E. von Däniken, *The Return of the Gods*, Element, Shaftsbury, Dorset, 1997; A.F. Alford, *Gods of the New Millennium*, Hodder and Stoughton, London, 1997; A.F. Alford, *The Phoenix Solution*, Hodder and Stoughton, London, 1998; L. Picknett and C. Prince, *The Stargate Conspiracy*, Warner Books, London, 2000, 116–254.

19. H. Brennan, *The Atlantis Enigma*, Judy Piatkus/Book Club Associates, London, 1999.

20. E. Spedicato, *Apollo Objects, Atlantis and the Deluge*, Instituto Universitario di Bergamo, 1990, 24–41.

21. A. Collins, *Gateway to Atlantis*, 230–333; P. Jordan, *The Atlantis Syndrome*, 257–274.

22. J. Allen, *Atlantis – The Andes Solution*, Windrush Press, London, 1998; S. Oppenheimer, *Eden in the East*, Weidenfeld and Nicolson, London, 1998; P. Jordan, *The Atlantis Syndrome*, 175–180; S. Hodge, *Atlantis*, Piatkus, London, 2000, 102–109, 125–131.

23. J. Copley, Lost and found – have we finally located the legendary island of Atlantis?, *New Scientist*, 22 September 2001, 17; Atlantis 'obviously near Gibraltar', see BBC News Sci/Tech web-site at http://news.bbc.co.uk/hi/english/sci/tech/newsid_1554000/1554594.stm.

24. C. Wilson, *From Atlantis to the Sphinx*, 114–120; R. Flem-Ath and R. Flem-Ath, *When the Sky Fell*, Weidenfeld and Nicolson, London, 1995.

25. C. Hapgood, *Maps of the Ancient Sea Kings*, revised edn, Turnstone Books, London, 1979, 62–95, 174–178.

26. R. Flem-Ath and R. Flem-Ath, *When the Sky Fell*, 73–136; R.M. Schoch, *Voices of the Rocks*, 95–106; P. James and N. Thorpe, *Ancient Mysteries*, 58–76; S. Hodge, *Atlantis*, 115–121; P. Jordan, *The Atlantis Syndrome*, 181–206.

27. R. Flem-Ath and R. Flem-Ath, *When the Sky Fell*, 53–71, 137–150; R. Flem-Ath and C. Wilson, *The Atlantis Blueprint*, Little, Brown and Co., London, 2000, 63–318.

28. G. Hancock, *Fingerprints of the Gods*, Heinemann, London, 1995, 62–92; P. James and N. Thorpe, *Ancient Mysteries*, 231–244.

29. R. Flem-Ath and R. Flem-Ath, *When the Sky Fell*, 53–71; G. Hancock, *Fingerprints of the Gods*, 42–52, 100–108, 268–269, 391–395; P. James and N. Thorpe, *Ancient Mysteries*, 241–244, 423–424; P. Jordan, *The Atlantis Syndrome*, 207–221.

30. R. Bauval and A. Gilbert, *The Orion Mystery*, Heinemann, London, 1994, 105–137. R. Bauval and G. Hancock, *Keeper of Genesis*, Heinemann, London, 1996.

31. P. James and N. Thorpe, *Ancient Mysteries*, 128–134; R.M. Schoch, *Voices of the Rocks*, 52–78; P. Jordan, *The Atlantis Syndrome*, 221–241; R. Chadwick, The so-called 'Orion Mystery', *KMT*, 7(3), 1996, 74–83; T. Palmer, Review of *Keeper of Genesis*, *Chronology and Catastrophism Review*, 1996:2, 44–46; P. Jordan, *Riddles of the Sphinx*, New York University Press, 1998, 127–143; I. Lawton and C. Ogilvie-Herald, *Giza – The Truth*, Virgin Books, London, 1999, 349–374.

32. R. Bauval and G. Hancock, *Keeper of Genesis*, 23–34.

33. G. Hancock, *Fingerprints of the Gods*, 135–136, 354–359, 411–428; R. Bauval and G. Hancock, *Keeper of Genesis*, 15–22; L. Picknett and C. Prince, *The Stargate Conspiracy*,

36–40, 255–302; J.A. West, *Serpent in the Sky*, Quest Books, Wheaton, Illinois, 1993, 184–220.

34. P. Jordan, *Riddles of the Sphinx*, 145–161; P. James and N. Thorpe, *Ancient Mysteries*, 213–226; I. Lawton and C. Ogilvie-Herald, *Giza – The Truth*, 313–335.

35. C. Reader, A geomorphological study of the Giza necropolis, with implications for the development of the site, *Archaeometry*, 43(1), 2001, 149–159; J. Dixon and C. Reader, The riddle of the sphinx – one geological solution, *Geoscientist*, 11(7), 2001, 4–6.

36. R.M. Schoch, *Voices of the Rocks*, 33–59, 74–78; R.M. Schoch, Redating the Great Sphinx of Giza, KMT, 3(2), 1992, 52–70; C. Reader, A geomorphological study of the Giza necropolis, with implications for the development of the site, *Archaeometry*, 43(1), 2001, 149–159; Astronomical alignments abound at Egyptian site, *Sky and Telescope*, 96(2), 1998, 18; R. Kunzig, Exit from Eden. *Discover*, 21(1), 2000, 84–91.

37. *The Times Archaeology of the World*, 80–83; *The Cassell Atlas of World History, Volume 1 – The Ancient and Classical Worlds*, section 1.08; P. James and N. Thorpe, *Ancient Mysteries*, 216–218; R.M. Schoch, *Voices of the Rocks*, 58–59; M. Magnusson, *BC – The Archaeology of the Bible Lands*, Book Club Associates/Bodley Head/ BBC, London, 1977, 9–14; R. Lewin, *Human Evolution*, Blackwell Science, Malden, Mass. and Oxford, 1999, 215–219; R. Kunzig, A tale of two obsessed archaeologists, *Discover*, 20(5), 1999, 84–92; R. Rudgley, *Lost Civilisations of the Stone Age*, Arrow Books, London, 1999, 13–22; R. Rudgley, *Secrets of the Stone Age*, Random House, London, 2000, 66–91.

38. J.A. West, *Serpent in the Sky*, 225–232; R. Bauval and G. Hancock, *Keeper of Genesis*, 15–21.

39. R. Bauval and G. Hancock, *Keeper of Genesis*, 15–22, 58–82, 254–267.

40. R. Bauval and A. Gilbert, *The Orion Mystery*, 105–161; G. Hancock, *Fingerprints of the Gods*, 353–358, 442–458; R. Bauval and G. Hancock, *Keeper of Genesis*, 66–78, 208–237.

41. P. James and N. Thorpe, *Ancient Mysteries*, 133–134; I. Lawton and C. Ogilvie-Herald, *Giza – The Truth*, 364–367.

42. R. Bauval and G. Hancock, *Keeper of Genesis*, 215–229 (quotation from pages 215–216).

43. This may be seen by comparing a large-scale map of the Memphis area of Egypt with a star-map. For the former, see, for example J. Baines and J. Malek, *Atlas of Ancient Egypt*, Andromeda, Oxford/Facts on File, New York, 1980, 135, revised edn, Andromeda, Oxford, 2000, 135. For a map of the appropriate part of the night sky see, for example, W. Tirion, *The Cambridge Star Atlas*, Cambridge University Press, 3rd edn, 2001, star chart 9.

44. R. Bauval and G. Hancock, *Keeper of Genesis*, 216.

45. R. Chadwick, KMT, 7(3), 1996, 74–83 (quotation from page 80).

46. R. Bauval and A. Gilbert, *The Orion Mystery*, 123–124.

47. J. Baines and J. Malek, *Atlas of Ancient Egypt*, 135; W. Tirion, *The Cambridge Star Atlas*, star chart 9; R. Chadwick, KMT, 7(3), 1996, 74–83; P. Jordan, *Riddles of the Sphinx*, 141–143; P. James and N. Thorpe, *Ancient Mysteries*, 131–133; I. Lawton and C. Ogilvie-Herald, *Giza – The Truth*, 357–364.

48. P. Jordan, *Riddles of the Sphinx*, 141–142; S. Hodge, *Atlantis*, 124–125; I. Lawton and C. Ogilvie-Herald, *Giza – The Truth*, 359–362; M. Lehner, *The Complete Pyramids*, Thames and Hudson, London, 1997, 106–107

49. G. Hancock, *Fingerprints of the Gods*, 454–458; R. Bauval and G. Hancock, *Keeper of Genesis*, 78–82, 182–186, 254–267.

50. P. Jordan, *Riddles of the Sphinx*, 133–138; P. James and N. Thorpe, *Ancient Mysteries*, 226–230; I. Shaw and P. Nicholson, *British Museum Dictionary of Ancient Egypt*, British Museum Press, London, 1995, 84–85.

51. R. Bauval and G. Hancock, *Keeper of Genesis*, 244–245.

52. R. Bauval and G. Hancock, *Keeper of Genesis*, 182–186, 252–254.

53. R. Bauval and G. Hancock, *Keeper of Genesis*, 215–216.

54. M. Lehner, *The Complete Pyramids*, 108–109, 200–239; A. Siliotti, *The Pyramids*, Weidenfeld and Nicolson, London, 1997, 40–45; D. Wildung, *Egypt*, Taschen, Cologne, 2001, 47–55; V. Morell, The pyramid builders, *National Geographic*, 200(5), 2001, 78–99.

55. R. Bauval and G. Hancock, *Keeper of Genesis*, 23–57; T. Palmer, *Chronology and Catastrophism Review*, 1996:2, 44–46.

56. R. Bauval and G. Hancock, *Keeper of Genesis*, 262–267.

57. R. Bauval and G. Hancock, *Keeper of Genesis*, 85–128, 290–297.

58. G. Hancock, *Fingerprints of the Gods*, 42–52, 100–108, 268–269, 391–395.

59. G. Hancock and S. Faiia, *Heaven's Mirror*, Michael Joseph, London, 1998; P. Jordan, *The Atlantis Syndrome*, 235–250. Note that, by this time, Hancock was becoming more cautious about claiming that a displacement of the Earth's crust took place in the Late Pleistocene, a concept he had strongly supported in *Fingerprints of the Gods*, 3–13, 464–470, 487–505, although he did not reject it altogether (see *Heaven's Mirror*, 208–212 and also G. Hancock, *Fingerprints of the Gods – The Quest Continues*, Century, London, 2001, xxxvii–xl). In *The Mars Mystery*, Michel Joseph, London, 1998, 292–294, Bauval, Hancock and their researcher, John Grigsby, wrote that they considered a cosmic impact to be the primary cause of the events at the end of the last Ice Age, and they found themselves 'in more or less complete agreement' with the scenario of coherent catastrophism put forward by Victor Clube and Bill Napier.

60. G. Hancock and S. Faiia, *Heaven's Mirror*, 20–55, 116–153.

61. P. James and N. Thorpe, *Ancient Mysteries*, 21–24; K.L. Feder, *Frauds, Myths and Mysteries*, 169–178; M. Lehner, *The Complete Pyramids*, 82–199; G.S. Stuart, *America's Ancient Cities*, National Geographic Society, Washington, D.C., 1988, 112–186; S. Baudez and C. Picasso, *Lost Cities of the Maya*, Thames and Hudson, London, 1992, 101–115, 160–163; S. Gruzinski, *The Aztecs*, Thames and Hudson, London, 1992, 46–71; B. Dagens, *Angkor*, Thames and Hudson, London, 1995, 82–127, 168–171.

62. G. Hancock and S. Faiia, *Heaven's Mirror*, 116–133, 222–254; B. Dagens, *Angkor*, 14–73; P. James and N. Thorpe, *Ancient Mysteries*, 244–262; L. Picknett and C. Prince, *The Stargate Conspiracy*, 50–57; P. Jordan, *The Atlantis Syndrome*, 241–247; K.P.N. Shuker, *The Unexplained*, Colour Library Direct, Godalming, Surrey, 1998, 197–199; M. Freeman and C. Jacques, *Ancient Angkor*, Thames and Hudson, London, 1999.

63. G. Hancock and S. Faiia, *Heaven's Mirror*, 29–32, 61; A. Gurshtein, When the zodiac climbed into the sky, *Sky and Telescope*, 90(4), 1995, 28–33; A.A. Gurshtein, The origins of the constellations, *American Scientist*, 85, 1997, 264–273.

64. G. Hancock and S. Faiia, *Heaven's Mirror*, 212–221; R.M. Schoch, *Voices of the Rocks*, 106–113; P. Jordan, *The Atlantis Syndrome*, 245–246.

65. R.M. Schoch, *Voices of the Rocks*, 120–127; R. Lewin, *Human Evolution*, 201–206; R. Rudgley, *Secrets of the Stone Age*, 94–147; M. Settegast, Plato prehistorian – 10000–5000 B.C., in *Myth and Archaeology*, Rotenburg Press, Cambridge, Mass., 1986; M. Settegast, *Plato Prehistorian*, Lindisfarne Press, New York, 1990; R. Lewin, *The Origin of Modern Humans*, Scientific American Books, New York, 1998, 119–121, 146; R. Rudgley, *Lost Civilisations of the Stone Age*, 82–85, 164–165; P. Jordan, *Neanderthal*, Sutton Publishing, Stroud, Gloucs., 1999, 215–227.

66. I. Lawton and C. Ogilvie-Herald, *Giza – The Truth*, 1–8, 283–503 (quotation from page 4); L. Picknett and C. Prince, *The Stargate Conspiracy*, 57–115.

67. R. Bauval, *Secret Chamber – The Quest for the Hall of Records*, Century, London, 1999; L. Picknett and C. Prince, *The Stargate Conspiracy*, 361–375; G. Hancock, *Underworld – Flooded Kingdoms of the Ice Age*, Michael Joseph, London, 2002. Note that, whilst some of the underwater structures which Hancock drew attention to in the latter book might indeed be man-made, it does not necessarily follow that they were produced by a metal-using culture, or that they dated from the Pleistocene–Holocene boundary. Hancock brought *Underworld* to an apparently triumphant conclusion by drawing attention to a recent announcement that wood from a structure under the Gulf of Cambay, India, has been shown by radiocarbon dating to be 9,500 years old (see also J. Leake, An Asian Atlantis?, *The Sunday Times*, 20 January 2002, 14; *New Scientist*, 26 January 2002, 13). Although that is a very interesting finding, it would, if confirmed, still date the wood to two thousand years after the start of the Holocene, and the structure (even if man-made) might be younger still.

68. P. James, *The Sunken Kingdom*, 66–77; A.G. Galanopoulos and E. Bacon, *Atlantis – The Truth behind the Legend*, Nelson, London, 1969, 134; R. Castleden, *Atlantis Destroyed*, Routledge, London and New York, 1998, 169–170.

69. P. James, *The Sunken Kingdom*, 77–84; K.L. Feder, *Frauds, Myths and Mysteries*, 165–168; P. James and N. Thorpe, *Ancient Mysteries*, 26–31; R. Castleden, *The Minoans*, Routledge, London and New York, 1990, 32–37; S. Logiadou-Platonos (transl. D. Hardy) and N. Marinatos, *Crete*, Mathioulakis, Athens, undated, 18–29.

70. Z. Kukal, *Earth Science Reviews*, 21, 1984, 1–224 (see pages 192–194); P. James, *The Sunken Kingdom*, 10–12; K.L. Feder, *Frauds, Myths and Mysteries*, 168–169; E. Zangger, *The Flood from Heaven*, 40–45, 48–49; P. Jordan, *The Atlantis Syndrome*, 31–50; I. Wilson, *Undiscovered*, Michael O'Mara, London, 1987, 21–29; Horizon: Helike – The Real Atlantis, shown on BBC Television 10 January 2002, see http://www.bbc.co.uk/science/horizon/2001/helike.shtml.

71. E. Zangger, *The Flood from Heaven*, 68–176, 217–219 (quotations from page 219); S. Hodge, *Atlantis*, 96–101; P. Jordan, *The Atlantis Syndrome*, 165.

72. E. Zangger, *The Flood from Heaven*, 91–131; N. Zhirov, *Atlantis – Basic Problems*, Progress Publishers, Moscow, 1970, 374–375.

73. E. Zangger, *The Flood from Heaven*, 108–115.

74. E. Zangger, *The Flood from Heaven*, 131–167; Homer (transl. E.V. Rieu), *The Iliad*, Penguin Books, Harmondsworth, 1950, 401.

75. P. James, *The Sunken Kingdom*, 200–207, 214–216; P. Jordan, *The Atlantis Syndrome*, 164–175.

76. P. James, *The Sunken Kingdom*, 207–216.

77. P. James, *The Sunken Kingdom*, 214–254; P. James and N. Thorpe, *Ancient Mysteries*, 34–41; P. James, Rock goddess, *Fortean Times*, 142, January 2001, 40–43.

78. P. James, *The Sunken Kingdom*, 255–265.

79. P. James, *The Sunken Kingdom*, 266–280.

80. P. James, *The Sunken Kingdom*, 96–102, 281–288.

81. R. Castleden, *Atlantis Destroyed*, 134–182.

82. W.L. Friedrich (transl. A.R. McBirney), *Fire in the Sea – The Santorini Volcano: Natural History and the Legend of Atlantis*, Cambridge University Press, 2000; T.F. Druitt, L. Edwards, R.M. Mellors, *et al.*, *Santorini Volcano*, Geological Society Memoir 19, Geological Society of London, 1999.

83. W.L. Friedrich, *Fire in the Sea*, 147–160 (quotations from page 157).

Chapter 29

1. R. Castleden, *Atlantis Destroyed*, Routledge, London and New York, 1998, 191–192; M. Baillie, *Exodus to Arthur*, Batsford, London, 1999, 51–55, 80–85, 93–98, 239–240; M.G.L. Baillie, Dendrochronology and Thera – the scientific case for a 17th century BC eruption, *Journal of the Ancient Chronology Forum*, 4, 1990/91, 15–28; P.M. Warren, The Minoan civilisation of Crete and the volcano of Thera, *Journal of the Ancient Chronology Forum*, 4, 1990/91, 29–39; R.M. Porter, Demise of the 'scientific' date for Thera, *Chronology and Catastrophism Review*, 1998:1, 27–29; G.A. Zielinski and M.S. Germani, New ice-core evidence challenges the 1620s BC age for the Santorini (Minoan) eruption, *Journal of Archaeological Science*, 25, 1998, 279–289, 1043–1045; R.M. Porter, Other news, *Chronology and Catastrophism Review*, 2001:2, 37; W.L. Friedrich (transl. A.R. McBirney), *Fire in the Sea*, Canbridge University Press, 2000, 82–93; T.F. Druitt, L. Edwards, R.M. Mellors, *et al.*, *Santorini Volcano*, Geological Society Memoir 19, Geological Society of London, 1999, 43–48; F.W. McCoy and G. Heiken, The Late-Bronze Age explosive eruption of Thera (Santorini), Greece, in F.W. McCoy and G. Heiken (eds.), *Volcanic Hazards and Disasters in Human Antiquity*, Geological Society of America Special Paper 345, 2000, 43–70; *Ancient Apocalypse*, shown on BBC Television 2 August 2001 (for a summary of this programme see J. Cecil, http://www.bbc.co.uk/history/ancient/apocalypse_minoan1.shtml).

2. I. Velikovsky, *Earth in Upheaval*, Sphere Books, London, 1973, 129–134, 164–167; I. Wilson, *The Exodus Enigma*, Guild Publishing, London, 1986, 86–188; G. Phillips, *Act of God*, Pan Books, London, 1998, 207–313; D.M. Rohl, *A Test of Time*, Random House, London, 1995, 173–185, 299–325; J.J. Bimson, *Redating the Exodus and Conquest*, Almond Press, Sheffield, 2nd edn, 1981; J.J. Bimson, Exodus and conquest – myth or reality?, *Journal of the Ancient Chronology Forum*, 2, 1988, 27–40; A. Nur, And the walls came tumbling down, *New Scientist*, 6 July, 1991, 45–48; P. James, *Centuries*

of Darkness, Jonathan Cape, London, 1991, 162–203; I. Wilson, *The Bible is History*, Weidenfeld and Nicolson, London, 1999, 10–14, 42–53, 64–77, 118–133; M. Steiner, Problems of synthesis, *Journal of the Ancient Chronology Forum*, 8, 1999, 12–13.

3. Joshua 10:10–14; J.J. Bimson, The nature and scale of an Exodus catastrophe reassessed, in *Proceedings of the 1993 Cambridge Conference*, Society for Interdisciplinary Studies, 1994, 33–44; P. James and N. Thorpe, *Ancient Mysteries*, Ballantine Books, New York, 1999, 135–153; W.J. Phythian-Adams, A meteorite of the fourteenth century B.C., *Palestine Excavation Quarterly*, 1946, 116–124.

4. V. Clube, Cometary catastrophes and the ideas of Immanuel Velikovsky, *Society of Interdisciplinary Studies Review*, 5, 1984, 106–111; V. Clube and B. Napier, *The Cosmic Serpent*, Faber and Faber, London, 1982, 153–189, 224–275; V. Clube and B. Napier, *The Cosmic Winter*, Basil Blackwell, Oxford, 1990, 123–127, 147–154, 284; P. James and N. Thorpe, *Ancient Mysteries*, 100–106; D. Steel, *Rogue Asteroids and Doomsday Comets*, Wiley, New York, 1995, 155–156; G.L. Verschuur, *Impact!*, Oxford University Press, 1996, 95–107; J. Gribbin and M. Gribbin, *Fire on Earth*, Simon and Schuster, London, 1996, 202–211; J.S. Lewis, *Rain of Iron and Ice*, Addison-Wesley, Reading, Mass., 1996, 1–2, 16; D.H. Levy, *Comets – Creators and Destroyers*, Touchstone, New York, 1998, 128–129; J. Man, *Comets, Meteors and Asteroids*, BBC Books, London, 2001, 8; D.K. Yeomans and T. Kiang, The long-term notion of Halley's comet, *Monthly Notices of the Royal Astronomical Society*, 197, 1981, 633–646.

5. V. Clube and B. Napier, *The Cosmic Serpent*, 224–272; D. Roth, G.J. Gammon and P.J. James, Velikovsky's history and cosmology, *Society for Interdisciplinary Studies Review*, 5, 1983, 72–74; J.B. Moore, J. Abery and P.J. James, Global catastrophes – new evidence from astronomy, biology and astronomy, *Society for Interdisciplinary Studies Review*, 6, 1984, 89–91; V. Clube and B. Napier, *The Cosmic Winter*, 164–204; W.M. Napier, Cometary catastrophes, cosmic dust and ecological disasters in historical times – the astronomical framework, in B.J. Peiser, T. Palmer and M.E. Bailey, *Natural Catastrophes during Bronze Age Civilisations*, Archaeopress, Oxford, 1998, 21–32; H. Sigurdsson, *Melting the Earth*, Oxford University Press, 1999, 11–33; S.L. Harris, Archaeology and volcanism, in H.Sigurdsson (ed.), *Encyclopedia of Volcanoes*, Academic Press, San Diego, Calif., 2000, 1301–1314; D. Smith and A. Dawson, Tsunami waves in the North Sea, *New Scientist*, 4 August 1990, 46–49; R. Mestel, Noah's Flood, *New Scientist*, 4 October 1997, 24–27; W. Ryan and W. Pitman, *Noah's Flood*, Simon and Schuster, New York, 1999, 229–252; I. Wilson, *Before the Flood*, Orion, London, 2001; E. Bryant, *Tsunami – The Underated Hazard*, Cambridge University Press, 2001, 200–209.

6. C.F.A. Schaeffer, *Stratigraphie comparée et chronologie de l'Asie occidentale (IIIe et IIe millénaires)*, Oxford University Press, 1948; G.J. Gammon, Bronze Age destructions in the Near East, *Society for Interdisciplinary Studies Review*, 4, 1980, 104–108; G.J. Gammon, Dr Claude Schaeffer-Forrer, 1898–1982 – an appreciation, *Society for Interdisciplinary Studies Review*, 5, 1983, 70–71; P. James and N. Thorpe, *Ancient Mysteries*, 10–11, 41–58; C. Burgess, Volcanoes, catastrophe and the global crisis of the late second millennium B.C., *Current Archaeology*, 117, 1989, 325–329; I. Wilson, *Undiscovered*, Michael O'Mara, London, 1987, 16–20; A. Nur, *New Scientist*, 6 July 1991, 45–48;

I. Wilson, *The Bible is History*, 30–31; *Ancient Apocalypse*, shown on BBC Television, 16 August 2001 (see J. Cecil, Ancient Apocalypse – the destruction of Sodom and Gomorrah, http://www.bbc.co.uk/history/ancient/apocalypse_gomorrah1.shtml).

7. H. Weiss, M.-A. Courty, W. Wetterstrom, *et al.*, The genesis and collapse of third millennium North Mesopotamian civilization, *Science*, 261, 1993, 995–1004; R.A. Kerr, Sea-floor dust shows drought felled Akkadian empire, *Science*, 279, 1998, 325–326; P. James and N. Thorpe, *Ancient Mysteries*, 56–57; *The Times History of the World*, Times Books, London, 1999, 54–55; *The Cassell Atlas of World History, Volume 1 – The Ancient and Classical Worlds*, Andromeda, Oxford, 2000, section 1.11; G. Leick, *Mesopotamia*, Allen Lane, London, 2001, 92–103.

8. M.M. Mandelkehr, An integrated model for an Earth-wide event at 2300 BC. Part I, The archaeological evidence, *Society for Interdisciplinary Studies Review*, 5, 1983, 77–95; M.M. Mandelkehr, An integrated model for an Earth-wide event at 2300 BC. Part II, Climatology, *Chronology and Catastrophism Review*, 9, 1987, 34–44; M.M. Mandelkehr, An integrated model for an Earth-wide event at 2300 BC. Part III, The geological evidence, *Chronology and Catastrophism Review*, 10, 1988, 11–22.

9. J.J. Bimson, The nature and scale of an Exodus catastrophe re-assessed, in *Proceedings of the 1993 Cambridge Conference*, Society for Interdisciplinary Studies, 1994, 33–44; R.M. Porter, Bronze Age multi-site destructions, in *Proceedings of the 1993 Cambridge Conference*, 45–50.

10. R.A.J. Matthews, The past is our future, in *Natural Catastrophes during Bronze Age Civilisations*, 6–9.

11. M.E. Bailey, Sources and populations of near-Earth objects – recent findings and historical implications, in *Natural Catastrophes during Bronze Age Civilisations*, 10–20; W.M. Napier, Cometary catastrophes, cosmic dust and ecological disasters in historical times – the astronomical framework, in *Natural Catastrophes during Bronze Age Civilisations*, 21–32.

12. D. Steel, Before the stones – Stonehenge I as a cometary catastrophe predictor, in *Natural Catastrophes during Bronze Age Civilisations*, 33–48; G.L. Verschuur, Our place in space, in *Natural Catastrophes during Bronze Age Civilisations*, 49–52.

13. W.B. Masse, Earth, air, fire and water – the archaeology of Bronze Age cosmic catastrophes, in *Natural Catastrophes during Bronze Age Civilisations*, 53–92.

14. M.-A. Courty, The soil record of an exceptional event at 4000 BP in the Middle East, in *Natural Catastrophes during Bronze Age Civilisations*, 93–108.

15. M.G.L. Baillie, Hints that cometary debris played some role in several tree-ring dated environmental downturns in the Bronze Age, in *Natural Catastrophes during Bronze Age Civilisations*, 109–116.

16. B.J. Peiser, Comparative analysis of Late Holocene environmental and social upheaval – evidence for a global disaster around 4000 BP, in *Natural Catastrophes during Bronze Age Civilisations*, 117–139.

17. A. Nur, The end of the Bronze Age by large earthqukes?, in *Natural Catastrophes during Bronze Age Civilisations*, 140–147; L. Franzén and T.B. Larsson, Landscape analysis and stratigraphic and geochemical investigations of playa and

alluvial fan sediments in Tunisia and raised bog deposits in Sweden, in *Natural Catastrophes during Bronze Age Civilisations*, 148–161; B. van Geel, O.M. Raspopov, J. van der Plicht and H. Renssen, Solar forcing of abrupt climate change around 850 calender years BC, in *Natural Catastrophes during Bronze Age Civilisations*, 162–168.

18. E.W. MacKie, Can European prehistory detect large-scale natural disasters?, in *Natural Catastrophes during Bronze Age Civilisations*, 169–171.

19. G. Heinsohn, The catastrophic emergence of civilization – the coming of blood sacrifice in the Bronze Age, in *Natural Catastrophes during Bronze Age Civilisations*, 172–186; D.W.Pankenier, Heaven-sent – Understanding cosmic disaster in Chinese myth and history, in *Natural Catastrophes during Bronze Age Civilisations*, 187–197; O.B. Duane and N. Hutchison, *Chinese Myths and Legends*, Brockhampton Press, London, 1998, 24–43, 50–71.

20. W. Mullen, The agenda of the Milesian school – the post catastrophic paradigm shift in Ancient Greece, in *Natural Catastrophes during Bronze Age Civilisations*, 198–218; I. Wolfe, The 'Kultursturz' at the Bronze Age/Iron Age boundary, in *Natural Catastrophes during Bronze Age Civilisations*, 219–231.

21. S.V.M. Clube, The problem of historical catastrophism, in *Natural Catastrophes during Bronze Age Civilisations*, 232–249.

22. B. Peiser (moderator), *Cambridge Conference Network* (electronic communication, see archive at http://abob.libs.uga.edu/bobk/cccmenu.html). Anyone interested in becoming a member of the network should contact the moderator, b.j.peiser@livjm.ac.uk.

23. B.J. Peiser (moderator), *Cambridge Conference Network*.

24. J. Malek, *In the Shadow of the Pyramids – Egypt during the Old Kingdom*, The American University in Cairo Press, 1986, 117–120; D. Steel, *Rogue Asteroids and Doomsday Comets*, Wiley, New York, 1995, 162–167; T. Radford, Pyramids seen as stirways to heaven – pharaohs used monuments as launch pads to the afterlife says scientist, *The Guardian*, 14 May 2001, 7; C. Wickramasinghe, Pyramid to paradise – were these enigmatic monuments air-raid shelters against cosmic missiles?, *The Daily Mail*, 15 May 2001, 11; *Ancient Apocalypse*, shown on BBC Television, 26 July 2001 (see F. Hassan, Ancient Apocalypse – the fall of the Egyptian Old Kingdom, http://www.bbc.co.uk/history/ancient/egyptians/apocalypse_egypt1.shtml).

25. B. Peiser (moderator), *Cambridge Conference Network*; D.J. Keenan, The three century climatic upheaval of c. 2000 BC, and regional radiocarbon disparities, http://freespace.virgin.net/doug.keenan.

26. P. James and N. Thorpe, *Ancient Mysteries*, 57.

27. M. Mandelkehr, The causal source for the climatic changes at 2300 BC, *Chronology and Catastrophism Review*, 1999:1, 3–10; M. Mandelkehr, The causal source for the geological transients at 2300 BC, *Chronology and Catastrophism Review*, 1999:1, 11–16; M. Mandelkehr, Geomagnetic effects of an Earthwide event in 2300 BC, *Chronology and Catastrophism Review*, 2001:1, 4–10; M. Mandelkehr, The ring around the Earth at 2300 BC, *Chronology and Catastrophism Review*, 2001:2, 8–17; M. Baillie, *Exodus to Arthur*, Batsford, London, 1999, 143–149.

28. S. Master, Possible Holocene impact structure in the Al 'Amarah marshes, near the Tigris–Euphrates confluence, southern Iraq, *Meteoritics and Planetary Science*, 36 (supplement), A124, 2001; R. Matthews, Meteor clue to end of Middle East civilisation, *The Sunday Telegraph*, 4 November 2001, 8; B. Peiser (mod.), *Cambridge Conference Network*.

29. M. Baillie, *Exodus to Arthur*, 48–88; A. Nur, The collapse of ancient societies by great earthquakes, in *Natural Catastrophes during Bronze Age Civilisations*, 140–147; E. Zangger, *The Flood from Heaven*, Book Club Associates/ Sidgwick and Jackson, London, 1992, 79–85; J.P. Grattan and D.D. Gilbertson, Prehistoric 'settlement crisis', environmental changes in the British Isles and volcanic eruptions in Iceland – an exploration of plausible linkages, in F.W. McCoy and G. Heiken (eds.), *Volcanic Hazards and Disasters in Human Antiquity*, 33–42.

30. M. Baillie, *Exodus to Arthur*, 122–139, 153–161, 192–197; M. Winterbotham, *Gildas – The Ruin of Britain and Other Works*, Phillimore, London, 1978; M.G.L. Baillie, Dendrochronology raises questions about the nature of the AD 536 dust-veil event, *The Holocene*, 4, 1994, 212–217; D. Keys, *Catastrophe*, Randon House, London, 1999, 109–137, 251–252; E. Jones, Climate, archaeology, history and the Arthurian tradition, in J.D. Gunn (ed.), *The Years Without Summer – Tracing A.D. 536 and its Aftermath*, Archaeopress, Oxford, 2000, 25–34; B.K. Young, Climate and crisis in sixth century Italy and Gaul, in *The Years Without Summer – Tracing A.D. 536 and its Aftermath*, 35–42; M. Baillie, The AD 540 event, *Current Archaeology*, 15(6), 2001, 266–269.

31. D. Keys, *Catastrophe*, 1–155.

32. D. Keys, *Catastrophe*, 159–190; M.S. Houston, Chinese climate, history, and state stability in A.D. 536, in *The Years Without Summer – Tracing A.D. 536 and its Aftermath*, 71–78.

33. D. Keys, *Catastrophe*, 193–248; H.R. Robichaux, The Maya hiatus and the A.D. 536 atmospheric event, in *The Years Without Summer – Tracing A.D. 536 and its Aftermath*, 45–53.

34. D. Keys, *Catastrophe*, vii.

35. D. Keys, *Catastrophe*, 251–295; V. Horwell, *Secrets of the Dead*, Channel 4 Television, London, 1999, 37–45.

36. N. Saunders, Blown away, *New Scientist*, 18 March, 2000, 48–49; E. James, Review of *Catastrophe*, *Medieval Life*, Issue 12, 1999/2000, 3–6.

37. D. Keys, *Catastrophe*, 264–269; M. Baillie, *Exodus to Arthur*, 79–88, 192–218, 230–248 (quotation from page 248); M. Baillie, in *Natural Catastrophes during Bronze Age Civilisations*, 109–116; C. Wickramasinghe, *Cosmic Dragons*, Souvenir Press, London, 2001, 97–125.

38. R.B. Gill, *The Great Maya Droughts*, University of New Mexico Press, Albuquerque, 2000 (quotation from pages 234–235). See also R. D'Arrigo, D. Frank, G. Jacoby and N. Pedersen, Spatial response to major volcanic events in or about AD 536, 934 and 1258, *Climatic Change*, 49, 2001, 239–246.

39. J.P. Sadler and J.P. Grattan, Volcanoes as agents of past environmental change, *Global and Planetary Change*, 21, 1999, 181–196; J.P. Grattan and F.B. Pyatt,

Volcanic eruptions dry fogs and the European palaeoenvironmental record – localised phenomena or hemispheric impacts?, *Global and Planetary Change*, 21, 1999, 173–179.

40. J.D. Gunn, A.D. 536 and its 300-year aftermath, in *The Years Without Summer – Tracing A.D. 536 and its Aftermath*, 5–20 (quotation from page 18).

41. D.G. Anderson, Epilogue, in *The Years Without Summer – Tracing A.D. 536 and its Aftermath*, 169–170 (quotation from page 169).

42. K.R. Briffa, T.S. Bartholin, D. Echstein, *et al.*, A 1,400-year tree-ring record of summer temperatures in Fennoscandia, *Nature*, 346, 1990, 434–439; K.R. Briffa, Annual climate variability in the Holocene, *Quaternary Science Reviews*, 19, 2000, 87–105; J.T. Palmer and T. Palmer, The catastrophic years around 850 AD, *Cambridge Conference Network Essay*, 5 December 2000; R. Rudgley, *Barbarians – Secrets of the Dark Ages*, Channel 4 Books, London, 2002, 111–277.

43. Introduction, in *Carolingian Chronicles*, University of Michigan Press, Ann Arbor, 1972, 1–33; *The Royal Frankish Annals* (translated by B.W. Scholz), in *Carolingian Chronicles*, 35–125; *Nithard's Histories* (translated by B.W. Scholz), in *Carolingian Chronicles*, 127–174; *The Annals of St. Bertin* (translated and annotated by J.L. Nelson), Manchester University Press, 1991; *The Annals of Fulda* (translated and annotated by T. Reuter), Manchester University Press, 1991; *The Annals of Xanten* (translated by S. Coupland), from a forthcoming volume about sources for the reign of Charles the Bald, to be published by Manchester University Press (copy made available by Dr S. Foot, University of Sheffield).

44. *Royal Frankish Annals*, years 800–811; *Annals of Xanten*, year 810.

45. *Royal Frankish Annals*, years 815, 817, 820, 821; *Annals of Xanten*, years 813, 817, 821.

46. *Royal Frankish Annals*, years 823, 824, 827, 829.

47. *Annals of St. Bertin*, year 834; *Annals of Xanten*, years 834, 836.

48. *Annals of Xanten*, years 837, 838.

49. *Annals of Fulda*, year 838

50. *Annals of St. Bertin*, year 839; *Annals of Fulda*, year 839; *Annals of Xanten*, years 839, 840, 841; *Nithard's Histories*, year 840.

51. *Annals of Fulda*, year 841; *Nithard's Histories*, years 841, 842.

52. *Nithard's Histories*, year 842.

53. *Nithard's Histories*, year 843.

54. *Annals of St. Bertin*, years 843, 845; *Annals of Xanten*, year 845.

55. *Annals of St. Bertin*, year 849; *Annals of Fulda*, year 850; *Annals of Xanten*, years 850, 852, 853.

56. *Annals of St. Bertin*, year 855; *Annals of Fulda*, year 855.

57. *Annals of St. Bertin*, years 856, 857; *Annals of Fulda*, year 857; *Annals of Xanten*, year 857.

58. *Annals of St. Bertin*, years 858, 859; *Annals of Fulda*, years 858, 859; *Annals of Xanten*, year 859.

59. *Annals of St. Bertin*, year 860; *Annals of Fulda*, year 860; *Annals of Xanten*, years 860, 861.

60. *Annals of Fulda*, year 868; *Annals of Xanten*, years 867–869.

61. *Annals of Fulda*, year 872; *Annals of Xanten*, year 872.

62. *Annals of Fulda*, years 873, 874; *Annals of Xanten*, year 873.

63. *Annals of Fulda*, years 875, 882.

64. D. Steel, *Rogue Asteroids and Doomsday Comets*, Wiley, New York, 1995, 109–136; W.M. Napier, in *Natural Catastrophes during Bronze Age Civilisation*, 21–32.

65. N. Henbest (ed.), *Halley's Comet*, New Science Publications, London, 1985, 16–19; B. Hetherington, *A Chronicle of Pre-Telescopic Astronomy*, Wiley, Chichester and New York, 1996, 95; D.H. Levy, *Comets – Creators and Destroyers*, Touchstone, New York, 1998, 132–134; G. Vanin, *Cosmic Phenomena*, Firefly Books, Buffalo, New York, 1999, 19–21.

66. N. Bone, Lights in the sky, *Astronomy Now*, 14(6), 2000, 28–31; K. Taylor, Auroras, *National Geographic*, 200(5), 2001, 48–63.

67. H.H. Lamb, *Climate – Present, Past and Future*, Volume 2, Methuen, London, 1977, 427; J.D. Gunn, A.D. 536 and its 300-year aftermath, in *The Years Without Summer – Tracing A.D. 536 and its Aftermath*, 5–20; P. James and N. Thorpe, *Ancient Mysteries*, 76–95; K.L. Feder, *Frauds, Myths and Mysteries*, Mayfield, Mountain View, Calif., 3rd edn, 2000, 271–275; R.B. Gill, *The Great Maya Droughts*, 318–320, 384–387; H. Weiss and R.S. Bradley, What drives societal collapse?, *Science*, 292, 2001, 609–610; *Ancient Apocalypse*, Shown on BBC Television, 9 August 2001 (see J. Cecil, Ancient Apocalypse – the fall of the Mayan civilisation, http://www.bbc.co.uk/history/ancient/apocalypse_maya1.shtml).

68. D.A. Hodell, M. Brenner, J.H. Curtis and T. Guilderson, Solar forcing of drought frequency in the Maya lowlands, *Science*, 292, 2001, 1367–1370; T. Radford, How the sun god frowned on the Mayan civilisation, *The Guardian*, 18 May 2001, 5; R.B. Gill, *The Great Maya Droughts*, 191–246; A.L. Martin del Pozzo, C. Córdoba and J. López, Volcanic impact on the southern basin of Mexico during the Holocene, *Quaternary International*, 43/44, 1997, 181–190; D. Webster, *The Fall of the Ancient Maya*, Thames and Hudson, London, 2002, 239–247 (quotation from page 247).

69. W.W. Fitzhugh and E.I. Ward (eds.), *Vikings*, Smithsonian Institution Press, Washington and London, 2000, 164; K. Grönvold, N. Óskarsson, S.J. Johnson, et al., Ash levels from Iceland in the Greenland GRIP ice core correlated with oceanic and land sediments, *Earth and Planetary Science Letters*, 135, 1995, 149–156.

70. R.B. Gill, *The Great Maya Droughts*, 191–246; K.R. Briffa, T.S. Bartholin, D. Eckstein, et al., *Nature*, 346, 1990, 434–439.

71. B. Hetherington, *A Chronicle of Pre-Telescopic Astronomy*, Wiley, Chichester, 1996, 87–112; U.Dall'Olmo, Meteors, meteor showers and meteorites in the middle ages – from European medieval sources, *Journal for the History of Astronomy*, 9, 1978, 123–134.

72. M. Baillie, *Exodus to Arthur*, 46–47, 151–165; V. Clube and B. Napier, *The Cosmic Winter*, 19–20, 274; K.R. Briffa, T.S. Bartholin, D. Eckstein, et al., *Nature*, 346, 1990, 434–439; K.R. Briffa, *Quaternary Science Reviews*, 19, 2000, 87–105; C. Officer and J. Page, *Tales of the Earth*, Oxford University Press, 1993, 98; R.B. Stothers, Far reach of the tenth century Eldgjá eruption, Iceland, *Climatic Change*, 39, 1998, 715–726; R.B. Stothers,

Volcanic dry fogs, climate cooling, and plague pandemics in Europe and the Middle East, *Climatic Change*, 42, 1999, 713–723; M.M. Naurzbaev and E.A. Vaganov, Variation of early summer and annual temperature in east Taymir and Putoran (Siberia) over the last two millennia inferred from tree rings, *Journal of Geophysical Research*, 105, 2000, 7317–7326; F. McDermott, D.P. Mattey and C. Hawksworth, Holocene climate variability revealed by a high-resolution speleotherm δ^{18}O record from SW Ireland, *Science*, 294, 2001, 1328–1331; R. D'Arrigo, D. Frank, G. Jacoby and N. Pedersen, *Climatic Change*, 49, 2001, 239–246.

73. C. Officer and J. Page, *Tales of the Earth*, 97–101; J. Gribbin and M. Gribbin, *Fire on Earth*, Simon and Schuster, London, 1996, 201–220; S. Huang and H.N. Pollack, Late Quaternary temperature changes seen in world-wide continental heat flow measurements, *Geophysical Research Letters*, 24, 1997, 1947–1950; H. Miller, *Secrets of the Dead*, Channel 4 Books, London, 2000, 83–116; Hekla web site, see *http://www.norvol.hi.is/hekla.html*; R.B. Stothers, Climatic and demographic consequences of the massive volcanic eruption of 1258, *Climatic Change*, 45, 2000, 361–374; R.B. Stothers, *Climatic Change*, 42, 1999, 713–723; D. Dahl-Jensen, K. Mosegaard, N. Gundestrup, *et al.*, Past temperatures directly from the Greenland ice sheet, *Science*, 282, 1998, 268–271; R.B. Gill, *The Great Maya Droughts*, 235–242; K.R. Briffa, *Quaternary Science Reviews*, 19, 2000, 87–105; B. Wagner and M. Mulles, A Holocene seabird record from Raffles So. Sediments, East Greenland, in response to climatic and oceanic changes, *Boreas*, 30, 2001, 228–238; S. de Silva, J. Alzueta and G. Salas, The socio-economic consequences of the A.D. 1600 eruption of Huaynaputina, southern Peru, in F.W. McCoy and H. Heiken (eds.), *Volcanic Hazards and Disasters in Human Antiquity*, 15–24; R. D'Arrigo, D. Frank, G. Jacoby and N. Pedersen, *Climatic Change*, 49, 2001, 239–246.

74. S.V.M. Clube, in *Natural Catastrophes during Bronze Age Civilisations*, 232–249; A. Walsham, Sermons in the sky, *History Today*, 51(4), 2001, 56–63.

75. C. Officer and J. Page, *Tales of the Earth*, 97–101; D. Webster, *The Fall of the Ancient Maya*, 217–348.

76. D.A. Hodell, M. Brenner, J.H. Curtis and T. Guilderson, *Science*, 292, 2001, 1367–1370; T. Radford, *The Guardian*, 18 May 2001, 5; A.M. Waple, The sun-climate relationship in recent centuries – a review, *Progress in Physical Geography*, 23, 1999, 309–328.

Index